THE IMPACT OF VERY HIGH S/N SPECTROSCOPY ON STELLAR PHYSICS

INTERNATIONAL ASTRONOMICAL UNION
UNION ASTRONOMIQUE INTERNATIONALE

THE IMPACT OF VERY HIGH S/N SPECTROSCOPY ON STELLAR PHYSICS

PROCEEDINGS OF THE 132ND SYMPOSIUM OF THE
INTERNATIONAL ASTRONOMICAL UNION
HELD IN PARIS, FRANCE
JUNE 29 – JULY 3, 1987

EDITED BY

G. CAYREL DE STROBEL

and

MONIQUE SPITE

Observatoire de Paris, Meudon, France

KLUWER ACADEMIC PUBLISHERS
DORDRECHT / BOSTON / LONDON

Library of Congress Cataloging in Publication Data

International Astronomical Union. Symposium (132th : 1987 : Paris, France)
  The impact of very high S/N spectroscopy on stellar physics : proceedings of the 132th Symposium of the International Astronomical Union held in France, June 29-July 3, 1987 / edited by G. Cayrel de Strobel and Monique Spite.
       p. cm.
    Includes index.
    ISBN 9027726957. ISBN 9027726965 (pbk.)
    1. Stars--Spectra--Congresses. 2. Astronomical spectroscopy--Congresses. 3. Astrometry--Congresses. I. Strobel, G. Cayrel de, 1920- . II. Spite, Monique, 1939- . III. Title.
  QB870.I58 1987
  523.8'7--dc19   88-3067
                                                                    CIP
ISBN 90-277-2695-7
ISBN 90-277-2696-5 (pbk.)

Published on behalf of
the International Astronomical Union
by
Kluwer Academic Publishers, P.O. Box 17, 3300 AA Dordrecht, The Netherlands.

Kluwer Academic Publishers incorporates
the publishing programmes of
D. Reidel, Martinus Nijhoff, Dr W. Junk and MTP Press.

Sold and distributed in the U.S.A. and Canada
by Kluwer Academic Publishers,
101 Philip Drive, Norwell, MA 02061, U.S.A.

In all other countries, sold and distributed
by Kluwer Academic Publishers Group,
P.O. Box 322, 3300 AH Dordrecht, The Netherlands.

All Rights Reserved
© *1988 by the International Astronomical Union*

No part of the material protected by this copyright notice may be reproduced or utilized in any form or by any means, electronic or mechanical including photocopying, recording or by any information storage and retrieval system, without written permission from the publisher.

Printed in The Netherlands

TABLE OF CONTENTS

Organizing committees	XIII
J.L. Greenstein	XV
Acknowledgments	XVI
The Scientific career of J.L Greenstein	XVII
List of participants	XXV
Introductory talk	XXXIII

I. SPECTROGRAPHS, DETECTORS, FOURIER TRANSFORM SPECTROSCOPY, AND RADIAL VELOCITIES

Ia SPECTROGRAPHS

S.S. VOGT
The Lick Observatory Hamilton Echelle Spectrometer   1
    H. MANDEL
High Resolution Spectroscopy with a Fiber-Linked
Echelle-Spectrograph   9
    S.Y. JIANG
The Most Efficient Tool for very High Resolution and very
High Signal to Noise Ratio Spectral Observation of Stars   15
    A.W. RODGERS, P. HARDING, G. BLOXHAM and M.S. BESSELL
High Signal to Noise Observations with a Photon Counting Array   23
    S. FRANDSEN
High Resolution Spectrograph Needs for Stellar Seismology 27
    P. FELENBOK and J. GUERIN
Fiber Fed Spectrograph for Line Variability Studies   31
    D.C. EBBETS, S.R. HEAP and D.J. LINDLER
Photometric Characteristics of Data from the Hubble Space
Telescope Goddard-High Resolution Spectrograph   35

Ib DETECTORS

    G.A.H. WALKER
Limitations and Future of Reticon Detectors   39
    J.L. COUTURES and G. BOUCHARLAT
TH 7832CDZ (TH X31513) Bilinear CCD Array for
Astronomy Applications   45
    B. FORT
On High Signal to Noise Spectroscopy with CCDs   49
    R. GRIFFIN
The Chromospheric Spectrum of the ζ Aur Binary HR 6902   55
    S.T. RIDGWAY and K.H. HINKLE
The Impact of Array Detectors on High Resolution
Infrared Spectroscopy   61

Ic   FOURIER TRANSFORM SPECTROSCOPY AND RADIAL VELOCITIES

    J.P. MAILLARD
Signal-to-Noise Ratio and Astronomical
Fourier Transform Spectroscopy   71
    B. CAMPBELL, G.A.H. WALKER and S. YANG
Precise Radial Velocities of F, G, K Dwarfs   79
    K.G. LIBBRECHT
A Search for Radial Velocity Oscillations in Procyon   83

II.   PRE-MAIN-SEQUENCE STARS

    I. APPENZELLER
High S/N Spectroscopy of Pre-Main Sequence Stars I   87
    J. BOUVIER
Rotation of Pre-Main Sequence Stars from High S/N Spectroscopy   95
    G. BASRI
High Quality Echelle Observations of T Tauri Stars   99
    C. CATALA, J. CZARNY and P. FELENBOK
Search for Rotational Modulation in Pre-Main Sequence
Herbig Ae/Be Stars (presented by F. Praderie)   105
    A. VITTONE, C. ROSSI, E. COVINO and F. GIOVANNELLI
High S/N Simultaneous Optical and IR Spectrophotometric
Observations of Herbig Ae/Be Stars   109

III.   STARS

IIIa   EARLY-TYPE STARS

    U. HEBER, K. HUNGER, T. RAUCH and K. WERNER
Improved Non-LTE Model Atmospheres for Subluminous O-Stars   117
    D. BAADE and L.B. LUCY
A Search for Coronal Emission Lines from Early-Type Stars   123
    B. BOHANNAN, S.A. VOELS, D.C. ABBOTT and D.G. HUMMER
Effective Temperatures and Gravities for O-type Stars
Determined from High Precision Line Profiles and
Wind-Blanketed Model Atmospheres   127
    A.M. HUBERT, H. HUBERT, B. DAGOSTINOZ and M. FLOQUET
Variability of Some Be Stars on High-Resolution, High S/N Spectra   131
    K. GESICKI
A Modelling of Line Transfer in Supergiant Envelopes   135
    A.A. CHALABAEV, J. BORSENBERGER, J.P. MAILLARD
    and F. PRADERIE
NLTE-Effects in Profiles of Paschen and Brackett Lines
of an AOIV Star γGem   139

IIIb LATE-TYPE STARS AND DEGENERATE STARS

    D. GILLET
Balmer Emission Profiles in Radially Pulsating Stars:
The Case of the Double Hα Emission   143
    B.H. FOING, F. CASTELLI, G. VLADILO and J. BECKMAN
Some Constraints on Chromospheric Modelling for
Solar-Type Stars with High S/N Spectra   149
    F. CASTELLI, P. GOUTTEBROZE, J. BECKMAN,
    L. CRIVELLARI and B. FOING
A Method for Calibrating, in Absolute Flux Units, Ca II H
Profiles of Late Type Stars Observed at ESO   153
    P.G. JUDGE
Spectroscopy of Cool Stars from IUE Data   163
    R. CROWE and D. GILLET
Shock Phenomena in β Cephei Stars   169
    J.L. GREENSTEIN
White Dwarfs as Observed at High Signal to Noise
(presented by K.G.Libbrecht)   175

IV. SPECTROSCOPIC PATTERNS OF STELLAR ATMOSPHERES

    D.F. GRAY
The Buying Power of High Signal-to-Noise Ratios in Spectroscopy   185
    D. BAADE
Doppler Imaging of Variable Early-Type Stars   193
    A.P. HATZES
Doppler Images of Rapidly Rotating Ap Stars   199
    P.L. COTTRELL, W.A. LAWSON and S.M. SMITH
High Time Resolution Spectroscopic Observations of
Stellar Shock Waves   205
    E. FOSSAT
Seismology of the Stellar Cores   209
    W. DAPPEN
Theoretical Constraints from Asteroseismological
High S/N Observations   211
    D. BAADE
Nonradial Pulsations and the Be Phenomenon   217
    J.E. NEFF
Ultraviolet Spectral Imaging   223
    J.L. LINSKY and J.E. NEFF
Applications of the Doppler Imaging Technique to the Analysis of
High Resolution Spectra of the 3 October 1981 Flare on V711 TAURI   231
    G. ALECIAN and M.C. ARTRU
Abundance Stratifications in the Atmospheres of Ap Stars :
The Case of Gallium   235
    D. DRAVINS
Stellar Granulation and Photospheric Line Asymmetries   239

J.P. MAILLARD and D. NADEAU
Observational Evidence by Fourier Transform Spectroscopy of
Convective Motion in Late-Type Stellar Atmospheres 249
    S.S. VOGT
Doppler Images of Spotted Late-Type Stars 253
    A.M. BOESGAARD
Abundance Signatures of Internal Stellar Structure in F Stars 273
    A. BAGLIN and P.J. MOREL
Light Elements as Probes of Stellar Interiors 279
    G. VLADILO, L. CRIVELLARI, F. CASTELLI,
    J.E. BECKMAN and B.H. FOING
Chromospheric Velocity Fields Diagnostics from
CaII and MgII Emission Profiles 283
    B.H. FOING, J.E. BECKMAN and G. VLADILO
Spectroscopic Variability in Late-Type Dwarfs
Using High S/N Spectra 287
    P.H. SMITH and R.S. McMILLAN
Short Period Oscillations in Acturus, Aldebaran, and Pollux 291

V. MAGNETIC FIELDS

    S.H. SAAR
Measurements of Magnetic Fields on Cool Stars 295
    G.W. MARCY and G. BASRI
Magnetic Field Measurements on Late-Type Stars: A New Technique 301
    D.A. BOHLENDER and J.D. LANDSTREET
Abundance and Magnetic Field Geometries of
Helium-Strong and Helium-Weak Stars 309
    P. DIDELON
Surface Magnetic Field Measurements in
Hot Chemically Peculiar Stars 313
    G. MATHYS and J.O. STENFLO
Spectropolarimetry of Magnetic Stars: HD 125248 317
    J.D. LANDSTREET
The Magnetic Field Geometry and Abundance Distributions of
Iron-Peak Elements in the Ap Star 53 Camelopardalis 321
    G. MATHYS and S.K. SOLANKI
Magnetic Field of Late-Type Stars: A New Approach 325
    C. MEGESSIER, T. LANZ and J.D. LANDSTREET
Magnetic Field and Silicon diffusion in Bp-Si Stars 329

VI. CHEMICAL COMPOSITION OF STARS

VIa METHODOLOGY

    B. GUSTAFSSON
Physical Input for the Determination of Stellar Abundances 333
    R. CAYREL
Data Analysis 345

B.H. FOING AND L. CRIVELLARI
Improved Data Reduction Techniques for the ESO CES
Plus Reticon Spectra   355
    M.C.E. HUBER
Precise Atomic Data   361
    R.J. RUTTEN
Oscillator Strengths from the High S/N Solar Spectrum   367
    D.K. DUNCAN
The Accuracy of High S/N Spectroscopic Measurements   373
    D.R. SODERBLOM
Lithium Abundances of Solar-Type Stars: A Critical Application
of High Signal-to-Noise, High Resolution Spectroscopy   381
    B. EDVARDSSON
Accurate Spectroscopic Surface Gravities for 8 Sub-Giants   387

VIb POP.I EARLY-TYPE STARS

    U. HEBER, K. HUNGER and K. WERNER
NLTE Analysis and Chemical Composition of Hot Low-Mass Stars   389
    D. GIGAS
The Chemical Composition of Vega   395
    C. BURKHART, M.F. COUPRY and C. VAN'T VEER
A New Look at the Am Stars   401
    S. GIRIDHAR
Spectroscopic Studies of 89 HER and HD 161796   407

VIc POP.I MID- AND LATE-TYPE STARS AND OPEN CLUSTERS

    H. HOLWEGER
Accurate Spectroscopy of Intermediate and Late Spectral Types   411
    C. ABIA, R. REBOLO, J.E. BECKMAN,
    L. CRIVELLARI and B. VILA
Abundances of Light Metals in Field Stars with
Metallicity Range $-1.2<[Fe/H]<+0.3$ 421
    U. PAULS, N. GREVESSE and M.C.E. HUBER
A Photospheric Solar Iron Abundance from Weak FeII Lines   425
    C. BENTOLILA and G. CAYREL DE STROBEL
Detailed Analyses of Four Solar Analogs Analysed
On High S/N CFH and ESO Spectra   429
    M.-N. PERRIN, G. CAYREL DE STROBEL and M.DENNEFELD
Well Determined Atmospheric Parameters fromHigh S/N Reticon
Spectra of Four G and K Dwarfs within 10 pc of the Sun   433
    I. FURENLID
Solar Flux Atlas from 296 to 1300 nm   435
    I. FURENLID and T. MEYLAN
A High S/N Spectroscopic Study of Alpha Centauri A   437
    J. ANDERSEN, B. EDVARDSSON, B. GUSTAFSSON
    and P.E. NISSEN
A High S/N Spectroscopic Survey of Chemical Abundances in
Disk Population F0-G2 Dwarfs   441

S. ARRIBAS and C. MARTINEZ-ROGER
Iron Lines and Surface Gravity Determination for τ Ceti ... 445

R. CAYREL, G. CAYREL DE STROBEL and B. CAMPBELL
The Iron Abundances, [Fe/H] in the Four Nearest Open Clusters:
Pleiades, Ursa Major Stream, Coma Berenices and Hyades ... 449

N. ARIMOTO and G. CAYREL DE STROBEL
Chemical Composition of the Four Hyades Giants ... 453

T.A. KIPPER
On Technetium Abundance in Late-type Stars ... 459

S. VAUCLAIR
Lithium Abundances, Diffusion and Macroscopic Motions
In Stellar Clusters ... 463

R. REBOLO and J.E. BECKMAN
Rotation, Activity and Lithium Depletion in the Hyades
Late Main Sequence ... 469

J.E. BECKMAN and R. REBOLO
Bimodality and Lithium Abundance on the Upper Main
Sequence of the Open Cluster NGC 752 ... 473

VId POP.II STARS

H.E. BOND and R.E. LUCK
Chemical Compositions of Population II Mid- and Late-Type Stars ... 477

P. MAGAIN
Non-LTE Effects and Abundance Analyses of Halo Stars ... 485

R.C. PETERSON
The Effect of Non-LTE and Atmospheric Perturbations on
Relative Abundance Determinations in Metal-Poor Giants ... 493

C. SNEDEN, C.A. PILACHOWSKI, K.K. GILROY and J.J. COWAN
Distribution of R- and S-Process Elements in the Galactic Halo ... 501

R. REBOLO, P. MOLARO and J.E. BECKMAN
Lithium Abundances in a New Sample of Metal Deficient Dwarfs ... 507

P. MOLARO
Boron Abundance in the PopII HD 76932 ... 511

B. BARBUY
Oxygen in Halo Giants ... 515

O. MORELL
The Chemical Composition of K Dwarfs ... 519

P. FRANCOIS
The Sulphur Abundance in Metal Deficient Dwarfs ... 521

VIe GLOBULAR CLUSTER STARS

R.G. GRATTON
Recent Advances in High Dispersion Spectroscopy of
Globular Cluster Stars ... 525

R. BUSER and R.L. KURUCZ
Line-Blanketing in Theoretical Model Atmospheres for
F, G, and K-Type Stars ... 531

F. SPITE, M. SPITE and P. FRANCOIS
Accurate Abundances in Some Giant Stars of the
Globular Cluster ω Centauri 537
    V.V. SMITH
The Chemical Composition of Giants in the Globular Cluster ω Cen 541

VIf MAGELLANIC CLOUD STARS

    S.C. RUSSELL, M.S. BESSELL and M.A. DOPITA
Abundances of Heavy Elements in the Magellanic Clouds 545
    M. SPITE, F. SPITE and P. FRANCOIS
High Signal to Noise Analysis of the Chemical Composition of
Stars in the Magellanic Clouds 551
    B. WOLF, O. STAHL and W. SEIFERT
High-Dispersion Spectroscopy of the Ofpe/WN9 Stars
R84 and S61 of the LMC 555
    O. STAHL and B. WOLF
The Spectral Evolution of the LMC S Dor Variable
R127 during outburst 557
    E.L. FITZPATRICK
"2-D Frutti" Spectra of B Supergiants in the Milky Way and the LMC 559

VII. ABUNDANCE CONSTRAINTS ON STELLAR EVOLUTION,
NUCLEOSYNTHESIS, AND COSMOLOGICAL THEORIES

    D.L. LAMBERT
Chemical Composition as a Signature of Stellar Evolution:
The Barium Stars 563
    J.W. TRURAN
Abundance Clues to Early Galactic Chemical Evolution 577
    K.G. BUDGE, A.M. BOESGAARD and J. VARSIK
Beryllium Abundances of F Dwarfs 585
    Y. CHMIELEWSKI and D.L. LAMBERT
The $^{12}C/^{13}C$ Ratio in Unevolved Cool Stars 589
    A. JORISSEN
The $^{12}C/^{13}C$ Ratio in Barium Stars 593
    J.E. BECKMAN and B.E.J. PAGEL
Cosmological Signatures in High S/N Spectra 597
    J. AUDOUZE and E. VANGIONI-FLAM
Chemical Evolution and Primordial Nucleosynthesis 603

VIII. CONCLUSION

    P.E. NISSEN
Concluding Suggestions for Future High S/N Spectroscopic
Work on Stars 613
    I. FURENLID
Final Remarks on IAU Symposium 132 621

INDEX 623

Scientific Organizing Committee: I. Appenzeller, M.S. Bessell, A.M. Boesgaard, B. Campbell, G. Cayrel de Strobel (Chair), D.K. Duncan, D.F. Gray, R.E. Griffin, B. Gustafsson, G.H. Herbig, H. Holweger, J. Jugaku, T. Kipper, D.L. Lambert, P.E. Nissen, F. Spite.

Local Organizing Committee: N. Arimoto, C. Bentolila, C. Catala, J. Czarny, W. Däppen, P. Felenbok (Chair), M. Spite.

IAU Symposium 132 was sponsored by Commission 29 (Stellar Spectra) and cosponsored by Commissions 30 (Radial Velocities), 36 (Theory of Stellar Atmospheres) and 37 (Star Clusters and Associations).

IAU Symposium N°132 :

The impact of very high S/N
Spectroscopy on Stellar Physics

is dedicated
to
Jesse L. Greenstein

whose contribution to stellar spectroscopy is essential
and whose pioniring work in quantitative spectroscopy
has been determining in our present concepts and studies
with the modern physical tools available today.

ACKNOWLEDGEMENTS

We thank the following Organisations and Institutions for giving their financial support:
International Astronomical Union
Observatoire de Paris
Centre National de la Recherche Scientifique
Digital Equipement France

We are most grateful to the members of the Scientific and Local Organizing Committees for contributing so much to the success of this Symposium

We thank the President of Paris Observatory, Prof. P. Charvin, for his Welcome Address, and for having made available the Meudon Observatory facilities.

The Symposium banquet was held at the Luxembourg Palais in the Senate dining room. We address our deepest thanks to Senator Paul Loridant for having allowed this social event at the same time, sumptuous and intimate.

A concert held at the auditorium "Marcel Dupré" of Meudon has been organized by Dominique Proust. The musical ensemble was composed by astronomers. Many thanks to all of them for such an enjoyable evening.

We aknowledge the efficient secretarial help of Geraldine Prieur during the first months of conference planning. We are very thankful to Sylvie Gordon, Suzanne Huille, Christiane Jouan, Mireille Petit and Véronique Thévenet for the perfect material organisation of the conference. Michel Poinse supervised very efficiently the technical organisation of the meeting, our best thanks to him too; all these friends and many others not named here have greatly contributed to the smooth running of this Symposium.

Major review and contributed papers are published as received from the authors, short contributions and all the discussions were done by Monique Michel: we extend to her our gratitude for a difficult job well done. Last not least Suzanne Huille helped very efficiently in the editional work of the Symposium.

THE SCIENTIFIC CAREER OF JESSE L. GREENSTEIN

Ann Merchant Boesgaard

Palomar Observatory and Department of Astronomy
California Institute of Technology

ABSTRACT. The scientific career of Dr. Jesse L. Greenstein covers nearly 60 years, a thousand nights on various telescopes, and several hundred papers. He has made substantial contributions in a number of major research areas including interstellar work, stellar abundances and nucleosynthesis, faint blue stars, white dwarf stars, comets, and quasars. He has combined productive research with distinguished service to our profession in a truly exemplary manner.

1. INTRODUCTION

I am delighted to present the scientific work of Jesse L. Greenstein, to whom this Symposium is dedicated. His extraordinary carrer is spread over nearly 60 years and has been very prolific. He has an average of $8\frac{1}{2}$ papers published per year throughout his career. Even now that he has retired he keeps up this rate: 9 papers in 1984, 8 in 1985, 6 in 1986. When Greenstein was 8 years old his grandfather gave him a brass telescope and he delivered lectures to his friends on astronomical topics from planets to nebulae. Later he had a basement lab complete with prism spectroscope and arc and tried to identify atomic lines – portents of things to come... His favorite subjects in high school were Latin and chemistry; he admits to finding high school physics to be dull. He entered Harvard by age 15, received his A. B. degree in 1929 and his Master's degree in 1930. He was expected to go into the family business, and so he did. However, he did *volunteer* work at Columbia University where he searched for RR Lyr variables in M 3 and produced light curves for 199 stars. Although he found the work monotonous, he knew he preferred to pursue graduate work in astronomy rather than to remain in business. He returned to Harvard in 1934 for his Ph. D. studies, undiscouraged by Shapley's remarks that the science had progressed too rapidly in those 5 years for him to be able to catch up. After his Ph. D. in 1937 from Harvard, he received a National Research Council Fellowship for 1937–1939 and went to Yerkes Observatory where he stayed on until 1948. The development of Palomar Observatory brought him to the California Institute of Technology in Pasadena in the leadership role in the pursuit of astronomical frontiers for Caltech. He built the Caltech Astonomy Department into the premier astronomy department in the United States and has stayed there through and beyond his formal retirement in 1980; he remains active as the Lee A. DuBridge Professor of Astrophysics, Emeritus.

2. INTERSTELLAR WORK

His first paper (1930), written when he was still an undergraduate at Harvard University, was on the colors of B stars. It deals essentially with the then still unrealized effects of interstellar reddening. His Ph.D. thesis work at Harvard was with Bart J. Bok, observing the spectrophotometric differences between reddened and unreddened B stars. He established that the absorption varied as $\approx \lambda^{-0.7}$ and determined the ratio of photographic absorption to reddening (on the International scale) as ≈ 4. He interpreted this by computation of the extinction coefficient of small particles from the theory of Gustav Mie. The extinction was found to arise from dielectric grains with a power-law mixture of sizes and a high albedo. Earlier pioneers in the study of dust had been Schönberg and Jung, in Germany, Schalèn in Sweden, who were interested in metallic grains.

After his Ph. D. work, he was able to obtain a National Research Council Fellowship, one of the few available in physical sciences, and chose to go to Yerkes Observatory of the University of Chicago in Williams Bay, Wisconsin. He had first learned atomic spectroscopy and the theory of emission lines in ionized gases from Donald H. Menzel. At Yerkes he came under the powerful influence of the Director, Otto Struve, who had invited Jesse to come there. At Yerkes from 1937–1948, he at first continued new studies of interstellar matter. In 1938 the invention of the nebular spectrograph, a highly efficient low-dispersion instrument permitted him to study the spectra and colors of reflection nebulae, of extended emission nebulae of low surface brightness and of comets. Among the most important results of collaboration with Louis Henyey on the theory of radiation transfer in dust clouds was the discovery of diffuse galactic light, dust scattering of starlight in the Milky Way. They also established the phase-function and high albedo of dust in reflection nebulae, confirming that the grains are ices of frozen gases and relatively non-absorbing silicates, rather than metallic compounds. Bengt Strömgren, with Struve and C. T. Elvey, found the diffuse line emission of Hα to be generally present in the Milky Way; Strömgren soon developed the beautiful theory of H II regions, beginning the modern physics of interstellar matter, with hydrogen the dominant constituent.

Only much later did Greenstein return to the theory of interstellar dust after interstellar polarization was discovered by W. A. Hiltner and J. S. Hall. At Caltech, after 1948, he worked with the theoretical physicist Leverett Davis. This pioneering work, referred to as the Davis–Greenstein mechanism in graduate astronomy textbooks, is still valid in large part. They devised a very elaborate theory (1951) of the interaction between spinning, non-spherical, dust grains and interstellar magnetic fields, which align the grains with their long axis perpendicular to lines of force. Since the polarization vectors had been observed to be parallel over large stretches of galactic latitude, the magnetic field lines were along spiral arms. The Davis-Greenstein mechanism requires dissipation of the spin energy within the grains, through the complex part of the paramagnetism. Although the magnetic fields required (about 30 microgauss) were larger than eventually observed from Zeeman splitting of the 21-cm line, the basic physical mechanism of grain alignment is correct.

After significant contributions to the study of interstellar matter, he switched his astrophysical interests to a completely different subject, the study of stellar spectra at high resolution (for those days). It is characteristic of his career that he explored new specialities, both to learn new astronomy and its related physics and to make use of newly available equipment on new types of objects.

3. STELLAR SPECTRA, ABUNDANCES, NUCLEOSYNTHESIS

Greenstein started in this new scientific direction, stellar spectroscopy, while he was still at Yerkes. In this he was much influenced by Struve and by Pol Swings (from Liège), a visitor to Yerkes just before the outbreak of WW II. The 82-inch reflector at the McDonald Obervatory of the University of Texas was designed and operated by the Yerkes Observatory. Dedicated in 1939, it had an excellent ultraviolet 3-prism cassegrain spectrograph; its coudé, with Littrow prisms, had dispersion of 2.5 Å mm^{-1} at Hγ. Simultaneously, Theodore Dunham was building new, improved coudé grating spectrographs with schmidt optics for Mount Wilson. The astrophysical interests of Struve, Swings and Unsöld (also a visitor in 1939), gave spectroscopy at McDonald an especially modern innovative flavor. Struve told Greenstein, who was to be one of the first observers at McDonald, that he might consider studying υ Sgr. He did, with fascination; its spectrum was extraordinarily interesting, dominated by ionized metals and neutral helium, lacking Balmer lines and jump. It was a helium-rich star; Greenstein's pioneering analysis in 1940 is second only to Louis Berman's 1935 study of R CrB. The latter is rich in carbon, and was eventually found to have a helium chromosphere. Stars with both composition and opacity dominated by helium could not provide absolute abundances in those primitive days. Good opacities, model atmospheres, line-broadening theory and transition probabilities for most of the lines were non-existent. Comparisons were made with the sun. But his studies of υ Sgr gave some ideas of composition; He was high and N I and Ne I were strong. The star later showed line-profile variations, and evanescent Hα, material presumably from the invisible companion. In the same observing run excellent coudé spectra of the B star, τ Sco, were obtained on high-contrast films for Unsöld, who used them for the first, modern quantitative analysis of H, He, C, N, O, Si. Several of Unsöld's students at Kiel re-analyzed these spectra with progressively better models, line-broadening theory and transition probabilities, beginning the trend to "fein-analyse" of the stars.

Greenstein (1942) next observed the bright giant α Car, so far south as to strain the pointing capability of the 82-inch. He used what Unsöld would have called a "grob-analyse" method, with a curve of growth; the main difficulties were in the lack of knowledge of f-values. He used LS-coupling transition arrays and empirical solar transition probabilities, which were then being derived by Menzel and Goldberg. Without stellar models, it was necessary to use different excitation and ionization temperatures. Abundance ratios relative to the Sun were normal in α Car. Following this general, but improving methodology, many stars were analyzed. Among the first was the metallic-line star, τ UMa, which showed numerous abundance deficiencies (notably Ca) as well as excesses; some of these anomalies were ascribed to non-LTE ionization. Greenstein thus initiated a series of coudé analyses of stars using the differential curve of growth method. K. O. Wright independently developed the method of differential curve of growth analysis in Canada. Collaborative efforts established generally useful techniques of calibrating plates and standard equivalent widths in stars.

World War II interrupted research at Yerkes and McDonald; immediately afterwards, opportunities blossomed. The 200-inch mirror was approaching completion when, in 1948, Greenstein left Yerkes to set up a department and assemble a new staff for the California Institute of Technology and its Palomar Observatory. The Mount Wilson and Palomar Observatories were jointly operated with remarkable resources. High-resolution spectroscopy, first of the Sun, and then the stars was a specialty of the older Mount Wilson staff. Paul Merrill discovered technetium in S stars soon after the first laboratory spectroscopy of that unstable element. Greenstein brought a more quantitative, astrophysical approach, and extended the work to intrinsically fainter stars. An interesting new topic was study of the light elements, subject to nuclear reactions at relatively low temperatures. The physicists of the Kellogg Radiation Laboratory specialized in energy levels and reaction rates of these relatively simple nuclei; Charles C. Lauritsen and William A. Fowler were interested in the astrophysical applications, notably energy generation. Greenstein worked on the changes in abundances produced by such reactions, devising astronomical tests; with collaborators, he searched for Li in the Sun, and the $^6Li/^7Li$ ratio. In the far ultraviolet region he looked for Be, neutral and ionized; B was inaccessible in the far ultraviolet. Lithium was depleted showing that solar surface material had mixed down into the envelope reaching above 10^6 K. Such solar spectroscopy, at least, qualifies for the subject of this conference, high S/N. Fred Hoyle had proposed that elements were synthesized in stellar interiors. Large He/C or C/H in evolved stars, high $^{13}C/^{12}C$ ratios, unstable Tc, elements with large neutron excesses such as rare earths and uranium, combined with the expanding knowledge of reaction rates all suggest the generalization that all elements heavier than He were synthesized in stellar interiors. Greenstein was active, pushing astronomical observations of these rare elements to compare with critical nuclear predictions; his first review of this comparison was at Utrecht in 1952. The elements Li, Be, B are not produced in stellar interiors, but by spallation by cosmic rays, according to Fowler, Greenstein and Hoyle. Energetics permit the process $^{13}C +^4 He$ to yield a neutron for each ^{13}C nucleus in red giants, thus leading to heavy elements in S stars. The theory of stellar nucleosynthesis culminates in the Burbidge, Burbidge, Fowler and Hoyle paper in *Reviews of Modern Physics* (1957).

From 1957 to 1970 Greenstein ran an "Abundance Project," under which numerous postdoctoral fellows came to Caltech from all over the world to use the Mount Wilson and Palomar telescopes and spectra. They produced, for the first time, a large number of quantitative studies of stellar composition. Many of those visitors have now become leaders in the field, including our hostess, Giusa de Strobel Cayrel. Some analyses were based on the differential curve-of-growth method, others used early model atmospheres. While few of the photographic spectra used would now qualify as having high signal-to-noise ratio, they were at high dispersion, with good instruments. They provided important information and tests of the new theory of stellar nucleosynthesis in the 1950's.

The composition of the oldest stars, born soon after the Galaxy formed, has great evolutionary significance. The metal deficiency of the unevolved sdG's was first determined by Aller and collaborators, including Greenstein; the weakness of the metallic lines is partly compensated for by the relative simplicity of the spectrum and ease of location of the continuum. Some abnormal ratios of heavy elements to Fe were suspected, and the abundances of C, N, O became targets of various investigations. A few, very long Palomar exposures, (up to parts of three nights), gave spectra of individual stars in globular clusters at 18 Å mm^{-1}. Similar halo giants were found among field stars, notably HD 122563, which was analyzed several times with more refined methods. The heavier elements appeared to have lower ratios to Fe in the most metal-deficient stars. Higher ratios were ascribed to nuclei containing integral numbers of α particles. An interesting observation was the 6300 Å line of [O I] in K giants at the highest possible signal to noise. Giusa and Roger Cayrel worked with the best possible data for ϵ Vir; they also provided tables facilitating relative analyses, allowing for different stratification depending on temperature and gravity, in model atmospheres with a "scaled" temperature distribution. Later, as a result of the accumulation of numerous abundance analyses, George Wallerstein found the useful correlation between ultraviolet excess, $\delta(U-B)$, and the metal deficiency $[Fe/H]$.

4. COMETS

Another field captured Greenstein's attention: during the 1950's there were several bright comets. So he used the new coudé facilities to obtain excellent spectra of Mrkos, Arend-Roland, Ikeya and Humason (which had unusually strong CO^+). The high resolution made possible detailed study of the intensity of the rotational lines, as well as their variation in the inner coma. The Swings excitation mechanism depends on the flux in the solar spectrum as seen velocity-shifted to the comet's rest frame; large intensity differences occurred between successive rotational lines, depending on the details of the solar line absorptions. Greenstein discovered a new phenomenon: changes of intensity of a rotational line along the slit, on opposite sides of the nucleus. Now called the "Greenstein effect" by cometary scientists, the phenomenon arises from internal motions of the cometary gas (probably from rotation of the nucleus) which shift the absorption of the molecules. This rotation was confirmed in the recent flybys of Halley. The Na emission was asymmetric with respect to the nucleus. An attempt at the cometary isotope ratio, $^{13}CN/^{12}CN$ was possibly vitiated by an overlapping NH_2 blend.

5. FAINT BLUE STARS, WHITE DWARFS

When Greenstein was 60 he changed fields again. The challenge of the availability of the 200-inch Hale reflector lead him to transfer his efforts to intrinsically fainter hot stars, both the "faint blue stars" (FB) and white dwarfs (WD), abandoning high dispersion spectroscopy. Most are apparently faint, demanding lower spectroscopic resolution. Only the newer generation of photoelectric sensors, with interferometric or echelle spectrometers, makes high resolution and good signal to noise possible for fainter stars. The faint blue stars and white dwarfs were the subject of an IAU Colloquium at Tucson in June 1987, and are therefor less fully reviewed here. By 1958 it was clear that a variety of types of stars populated the lower-left-hand quadrant of the HR diagram; Guido Münch and Greenstein found that blue stars on the horizontal branch of globular clusters had weak He I lines, as did the halo sdB's. Sargent and Searle developed a rapid quantitative method to obtain gravity and temperature from UBV colors and the widths of the Balmer-line profiles based on the available LTE models. Baschek, Newell, Norris studied field and cluster FB's. Finally, in their extensive survey of FB stars, Greenstein and Anneila Sargent (1974) found the HBB and sdB stars to have constant $log\ g\theta^4$, which measures the M/L ratio. They derived the mass of the halo field stars as $0.56M_\odot$. Some FB's are apparently normal B's, with characteristics of population I, but at distances and travel times from the galactic plane exceeding their nuclear lifetimes. Most FB's have only slow rotation, and are highly evolved. Quantitative analysis has since revealed helium-rich, chemically-peculiar, often variable stars among the FB's. Convection, diffusion, radiation pressure and mixing complicate the subject. The Kiel group has been extensively observing the sdB, sdOB and sdO stars at ESO, with great success.

Observing high quality spectra of white dwarfs is not easy; they are faint and have weak lines at high (or low) temperature. Starting with stars selected by blue color and proper motion, Eggen and Greenstein (1965-67) confirmed and published colors and spectra of 200 WD's; in total, Greenstein spent about 300 nights with the 200-inch on this problem. He published data on 550 white dwarfs from the prime-focus spectrograph, Oke's multichannel spectrophotometer and the double CCD spectrograph (1965–1986). In 1983, after his retirement he observed 200 stars at 5 Å resolution in 8 nights; he could achieve signal-to-noise > 100 in 10 minutes for objects at V = 16. Some observations are high signal-to-noise spectroscopy, with counts $\approx 10^5$ in the low resolution multichannel photometry and with the CCD yielding 30000 photons per pixel; for example, a composite hydrogen-line profile in ten DA's is based on nearly 10^6 detected photons. High S/N is essential to reveal the weak, shallow and very diffuse absorptions found (1986) in the strongly magnetic WD Grw +70°8247; four had been found (1957) by superposition of tracings of numerous photographic spectra. These are Zeeman components of Hα, Hβ and Hγ in a 350 megagauss dipole field. By superposition of photographic spectra he discovered and first identified a band of C_2 200 Å wide and 10 % deep in a carbon-rich WD. Quantitive analyses of composition of WD's with metallic lines were carried out first by Weidemann on van Maanen 2, and in others by Greenstein and his collaborators. Spectroscopy of the ultraviolet with IUE has become possible; it provided a confirmation of the ground-based temperature scale, and also revealed an unidentified broad feature at 1400Å. Koester and Weidemann later identified it as a quasi-molecular transition in H_2. The extraordinary transparency of an atmosphere dominated by helium made it possible to detect metals even when they were 10^{-5} or less as abundant as in the sun; the same is true in the few WD's showing carbon.

A general picture of WD atmosphere emerged: the DA's have negligible helium and only rarely detectable metals. The non-DA's have negligible abundance of hydrogen, and often traces of carbon and metals. The WD's have very stable atmospheres, and the hydrogen even in DA's is concentrated in a surface layer of less than 10^{-5} M $_\odot$. The metals and helium diffuse rapidly downwards, and there may be some accretion (which should, magically, exclude hydrogen). Some pre-WD's may lose all their hydrogen, resulting in the non-DA's; alternatively convective mixing could dilute the residual hydrogen.

Greenstein's study of WD's over the years produced abundance analyses and extensive quantitative results on their physical state. A temperature scale came from fitting models to his spectrophotometry, with energy distributions and line profiles in both the hydrogen-rich and helium-rich degenerates. Results obtained from his extensive data include the fact that the HR diagram is narrow; parallaxes give the $\sigma(M_V) \approx \pm 0.3$ mag, i.e. only small range of radii and surface gravities. The WD's rotate only slowly (≤ 100 km s^{-1}); their red-giant predecessors have lost proportionally more of their angular momentum than mass. In a series of papers with Virginia Trimble and others, the Einstein gravitational redshift was found to agree with that expected from the mass-radius diagram with a C, He core. After the rapid variability of the ZZ Ceti type DA's was discovered, multichannel colors established that oscillations are confined to a narrow range of temperature.

6. RADIO ASTRONOMY

Greenstein had a long enthusiasm for radioastronomy; with F. L. Whipple he had written a premature theoretical explanation of Jansky's observations in 1937. In

1956 he helped organize the start of Caltech's radio observatory. Identification of many Cambridge 3C sources with active galaxies left unexplained some that remained unresolved by the best radio-interferometric measurements. These were identified with "blue stellar objects" as apparently stars with peculiar spectrum and colors, by A. R. Sandage and Tom Matthews; the accurate lunar-occultation position of 3C273 permitted Maarten Schmidt to identify it with a bright stellar object and recognize its redshift as 16 %. Greenstein had observed extensively weak emission lines in the fainter blue stellar object 3C48. When Schmidt discovered the redshift of 3C273, Greenstein immediately recognized that 3C48 was also redshifted, by 37 %. Greenstein and Schmidt (1964) analyzed the physical conditions in the line-emitting regions of these first two quasars recognized. The methodology was Menzel's for planetary nebulae but essential new concepts include the implausibility of a gravitational origin of the redshift, the high density from observed collisional de-excitation, line broadening by electron scattering, the spatial separation of the emission-line and radiofrequency emitting regions. The energy required to maintain the high luminosity at cosmological distances suggested that nuclear energy did not suffice. With the mystery of the central source energy source unsolved, Greenstein abandoned that field, happy that the cosmological redshift left an exciting difficult question for the next generation.

7. CONCLUDING REMARKS

During all this time, he served as department head for 24 years, observed over a thousand nights at large telescopes, and frequently advised the U.S. government on the requirements for financial support of astronomy. He served on some 50 important national and international committees on various aspects of astronomy. Among the most significant was his chairmanship of the "Greenstein Report," the study of the priorities and prices of *Astronomy and Astrophysics for the 1970s* for the National Academy of Sciences of the U.S.A. We retain his name on some of his discoveries: the Henyey-Greenstein camera, the Davis-Greenstein mechanism, the Greenstein effect.

Among many honors are the the California Scientist of the Year award (with Schmidt) for work on quasars, the Bruce Gold Medal of the Astronomical Society of the Pacific, the Gold Medal of the Royal Astronomical Society, NASA's Distinguished Public Service Medal and an honorary D. Sc. from the University of Arizona this June.

Having been named the Lee A. DuBridge Professor of Astrophysics in 1970, he retired to the Emeritus status of that Professorship in 1980; he stopped observing in 1983, but still works, currently on the molecules that may serve as population discriminants in M dwarfs. He is to be greatly revered for his significant scientific achievements and his distinguished professional service. Many of us here at this conference eagerly and fortunately learned much about stellar spectroscopy from Jesse Greenstein. His influence is far-reaching. It is a very fitting tribute to dedicate this Symposium to him.

List of participants

DR. C. ABIA	INSTIT. DE ASTROFIS. DE CANARIAS 38200 LA LAGUNA, TENERIFE, SPAIN
DR. G. ALECIAN	OBSERVATOIRE DE PARIS-MEUDON 92195 MEUDON Cedex, FRANCE
DR. J. ANDERSEN	SMITHSONIAN ASTROPHYS. OBSERVATORY 60 GARDEN ST., CAMBRIDGE, MA 02138 U.S.A.
PROF. I. APPENZELLER	LANDESSTERNWARTE, KOENIGSTUHL D-6900 HEIDELBERG 1 F.R.G.
DR. N. ARIMOTO	OBSERVATOIRE DE PARIS-MEUDON 92195 MEUDON Cedex FRANCE
DR. M.C. ARTRU	OBSERVATOIRE DE PARIS-MEUDON 92195 MEUDON Cedex FRANCE
DR. J. AUDOUZE	INSTITUT D'ASTROPHYSIQUE DE PARIS 98 BIS BOULEVARD ARAGO, 75014 PARIS, FRANCE
DR. D. BAADE	E C F / ESO, KARL SCHWARZSCHILD STRASSE 2 D 8046 GARCHING BEI MUNCHEN F.R.G.
DR. A. BAGLIN	OBSERVATOIRE DE NICE, BP139 06003 NICE Cedex FRANCE
DR. B. BARBUY	IAG USP DEPTO ASTRONOMIA C.P. 30627, 01051 SAO PAULO, BRASIL
DR. G. BARATTA	OSSERVATORIO ASTRO. DI ROMA VIALE DEL PARCO MELLINI, I-00136 ROMA, ITALY
DR. G. B. BASRI	ASTRON. DEPT., UNIV. OF CALIFORNIA BERKELEY, CA 94720 U.S.A.
PROF. J. E. BECKMAN	INSTITUTO DE FISICA DE CANARIAS LA LAGUNA, 38071 TENERIFE, SPAIN
MME C. BENTOLILA	OBSERVATOIRE DE PARIS-MEUDON 92195 MEUDON Cedex FRANCE
DR. M. S. BESSELL	MT STROMLO OBSERVATORY WODEN PO ACT 2606, AUSTRALIA
PROF. A. M. BOESGAARD	ASTRON. 105-24, CALIFORNIA INST. OF TECHNOLOGY PASADENA, CA 91125, U.S.A.

DR. B. E. BOHANNAN	ASTROPHYS. PLANET. and ATMOS. SCIENCES UNIV. OF COLORADO, BOULDER, CO 80309-0391, USA
DR. D. BOHLENDER	DEPT OF ASTRONOMY, UNIVERSITY OF WESTERN ONTARIO LONDON, ONTARIO CANADA N6A 3K7
DR. H. E. BOND	SPACE TELESCOPE SCI. INST., HOMEWOOD CAMPUS 3700 ST MARTIN DR., BALTIMORE, MD 21218 U.S.A.
DR J. BOUVIER	INSTITUT D ASTROPHYSIQUE DE PARIS 98bis BD ARAGO, 75014 PARIS FRANCE
DR J. BRULS	STERREWACHT "SONNENBORGH", Zomenburg 2 3512 NL UTRECHT, HOLLAND
DR. K. BUDGE	105-24, CALIFORNIA INST. OF TECHNOLOGY PASADENA, CA 91125 U.S.A.
DR. C. BURKHART	OBSERVATOIRE DE LYON F-69320 SAINT GENIS LAVAL FRANCE
DR. R. BUSER	ASTRONOMISHES INST., UNIVERSITAET BASEL VENUSSTRASSE 7, CH-4102 BINNINGEN, SWITZERLAND
DR. F. CASTELLI	OSSERVATORIO ASTR., VIA TIEPOLO 11 I-34131 TRIESTE ITALY
DR. G. CAYREL DE STROBEL	OBSERVATOIRE DE PARIS-MEUDON 92195 MEUDON Cedex FRANCE
DR. R. CAYREL	OBSERVATOIRE DE PARIS 61 AV. DE L'OBSERVATOIRE, 75014 PARIS FRANCE
DR. A. CHALABAEV	OBS. DE HTE PROVENCE 04870 ST MICHEL-L'OBSERVATOIRE, FRANCE
DR. Y. CHMIELEWSKY	OBSERVATOIRE DE GENEVE, CHEM. DES MAILLETTES 51, 1290 SAUVERNY, SWITZERLAND
DR. R. COLUZZI	VIA DELLA STAZIONE 205 04013 LATINA SCALO, ITALY
DR. P. L. COTTRELL	DEPT OF PHYSICS, UNIVERSITY OF CANTERBURY CHRISTCHURCH 1 NEW ZEALAND
Mr. J.L. COUTURES	THOMSON-CSF Div: TUBES ELECTRONIQUES 38 RUE VAUTHIER BP305 92102 BOULOGNE-BILLANCOURT Cedex FRANCE

Mme M.F. COUPRY	OBSERVATOIRE DE PARIS 61 AV. DE L'OBSERVATOIRE, 75014 PARIS FRANCE
DR. L. CRIVELLARI	OSSERV. ASTR. DI TRIESTE VIA G.B. TIEPOLO 11, 34131 TRIESTE, ITALY
DR. R. CROWE	CFHT CORPORATION, P.O. BOX 1597 KAMUELA, HAWAII 96743 U.S.A.
MR. J. CZARNY	OBSERV. DE PARIS-MEUDON 92195 MEUDON Cedex FRANCE
DR. W. DÄPPEN	OBSERV. DE PARIS-MEUDON 92195 MEUDON Cedex FRANCE
DR. M.R. DESPHANDE	PHYS. RES. LAB. NAVRANGPURA, AHMEDABAD 38009, INDIA
DR. P. DIDELON	OBSERVATOIRE DE STRASBOURG 11 RUE DE L'UNIVERSITE, 67000 STRASBOURG, FRANCE
DR. V. DOAZAN	OBSERVATOIRE DE PARIS 61 AV. DE L'OBSERVATOIRE, 75014 PARIS FRANCE
DR. J.D. DRAKE	UNIVERSITY OF OXFORD, DEPARTMENT OF ASTROPHYSICS SOUTH PARK ROAD, OXFORD 3RQ, GREAT BRITAIN
PROF. D. DRAVINS	LUND OBSERVATORY, BOX 43 S -22100 LUND SWEDEN
DR. D. DUNCAN	SPACE TELESCOPE SCIENCE INSTITUTE 3700 SAN MARTIN DRIVE HOMEWOOD CAMPUS BALTIMORE MD 21218 U.S.A.
DR. D. EBBETS	INS. SUPP. BRANCH, SPACE TELESCOPE Sci. Inst. HOMEWOOD CAMPUS, BALTIMORE, MD 21218 U.S.A.
DR. B. EDVARDSSON	UPPSALA ASTRONOMICAL OBSERVATORY BOX 515, S-751-20 UPPSALA SWEDEN
DR. P. FELENBOK	OBSERVATOIRE DE PARIS-MEUDON 92195 MEUDON Cedex FRANCE
DR. E. FITZPATRICK	JOINT INST. FOR LAB. ASTROPHYSICS UNIV. OF COLORADO, BOULDER CO 80309-0440 U.S.A.
DR. M. FLOQUET	OBSERVATOIRE DE PARIS-MEUDON 92195 MEUDON Cedex FRANCE
DR. B. FOING	LAB. DE PHYS. STELLAIRE ET PLANETAIRE B.P. 10 91371 VERRIERES-LE-BUISSON Cedex, FRANCE

DR. B. FORT	O.P.M.T., 14 AVENUE EDOUARD BELIN 31400 TOULOUSE FRANCE
DR. E. FOSSAT	OBSERVATOIRE DE NICE B.P. 139 06003 NICE-Cedex FRANCE
DR. P. FRANCOIS	ESO, KARL SCHWARZSCHILD STRASSE 2 8046 GARCHING BEI MUNCHEN, F.R.G.
DR. S. FRANDSEN	INSTITUTE OF ASTRONOMY, UNIVERSITY OF AARHUS DK 8000 AARHUS C DENMARK
DR. I. FURENLID	DEPT OF PHYSICS AND ASTRONOMY GEORGIA STATE UNIV., ATLANTA GA 30303 U.S.A.
DR. K. GESICKI	CENTRUM ASTRONOMI. im. M.KOPERNIKA PRACOWNIA ASTROFIZYKI, ul. CHOPINA 12/18 TORUN, POLAND
PROF. D. GIGAS	INST THEOR PHYS & STERNW, UNIVERSITAET KIEL OLSHAUSENSTRASSE, D-2300 KIEL F.R.G.
DR. D. GILLET	OBSERVATOIRE DE HAUTE PROVENCE 04870 SAINT MICHEL L'OBSERVATOIRE, FRANCE
DR. S. GIRIDHAR	INDIAN INSTITUTE OF ASTROPHYSICS KORAMANGALA, BANGALORE 34 INDIA
DR. R. GRATTON	INSTITUTO ASTROFISICA SPAZIALE C P 67 I-00044 FRASCATI ITALY
PROF. D. F. GRAY	DEPT OF ASTRONOMY, UNIV OF WESTERN ONTARIO PHYS.-ASTRONOMY BLDG, LONDON ONT N6A 3K7, CANADA
DR. R.E.M. GRIFFIN	THE OBSERVATORIES MADINGLEY ROAD, CAMBRIDGE CB3 OHA, GREAT BRITAIN
PROF. B. GUSTAFSSON	STOCKHOLM OBSERVATORY 13300 SALTSJOBADEN, SWEDEN
DR. A. HATZES	LICK OBSERVATORY, UNIVERSITY OF CALIFORNIA SANTA CRUZ, CA 95064 U.S.A.
DR. U. HEBER	INST THEORET & STERNW, NEUE UNIV PHYSIK ZENTRUM OLSHAUSENSTRASSE 40 N61C, 2300 KIEL 1 F.R.G.
DR. H.F. HENRICHS	ASTRONOMICAL INSTITUTE, UNIVERSITY OF AMSTERDAM ROETERSSTRAAT, 15 1018 W B AMSTERDAM THE NETHERLANDS.

PROF. H. HOLWEGER	INST. THEO. PHYS. STERNWARTE, UNIVERSOTAET KIEL OLSHAUSENSTRASSE, 2300 KIEL F.R.G.
DR. A.M. HUBERT	OBSERVATOIRE DE PARIS-MEUDON 92195 MEUDON Cedex FRANCE
MR. H. HUBERT	OBSERVATOIRE DE PARIS-MEUDON 92195 MEUDON Cedex FRANCE
DR. M.C.E. HUBER	INSTITUTE OF ASTRONOMY, ETH-ZENTRUM CH-8092 ZURICH SWITZERLAND
DR. JIANG SHI-YANG	BEIJING ASTRONOMICAL OBSERVATORY BEIJING CHINA
DR. A. JORISSEN	UNIVERSITE LIBRE DE BRUXELLES INST. D'ASTRON. ASTROPHYS. GEOPHYS. B.P. 165, B 1050 BRUXELLES BELGIUM
DR. P.G. JUDGE	DEPT THEORETICAL PHYSICS, OXFORD UNIVERSITY 1 KEBLE ROAD, OXFORD OX1 3NP GREAT BRITAIN
DR. J. JUGAKU	TOKYO ASTRONOMICAL OBS OSAWA MITAKA, TOKYO 181 JAPAN
DR. T. KIPPER	TARTU ASTROPHYS.OBS., ESTONIAN ACAD. OF SCIENCES TORAVERE, 202444 TARTU U.S.S.R.
DR. J. KOPPEN	OBSERVATOIRE DE STRASBOURG 11 RUE DE L'UNIVERSITE, 67000 STRASBOURG FRANCE
PROF. D. L. LAMBERT	UNIVERSITY OF TEXAS R L MOORE HALL AUSTIN TX 78712 U.S.A.
PROF. J.D. LANDSTREET	DEPT OF ASTRONOMY, UNIV OF WESTERN ONTARIO LONDON ONTARIO N6A 3K7 CANADA
DR. T. LANZ	UNIVERSITY DE LAUSANNE, INSTITUT D'ASTRONOMIE CH 11290 CHAVANNES DES BOIS SWITZERLAND
DR. Y. LEBRETON	OBSERVATOIRE DE PARIS-MEUDON 92195 MEUDON Cedex FRANCE
DR. K. LIBBRECHT	ASTRONOMY 264-33 CALTECH PASADENA, CA 91125 U.S.A.
PROF. J.L. LINSKY	JILA, UNIV. OF COLORADO BOULDER, CO 80309 U.S.A.
DR. M.C. LORTET	OBSERVATOIRE DE PARIS-MEUDON 92195 MEUDON Cedex FRANCE

xxx

DR. P. MAGAIN	E.S.O. CASILLA 19001 SANTIAGO 19 CHILE
MR. J.P. MAILLARD	INSTITUT D'ASTROPHYSIQUE DE PARIS 98 BIS BOULEVARD ARAGO, 75014 PARIS FRANCE
DR. H. MANDEL	LANDESSTERNWARTE D-6900 HEIDELBERG 1 F.R.G.
DR. G.W. MARCY	DEPT. OF PHYS. AND ASTRON. SAN FRANCISCO STATE UNIV SAN FRANCISCO, CA 94132 U.S.A.
DR. G. MATHYS	OBSERV. DE GENEVE, CHEMIN DES MAILLETTES 51 1290 SAUVERNY SWITZERLAND
DR. C. MEGESSIER	OBSERVATOIRE DE PARIS-MEUDON 92195 MEUDON Cedex FRANCE
DR. P. MOLARO	OSSERVATORIO ASTR., VIA TIEPOLO 11 I-34131 TRIESTE ITALY
DR. P. MOREL	OBSERVATOIRE DE NICE BP139 06003 NICE Cedex FRANCE
DR. O. MORELL	ASTRONOMICA OBSERVATORIET BOX 515, S-751-20 UPPSALA SWEDEN
DR. C. MOROSSI	OSSERVATORIO ASTRONOMICO, VIA TIEPOLO 11, I-34131 TRIESTE ITALY
DR. D. NADEAU	DEPT DE PHYS, UNIV DE MONTREAL, CP 6128 SUCC A MONTREAL PQ H3C 3J7 CANADA
DR. J.NEFF	JILA, CAMPUS BOX 440, UNIV. OF COLORADO BOULDER , CO 80309-0440 U.S.A.
PROF. P.E. NISSEN	INSTITUTE OF ASTRONOMY, UNIVERSITY OF AARHUS LANGELANDSGADE DK-3000 AARHUS C, DENMARK
PROF. L. PASINETTI	DIPART. DI FISICA, UNIVERSITA DI MILANO VIA CELORIA 16, I-20133 MILANO ITALY
DR. U. PAULS	INSTITUTE OF ASTRONOMY, ETH-ZENTRUM CH-8092 ZURICH SWITZERLAND
DR. M.N. PERRIN	OBSERVATOIRE DE PARIS 61 AV. DE L'OBSERVATOIRE, 75014 PARIS FRANCE
DR. R.C. PETERSON	WHIPPLE OBSERVATORY, STEWARD OBS. OFFICES UNIV. OF ARIZONA, TUCSON AZ 85721 U.S.A.

MR. B. PLEZ	CERGA OBSERV. DE CALERN, CAUSSOLS 06460 ST VALLIER DE THIEY FRANCE
DR. F. PRADERIE	OBSERVATOIRE DE PARIS-MEUDON 92195 MEUDON Cedex FRANCE
DR. R. REBOLO	INST. DE ASTROFISICA DE CANARIAS CAMINO DE LA HORNERA, LA LAGUNA, TENERIFE SPAIN
DR. T. RICHTLER	UNIV. STERNWARTE, AUF DEM HUEGEL 71 D 5300 BONN F.R.G.
DR. S. RIDGWAY	KPNO P.O. Box 26732 TUCSON, AZ 85726 U.S.A.
DR. G. ROBERTI	OSSERV. ASTRO. DI CAPODIMONTE 80131 NAPOLI ITALY
DR. C. ROSSI	OSSERV. ASTRO. DI ROMA, VIALE DEL PARCO MELLINI I-00136 ROMA ITALY
DR. R. RUTTEN	STERREWACHT "SONNENBORGH", ZONNENBURG 2-3512 UTRECHT THE NETHERLANDS
DR. S.H. SAAR	JILA, UNIV. OF COLORADO BOULDER, CO 80309 U.S.A.
DR. SCHERBAKOV	CRIMEA ASTROPHYSICAL OBSERVATORY, ACAD. OF SCI., p/o NAUCHNY, 334413 CRIMEA U.S.S.R.
DR. P.H. SMITH	LUNAR AND PLANETARY LABORATORY UNIV. OF ARIZONA TUCSON AZ 85721 U.S.A.
DR. V. SMITH	DEPT OF ASTRONOMY, UNIVERSITY OF TEXAS AUSTIN, TEXAS 78712 U.S.A.
DR. C. SNEDEN	DEPT. OF ASTRONOMY, UNIVERSITY OF TEXAS, AUSTIN TX 78731 U.S.A.
DR. D.R. SODERBLOM	SPACE TELESCOPE SCI. INST. HOMEWOOD CAMPUS BALTIMORE, MD 21218 U.S.A.
DR. F. SPITE	OBSERVATOIRE DE PARIS-MEUDON 92195 MEUDON Cedex FRANCE
DR. M. SPITE	OBSERVATOIRE DE PARIS-MEUDON 92195 MEUDON Cedex FRANCE
DR. O. STAHL	ESO, KARL SCHWARZSCHILD STRASSE 2 D-8046 GARCHING BEI MUNCHEN F.R.G.

DR. A. TALAVERA   IUE OBSERVATORY, VILLAFRANCA SATELLITE STATION
                       PO Box 54065   MADRID   SPAIN

DR. F. THEVENIN   CERGA, OBSERV. DE CALERN, CAUSSOLS
                       06460 ST VALLIER DE THIEY   FRANCE

DR. F. TRAN MINH   OBSERVATOIRE DE PARIS-MEUDON
                       92195 MEUDON Cedex   FRANCE

DR. J.W. TRURAN   DEPT. OF ASTRONOMY, UNIV. OF ILLINOIS
                       1011 W.SPRINGFIELD AV. URBANA IL 61801 U.S.A.

DR. J. VAN SANTOORT   IUE OBSERVATORY, VILLAFRANCA SATELLITE STATION
                       PO Box 54065   MADRID   SPAIN

DR. C. VAN'T VEER MENNERET   OBSERVATOIRE DE PARIS
                       61 AV. DE L'OBSERVATOIRE, 75014 PARIS   FRANCE

DR. S. VAUCLAIR   O.P.M.T., 14 AVENUE EDOUARD BELIN
                       31400 TOULOUSE   FRANCE

DR. A.A. VITTONE   OSSERVATORIO ASTR. DI CAPODIMONTE
                       80131 NAPOLI   ITALY

Dr. G. VLADILO   OSSERVATORIO ASTR., VIA TIEPOLO 11
                       I-34131 TRIESTE   ITALY

DR. S.S. VOGT   LICK OBSERVATORY, UNIVERSITY OF CALIFORNIA
                       SANTA-CRUZ CA 95064   U.S.A.

PROF. G.A.H. WALKER   DEPT. GEOPHYS. AND ASTRON.
                       UNIV. OF BRITISH COLUMBIA   2075 WESBROOK PLACE
                       VANCOUVER BC V6T 1W5   CANADA

INTRODUCTORY TALK: AIM OF THE SYMPOSIUM

G. Cayrel de Strobel

The idea to organize a Conference on high S/N spectroscopy came to me several years ago, in the beginning of the eighties, when the first tracings of Reticon spectra of 8 and 9 magnitude stars were published. I suddendly realized that the quality of those spectra was comparable to those we find in the atlasses of the Sun, Procyon, Arcturus and of a very few other very bright stars. I thought at that epoch, probably at the top of Mauna Kea, that when high-resolution spectroscopists will have collected enough high S/N results, then, time would be ripe to discuss the impact of these results on our knowledge of Stellar Physics.

The time is now ripe, and here we are with our collected material opening this 132 IAU Symposium. The material is so vast, that I fear, that many of us will be afraid that we will not have time to discuss all the items contained in the program.

This program is centered along two main axes. The first one concerns some important advances in a better understanding of the physical structure of the stars in the light of new results of high S/N spectroscopy. The second one deals with the interpretation of the chemical composition of stars belonging to different populations and to... different galaxies (for the time being belonging to our Galaxy and to the Magellanic Clouds).

These two main topics are preceded by short sessions on spectrographs detectors, Fourier transform spectroscopy and radial velocity measurements, in the aim to give us the necessary background for a better understanding of high S/N observations. They are followed by a session in which abundance constraints on stellar evolution, nucleo-synthesis and cosmological theories will be very briefly discussed.

A week ago in the "Liège Conference" spectrographs and detectors have been discussed much more in detail, and during this week the most remote objects will be on the grill at a Conference of the Institut d'Astrophysique de Paris.

Another point of interest of this meeting is that, at my knowledge, it is the first time that almost all the teams interested in lithium abundance in stars will gather together. Unhappily, our dear friend George Herbig, who was the first to discover the large variation of Lithium in F and G stars, was not able to attend the Conference.

Poul Erik Nissen will conclude the Conference and will have the hard task to be the most present person in the conference-room during this week.

THE LICK OBSERVATORY HAMILTON ECHELLE SPECTROMETER

Steven S. Vogt
Lick Observatory and Board of Studies in Astronomy
and Astrophysics
University of California, Santa Cruz, CA 95064

ABSTRACT. The Hamilton Echelle Spectrometer, recently installed at the coudé focus of the Shane 3-m telescope, is a high dispersion spectrograph optimized for use with large format CCD's. It was designed primarily for high resolution (R=50,000) wide bandpass spectroscopy of point-like sources down to a limiting magnitude of about V=16.5, over the 0.34 μm to 1.1 μm spectral region. Its design features a relatively large collimated beam size, the use of prisms rather than gratings for cross dispersion, minimum order separation, the use of protected silver mirror coatings throughout the system, and a fast (f/1.67) folded Schmidt camera with a flat external focal plane. Together, these features yield a very powerful spectrometer for high resolution stellar spectroscopy. This paper gives a brief description of the Hamilton spectrometer and gives several examples of its performance on astronomical objects.

DISCUSSION

The Hamilton Echelle is a prism-cross-dispersed echelle spectrometer recently commissioned at coudé of the 3-meter telescope. It was optimized for today's largest format CCD detectors and for the even larger format CCD's anticipated in the next few years. This paper presents a very brief description of the spectrometer. A more detailed discussion of the instrument and the philosophy behind its design will appear elsewhere.

Figure 1 shows the layout of the Hamilton in top and side views. The f/36 beam from the 3-m telescope enters the slit (**1**) at upper right. The beam is collimated into a 20.4 cm beam by an off-axis parabolic collimator (**2**) and deviated 11° to an echelle grating (**3**). The echelle is a 204×408 mm R-2 with 31.6 gr/mm, purchased from Milton Roy Inc. (formerly Bausch and Lomb). It is used in-plane ($\gamma = 0°$) with $\theta = 5.5°$. The light is then cross dispersed by two large UBK-7 prisms (**4**) and (**5**). Prisms were chosen over gratings since they provide higher throughput, lack of blaze fall-off, and much more uniform order separation. The latter is particularly important since real estate on the CCD detector is precious. Order separation with gratings varies substantially with wavelength, resulting in much wasted detector area between orders. Prisms yield approximately constant order separation over the entire visual and near IR spectrum. The Hamilton was designed for minimum order separation, thus maximizing spectral coverage per observation.

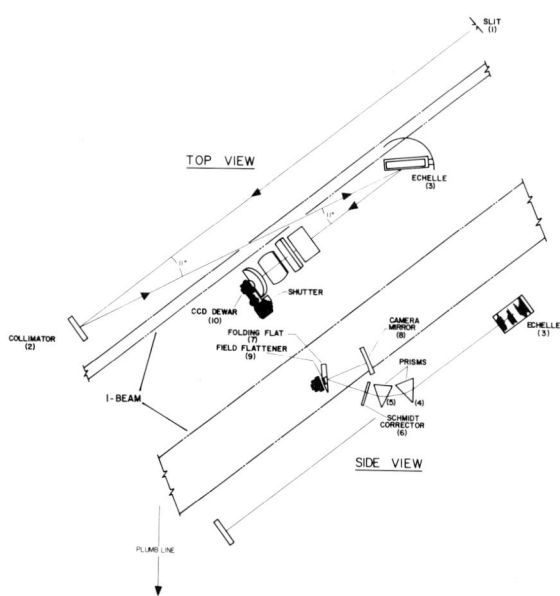

Figure 1. Hamilton spectrometer in top and side views.

After cross dispersion, the light is brought to focus by an f/1.67 (polychromatic) folded Schmidt camera designed by Dr. Harlan Epps at the University of California at Los Angeles. This camera features a flat, external focal plane located 2.3 cm beyond the last optical surface. As such, it reaches easily into a typical CCD dewar. The focal surface is flat, and no refocussing is required over the entire 3400 Å to 11,000 Å range. The camera consists of a Schmidt corrector (**6**), a folding flat (**7**), a spherical camera mirror (**8**), and a field flattener (**9**) mounted in the hole in the folding flat. At present, the principle detector for the Hamilton is a thinned backside illuminated Texas Instruments 800×800 CCD with readout noise of about 7 electrons. A large mounting stage behind the camera is provided to allow other detectors (MAMA's, PAPA's, Ranicons, etc.) to be used with the Hamilton.

The Hamilton's relatively large beam size and high blaze angle echelle ($\theta_b = 64.7°$) allow a relatively large slit to be used, even at high resolutions. A

1.25 arc-sec slit projects down to two 15 micron CCD pixels. The 'two-pixel' resolution equivalent is 60,000 and the true effective resolution with a 'two-pixel projected slit' is about 40,000 or 0.125 Å at 5000 Å.

All the reflective optical surfaces of the Hamilton have been optimized for highest efficiency in the visual/near-IR region by using protected silver coatings. The silver coatings were done in the Lick Coating Laboratory. The silver is laid down 1000 Å thick on a protective underlayer of 400 Å of copper. A 1000 Å thick overcoat of sapphire is then applied by e-beam deposition to seal the silver from exposure to the atmosphere, thus preventing tarnishing and protecting against mechanical abrasion. All transmitting optical surfaces are coated with single layer MgF anti-reflection coatings.

Figure 2 shows the optical format of the Hamilton. Each horizontal line in the format represents a single order, and the length of each line represents one free spectral range. A few order numbers are shown at the left of each order, and blaze wavelengths are at the right. Our present TI CCD (format shown by the inner box) does not cover the entire format, but it nevertheless spans typically 800 Å to 3600 Å in a single exposure. The addition of a Tektronix 1024×1024 (format shown by the outer border) will bring the Hamilton to full maturity, providing complete spectral coverage from 3400 Å to 9100 Å in a single exposure with no interorder gaps, a glorious match between echelles and CCD's.

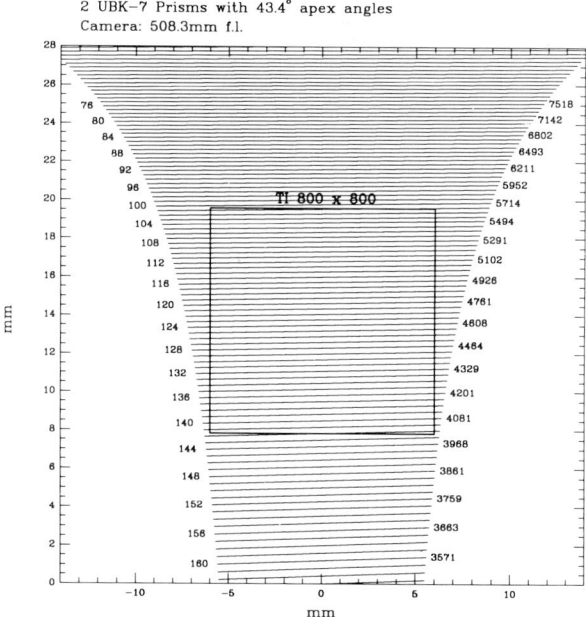

Figure 2. Hamilton spectral format.

Figure 3 shows a Hamilton spectrum of the V=12.3 magnitude globular cluster star M71-A4, obtained with the 800×800 CCD. This star was selected as a good benchmark test of the system performance since it had already been studied with echelles or other spectrographs on several different large telescopes in recent years. The horizontal white bars are the spectral orders of the echelle. The spectrum runs from 4530 Å at the bottom to 7320 Å at the top. The atmospheric B-band is visible near row 80. Hα is the prominent absorption feature near row 135. The strong pair of lines at row 285 are the sodium D lines. The strong line to the left at row 495 is part of the Mg b feature. This 30 minute exposure yielded a resolution of R=30,000 (0.22 Å) and a S/N of 150 per resolution element at Hα.

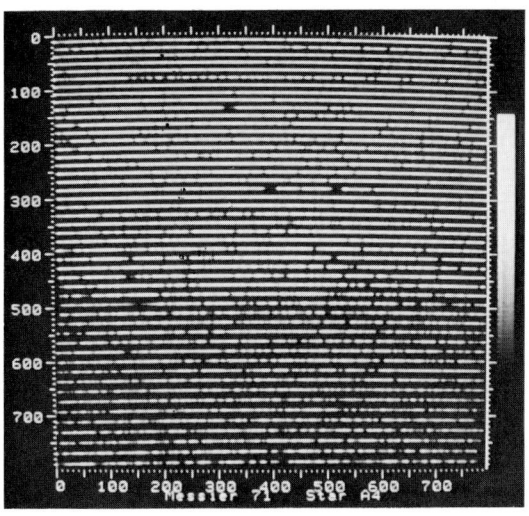

Figure 3. Hamilton spectrum of M71-A4.

Figure 4 shows a performance comparison between the Hamilton and telescope/spectrometer combinations at Kitt Peak and Palomar. The object is again the globular cluster star M71-A4 in the region around 6200 Å. The top spectrum (Cohen 1983, *Ap.J.*, **270**, 654) shows a 2.3 hour exposure obtained with the Kitt Peak 4-m telescope Cassegrain Echelle Spectrograph with the Singer image tube camera and baked IIIa-J plates. A resolving power of about 12,500 was achieved here. The middle spectrum (from the same reference) shows a 1 hour exposure with the Palomar 5-m telescope using the red side of the Double Spectrograph, and a TI 800×800 CCD. A resolution of 9,500 was achieved here. The lower spectrum shows the corresponding spectral region extracted from the 30 minute Hamilton exposure of Figure 3. The Hamilton data have been smoothed to a resolution of about 20,000 here. The Hamilton thus obtained 2 to 3 times the resolution in one-half to one-quarter the observing time, using a telescope with only one-half to one-third the collecting area. This comparison illustrates the great strides in performance that are now achievable in high resolution spectroscopy because of technological advances in detectors, coatings, and optical glasses.

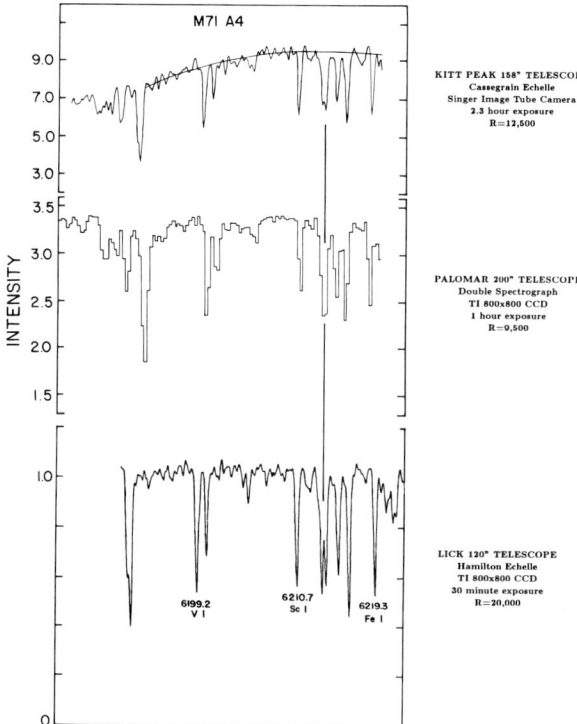

Figure 4. Performance comparison

As an example of the Hamilton performance near its present limiting magnitude, Figure 5 shows a 240 minute observation of the B=16.0 magnitude QSO 1331+170. This is the sum of two 2-hour exposures taken under very good conditions on 4/17/87 UT. The CCD was binned 2×2 on-chip to reduce the readout noise contribution. Shown here is only a subsection of the original 400×400 pixel binned image (the areas not shown contain hot pixels and hot columns which became heavily saturated in the long exposures). Most of the cosmic ray hits have been removed by median filtering. The 2.25 arcsecond entrance slit projected to 1.8 binned pixels yielding an effective resolution of about 2.5 pixels (0.2 Å) or resolving power of 24,000. The section shown covers the wavelength range from about 3985 Å at the bottom to 5020 Å at the top. At the blue end the wavelength coverage is about 75% complete, dropping to 60% complete at the red end. The effective S/N is about 8-12 per pixel or 13-19 per resolution element. The strong night sky line just below center at row 202 is Hg I λ4358 Å, and the two weaker night sky lines near the bottom of the chip are Hg I $\lambda 4078$ Å (row 299), and Hg I λ 4047 Å(row 310). Numerous absorption features are visible in the QSO spectrum. The very strong complex of absorption features to the right of center near row 221 is the C IV $\lambda\lambda$ 1548, 1551 Å doublet at z=1.7755. Two other strong absorption features from the same system are visible; Si II λ 1527 Å just below the C IV complex at row 241, and Al II λ 1671 Å left of center near row 124. The complex of absorption features near row 69 is the Mg II $\lambda\lambda$ 2795, 2803 Å doublet at a redshift of z=0.7441.

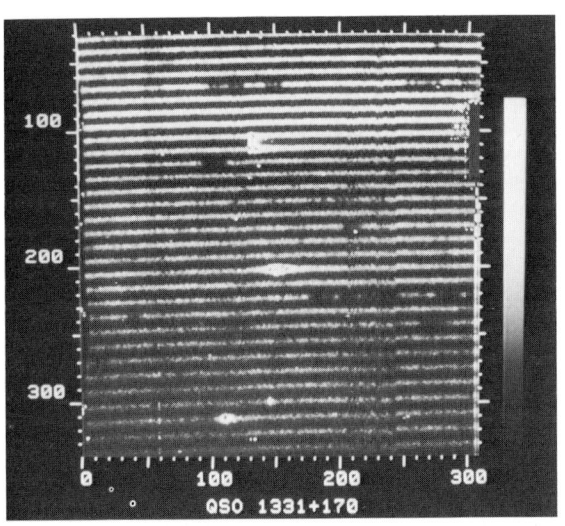

Figure 5. Hamilton spectrum of QSO 1331+170.

Figure 6 shows the reduced spectrum from echelle order 117 near row 69 of Figure 5 which contains the Mg II doublet. At the present resolution of 0.20 Å, the Mg features are resolved into at least 8 separate components which span a velocity range of some 260 km s^{-1}. These different components probably arise from discrete absorbing clouds moving with different velocities in a distant intervening galaxy along the line-of-sight to the QSO. Data such as these will make detailed metal abundance studies possible in very distant galaxies over a large range of look-back times. Here, high resolution is crucial for distinguishing abundance effects from multiple velocity shifted cloud components, both of which contribute to the total equivalent width.

As performance benchmarks for the Hamilton and our TI 800×800 CCD, one obtains S/N=45 per 0.15 Å resolution element at blaze center near 5500 Å in a 1 hour exposure at V=13.0 under excellent conditions and gray time. At the faint end, under similar conditions, the Hamilton delivers S/N=10 per 0.2 Å resolution element in 4 hours at V=17.0. At this level, dark current and CCD readout noise become major contributors to the noise.

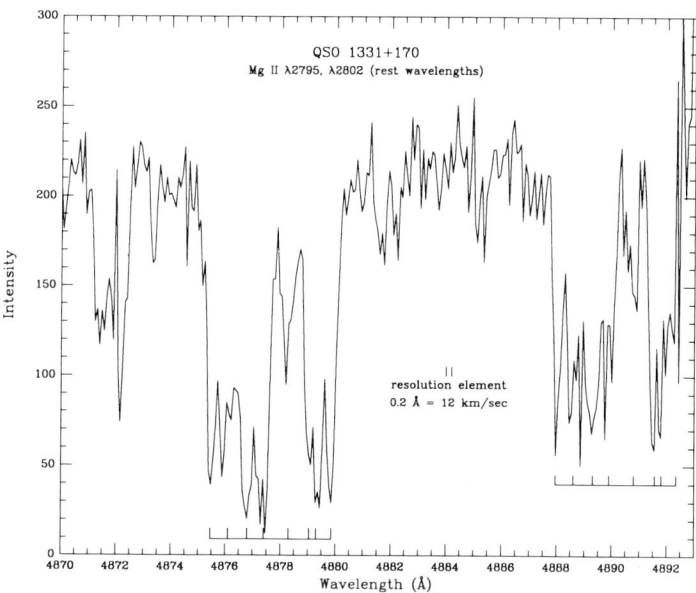

Figure 6. Reduced spectrum of Mg II doublet.

DISCUSSION

SMITH: Do you need different exposure times for different wavelength regions because of the large flux gradients?

VOGT: Yes, of course. When working on cool stars, there is an enormous flux gradient between the UV/blue and the near-IR regions. The dynamic range of the CCD is quite large, but to optimize exposures at such wavelength extremes, two exposures are advised.

EBBETS: Does the close order spacing create difficulties with measuring and removing background and scattered light?

VOGT: Not really. To first order, the orders are just far enough apart where background correction and scattered light removal are straightforward. However, we have not yet completed our software routines for data reduction. I am sure that, as we push to high accuracy, and study the effect more carefully, we will find it necessary to implement some more sophisticated approaches to the task.

HIGH RESOLUTION SPECTROSCOPY WITH A FIBER-LINKED ECHELLE-SPECTROGRAPH

H. Mandel
Landessternwarte
Heidelberg-Königstuhl
6900 Heidelberg
West Germany

ABSTRACT. A compact fiber-linked echelle-spectrograph has been designed and constructed at the Landessternwarte Heidelberg-Königstuhl and was successfully tested during its first observation period in the end of May 1987. The optical design of the instrument is discussed in general terms and preliminary results of the first observations are presented. The reduction of several CCD frames has shown that the real properties of the spectrograph are within a few percent of the calculated ones.

1. Introduction

The concept to use single fused silica fibers for astronomical applications allows physical disjointment from the telescope and a simplified spectrograph design. One also eliminates telescope related flexure in the spectrograph while being capable of housing the instrument in a temperature and humidity controlled room. The light scrambling properties of the fiber results in a constant spectrograph illumination independent of guiding errors and seeing variations. To use the fiber technology, however, two important problems must be solved: the input beam focal ratio degradation and the signal attenuation. Therefore each new fiber must be tested carefully and only a few fibers are useful for astronomical observations.

For a given telescope the $\lambda/\Delta\lambda$-resolution depends on the core-diameter of the fiber which represents the slit of the spectrograph. At the Cassegrain focus of the 0.75 m telescope of the Landessternwarte and slits between 1" and 5" diameter the $\lambda/\Delta\lambda$-resolution changes between 70.000 and 15.000.

2. The optical system and its calculated properties

The major components of the spectrograph and the TV-guiding-system are shown in Fig. 1 and Fig. 2. Fig. 3 shows the calculated properties of the instrument as a function of the echelle-order number. All calculations are done with respect to a blaze angle of $65°$ and a separation between incident and diffracted beam of $12°$ (see references (1)-(5)).

In the TV-guiding-system the infalling f/8 beam from the telescope is reflected by several mirrors and focussed on the photo-cathode of the TV-camera. The field of view using the 0.75 m telescope of the Landessternwarte is 440" x 330". A motor driven mirror and prism allows to change between starlight and light from the ArTh-comparison-lamp or the flat-field-lamp.

The fiber coupler is a mirror with a pinhole of 150 microns in it. The glued in fiber has an outer diameter of 140 microns and a core-diameter of 100 microns which represents the slit of the instrument. At variable seeing conditions it is possible to exchange the fiber in less than 10 minutes.

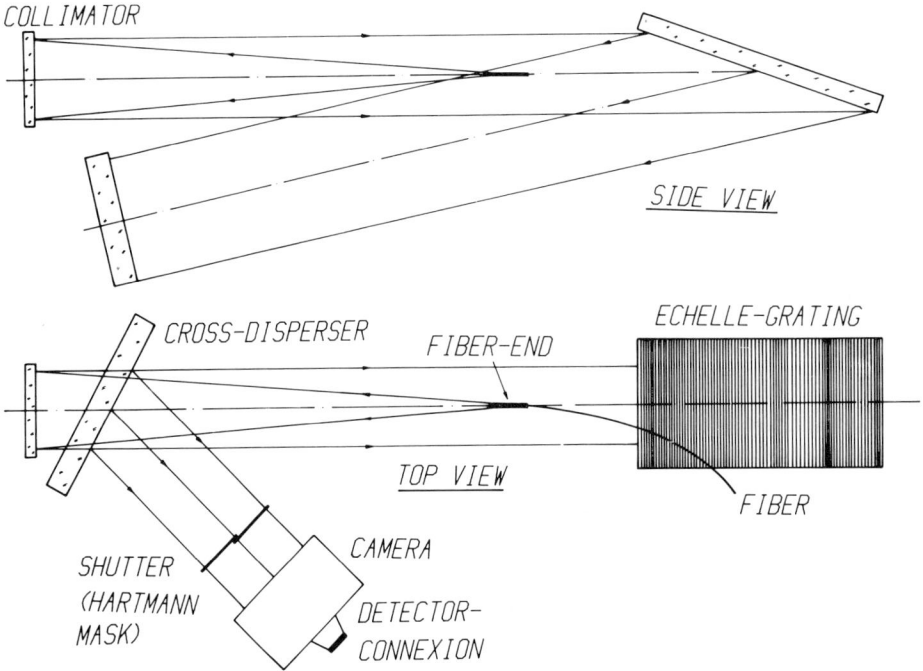

Figure 1. Optical lay-out of the echelle-spectrograph:
collimator: f/5, f=400 mm; echelle-grating: 110 x 220 mm, 31.6 grooves/mm, blaze: $65°$; cross-disperser: 165 x 135 mm, 300 grooves/mm, blaze: $4°.3$; camera: f/2.8, f=300 mm.

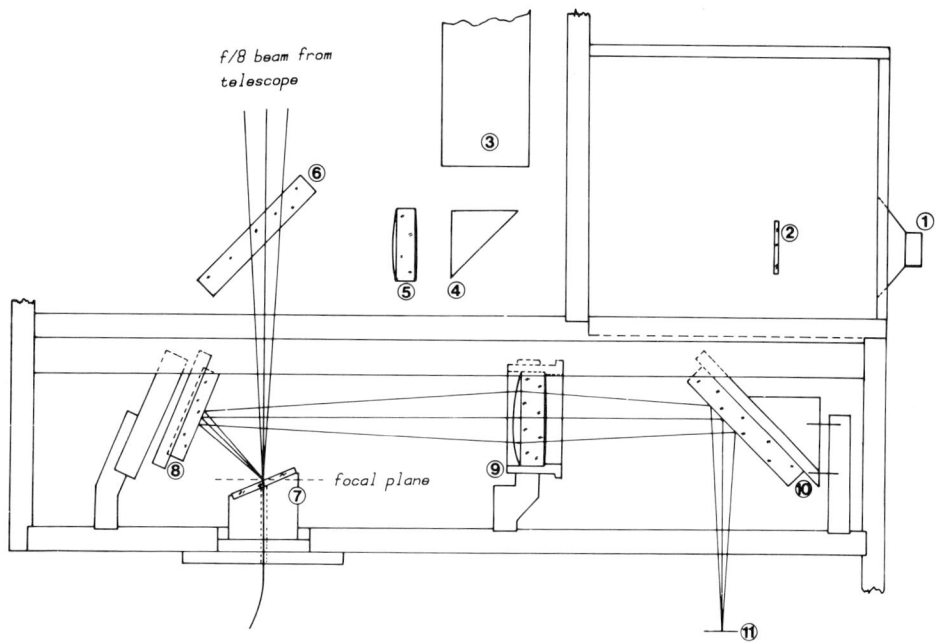

Figure 2. Optical lay-out of the TV-guiding-system:
(1) flat-field-lamp; (2) ground glass screen; (3) comparison-lamp (ArTh); (4) motor driven prism to select the lamp; (5) lens f/2.5, f=80 mm; (6) motor driven mirror; (7) fiber coupler with fiber AS 140/100 from Quartz & Silice; (8) mirror; (9) lens f/2.25, f= 90 mm; (10) mirror; (11) focal plane of the TV-camera.

3. Results

The spectrograph was tested at the 0.75 m Cassegrain telescope of the Landessternwarte. As detector a GEC chip type P8603/A was used. A 530 second exposure of α Boo (exposed through clouds) yielded a spectrum with a signal-to-noise ratio of 235 per pixel at λ= 5500Å. The estimated overall efficiency (including air mass, telescope, spectrograph and detector) at the same wavelength is $\geq$ 0.6 %. At better conditions the instrument reached 1.1 %.

As an example the 96th echelle-order of an one hour exposure of P Cygni is shown in Fig. 4. In the continuum the signal-to-noise ratio differs between 50 and 90. At HeI-wavelength the signal-to-noise ratio is 160.

Figure 3. Calculated properties of the spectrograph.

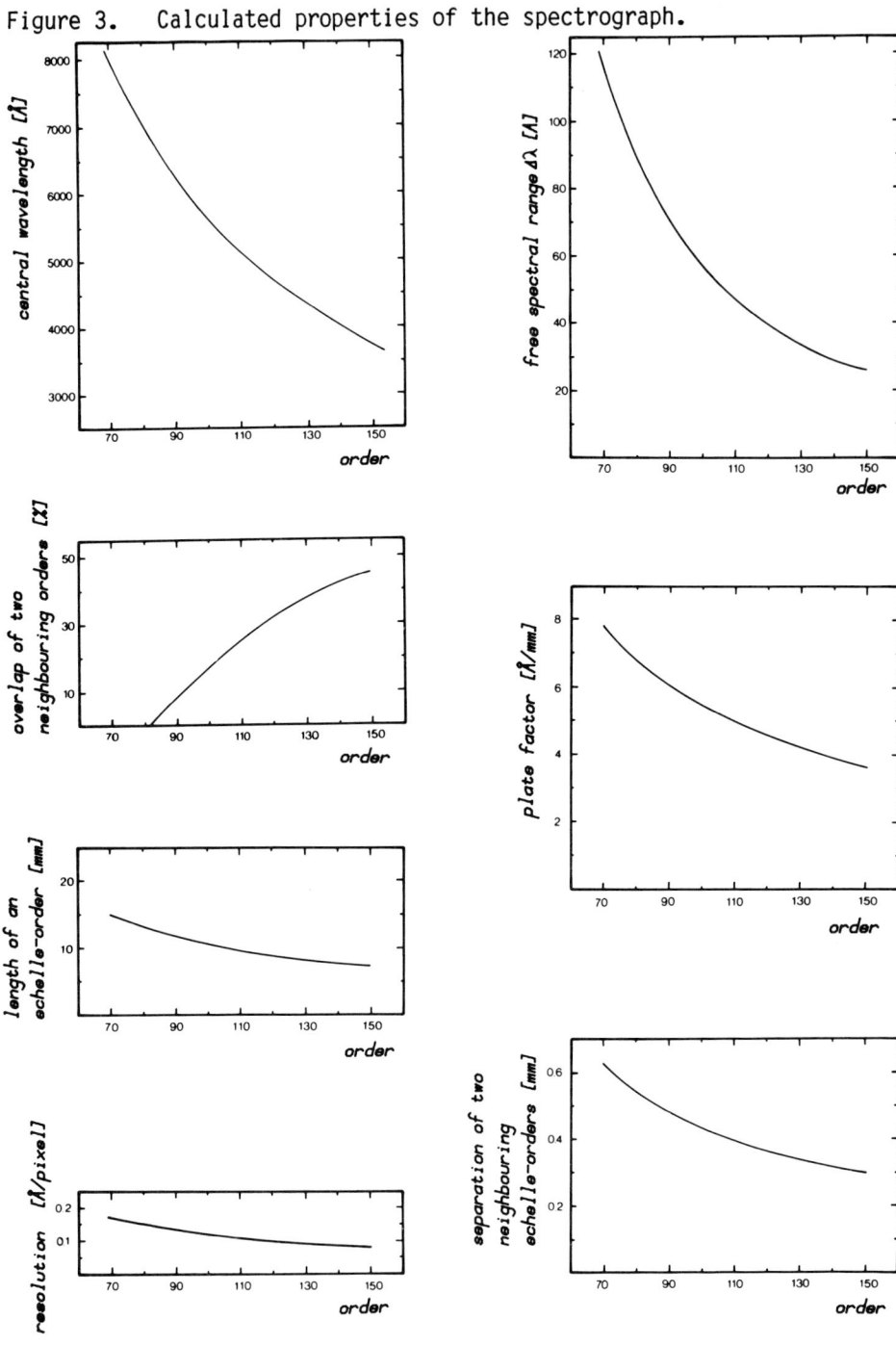

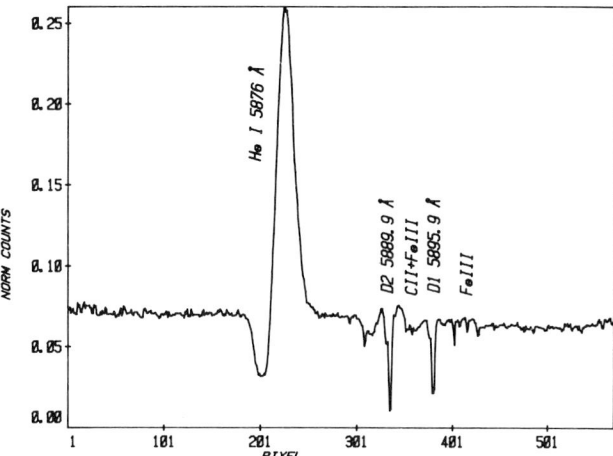

Figure 4.
Bias subtracted and flat-fielded echelle-order Nr. 96 of an one hour exposure of P Cygni.

4. References

(1) Bottema, M., 1981, Applied Optics 20, 528.
(2) Chaffee, F.H., Schroeder, J.D., 1976,
    Ann. Rev. Astr. a. Astrophys. 14, 23.
(3) Harrison, G.R., 1949, J. Opt. Soc. Am. 44, 761.
(4) Schroeder, J.D., 1967, Applied Optics 6, 1976.
(5) Schroeder, J.D., 1970, PASP 82, 1253.

DISCUSSION

SODERBLOM What improvment in S/N has your optical fiber achieved? Does it significantly improve calibration?

MANDEL The illumination of the collimator is exactly the same by starlight, flat-field-light and ArTh-comparison-light. The only link to the telescope is a 5-10m long fiber. In fact of the possibility of housing the spectrograph temperature and humidity controlled you have no bending and temperature effects in the instrument and therefore no blurred spectral lines. The improvement in S/N has not been calculated, but if there is a pixel-shift of 0.5-1 pixel during long exposures for telescope mounted spectrographs you have an improvement in S/N of more than 10 %.

VOGT Do you know how much efficiency you are loosing from polarisation effects with your crossed grating geometry?

MANDEL No, these effects have not been calculated separatly. The overall efficiency of the intrument at 5500Å was estimated to 1.1 %. If I calculate the efficiency with the optical elements I reach a value of 2.8%. The difference of 1.7% is due to guiding-errors, seeing variations and polarisation effects.

BAADE Is one of the advantages of using a fiber link that the internal flat-field lamp provides much better correction for fringing and low spatial frequences distortions than a conventional coupling between telescope and spectrograph would do?

MANDEL Yes, it is.

LINSKY What is the maximum spectral resolution possible with your instrument setup?

MANDEL The maximum $\lambda/\Delta\lambda$-resolution depends on the core-diameter of the fiber which represents the slit of the spectrograph. At best conditions I can reach a $\lambda/\Delta\lambda$-resolution of $\simeq$ 70.000.

THE MOST EFFICIENT TOOL FOR VERY HIGH RESOLUTION AND VERY HIGH SIGNAL TO NOISE RATIO SPECTRAL OBSERVATION OF STARS

S. Y. Jiang
Beijing Astronomical Observatory
The Academy of Sciences of China
Beijing 100080
The People's Republic of China

ABSTRACT. Today even for the most efficient spectrograph combined with a large telescope the light efficiency is only about 0.01 to 0.1 for spectral resolving power R larger than 10000 in optical wavelength band (OWB). Consequently for a very high signal to noise ratio spectral observation of rather bright stars still needs very large telescope. The main reason is that there are too many optical surface with rather low light efficiency and serious light loss at the limited slit width. In this paper we suggest a very high efficiency telescope-spectrograph system which will give an overall light efficiency varied from 0.21 at 400 nm to 0.44 at 700 nm, four fold higher than before. Using this system for R = 100000, S/N larger than 100 the limiting magnitude will be about 15.

1. INTRODUCTION

Today even for the most efficient spectrograph combined with a large telescope the light efficiency is only about 0.01 to 0.1 for spectral resolving power R larger than 10000 in OWB (Oke and Gunn, 1982). Consequently, for a very high S/N spectral observation of rather bright stars still needs very large telescope. So recent year there are many New Generation Telescope Programs in western countries. Because they are too expensive and also not very efficient, only very few mait be able stood to a real result. China have the largest population but still rather poor. We will not join such a competation but try do the best to promot the developing of astrophysics. After recognize that the main reason for low efficiency for a telescope-spectrograph system is that there are too many optical surface with rather low light efficiency and serious light loss at the limited slit width, we suggest a very high light efficiency telescope-spectrograph system as the next step for chinese astronomers after the 2.16 metres telescope.

2. THE IMPORTANT FACTORS FOR CHOICE OF SYSTEM

2.1. The Scientific Requirements

Recent astrophysics are mainly concentred on two large fields: The stellar physics and The extragalactic objects physics.

In stellar physics, to solve the evolutional problem, we must know the accurate chemical composition and atmospher condition of stars with different age, in different regions of different galaxies and stellar systems. The only reliable way of deriving atmospheric abundances of stars is the comparison of detailed theoretical models with high resolution spectrograms. We also need to know the constructions, the properties, the kinematics and the variations of the stellar winds and circumstellar mass flows for different stars. The rotation and magnetic fields also play important roles. All these can be determined from a careful analysis of its complex emission and absorption contribution to the stellar line spectrum. In some rather far-away star's spectrogram, we can find some narrow absorption lines caused by interstellar gas. We need to research there composition and kinematics. To determine the stellar radial velocities of different line system is also a common work. For all these work, we need spectral resolutions between 10000 to 100000, signal to noise ratio between 100 to 300, a broad wavelength coverage extending from 365 nm to 1100 nm, and a limiting magnitude better than 15 with a reasonable integration time.

For quasar absorption lines, we also need high spectral resolutions and high signal to noise ratio, but the limiting magnitude should be better than 20, which is very difficult to attain.

2.2. The Sites and Seeing

Good sites is very important for high light efficiency high spectral resolution instruments. If the diameter of the seeing disk change from 2 arcseconds to 1 arcsecond the light efficiency will be double. Although there are some good sites with seeing near 1 arcsecond, for China, we will use 2 arcseconds as design parameter, and the entrance slit width will be 2.3 arcseconds to accept more than 95% light collected by telescope.

2.3. The Dispersion Way and the Size of Echelle

For such a high resolution and wide wavelength coverage, the best dispersion way is to use echelle with prism as cross disperser. Up to now the echelle ruled by conventional techniques are limited in size up to 300 mm X 600 mm ruled area. To match such a large echelle, the cross dispersion prism will be very large and will absorb a large parts of light. So, if possible, we favour to use echelle with ruled area of 260 mm X 408 mm.

2.4. Detector and its Size

Up to now the best detector for echelle spectrograph is CCD. The largest CCD chip will be produced in recent year will be 2000 X 2000 pixels with pixel size of 27 microns. two of them side by side will cover a field of view about 6.°2 for a camera with focal length about one metre.

2. 5. The Camera

The best camera still is the classical Schmidt type camera. To match the detector pixel size with spectral resolution larger than 100000, the focal length will be about one metre. Of course, for different objects and different work, we would better to have at least 2 camera with different focal length, and easy to change.

3. THE OPTICAL ARRANGEMENT

3. 1. The Telescope

From the theoretical point of view, for a given pixel size and total pixels of detector, under the seeing limited condition, the spectral resolution times the luminosity and the information unit number attainable in one single exposure is almost the same for any aperture of telescopes. The only problem is that if the aperture of the telescope is too small, the integration time will be too long to get enough signal to noise ratio for the faint objects. So we must make some compromise between the aperture of the telescope and the information unit number attainable in one single exposure.

If there is no any light loss within the telescope-spectrograph system, in the OWB, the monochromatic photon rates within 6 hours for a 15th magnitude A0V star will be about 40 to 300 nm cm for different wavelength. If the resolving power is 100000, the signal to noise ratio is 100, we must collect at least 2 million photons in 1 nm. So the diameter of the telescope must larger than 2.6 metres. On consider about the light loss, the smallest diameter of the telescope is 3 metres. For such a telescope, the only usable focus for high light efficiency is the F/3 prime focus of a paraboloid mirror coated with a protected silver coating which will give better than 95% reflection for wavelength longer than 400 nm and still reasonably good till 365 nm. In principle we can directly attache a very high light efficiency, high spectral resolution wide band spectrograph to the prime focus (Jiang, 1978), but on consider about the technical improvement and convenience of optical fibre, we tends to use optical fibre linking from the prime focus to a spectrograph on the stable floor.

3. 2. The Dichroic Filter and Beam Spliting

The overall wavelength coverage we intended is from 365 nm to 1100 nm. To get high light efficiency for such a wide wavelength range, the light beam coming from the primary must be splited to two beams by a dichroic filter just before the prime focus. The best dividing wavelength is about 630 nm on considering equal orders in two beams. So we have two foci: one for the wavelength shorter than 650 nm (SWB) , another for the wavelength longer than 610 nm (LWB). Putting the dicroic filter before the focus can ease the atmospheric dispersion and can make use of specific optical fibre for different wavelength band to improve the light efficiency. Because there are 40 nm overlaping on wavelength in two beams, so we can add the signals in two beams to one single signal for the same wavelength to get very high light efficiency for those wavelengths near the separation.

3.3. The Optical Fibre Linking and Image Slicer

For the F/3 prime focus, it is very easy to match with any kind of optical fibres. For the SWB, a 10 metres long FHA type fibre can give a better than 90% transmition; for the LWB, a 10 metres long QSF-ASW type fibre can give a better than 95% transmition. The field scale is about 23 arcseconds per millimeter, so a core diameter 100 microns can accept 2.3 arcseconds on the sky. For higher resolution, a special optical fibre bundle made by fibres with 20 microns core diameter is needed to act as an image slicer. If the end shape is specially made (such as square shape) for each fibre, and is thin cladding, the whole stellar image will be accepted with very low loss except the image height on the detector become about 30% higher. A Walraven type image slicer (Walraven, 1972) after a single 100 microns fibre with focal ratio enlarge micro lens is also very attractive.

3. 4. The Echelle Spectrograph

To keep the image height of the entrance slit on the CCD chip smaller than the smallest distance of two nearby orders, we suggest a ruling height of the echelle larger than half of the width of the ruled area of the echelle. For equal orders and smaller angular field of view, the rulings are 55 per mm and 31.5 per mm respectively in SWB and LWB. Both echelle are blazed at $63°.433$. To ease the arrangement and changing of cameras with different focal length, it is better to choose a larger angle of construction (the angle between the incident beam and the blazing diffraction beam) which will make the blaze efficiencies and angular dispersions lower than Littrow configuration (Schroeder and Hilliard, 1980), but we can get rather smaller collimating beam size, so can use smaller prism for cross dispersion and beam widening which will have smaller light absorption. To find back the blaze efficiencies we can double the field of view in the main dispersion direction to include all the light of each wavelengths in three nearby orders and add the signal from corresponding CCD pixels to one single signal, so that for all wavelengths we get the same light efficiency as the peak efficiency in Littrow configuration. By the way we can use wider entrance slit to get higher light efficiency, and can use several pixels to share the whole light of the same wavelength to ease the saturation problem for high signal to noise ratio.

If the angle of incidence is $73°.433$, then the blazed angle of diffraction is $53°.433$, for a camera focal length of 1000 mm, two 2000 X 2000 pixels CCD chips with pixel size 27 microns square can cover a $6°.18$ X $3°.09$ angular field of view, which will include the whole wavelength band for each beam and the all light of each wavelengths within 3 nearby orders. Matching with 27 microns pixel size, the angular entrance slit width on the sky is 0.45 arcseconds, and the spectral resolving power is just 100000. Of course we need an image slicer with 5 slicers to make full use of the whole stellar light collected by the telescope. Then the total angular height of the entrance slit on the sky must larger than 12 arcseconds.

The collimeting beam size is 116 mm in diameter. For the cross dispersion of SWB, we choose a 55° apex angle prism made by KF6 glass, which will give an angular dispersion of 2°.25, and have a size about 150 mm X 150 mm X 120 mm. The output monochromatic beam size in the dispersion direction will be 120 mm which is not enough to fully use the groove height of the echelle, so we should add another KF6 prism for beam widening. If this prism is a 90° prism with an apex angle of 37°, and the incidence angle on the longest side is 67°, the output beam size for monochromatic light in the dispersion direction will be 240 mm. This prism will also produce an angular cross dispersion of 0°.9, so the total angular cross dispersion can easyly adjusted to 3°.09 which will make fully use of the CCD chip. The size of this prism will be 185 mm X 250 mm X 120 mm. The average light path is about 180 mm for the two prism on sum, so even for 365 nm, the absorption will be less than 10%. For the LWB, we choose ZF6 glass prism to get the same cross dispersion and the same beam width. The average light path is about 150 mm, so the absorption will be smaller than 7% for any wavelengths.

The monochromatic beam size of the camera will be 243 mm in diameter. The CCD chip on the focal plan with cold finger will obstruct less than 7% incident light. The total orders for both beam are 39, so the average separation is 51 pixels on CCD chip. The smallest separation is 34 pixels. For a 2".0 seeing, we choose the real image height as 2".3, which will give an image height on the CCD chip of 5 pixels. With 5 slicers for the whole image, the total image is about 26 pixels on height, so there are still 8 pixels for sky substraction.

Some time we only need lower spectral resolution but shorter integration time or simply for fainter objects, we can change by use a camera with focal length of 250 mm, equiped with two 512 X 512 pixels CCD chips. In this case, we do not need an image slicer, so we can use 5 independent optical fibres for 5 stars to make multi objects spectral observation.

4. CONCLUSION

Our design idea is to fully use the light collected by the primary of a simple telescope. We are not only try to make the image width of the entrance slit but also the image height as small as possible. We also try to fully use the beam separation and almost loss nothing for any wavelengths. We also can fully use the diffraction light for any wavelengths fallen in different orders. The light absorption in cross dispersion prisms also is very low, so the whole light efficiency for 400 nm is 0.21, for 700 nm is 0.44. The photon events rate for all the wavelengths between these two wavelengths within 6 hours will be larger than 2 million per nm. So the limiting magnitude will be 15 for S/N = 100. This is 10 times higher than that for the coude spectrograph of AAT (Walker and Diego, 1984).

REFERENCES

Jiang, S. Y., 1978, Acta Astron. Sinica, 19, 204.
Oke, J. B., Gunn, J. E., 1982, Publ. Astron. Soc. Pacific, 94, 586.
Schroeder, D. J., Hilliard, R. L., 1980, Appleid Optics, 19, 2933.
Walker, D. D., Diego, F., 1984, in Proceeding of the IAU Colloquium No. 79: Very Large Telescopes, their Instrumentation and Programs, eds. M. H. Ulrich, K. Kjar, P. 499.
Walraven, J. H. and Th., 1972, in ESO/CERN Conference on Auxiliary Instrumentation for large telescopes, eds. S. Laustsen and A. Reiz, P. 295.

DISCUSSION

VOGT How much order separation do you get ? My experience with prismatic cross-dispersion is that you cannot get near enough order separation with prism to have multiple objects (or multiple fibers) packed between adjacent orders.

YANG We use KF6 type glass for blue band. If the top angle is $60°$. The cross angular dispersion is :
$\Delta\beta = 2*\Delta n / (\sqrt{(4-\hbar^2)}) = 0.04649 = 2°66$ between $\lambda = 3300$ Å to $\lambda = 5461$ Å with 27 orders.
After this prism, we added another prism with top angle of $45°$, so the total angular separation is about $3°8$. With a focal length about 200mm, the linear separation will be 13.2mm. The smallest order separation will be about 240 micron. I think it is large enough for about 2 or 3 image with seeing disk of 2" (for more slices, you add another prism).
For red beam, we use 2F6 (SF4) type prism, it is also easy to get $3°5$ angular cross dispersion. It is very important to have two or 3 beams.

RUTTEN I want to point out a major difference between solar and stellar échelle spectrometry. In solar work one does not use a crossed presisperser to put the orders side by side, but instead a parallel predisperser with slits in the predispersed spectrum, to select short segments of specific orders to be projected linearly adjacent on the detector. The best example is the Sacramento Peale échelle spectrograph.
The advantage is two fold :
(i) - data compression. You only select those spectral elements you are interested in, to fall at high dispersion on the detector wherever they are in the spectrum. For example, in a cool-star activity program we would combine the Ca II H and K lines, the Ca II IR triplet and Hα, getting the important diagnostics while not registering lots of excess data.
(ii) - multiplexing. The other detector dimension is yet available. In solar physics it is employed to measure monochromatic image detail spatially along the slit. In stellar physics it might accomodate multiple input from an image-plane fiber-coupled aperture array.

YANG Yes, you are right. But, for solar physics, the object is an extended source, and only the sun, and it is very bright. Many people observe on it frequently. So it is no special need for very wide wavelengh range record. But in stellar physics, there are too many objects, only few people observe occasionally on some specific stars, and the exposure time usually is very long. So for easily find new things, you must take a wide wavelength range record. That is the difference.

HIGH SIGNAL TO NOISE OBSERVATIONS WITH A PHOTON COUNTING ARRAY

A.W.Rodgers, P.Harding, G.Bloxham and M.S.Bessell
Mount Stromlo and Siding Spring Observatories
The Australian National University
Private Bag Woden ACT 2606
AUSTRALIA

ABSTRACT. A simple high speed image widener is described which allows effective count rates of up to 100hz to be achieved with photon counting detectors thus allowing rapid accumulation of high signal to noise data in bright star spectroscopy.

1. INTRODUCTION

A significant limitation of photon counting detectors arises in saturation effects associated with high photon rates per pixel incident of the system.

While current versions of photon counting systems are able to provide detective quantum efficiencies, at low counting rates (below ca. 0.6Hz), which are equal to the diodic quantum efficiencies of the first cathode in the system, they compare unfavourably with analogue devices at high incident illumination levels. In the case of the Mount Stromlo photon counting array (Stapinski et al, 1981) and previous generation system such as the IPCS (Boksenberg et al, 1972) photon counting is achieved by optical detection of an intensified photon event using a TV type device such as a video camera or CCD operated in the line transfer mode. Among other properties, the basic limit is set on photon detection at high incidence rates when a significant proportion of the photons fall on the same pixel within the frame time of the CCD. Our photon counting arrays operate at up to 15 MHz clock frequencies and have frame subtraction circuitry to eliminate effects due to phosphor decay. Nevertheless, on point sources, count rates much above 1Hz show co-incidence effects which limit the use of the system for accurate photometric work.

Even with the highest resolutions available to us (up to 500,000) normal échelle spectroscopy of bright stars becomes inefficient when high signal to noise data is required due to the long exposure times required to accumulate the 10^4 to 10^6 counts necessary. This is additionally true for the acquisition of flat fields, which if they are to be useful in correcting pixel to pixel variations in high S/N data, must contain a significantly greater number of recorded photons.

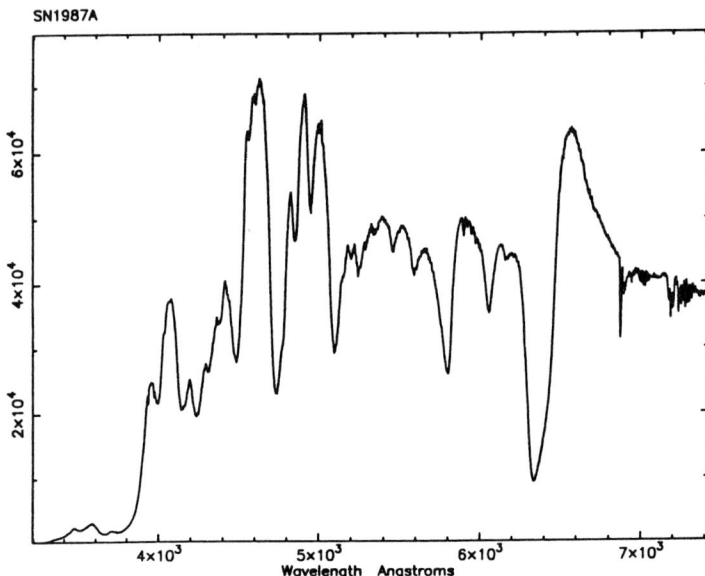

Figure 1. Spectrum of SN 1987A, the supernova in the Large Magellanic Cloud, obtained on the 74-in telescope with a resolution of 40 kmsec^{-1} with a S/N > 100 which was obtained in an exposure time of 600 seconds.

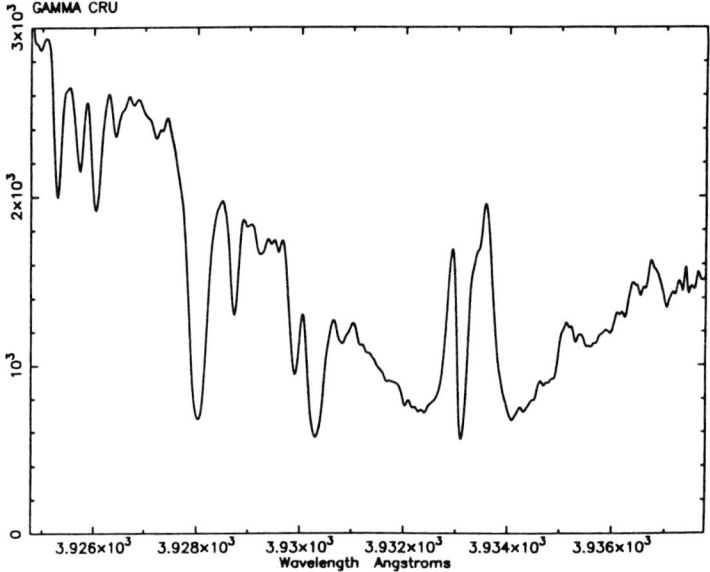

Figure 2. The region of the Ca II K line reversal in the M3 giant γ Crucis. The exposure was 40 min and in the wings of the line and in the other orders of the échelle image the effective count rate was 25Hz.

We describe below a simple and very effective device allowing high count rates on bright stars to be achieved.

2. THE RAPID WIDENER

In the Mount Stromlo photon counting arrays typical frame times lie between 7 and 20 msec. In typical seeing, a stellar image falls on 1 to 2 pixels perpendicular to the dispersion of the spectrum: essentially all the light is concentrated in one or two rows of the array. It is therefore clear that by spreading the light over the width of the array, typically 200 pixels, the total effective count read can be increased by between 100 and 200. Inefficient techniques, such as insertion of neutral filters and defocusing, can of course be used to keep the spectrum intensity below the system limit; however, as discussed above, it becomes logistically impossible to collect high signal to noise data in this way. Therefore we determined to widen the spectrum in a fashion that was analogously done in the era of photographic stellar spectroscopy. It is, of course, true that the widening rates used then would not spread the light along the slit in times small compared with the CCD frame time. We have therefore placed in front of the slit a hexagonal prism which is rotated at a speed of 6000 rpm which, through its rotation, wipes the star along the slit as each face of the hexagon is in turn presented to the incoming beam. Because of the hexagonal nature of the prism, on 6000 rpm it displaces the star image along the slit each 3 msec, a time which is short compared with the frame time of the CCD in the photon counting array system so that the photon counting array sees the spectrum as a uniformly illuminated strip.

In the case of the 74-in telescope coudé spectrograph the hexagon has a distance between its parallel faces of 42 mm and produces a widening at the slit of 22 mm. With the 32-in coude camera with 6-in beam size an image of the spectrum 3 mm or 100 pixels wide is recorded by the detector. Thus with a count rate per pixel of 0.5Hz a 1,000 second exposure we can record a total of 5×10^5 counts.

Since the coudé focus of the 74-in telescope has a focal ratio of f/32, negligible aberrations are introduced into the beam. Secondly, and most importantly, guiding from light reflected from the slit back through the hexagon to the video guider is possible since the guider sees a stationary star and the slit image behind it being oscillated backwards and forwards at 300Hz.

We present in Figures 1 and 2 examples of recent spectra obtained using the rapid widener. It is clear that such a device as the rapid widener will allow high signal to noise ratio spectra to be obtained with reasonable efficiency in the blue spectral region of many bright southern stars in wavelength regions of interest.

REFERENCES

Boksenberg, A. and Burgess, D., 1972 *Electronics and Electron Physics*, 33B, 835.
Stapinski, T.E., Rodgers, A.W. and Ellis, M.J., 1981 *P.A.S.P.*, 93, 242.

HIGH RESOLUTION SPECTROGRAPH NEEDS FOR STELLAR SEISMOLOGY

Søren Frandsen
Astronomisk Institut, Aarhus Universitet
DK-8000 Aarhus C
Denmark

1. INTRODUCTION.

There are two ways to become a famous observing astronomer: 1) by discovering something new and exciting, 2) by detecting a phenomena, which has been predicted, but not observed before.
Observing stellar oscillations belongs to the second category. Indications are around, that solar type stars do oscillate, but there is still a lot of uncertainty, so that nobody can claim to have a 100% solid identification of oscillations in a solar-type star.
The prospects for improving the understanding of stellar structure and evolution have been discussed by Christensen-Dalsgaard (1987) and by Däppen (1988).

2. WHAT ARE WE LOOKING FOR?

The signals that one should observe are shortly described in Table I.

TABLE I

Characteristic amplitudes of acoustic oscillations.

	Sun	F0 V star
Velocity	15 cm/s	40 cm/s
Intensity	3.5×10^{-6}	6.5×10^{-6}
Typical Period	300 s	350 s
Frequency Spacing	137 μHz	96 μHz

Clearly we are dealing with very small amplitudes, which can only be measured with carefully designed instruments.
Let me concentrate on techniques for observing oscillations from the ground. It might be possible to do such measurements with a photometer using differential photometry (Harvey et al.,1987), but generally the instruments

proposed for doing efficient observations integrate a high resolution spectrograph (HRS) into the design. One can think of a stellar seismometer as consisting of three parts:

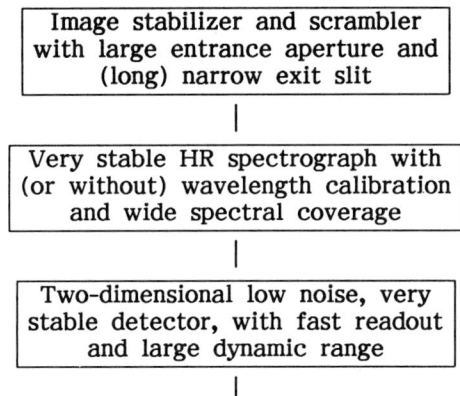

The first critical element brings the light from the telescope to the HRS. It will almost certainly consist of one or more optical fibers, which scramble the light and bring the light to a more convenient location for a spectrograph than close to or attached to the telescope. Some additional image stabilization could be needed and reimaging might be necessary to match the entrance aperture of the spectrograph.

Secondly one must have a very stable spectrograph, either inherently by eliminating internal turbulence and temperature drifts, or by calibration by a very stable reference source. To reach as low as the amplitudes observed in the Sun, a reference source is probably a necessity.

The detector must have low noise, so that readout noise does not dominate. It must be extremely stable with no change in sensitivity or zero point drift. The Reticon detector at the CES at ESO does not have that stability (Frandsen, 1987). A two dimensional detector is required to have access to several echelle orders and cover a wide spectral range. One needs to record as many photons as possible. Otherwise measurements are restricted to a few very bright stars.

3. TIME REQUIRED TO GET SATISFACTORY S/N RATIOS.

In one echelle order there are close to 20 useful lines, that are not awfully blended (Frandsen, 1987). The gradient of f.ex the K D1 line has a large value only over a range of 0.05Å, so that one can define a typical spectral element of that size. The maximum gradient will for a non-rotating star be given in velocity units by

$$\frac{d\log I}{dv} \approx 0.4 (km/s)^{-1}$$

This means, that a velocity amplitude of 1 m/s corresponds to an intensity change of

$$\frac{\Delta I}{I} \approx 4 \times 10^{-4}$$

Let us now assume, we have a 3.6m telescope at our disposal and the total instrumental efficiency is 1%, and we are observing a $m_V=0.0$ star. Then the number of photons recorded is

$$N \approx 4 \times 10^4 photons/s$$

If more than one line is used simultaneously, the equivalent bandwidth will be ≈ 1Å for one echelle order. The observing times required for a moderate S/N≈ 4 is listed in Table II.

TABLE II

Observing time(seconds) for a S/N ~ 4.

Technique	t(1m/s)	t($V_\odot$)
Doppler shift 1 linewing	150	5400
Doppler shift 10 lines	35	1200
Intensity 60Å band	150	5400

If we instead try to observe the intensity change in the line cores, the typical equivalent bandwidth is 5Å. The signal one expects for the Sun is two times the continuum changes (Frandsen, 1984, Stebbins and Goode, 1987). The amplitude will be around 7×10^{-6} for the Sun. The number of photons recorded within the line cores is

$$N \approx 4 \times 10^6 photons/s$$

The last row in Table II is for this case.

A velocity measurement is the most demanding technique considering the stability of HRS and the entrance element. Intensity studies of line cores is more vulnerable to detector flaws. Roughly speaking the accuracy required on time scales appropriate for stellar oscillations can be defined as in Table III.

TABLE III

Requirements for a seismometer.

	velocity	intensity
Velocity stability	10 cm/s	1 m/s
Detector/bias stability	10^{-3}	10^{-4}

The stability is for timescales around 5 minutes.
Bias stability describes zero point drifts.
Detector stability problems are changes in sensitivity, nonlinearities, hysteresis etc.

4. CONCLUSION.

If a very stable spectrograph can be constructed, it is more efficient to observe Doppler shift, because the related intensity changes will be the largest. To use telescope time efficiently several echelle orders should be observed with a two-dimensional detector, and to secure a high duty cycle fast readout should be possible. If a very stable detector is available, but no spectrograph, that satisfies the requirements of Table III for measuring Doppler shifts, a reasonable decision could be to measure line core changes. Ideally one should be able to derive both velocity and intensity oscillations from the observed spectra, which would give some additional information about amplitude ratios and phase delays of the oscillations.

REFERENCES.

Christensen-Dalsgaard, J., 1987a, *Proc. of IAU Symposium 123: Advances in Helio- and Asteroseismology*, (Eds. J. Christensen-Dalsgaard and S. Frandsen), Reidel, Dordrecht, to be published
Däppen, W., 1988, *These Proceedings*
Frandsen, S., 1984, *Proc. of the 25th Liege Int. Astroph. Coll.*, 303
Frandsen, S., 1987, *Astron.Astrophys.*, 181, 289
Harvey, J.W., 1987, *Proc. of IAU Symposium 123: Advances in Helio- and Asteroseismology*, (Eds. J. Christensen-Dalsgaard and S. Frandsen), Reidel, Dordrecht, to be published
Stebbins, R., and Goode, P.R. 1987, *Solar Phys.*, to be published

DISCUSSION

FOING How does the stability of such a spectroseismometer compare to that of a resonance cell? Or of a CORAVEL type instrument? What is the gain in S/N of a v measurement brought by the image stabilization, apart from the photon collection?

FRANDSEN A seismometer would probably use a resonance cell or Fabry-Perot filter to calibrate instrumental drifts, so the stability would be comparable to a resonance cell. A CORAVEL has no scrambling of the image and not the stability to be used as a seismometer. Without elimination of the effects of seeing, oscillation measurements are impossible at solar amplitude levels.

FIBER FED SPECTROGRAPH FOR LINE VARIABILITY STUDIES

P. Felenbok and J. Guérin
Observatoire de Paris, Section de Meudon

Introduction : In our studies of activity in pre-main sequence Herbig Ae/Be stars we are mainly interested in searching rotational modulation of line profile. If the period of star rotation is of the order of one or two days, the data collected from a single site is insufficient. This led us to start correlated observations from two or three sites spread as much as possible in longitude. Our first bi-site observations started in 1982 on AB Aur from two observatories located 11 hours apart.: CFHT in Hawaï and OHP in France. To achieve a high flexibility and to gain access to telescopes without attached spectrographs. We built an instrument that is mobile and specially designed for line profile studies.

The spectrograph (ISIS) : The optical design is of the Czerny-Turner perpendicular type which is a pure Littrow mount giving a high luminosity in the blaze angle. Fig. 1 shows the spectrograph and his link to the telescope. This instrument is coupled to any kind of telescope by an optical fiber. We use step index silica-silica fibers with 133 µm or 200 µm core. In order to take into account the fiber aperture degradation the fiber is fed with a telescope beam at f/5 and the spectrograph collimator is a parabola on axis at f/3.7. Two gratings are mounted in a flipping barrel, one for a resolution $R = 3.5 \ 10^4$ and the other one for $R = 10^4$. The optical quality of the instrument, with the slit parallel to the grating grooves was checked on photographic plates in a 36 mm field, far bigger than the CCD's surface. Fig. 2 shows the test spectrum taken with a thorium electrodeless discharge lamp. As it is seen on fig. 2, our optical design is generating tilted spectral lines with respect to the direction of dispersion. This is handled quite well with standard reduction software and cannot be avoided if the uttermost resolution is required.

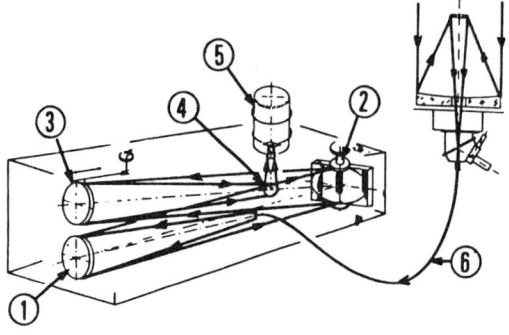
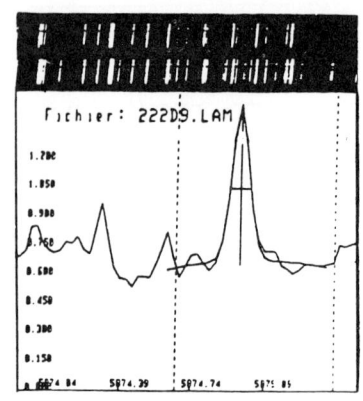

Fig. 1 : The optical design of the ISIS spectrograph
1) Parabolic collimator Ø 160 mm   f = 600 mm
2) Gratings -a 600 gr/mm B L   blaze = 48°
           -b 300 gr/mm J Y   blaze = 17°
3) Spherical camera Ø 200 mm   f = 1000 mm
4) Flat Newton mirror
5) Detector : CCD RCA or Thomson
6) Optical fiber

Fig. 2 : Test spectrum
Detector   : Photographic plate
Slit   : 21 μm
Grating   : 600 gr/mm 4th order
Dispersion : 2.9 Å/mm at NaD
Resolution : $R = 6\ 10^4$

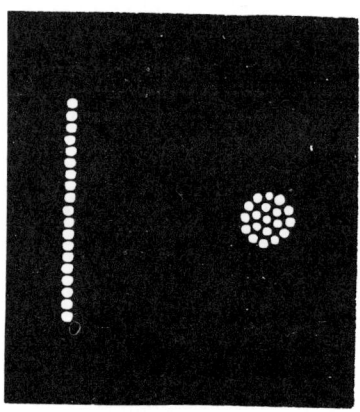
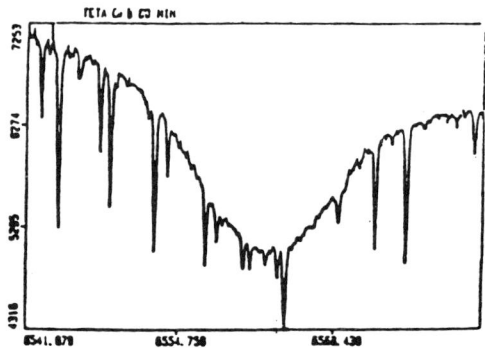

Fig. 3 : Fiber image slicer
19 glass fibers. Ø 45-50μm
l = 30mm

Fig. 4 : Resolution test on θ Cra
$R = 3.5\ 10^4$

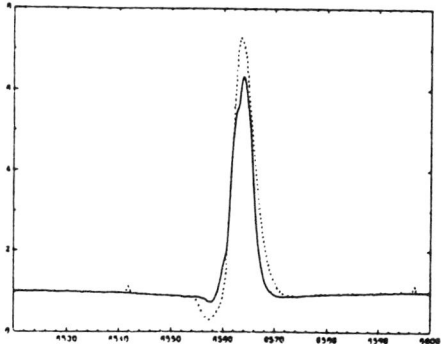
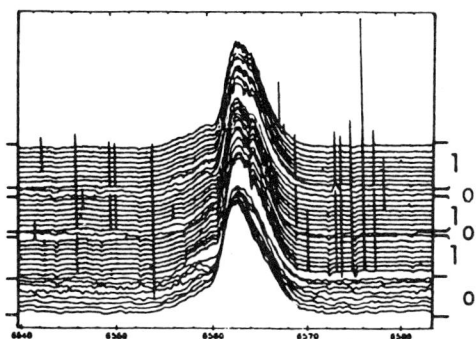

Fig. 5 : Hα variability of AB Aur (m_v = 7.2)   Fig. 6 : Hα bi-site observations of BD+46°3471

However, when a CCD with 30μm pixels is used, it is possible to set the lines perpendicular to the dispersion by rotating the slit with an adequate angle, thus simplifying the reduction procedure.This is the actuel set-up used by observing visitors on ISIS at OHP. The dotted aspect of the spectral lines seen on Fig. 2, is due to the use of an fiber image slicer instead of a slit.The shape of the image slicer is presented on Fig. 3. The image slicer must be made of glass because silica fibers have a thick cladding leading to a poor packing efficiency.

The observations : The spectral capability of ISIS was checked on the sky in association with the 193cm telescope at OHP. Fig. 4 shows a spectrum of θ Cra . The real spectral resolution is derived from this spectrum.The resolution is 3.5 10^4 and the water vapor lines are clearly visible. This is a very important point, because when searching for spectral variability one has to be cautious about water vapor variation if insufficient resolution is employed. Fig. 5 shows the Hα variation in AB Aur, a PMS Herbig Ae star with a resolution of 10^4 . ISIS was also used in a bi-site campaign on a rotational modulation search of the Herbig Ae star, BD+46°3471. Fig. 6 shows the Hα variations for this star as seen from OHP(o) and LICK(l). A Fourier analysis of the data showed no periode in these variations.

Acknowledgments:We are pleased to thank G. Rousset and I. Raulet who where deeply involved in ISIS construction as well as D. Gillet who is the astronomer in charge of ISIS at OHP. H. Hubert provided us with the θ Cra spectrum and M. Dreux and T. Fauconnier participated in the BD+46°3471 observation at OHP.

Question by : I. Appenzeller

Could you give a number on the total efficiency of your spectrograph when working in the "image slicer mode", i.e. when observing stars with a matrix of many small diameter fibers?

Answer : With our present image slicer (19 fibers, Ø 45-50µm, 30 mm long) the gain over a slit is 35%. We are making a new one, with only 7 fibers that is supposed to increase this gain.

Question by : Deshpande

Do the fibres introduce any polarization ?

Answer : Our step index fibers do not introduce any polarization. To preserve polarization of propagated light, single mode fiber with special cladding should be employed.

Question by : G. Marcy

The nonuniform illumination of the input to the fiber at the telescope focal plane may cause nonuniform illumination along the slit. The nonuniform illumination is due to the stellar image not being as large as the fiber. Does this affect flat fielding ?

Answer : When using a long optical fiber, its output illumination is uniform and independent from the input position of the beam at the entrance face. This leads to an uniform illumination of the slicer whatever the position of the star at the entrance of the single fiber.

PHOTOMETRIC CHARACTERISTICS OF DATA FROM THE HUBBLE SPACE TELESCOPE GODDARD-HIGH RESOLUTION SPECTROGRAPH

Dennis C. Ebbets
Ball Aerospace Systems Division
P.O. Box 1062   JWF-3
Boulder, CO 80306   USA

Sara R. Heap
NASA Goddard Space Flight Center
Greenbelt, MD 20771   USA

Don J. Lindler
Advanced Computer Concepts
Potomac, MD 20854   USA

ABSTRACT. The G-HRS is one of four axial scientific instruments which will fly aboard the Hubble Space Telescope (ref 1,2). It will produce spectroscopic observations in the 1050 A $\leq \lambda \leq$ 3300 A region with greater spectral, spatial and temporal resolution than has been possible with previous space-based instruments. Five first order diffraction gratings and one Echelle provide three modes of spectroscopic operation with resolving powers of R = $\lambda/\Delta\lambda$ = 2000, 20000 and 90000. Two magnetically focused, pulse-counting digicon detectors, which differ only in the nature of their photocathodes, produce data whose photometric quality is usually determined by statistical noise in the signal (ref 3). Under ideal circumstances the signal to noise ratio increases as the square root of the exposure time. For some observations detector dark count, instrumental scattered light or granularity in the pixel to pixel sensitivity will cause additional noise. The signal to noise ratio of the net spectrum will then depend on several parameters, and will increase more slowly with exposure time. We have analyzed data from the ground based calibration programs, and have developed a theoretical model of the HRS performance (ref 4). Our results allow observing and data reduction strategies to be optimized when factors other than photon statistics influence the photometric quality of the data.

1. THE ANALYTICAL MODEL

Counts are collected into spectrum "bins" and background "bins". The spectrum bins measure the gross signal of spectrum + background + dark count. The background bins measure background + dark count. There are three sources of noise; photon noise in the spectrum bins, photon noise in the background bins, and fluctuations in the spectrum counts due to granularity. If the noise sources are random and independent of each other their variances should add.

Let: S = signal count rate - counts per diode per second - product of stellar flux and G-HRS sensitivity (ref 2)
bS = scattered light background count rate
b = background fraction relative to the adjacent spectrum
d = detector dark count rate - counts per diode per second
t = total observing time - seconds
f = fraction of time spent measuring the spectrum
ft = spectrum exposure time - seconds
(1-f)t = background exposure time - seconds
σ^2 = variance of detector granularity
n_s = number of spectrum diodes per resolution element
n_b = number of diodes smoothed over in background

The signal to noise ratio of the net spectrum is:

$$\left\{\frac{S}{N}\right\}^2 = \frac{S^2 t}{S\left\{\frac{1+b}{n_s f} + \frac{b}{n_b(1-f)}\right\} + \sigma^2(1+b)^2 St + d\left\{\frac{1}{n_s f} + \frac{1}{n_b(1-f)}\right\}} \qquad (1)$$

This model has been compared to experimental data obtained during the ground-based calibration, and all of its significant scaling relations have been confirmed. Rarely will all of these effects be important during a single observation. Each influence can be seen more clearly by considering the following limiting cases.

2. IF ONLY PHOTON NOISE IN THE SPECTRUM IS IMPORTANT

If the signal is much greater than the dark count, scattered light is not significant, and the granularity can be ignored, then d=0, b=0, σ^2=0, and f=1. Equation (1) reduces to the familiar result:

$$\left\{\frac{S}{N}\right\}^2 = n_s St \qquad (2) \qquad n_s St = \text{net counts per resolution element}$$

We have found that for observations made with any of the first order gratings, (G140M, G160M, G200M, G270M, G140L), equation (2) describes the observed S/N for up to 5000 or so counts per resolution element.

3. THE EFFECT OF DARK COUNTS

Retaining terms which include d, but ignoring scattered light and granularity we find:

$$\left\{\frac{S}{N}\right\}^2 = \left[\frac{S/d}{S/d + 1}\right] n_s St \qquad (3)$$

Both detectors have been extremely quiet during the ground-based development; d < $3*10^{-4}$ cts/diode sec (<2 cts/orbit). Unless they deteriorate significantly, or prove to be unexpectedly sensitive to radiation, dark counts should rarely affect the data quality.

4. THE EFFECT OF INSTRUMENTAL SCATTERED LIGHT

In the Echelle modes scattering from both the Echelle and the cross-disperser causes an elevated background level which is counted by the diodes. The fraction b is calibrated as a function of order, and ranges from 0.02 to over 0.5 at the shortest wavelengths. This scattered light source is not significant in any of the first order grating modes, where b < 0.01. For a fixed observing time, t, the S/N of the net spectrum will be maximized if $f \simeq 0.90 \pm 0.05$. The "standard substep patterns" implemented at the STScI reflect this fact (ref 2).

$$\left\{\frac{S}{N}\right\}^2 \simeq \left(\frac{f}{1+b}\right) n_s St \quad (4)$$

Typically, $f \simeq 0.9$ and $b \simeq 0.2$, (ref 2) and S/N is reduced by about 15% compared to the ideal case. The exposure time must be increased by about 33% to compensate for the noise in the background bins.

5. THE EFFECT OF DETECTOR RESPONSE GRANULARITY

$$\left\{\frac{S}{N}\right\}^2 = \frac{1}{\frac{1}{n_s St} + \sigma^2} \quad (5) \qquad \text{as} \quad n_s St \to \infty, \quad \frac{S}{N} \to \frac{1}{\sigma} \quad (6)$$

σ is typically 0.01 or so. At low count levels the noise in the signal dominates, and equation (5) gives the same result as (2). As the collected signal counts increase, the photon noise decreases, and eventually becomes negligible compared to the granularity. Equations (1) and (5) describe the statistical effect of granularity, averaged over a wide spectral interval. At high S/N levels it will be desireable to actually remove the detector irregularities using an as yet to be defined flat-fielding type of procedure.

6. REFERENCES

1. Brandt, J.C. et al., "The High Resolution Spectrograph For The Space Telescope", in The Space Telescope Observatory, D.N.B.Hall ed., NASA CP2244, p. 76, 1982.

2. Ebbets, D.C., "High Resolution Spectrograph Instrument Handbook", Space Telescope Science Institute, 1985.

3. Ebbets, D.C. and Garner, H.W., "Dead-time Effects in Pulse Counting Digicon Detectors", Proceedings of SPIE, Vol. 627, Instrumentation in Astronomy VI, 1986, p 638.

4. Ebbets, D.C., "The Effects of Scattered Light and Detector Response Non-Uniformities on Signal to Noise Characteristics of HRS Data", Proceedings of the Eighth Workshop on the Vacuum Ultraviolet Radiometric Calibration of Space Experiments, Boulder, 1987.

DISCUSSION

BAADE The S/N at high spatial frequencies is one but not the only quality indicator. As the internal flat field lamp illuminates the detectors directly (the dispersive elements are by-passed), how do you correct for low spatial frequencies ? In view of the very limited length of single spectra and the continuum of grating angles this appears a very important point.

EBBETS The low frequency variations will be included in the absolute radiometric calibration. Standard stars will be observed at approximately 25Å intervals in the medium resolution modes, and 5Å intervals in the Echelles. Observers who require this kind of flat fielding will either have to use the calibrated wavelength settings, or arrange for new calibration observations to be made. It may be possible to interpolate between calibrated positions, but we have not yet developed such procedures.

SODERBLOM What is (are) the best achievable S/N with the HRS ? Your curve for the echelle mode suggest an asymptote value of around 150.

EBBETS With the medium resolution gratings (R=20000) S/N should be limited by Poisson statistics. It should be possible to achieve S/N > 100 without extraordinary effort, and perhaps 200 or more with careful observing and data reduction techniques. In the Echelle modes, S/N up to 50 or so will be determined by photon statistics. Observing techniques to remove the effects of scattered light and detector irregularities should allow S/N 700 to be achieved. We have plans to develop these techniques in the post-launch calibration program.

SNEDEN Have you had any problems with the lifetimes of Digicon detectors, and do you expect any additional problems after launch ?

EBBETS Both HRS detectors are now three years old, and continue to perform flawlessly. Tests are being run every few months while HST awaits launch. We measure dark count rates, resolution, geometrical stability, sensitivity, "dead-time" effects, and anomalous channels. A "pulse height analysis" sets the discriminator thresholds and searches for ion counts, which could be indicative of a leak in the vacuum seals. We expect the detectors to perform well for many years in orbit, after the 1 atmosphere pressure differential is removed.

FITZPATRICK What is the spatial scale of the photocathode granularity in the short wavelength detector ; and will the features be mapped ?

EBBETS The most troublesome structure has a spatial scale of 200 μm or so, which is comparable to the 50 μm resolution element. The irregularities will be mapped at a small number of wavelengths during the post launch "science verification" calibration program. These wavelengths have been selected to coincide with important interstellar lines. More comprehensive mapping may be undertaken as a long term calibration project by the Space Telescope Science Institute.

LIMITATIONS AND FUTURE OF RETICON DETECTORS

Gordon A. H. Walker
Geophysics and Astronomy Department
University of British Columbia
Vancouver, B.C., V6T 1W5, Canada

ABSTRACT. Linear arrays of self-scanned silicon diodes have been used in astronomical spectroscopy for over a decade. With care in the flat-fielding and data reduction they can be calibrated to better than 0.1%. They are still the best detector for signal to noise levels >100 when continuous wide-band coverage is needed. CCD's should be capable of this spectrophotometric performance but, for the forseeable future, the lack of a large format and their high cost only make them competitive for spectroscopy of single spectral features or multiple echelle spectra.

1. INTRODUCTION

"It is quite apparent that from now on the photographic plate will be considered a specialised detector. Arrangements for its use will have to be made well in advance much the same as was done in the past for the electronic detectors." J. M. Fletcher DAO, January 7. 1987 on finding that <u>no</u> plates were taken with the 1.8m telescope in 1986.

While the Reticon is unlikely to have as long a reign as the photographic plate, it is clear from the wealth of results being presented at this meeting that it will remain part of the equipment at most observatories for at least the rest of this century. At CFHT it is used for one third of the observations.

Self-scanned, linear-arrays of silicon diodes were first introduced by the Reticon Corporation (now E. G. and G. Reticon) in 1971 for optical character recognition, facsimile production, and non-contact control. The first devices had 64 diodes and they are now available with up to 4096. Tens of thousands are produced each year, mostly for optical character recognition at check-out counters. The major scientific use is in multichannel spectrometers for chemistry with astronomy only taking a tiny fraction of the total production. Unlike the situation with CCD's, astronomers seem assured of a secure supply of quality diode arrays at a reasonable price (a selected device costs only a few thousand $).

Their geometrical stability, high panchromatic responsive quantum efficiency, high storage capacity, and almost total lack of lag make diode arrays superior to other signal generating detectors for high signal to noise spectroscopy. Tull and Nather (1973) were the first to report the use of a refrigerated diode array for astronomical spectroscopy and they are now in routine use for analog integration at most major and many smaller observatories. I counted over 200 papers based on their use when preparing this review. Reticons are widely used in astronomy in other than an integrating mode e.g. photon-counting detecting phosphor scintillations or electron bombarded (Digicon), and in radio-frequency acousto-optic spectrometers. (E. G. and G. Reticon are planning to build a 512x512 CCD suitable for astronomy.)

2. RETICON vs. CCD

Reticons have a highly linear response because each photon detected produces a single charge-carrier pair but the lack of internal gain introduces a large read-out noise. Detective Quantum Efficiency, DQE, measures the efficiency with which a detector reproduces the $N^{0.5}$ signal to noise in an incident beam of N photons. For a single diode:
DQE = $Q/(1 + (\text{read-out noise})^2/NQ) = ((S/N \text{ out})/(S/N \text{ in}))^2$,
where Q is the responsive quantum efficiency, and the (read-out noise)^2 is measured in equivalent detected photons. Q for the Reticon is almost optimal for a silicon device as can be seen in Figure 1 where Q is shown as a function of wavelength for several front and rear illuminated CCD's, a semi-transparent S-20 photocathode, and the RL 1872 Reticon. The dashed line indicates Q enhancement using a fluorescent coating on the thick CCD's (Cullum, Deires, D'Ordico, and Reiβ (1985)).

The limiting read-out noise for both Reticons and CCD's is KTC switching noise (=$2.5 \times 10^9 (KTC)^{0.5}$ electrons). For diode arrays most of the capacitance is in the diffusion associated with the shift registers, while for the CCD it is largely the input node capacitance of the on-chip amplifier. For the Reticon C is of the order of 1 pF while for a CCD it is only about 1 fF (i.e. about 10^{-3} less).

Figure 2 shows the improvements which we have made in the read-out noise of our RL 1872 F/30 systems (Walker, Johnson and Yang (1985)) based on the pioneering work of Geary (1979), Livingstone, Harvey, Slaughter, and Trumbo (1976), and Vogt (1981). With the three systems we have built for CFHT, DAO, and UBC we regularly achieve read-out noise values of less than 400 electrons rms which is close to the expected $(KTC)^{0.5}$ limit. In the following discussion

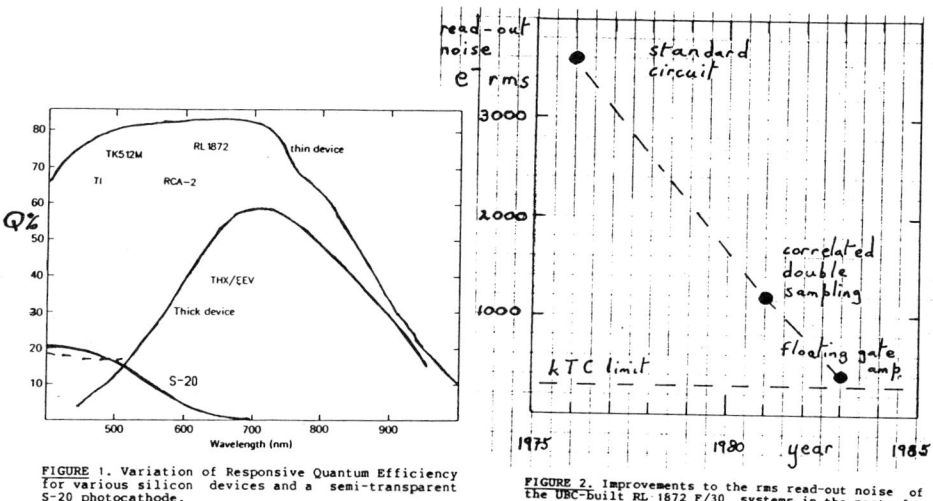

FIGURE 1. Variation of Responsive Quantum Efficiency for various silicon devices and a semi-transparent S-20 photocathode.

FIGURE 2. Improvements to the rms read-out noise of the UBC-built RL 1872 F/30 systems in the past few years.

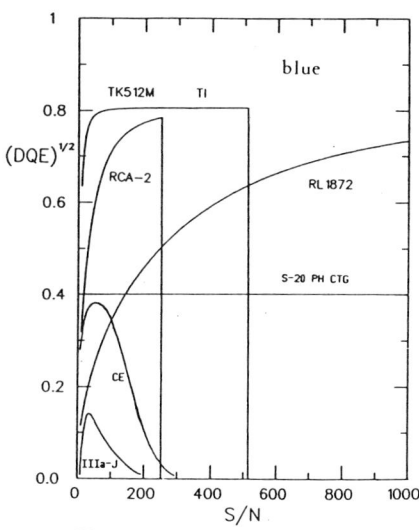

FIGURE 3. Variation of DQE with signal to noise level in the blue for four CCD's, an RL 1872, a IIIa-J emulsion, and an electronic camera (CE). The cutoffs for the CCD's correspond to saturation of the horizontal shift registers.

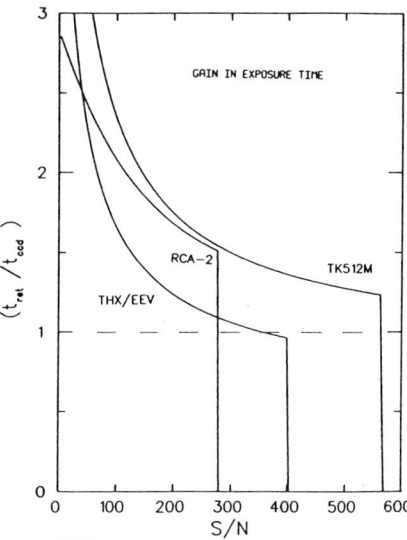

FIGURE 4. The ratio of exposure time for a given S/N with the 1872 Reticon to that with different CCD's.

I adopt the very conservative value of 500 electrons rms.

Figure 3 shows how $(DQE)^{0.5}$ varies with detected signal to noise for various detectors in the blue. Binning is assumed for the CCD's and the cut-off at high S/N corresponds to saturation of the horizontal shift register. The curve for the IIIa-J emulsion is for an area of 1000 square microns and was taken from Hoag (1976) while the curve for the Electronic Camera is an educated guess, again for 1000 square microns. It is clear that within their range of operation the silicon devices are markedly superior, with the CCD's being even better than the Reticon.

Of more importance to the astronomer than these technicalities is just what is the difference in exposure times implied by Figure 3. This is illustrated in Figure 4 where the the exposure time required with the Reticon is compared with that needed with the four different CCD's to achieve a given signal to noise ratio. In all cases the exposure is longer for the Reticon and catastrophically so for S/N<100. However, this does not take account of the simultaneity gain of the Reticon which has more pixels in-line than the CCD's. This is remedied in Figure 5 where the curves of Figure 4 have been multiplied by the ratio of the number of pixels in each device. An 1872 Reticon is assumed. In this case the Reticon is superior above S/N = 50. This advantage is ofcourse regained by CCD's when used in an echelle spectrograph but, by the same token, some groups are implementing the 4096 Reticon array and at the University of Texas Tull and Young now have the 'Octicon', eight 1872 Reticons, operating successfully, to give 1.5×10^4 diodes in line! It is worth remembering that the proposed 2048x2048 Tek CCD will cost $>\$10^5$!

The above analysis suggests that for:
(i) S/N < 250, use a CCD for single spectral features or multiple echelle spectra,
(ii) S/N > 100, use a Reticon for wide continuous spectral coverage. (For S/N < 100, unless exposure times are short, spectra become progressively contaminated by cosmic ray muon events see for e.g. Walker (1987)).
(iii) S/N > 250, only use a Reticon unless the reported photometric calibration of CCD's can be improved by an order of magnitude from the current 1% level.

3. CALIBRATION

I am confident that CCD calibrations will eventually improve although it would be interesting to know what limitations are introduced by scattered light and other problems in echelle spectrographs. It is certainly possible with care to calibrate Reticons to better than 0.1%.

As many groups seem to have difficulty achieving the high photometric performance of which the Reticon is capable I briefly outline below our flat fielding procedures.

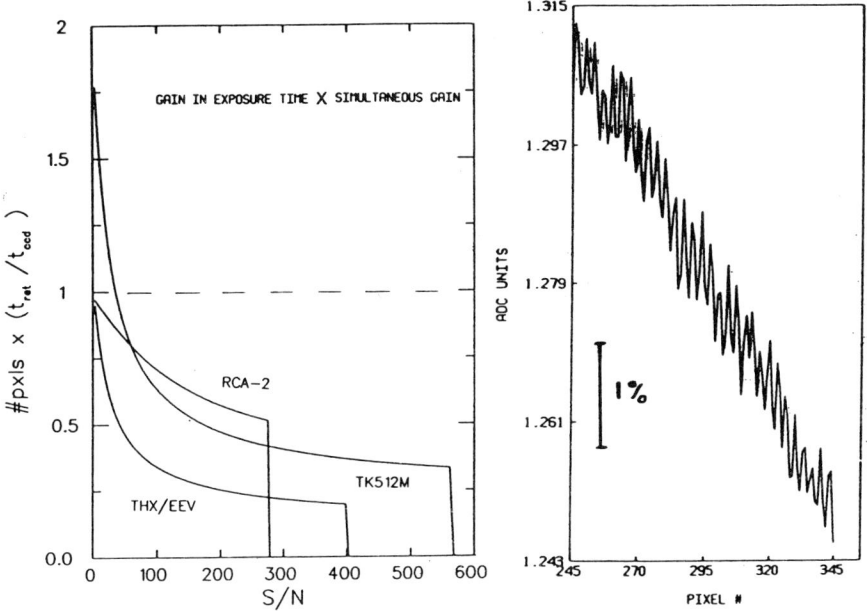

FIGURE 5. The same as Fig. 4 but for the same number of spectral resolution elements for each detector.

FIGURE 6. 100 diode portion of a flat field spectrum corrected for baseline which shows the 10 to 15 quasi-periodic variation of Q with diode number.

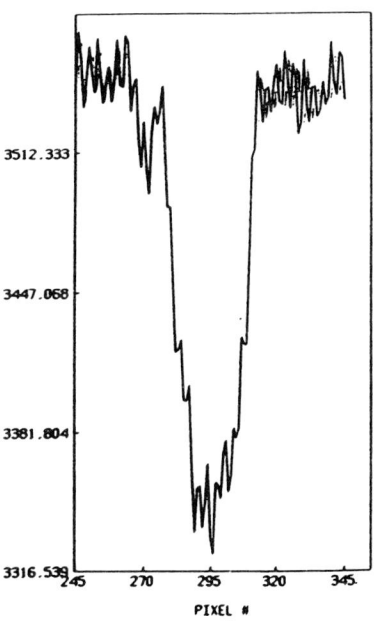

FIGURE 7. 100 diode portion of the spectrum of Spica before normalisation with the flat field of Fig. 6.

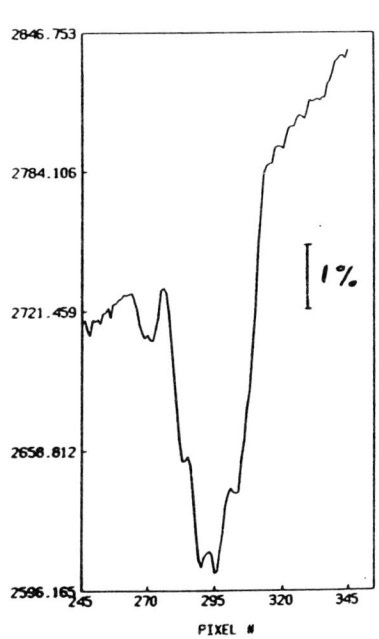

FIGURE 8. As Fig.7 but following normalisation with the flat field in Fig. 6.

Diode to diode variations in Q show a quasi-periodicity of ten to fifteen diodes at a level of about half of one percent. This can be very clearly seen in Figure 6 which shows a 100 diode portion of a flat field spectrum after subtraction of the baseline. Unless star and flat-field light fill exactly the same area of the array the quasi-periodic bumps will appear at some level in the final spectra where they can be mistaken for real features as well as confusing radial velocity measurements based on small shift techniques.

Spectra at CFHT, DAO, and UBC are taken with a Richardson image slicer (Richardson, Brealey, and Dancey 1971) which focusses a one dimensional image of the telescope exit pupil on the detector, normal to the direction of dispersion. Each of the coudé spectrographs is fed by a small-mirror coudé train which operates at a large intermediate f/ratio with a focal reducer before the slicer (Richardson *et al.* 1971). The reducing lens provides an accessible exit pupil between lens and slicer which can be located by eye.

To eliminate residual collimation errors, the exit pupil is isolated by adjusting the position of an iris diaphragm (with a central disc to cover the secondary mirror shadow) for each star. It is sometimes even necessary to move it during a long spectral series. The iris is illuminated by a lamp and diffuser to give the flat field. The array is aligned 'vertically' on the spectrum of the flat field.

Figures 7 and 8 show 100 diodes of the spectrum of Spica unnormalised and normalised by the flat-field shown in Figure 6, respectively. It is clear from Figure 8 that the calibration is good to better than the 0.1% level. For complete processing the spectrum should be filtered to 50% of the Nyquist frequency if the line spread function has been properly sampled by the diode spacing.

REFERENCES

Cullum, M., Deires, S, D'Ordico, S., Reiß, R., 1985, *Astron. Astrphys.*, **153**, L1.
Geary, J. C., 1979, *Proc. S.P.I.E.*, **172**, 82.
Hoag, A.A., 1976, *AAS Phot. Bull.*, **13**, 14.
Livingstone, W.C., Harvey, J., Slaughter, C., Trumbo, D., 1976, *Appl. Opt.*, **15**, 40.
Richardson, E.H., Brealey, G.A., Dancey, R., 1971, *Publ. Dom. Ap. Obs.*, **14**, 1.
Tull, R.D., Nather, R.E., 1973, *Astr. Obs. with TV Sensors*, UBC, ed. Glaspey and Walker, 171.
Vogt, S.S., 1981, *Proc. S.P.I.E.*, **290**, 70.
Walker, G.A.H., Johnson, R., Yang, S., 1985, *Adv. Electronics Electron Phys.*, **64A**, 213.
Walker, G.A.H., 1987, *Astronomical Observations*, Cambridge University Press.

TH 7832CDZ (TH X31513) BILINEAR CCD ARRAY FOR ASTRONOMY APPLICATIONS

J.L. COUTURES and G. BOUCHARLAT
THOMSON-CSF Division Tubes Electroniques
38, rue Vauthier BP 305
92102 BOULOGNE-BILLANCOURT CEDEX France

ABSTRACT. Photodiode linear arrays are perfectly adapted for spectral analysis. The TH 7832CDZ bilinear array is a new device specially adapted to low level detection (exposure $\lessdot 7$ nJ/cm^2) with a reading efficiency of the photodiode signal better than 97 % on all the dynamic range ($\gtrdot$ 70 dB).

1. GENERAL DESCRIPTION

The aperture of the sensitive line is 750 μm with 13 μm pixel pitch. The gap between the two sensitive lines is 500 μm. There are two identical sensitive lines in order to have the possibility of processing two pieces of information at the same time, e.g. for background signal suppression. Two CCD registers are associated to each photodiode line, one to read information from odd pixels and the other one from even pixels. One of the particularities of this device is an electrical input on each CCD register which generate bias charges.

2. EFFECTS OF THE BIAS CHARGE

2.1. Reading Efficiency

The bias charges, added to the signal charges on the photodiode, increase the reading efficiency of the signal. This effect is rendered evident by the response to a transient light versus the time. Fig 2a shows the transient response on one output (odd or even pixels) without the bias charge. The light has a quasi uniform level along the sensitive line after the transient. The output signal level reaches the correct value after several integration times (Ti). If the light has a short duration, just during one integration time (as one event for example), the reading efficiency is the ratio of the first step value to the asymptotic value. On the photograph fig 2a, the reading efficiency should be about 15 %.

The photograph fig. 2b shows the response with an added bias charge. In this case, the output signal has the correct value after the first integration time when the light is on.

Usually, for large length photodiodes read by CCD register or by switches to a bus, the signal reading efficiency decreases with low level signals. The bias charge reduces this effect. With a bias charge equal to 1 volt (bias voltage read at the output), and for signal levels on 1.5 % to 50 % of the saturation level, the reading efficiencies are respectively 98 % and 99 % (see fig 3).

2.2. Noise Level of the Device

In order to have a large dynamic range, the detector must have the lower detection noise than possible. The following table gives all the noise source values (in electrons) of the bilinear detector, from experimental results, at room temperature, 5 ms integration time and 1 MHz output sample frequency :

Reset noise for :
- optical input stage .... $295e^-$
- electrical input stage .... $70e^-$
- reading output stage .... $110e^-$

Shot noise for leakage current :
- on photodiode .... $53e^-$
- on CCD register .... $11e^-$

Transfer noise on CCD register .... $45e^-$
Output amplifier noise .... $100e^-$
RMS total noise .... $344e^-$

Among all the noise sources, the optical input stage has the highest noise. This noise is proportional to the square root of the photodiode capacitance. So the larger the photodiode is, the higher the capacitance is and the stronger the optical stage reset noise is.

Fig 4 shows experimental results on noise measurement with a variation of the added bias charge. The solid curve gives the noise variation on the standard transfer conditions. The noise level increases as the bias charge level up to an asymptotic value about 330 electrons. In order to prove that the noise variation is not due to the using of a bias charge, the dashed curve represents the noise variations when the bias charge is transferred from the electrical input to the output stage, without a transfer between photodiode and register. We can verify that the two variations for reading efficiency (fig 3) and noise level (fig 4) versus the bias charge level are the same pattern. Then a bad transfer efficiency from photodiode to CCD register acts as a low pass filter for signal and noise. When the reading signal efficiency increases with the bias charge, the noise level increases as the cutoff frequency of the filtering.

3. GENERAL ELECTROOPTICAL CHARACTERISTICS

The following table gives the typical electrooptical characteristics of the bilinear CCD array, at room temperature, 5 ms integration time and 1 MHz output sample frequency :

Output conversion factor .... 1.6 µV/e⁻
Response to a light source (tungsten filament lamp
(2854K) + infrared filter) .... 300 V/µJ/cm^2
Saturation output signal (with 0.8 V added bias charge) .... 2 V
Saturation exposure .... 7 nJ/cm^2
Average dark current (with respect to the integration time) .. 1 mV/ms
Quantum efficiency :
- at 400 nm wavelength .... 55 %
- at 700 nm wavelength .... 75 %
Contrast transfer function (at 500 nm wavelength
and Nyquist point) .... 68 %
Minimum integration time .... 250 µs

4. CONCLUSION

The bilinear CCD array TH 7832CDZ is packaged in dual in line integrated circuit package with quartz window or coupled with fiber optical window. The sensitive line geometrical ratio is suitable for coupling to spectrographs. An integration time reduction is allowed owing to the low level detection possibilities with a good reading efficiency in the full 70 dB dynamic range. Just one integration time is necessary to suppress background level from signal with the process of the two sensitive line readings. According to the application the bias charge could not be used in two cases. First, when the signal has slow variations with respect to the frequency of the integration time repetition (200 Hz maximum signal frequency with the minimum integration time). Then, the noise detection is filtered and his level falls to 110 electrons. Second, when the usefull signal is added with a constant background level which acts as the bias charge (level higher than 0.8 V).

DISCUSSION

VOGT Presumably, by going to the CCD register readout to achieve lower noise, you have sacrificed a lot of dynamic range, as compared to the RETICON. RETICONs have saturation exposure levels of ≃ 2x10⁷, whereas your device looks more like 2x10⁶ electrons. However, your CCD register has not yielded a significantly lower readout noise than the level of ≃ 350 electrons which Gordon Walker's RETICON devices achieve. Therefore, it seems that your array has succeeded only in lowering the dynamic range of the diode array approach.

COUTURES Yes the saturation exposure level is 2 10⁶ e- but the purpose is to detect low level signal with a correct reading efficiency. What is the reading efficiency for RETICON devices ? 70 % ? Is a part of photodiode noise filtered on RETICON devices as the signal is ?

COTTRELL What is the cost of one of these devices ?

COUTURES Not known, possibly not very expensive.

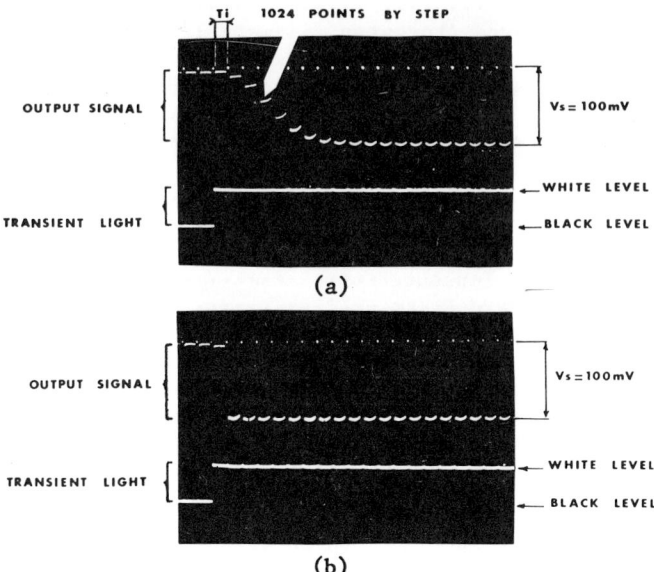

Fig. 2 - Response to a transient light, (a) without a bias charge, (b), with a bias charge. Each step represents the read signal on one output at the integration time repetition.

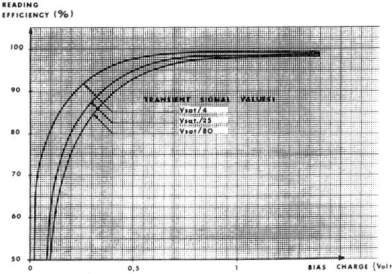

Fig.3 - Reading efficiency for several signal levels v.s. the bias charge

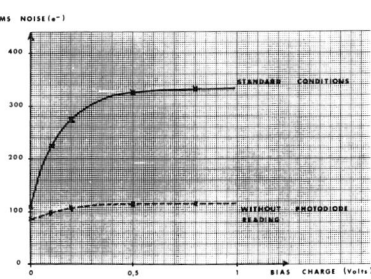

Fig. 4 - RMS noise variation v.s. the bias charge

ON HIGH SIGNAL TO NOISE SPECTROSCOPY WITH CCDs

Bernard Fort
Observatoire de Toulouse
14, avenue Edouard Belin
31400 Toulouse

ABSTRACT : This communication presents a very quick overview of major trends in CCD development and somes ideas on how to use a CCD for spectroscopy with H.S.N. ratio.

1. INTRODUCTION

A lot of reviews have been devoted to CCDs in the past 10 years. The two most recent and complete contributions on the subject are probably the publication of the ESO/OHP Workshop on the optimization of the use of CCD Detectors in Astronomy [1] and the three new books on CCDs : theory, manufacturing and application by J. Janesick [2. 3. 4]. Reading these publications immediatly shows that CCDs were up to now rarely used for spectroscopy with S/N ratio above 200. The reason is quite evident : for such observations the read-out noise of the detector is negligible compared to the photoelectronic shot noise. The main performance of the detector becomes the maximum capacity of charge storage per pixel. In this respect, good Reticon cameras with a read-out noise below 400 electrons : $pixel^{-1}$ and a storage capacity of several billions of charge definitively out perform CCD cameras. (cf. paper by M. Walker in this conference).

The advantage of CCDs in this field of application may only come from their twodimensional format necessary for long slit spectroscopy of extended objects or cross-dispersion spectroscopy.

In this respect, in this -very short- communication we have chosen to briefly list the new trends in CCD development and some ideas on how to use a CCD for H.S.N. spectroscopy. Although the author is familiar with CCD, he has only used them in extragalactic spectroscopy. Therefore this paper reflects more the theoretical view of a CCD camera maker than that of an expert astronomer in H.S.N. Spectroscopy.

II. MAJOR TRENDS ON CCD DEVELOPMENTS FOR THE VISIBLE SPECTRAL RANGE

The possibility to run CCDS with very low read-out noise, 6 to 10 electrons r.m.s. per pixel and per reading, has now been successful in many laboratories, and on telescopes. Various small photometric defects are also being cured, that we do not comment here because most of them are significant at very low flux [4]. Finally the two major efforts which are being made concern the increase of quantum efficiency in the UV and the attempt to have large pixel formats.

Obtaining a high UV response

In the UV, the opacity of Silicium is high and the photocharges are created very close to the surface, where they have a great chance to be recombined and trapped by impurities before being collected in the depletion layer. One possibility to collect these charges efficientily is to use a thin CCD with a potential curve which makes the charges drift to the potential pit.

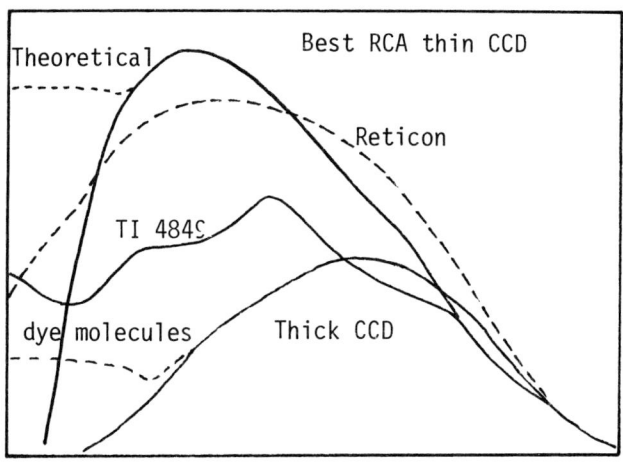

Fig. 1 : Typical responsive efficiency of CCDs.

Many solutions are investigated and discussed by Janesick et al. [3].
   a) coating with dye molecules which convert UV photons to red photons detected by a thick CCD [5].
   b) P+ diffusion in a thick epitaxial SiO_2 layer followed by an optimum thinning of about 200 nm [2].
   c) direct ionic implantation of Boron P+ on an optimum thin CCD [6].
   d) Backside charging of the thin CCD by gas absorption corona discharge or UV flooding [2.3].

e) the so-called "flash gate technology" which consist in coating the backside with a thin platinum layer [2. 3].

Solution a) is presently working very well but should be supplanted within a few years by one of the promising technologies b, c, d or e. (Fig. 1)
We should also note that such an effort is accompagned by an attempt to minimize fringing effect on thin CCDs, and Fresnel losses at the SiO_2/ air interface.

- Increasing the size of frame-transfer CCDs

Up to now the standard size of CCDs does not allow us to use the full field of a telescope without a field segmentation. This often produces an intolerable waste of large telescope time. Several efforts are being made to develop large size CCDs for 1988. Two approaches are considered : the first is to build large monolithic CCDs, [2] the second is to use buttable CCDs for making mosaics [2] (see Fig. 2).

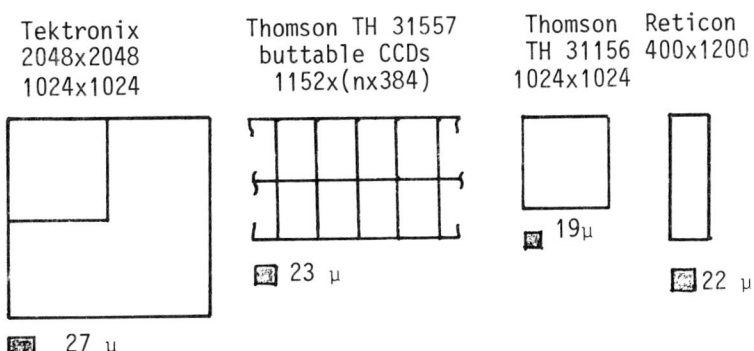

Fig. 2 : Relative size of futur large size CCDs. The shaded area represent the relative pixel size (ref. 9).

III. HOW TO HAVE HIGH S/N WITH A CCD

In this paragraph we summarize some steps that should be followed if we want to get a spectrum with signal to noise higher than 200.
1) The observer has to choose carefully the best place where he can record the spectrum. Even with the best scientific CCD, there always remains blemish defects, trapping effects etc..., which affect some pixels, on a chip.

2) The long exposure necessary to get the total number of charge per pixel (say $5 \cdot 10^5$ charge pixel^{-1}) has to be segmented in several shorter exposures which do not rise the continuum above a level of about 80 000 charges/pixel. This point is very important and needs to be quickly developed here.

For many CCDs, there is an optimum charge capacity close to 10^5 above which a small but real degradation of the MTF occurs. Some charges in a pixel spread to adjacent pixel. This is for example the case for the Thomson CCD [7]. A non linearity results for high spatial frequencies as well as a smoothing of the Flat Field response which can modify the line profile of absorption or emission lines, if poorly sampled. As the read-out noise is rapidly negligible as compared to the photoelectron shot noise it is then an imperative rule to stop exposures at the optimum charge level and then to co-add each of them. This also allows a better discrimination of cosmic rays glitches on the spectrum.

This rule of not trying to use a CCD above its optimum charge level, also holds for the Flat Field exposure. It has also to be done with a HSN level by co-adding many short exposures.

3) Even when working in this careful way it is difficult to avoid systematic errors, many of them coming from the residual in fringing effect correction. As an example the distribution of light through the slit is different for the calibration lamp and the star, and correspond to different flat field responses.

It is then of great importance to move slightly the grating between each individual exposure in order to average some systematic residual errors. The fringing effect is probably the most pervers one which prevents to get HSN in the far red. But in the green it is already possible to get S/N as high as 500 to 700 with a CCD. This is in fact illustrated by a curve derived from a table given by Bohannan et al. [8]

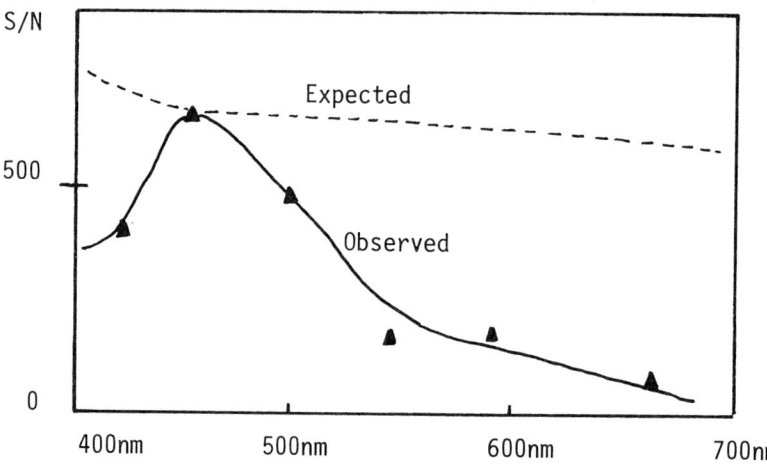

Fig. 3 : High S/N with the KPNO coudé feed spectrograph and a RCA2 thin CCD. 3 to 4 exposures with an optimum pixel charge : 87000e (ref. 8).

REFERENCES

[1] ESO-OHP Workshop on the optimization of the use of CCD detectors in Astronomy 16-19 June 1986. ESO Proceeding n° 25, Edited by J.P. Baluteau and S. D'Odorico

[2] Janesick J. Optical Engineering August 1987, Vol 26 n° 8, Charge Coupled Devices : Theory and Characterization

[3] Janesick J. Optical Engineering September 87, Vol 26 n° 9, Charge Coupled Devices : Manufacture and Applications

[4] Janesick J. Optical Engineering October 87, Vol 26 n° 10, Charge Coupled Devices : Applications

[5] Cullum, M., Deiries, S., D'Odorico, S., Reiss, R., A & A, 153, L1, 1985

[6] Thomson CSF, 1987 private communication

[7] Mellier, Y., Cailloux, M., Dupin, J.P., Fort, B., Lours, C., Picat, J.P., Tilloles, P., A & A janvier 1986, 157, p. 96

[8] Bohannan, B., Abbott, D., Voels, S., Hummer, D., Ap. J., 308, p. 728

[9] Cullum M., 1986 ESO'S Very Large Telescope ESO proceeding n° 24, Edited by S. D'Odorico and J.P. Swings, p. 331

THE CHROMOSPHERIC SPECTRUM OF THE ζ AUR BINARY HR 6902

R. Griffin
The Observatories
Madingley Road
Cambridge
U.K.

ABSTRACT. Observations of the composite-spectrum binary HR 6902 around the time of total eclipse reveal absorption features that are due to the chromosphere of the G9 II primary star. By the application of our method of digital subtraction we have succeeded in isolating the spectrum of the stellar chromosphere.

Binary stars exhibiting composite spectra form a particular class of double-lined binary systems. In the present application a composite spectrum is one which arises from a binary or multiple stellar system whose components have a very small angular separation and dissimilar spectral types; composite-spectrum binaries of interest here consist of a late-type giant and a main-sequence star of type A or B. Spectra of composite systems are rather confusing; spectral classification presents considerable difficulties, and so does the measurement of the features of the spectrum of one component in the presence of those of the other. If, however, the two spectra can somehow be separated then we can learn much more about the system. In particular, if the relative radial velocities of the two components can be measured, the mass ratio of the two stars follows immediately as the inverse of the velocity ratio; and since the masses of main-sequence stars are known tolerably well we can determine the mass of the cool giant with reasonable confidence. Our method avoids any reliance on spectroscopic gravities, which have been responsible for the notorious uncertainty associated with giant star masses, and in applying it to composite-spectrum binaries we are exploiting a hitherto neglected class of object.

We have developed a technique of digital subtraction which enables the components of composite spectra to be separated. As the details of our technique have been described fully elsewhere (Griffin, 1986) we will only summarize here the main features of the procedure.

First, a digital record of the composite spectrum has to be prepared and linearized in wavelength in the rest-frame of the primary star. From a library of standard single-star spectra, similarly linearized in wavelength in their respective rest-frames, we then select by trial and error one that matches the primary, and subtract it point-by-point from the composite spectrum. When a good match is achieved and the right proportion of it is subtracted, the spectrum of the secondary is uncovered. The procedure is optimized iteratively until there is no trace of the spectrum of the primary left to contaminate that of the secondary; to achieve that requires careful control, since both spectra will contain some of the same spectral features unless the secondary is appreciably earlier than type A0. The secondary spectrum thus revealed can then be used for any purpose that a spectrum would normally serve, including spectral classification and the measurement of radial velocities and equivalent widths.

Studies of this nature require very high S/N, because in the process of subtraction the signal is subtracted but the noise is added. In addition, attempts at spectral classification are aided by the availability of long regions of spectra, as also are the assessments of the subtraction procedure itself. Furthermore, spectra of normal late-type giants have very narrow lines, and in order to subtract away the primary spectrum cleanly without leaving traces of apparent P-Cygni profiles, such as occur if the two spectra are not aligned accurately enough, it is imperative to achieve adequate resolution per pixel element of the detector, and to sample the data sufficiently frequently. The photographic plate is a detector which satisfies all of these stringent requirements: it can cover a large wavelength region in one exposure and it gives a virtually continuous record; S/N ratios can be increased at will by co-adding digitized spectra, and as a matter of routine we also widen the spectra as far as conditions will permit by trailing the star image along the slit of the spectrograph. Suitable photographic facilities were made available to us for this project at the coudé focus of the 100-inch telescope on Mt. Wilson. We have chosen to work at a dispersion of 10 Å/mm throughout, in order to handle spectra of composite stars down to 7^m in the same way as those in the library of standards; the spectra are digitized every 5μ on the plate, and the intensities are computed at intervals of 50 mÅ.

The first system which we analyzed, HR 6902, was found to have components of types G9 II ad B8 V; part of the results of the subtraction procedure are shown in Fig. 1. As the B8 V star has very narrow lines we were able to measure accurately (by cross-correlation with a B8 V standard) the difference in radial velocity between the two components, and to derive for the system a double-lined orbit. The orbit solution gave a mass ratio of 1.31 ± 0.021, the individual masses being

$m_1 \sin^3 i = 3.86 \pm 0.13\ M_\odot$ (G9 II) and

$m_2 \sin^3 i = 2.95 \pm 0.09\ M_\odot$ (B8 V).

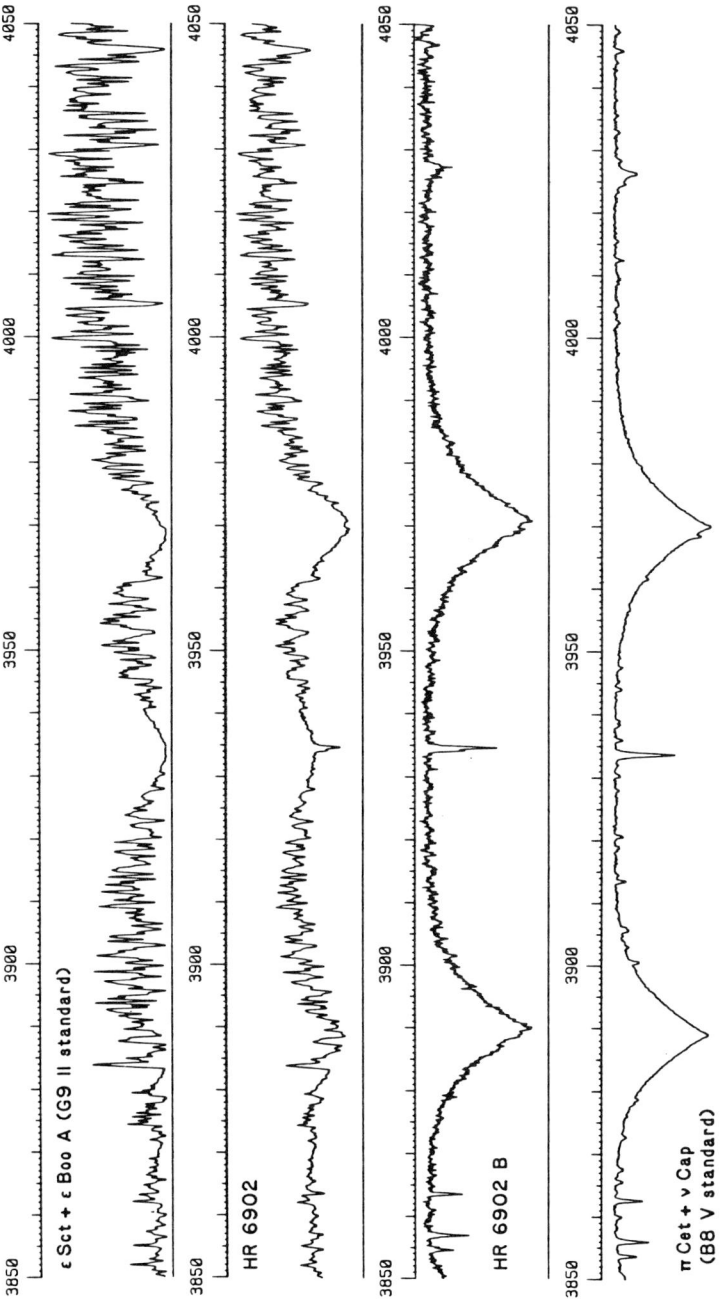

Figure 1. Subtraction of spectra. The top spectrum is adopted as the analogue of the late-type component of **HR 6902**. Subtraction of an optimal proportion of it from the spectrum of HR 6902 (second tracing down) isolates the early-type component (third tracing), which may be compared with the standard spectrum at the bottom. The horizontal line below each tracing is the zero-intensity level.

However, according to Popper (1980) the mass of a B8 or B9 dwarf is near 3 $M_\odot$, and since our value of m_1 has a minimum of 2.95 $M_\odot$ then the angle i must be very close to 90°, i.e. the system is likely to undergo eclipses.

In order to verify the occurrence of eclipses, photographic spectra of the system were obtained with the D.A.O. 48-inch telescope and coude spectrograph during the week following the predicted date of mid-totality. An eclipse did take place, and moreover the spectra showed clear evidence of an atmospheric eclipse which is characteristic of ζ Aur binaries: immediately before and after totality the hot dwarf shines through the eclipsing giant's extended atmosphere, which causes additional absorption lines to appear in the composite spectrum (see, for example, the sequence of events shown by Wright (1952) during the atmospheric eclipse of 32 Cyg). Since the giant component in HR 6902 is less luminous than the primary stars of the classical ζ Aur binaries and therefore has a less extended atmosphere, the atmospheric eclipse phase lasted only two or three days and little height-resolution in the chromosphere could be obtained.

The atmospheric-eclipse spectra were first processed like normal composite spectra, by subtracting the spectrum of the system in totality to reveal the B8 V spectrum with additional chromospheric lines superimposed. We removed the B8 V spectrum by dividing it by our standard B8 V spectrum, and thereby isolated the G9 II's chromospheric lines. They include strong, narrow Ca II H and K lines that are black in the line cores, narrow Balmer lines, and a few resonance or low-E.P. metallic lines such as Mg I, Al I, Ca I, Ti II, Cr I, Mn I, Fe I and Sr II.

We hope to be able to improve upon our result during a future eclipse of HR 6902, and also to apply the method to some of the classical ζ Aur binaries in due course; in the latter events we expect to be able to obtain spectra of the chromospheric material at different heights above the photosphere of the giant star. An observational analysis of this kind has apparently been attempted once before (Morbey et al., 1975; Wright, 1975) for 31 Cyg, but was unfortunately left in a preliminary stage and not fully discussed in the literature.

References

Griffin, R. & R.: 1986, J.Astrophys.Astr. 7, 195.
Morbey, C.L., Wright, K.O., Carlberg, R.G.:
    1975, J.Roy.Astron.Soc.Canada 69, 40.
Popper, D.M.: 1980, Ann.Rev.Astr.Ap. 18, 115.
Wright, K.O.: 1952, Publ.Dom.Astrophys.Obs. 9, 189.
Wright, K.O.: 1975, J.Roy.Astron.Soc.Canada 69, 277.

DISCUSSION

COTTRELL   You want to be able to monitor effects which have time scales of less than a day and yet you only obtain one photographic spectrum per day. Surely you need to go to a solid-state system to do this?

GRIFFIN   The geometry of the system is such that we cannot get height- resolution; we still need large wavelength coverage for classification purposes.

SODERBLOM   What S/N ratio were you able to achieve with your photographic plates? To what extent was this limited by horizontal streaks (guiding errors?) on the original spectra?

GRIFFIN   In our spectra at very high resolution (1 Å/mm or better) we can reduce the noise level to a fraction of 1 %. The work on composite spectra employs 10 Å/mm, though the S/N achieved will depend upon the opportunity to co-add spectra. Horizontal streaks are caused by imperfections in the entrance slit as well as by guiding errors during short exposures and are to be avoided where possible. If the continuum density lies on the linear region of the characteristic curve the relative error will be small, but the effect in absolute terms may cause some anxiety.

MOROSSI   How did you make the actual subtraction of the standard spectrum from the composite one? I refer to consistency of the two wavelength scales.

GRIFFIN   There is no problem. A wavelength scale is established for each spectrum in the rest-frame of the star, by identifying stellar lines. We do not use a reference spectrum.

THE IMPACT OF ARRAY DETECTORS ON HIGH RESOLUTION INFRARED SPECTROSCOPY

Stephen T. Ridgway and Kenneth H. Hinkle
Kitt Peak National Observatory [1]
P.O. Box 26732
Tucson, Arizona 85726
U.S.A.

ABSTRACT. Infrared detector arrays implemented for astronomical use during the past few years achieve performance gains which have profound implications for infrared spectroscopy. Arrays are now available with $\sim few \times 10^3$ pixels, each of which is $\sim 10^2$ times more sensitive than previous single element detectors. Depending on the spectral regime, it is now possible to construct infrared spectrometers with limiting sensitivities $10 - 500$ times fainter than in current use.

1. INFRARED DETECTOR ARRAYS, LARGE TELESCOPES AND ADAPTIVE OPTICS

The short history of infrared detector arrays, like the longer history of visible detector arrays, is filled with dead ends, false hopes, and unmarketed one-of-a-kinds. Therefore, this projection of the impact of infrared arrays on spectroscopy will be based on an array which has been manufactured commercially for astronomy and is now available at several observatories. This detector is a 58×62 element MOS Direct Readout InSb array from Santa Barbara Research Center. It is a detector of the hybrid type, with photovoltaic InSb detector material and a bonded silicon readout layer (Orias et al. 1986). Laboratory and telescope tests have been reported by Fowler et al. (1987a,b).

The detector characteristics appearing in Table 1 have been taken from these sources and are based on detectors actually delivered and tested. In addition to the characteristics reported in Table 1, the SBRC InSb array detectors exhibit some non-linearity. Very preliminary tests show that linearity can be recovered to 5% in a flux ratio of 1700:1, with further improvement expected. Approximately one cosmic ray event is observed per frame and per 500 seconds of integration.

The InSb detector arrays may be compared with single element InSb detectors which are the most sensitive near-IR detectors in general use. These have quantum efficiency similar to the array elements. Normally employed in an analog mode, the performance of these detectors is usually characterized by the Noise Equivalent Power. Except for special and limited situations, the best performance obtained is $\simeq 2 \times 10^{-16}$ W Hz$^{-\frac{1}{2}}$. As an analog detector, the S/N of a measurement improves

---

[1] National Optical Astronomy Observatories, operated by the Association of Universities for Research in Astronomy, Inc., under contract to the National Science Foundation

Table 1: SBRC 58×62 InSb Detector Characteristics

Format	62×58 (3596 pixels)
Pixel Spacing	76×76 μm
Active Area	$\simeq$100%
Readout Noise	400 e$^-$
Quantum Efficiency	>60%
Well Capacity	>1 × 10^6 e$^-$
Dark Current	100 e$^-$/pixel-sec
Good Pixels	>97%
Response uniformity	±25%
Dynamic Range	$\simeq 10^3$

$\propto \sqrt{t}$, where t is the integration time. This differs from the array detector, an integrating device, for which, as long as the measurement is detector noise limited, the S/N improves $\propto t$. Thus a comparison of performance between discrete and array detectors requires an assumed integration time. A single pixel of the array described above has an equivalent NEP of $\simeq 2 \times 10^{-16}$ W Hz$^{-\frac{1}{2}}$ for an integration time of 0.1 seconds. For longer integration times, the array pixel outperforms the discrete detector $\propto \sqrt{t}$. With an integration time of 1000 seconds, the improvement in limiting sensitivity is $\simeq 100$.

The development of large telescope technology offers much to infrared astronomy. In the near infrared, 1–2.5 μm, where the faint limit for high resolution spectroscopy will be set by detector noise for some years to come, the limiting flux improves $\propto (Dd)^{-1}$, where D is the aperture diameter. Several groups (e.g. Beckers et al. 1986; Merkle 1987) are developing adaptive optical systems for large telescopes. The visible light from a point source will be used to measure the wavefront errors, and with this information an active optical element will be adjusted to correct the errors in real time. The correction should suffice to render the telescope diffraction limited at infrared wavelengths $\simeq$ 3–5 μm. Then a smaller acceptance aperture will be selected appropriate to the smaller, diffraction limited images produced by larger telescopes. In this case, the limiting flux also improves $\propto 1/D^2$.

Therefore, for high resolution infrared spectroscopy (and many other types of infrared observations) the gains in limiting flux for a 10 m (TMT) or 22 m (NNTT) aperture relative to a 4 m aperture will be $\simeq$ 6 and $\simeq$ 22, respectively (the NNTT is an unfilled aperture, hence the gain is $\propto 1/Dd$, where d is the equivalent filled aperture diameter, and D is the unfilled aperture diameter).

2. PROJECTED SENSITIVITY GAINS FOR INFRARED SPECTROSCOPY

Figure 1 presents the expected gains in limiting magnitude associated with available detectors, expected telescope and optics developments, and possible future detector improvements. Some simplifying assumptions have been made, but the implication of major improvements in limiting magnitude is very accurate.

A gain of more than 2 magnitudes is expected at all wavelengths from the number of pixels across the width of the array (assuming that at least 62 pixels

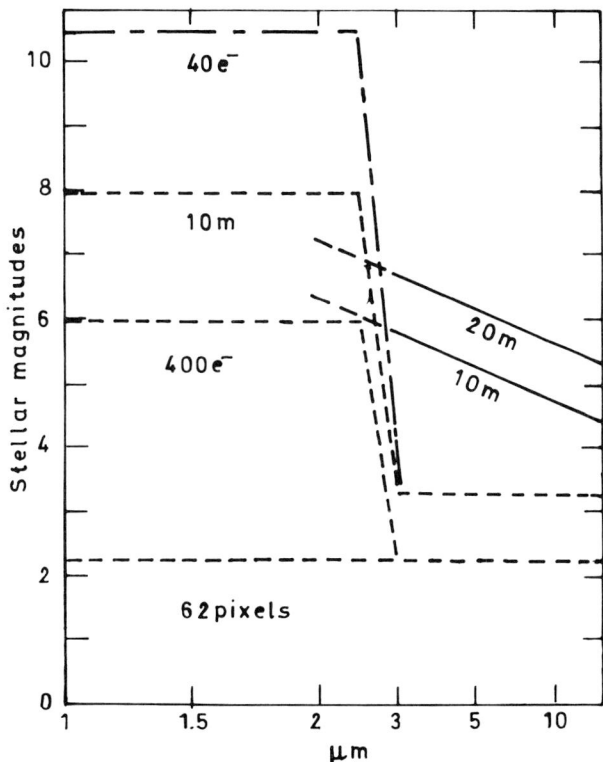

Figure 1. The relative sensitivity gain in magnitudes vs wavelength in μm, arising from use of an integrating InSb array 62 pixels wide instead of a single discrete detector, with integration time 100 seconds, and readout noise of 400 and 40 e$^-$, and the gains associated with a 10 m effective aperture for flux collection, and a 10 or 20 m effective aperture for diffraction limited imaging.

of spectrum are needed). The available readout noise of 400 e$^-$ provides a gain of more than 3 magnitudes in the near-IR. A 10 m telescope aperture, relative to a 4 m aperture, provides 2 magnitudes gain in the near-IR. It provides less in the longer wavelengths, unless adaptive optics are available, in which case a substantial additional gain is achieved. Reduction of detector read noise to levels associated with CCD's (possible in the not too distant future) would provide another 2.5 magnitude gain in the near-IR.

3. EXPLOITATION OF THE IMPROVED INFRARED DETECTORS

High resolution infrared spectroscopy has brought abundant and detailed astrophysical results during the last decade (e.g. reviews by Ridgway and Brault 1984; Ridgway 1987). Areas of particularly impressive achievements from high resolution infrared spectroscopy have been non-equilibrium chemistry in outer planet atmospheres, atomic and isotopic abundances in cool stars, dynamical structure of pulsating stellar atmospheres, kinematics of mass loss, chemistry in stellar envelopes, mass-motions in molecular clouds, and young stars and star formation.

The most spectacular improvements projected in Figure 1 are for detector noise limited measurements. As it is in high resolution spectroscopy that this condition is most commonly achieved, implementation of array detectors in a high resolution spectrograph is a natural application of the new technology. The required resolution is determined from the astrophysical objectives. For a high resolution spectrograph the scientific goals normally require that the lines be resolved. The lines in interstellar or circumstellar clouds have intrinsic widths of < 2 km s^{-1} and in addition commonly have multiple components with spacings of a few km s^{-1} (Bernat 1981; Black and Willner 1984). For the photospheres of late-type stars line widths are typically 5 km s^{-1} (Hinkle 1978). High resolution (10 km s^{-1}), long slit spectra of optical emission in the nuclei of a variety of nearby spiral and active galaxies reveal complex line profiles of total widths $\simeq 200$ km s^{-1} but with considerable finer structure (Goad and Gallagher 1985).

Additional insight into selecting resolution can be obtained from experience with the FTS at the KPNO 4 m telescope. The FTS is an excellent learning device for this purpose since the resolution is readily adjusted at the telescope. A majority of observations with the 4 meter FTS are of the photospheres of stellar objects. Here experience with the FTS indicates that resolutions between 40000 and 80000 are requested. Most stellar observations are in the 2.0 to 4.0 μm region, although the oxygen isotopes are best measured using the 4.6 μm CO fundamental. Perhaps 20% of FTS observations are of interstellar or circumstellar clouds. Requested resolutions have been as high as 150000, where for best results in detecting weak lines the resolution should be an approximate match to or exceed the intrinsic width of the line.

In circumstellar or interstellar clouds, not only does the ability to detect weak features depend on high resolution, but important information on the source is gained by resolving the line profiles. For instance, in IRC+10216 CO first overtone lines are obviously present at a resolution of 40000 but at a resolution of 100000 the line profiles reveal that several outflow regions at different temperatures and velocities exist along the line of sight. FTS observations of the BN object in Orion provide a limit on how low the resolution can be for a narrow lined source and still provide useful results. Scoville et al. (1983) were able to use observations at resolution 54000 at 4.6 μm.

Of course, a caveat is that cryogenic echelle projects certainly will be different from those done on the FTS due to the factor of 100 difference in sensitivity between these instruments. While virtually all infrared spectroscopy will enjoy the benefits of the 3–10 magnitude gain in sensitivity, the following areas seem to us particularly promising for rapid progress with improved sensitivity. (Typical spectral resolution in km s^{-1} required by previous similar observations is given in parentheses): molecular clouds (1-2 km s^{-1}), young stellar objects (2-5 km s^{-1}), galactic nuclei (20-50 km s^{-1}), abundances of stars in clusters (2-5 km s^{-1}), and nebulae and novae (5-10 km s^{-1}). We note that all of these areas profit from a long-slit spectroscopic capability, either because the sources are extended, or because several stars may be placed on the slit simultaneously (possibly with fibers, rapidly becoming available in the IR).

We conclude that the scientific goals require that at the longest end of the wavelength range (4.6 μm) the grating size provide a resolution of at least 50000. At resolution 50000 the 2.1 m with cryogenic echelle will reach $\simeq 8$ magnitudes fainter than the 4 m with FTS. (At resolution 50000 the FTS is imited by source or background photon noise at all wavelengths, hence the dispersed spectrometer has as additional advantage of $\sqrt{N_{sp}}$ in addition to the gains in Figure 1, where N_{sp} is the number of spectral elements in the FTS bandpass.) Recognizing the great gains to be achieved, a number of observatories are considering the design of a new generation of infrared spectrographs. At NOAO, design studies have been carried out for the 2.1 m telescope (results described below) and are in progress for the NNTT.

4. THE NOAO INFRARED CRYOGENIC ECHELLE DESIGN

What follows is a discussion of the parameters for an echelle spectrograph matched to a two-dimensional infrared array. We first discuss detector considerations, the wavelength coverage, and the size telescope best suited to the scientific applications. We then outline the arguments specifying the size of the envisaged instrument. Grating size, slit size, and resolution will be reviewed. The size of the optics package is critical because the entire optics assembly must be placed in a dewar and chilled to LN_2 temperature. Finally, we review the various order separation schemes.

4.1. Telescope, Wavelength Coverage and Detector

For the detector we adopt the confirmed performance of the SBRC array described above. The SBRC arrays are relatively small arrays, having only a few thousand elements. We propose designing the spectrograph for a 64 x 256 element array. Owing to thermal stress in detector material, it is unlikely that similar arrays will be much larger than 256 pixels wide for some time. Foreseeable detector improvements should result in pixels no smaller than 50 μm in the lifetime of the cryogenic echelle. Two of the currently available SBRC arrays would take a similar amount of space in the focal plane as a 64 x 256 array and would provide acceptable spectral coverage.

The two dimensional nature of the detector is an important aspect of the operation of the instrument. We propose a slit length of 0.5 to 1 arc-minute which is roughly matched to the height of the current array at 1 arc-second per pixel at the 2.1 meter telescope. This will allow long slit operation on extended sources. On point sources the sky background will be determined by observing the source at various positions along the slit, saving a factor of two in observing time compared with conventional beam-switching. Arguments presented below will demonstrate why we do not favor a cross dispersed output format.

Several considerations dictate the wavelength range of the instrument. At NOAO there is considerable experience with detector arrays for the 1-5 μm region. The 1-5 μm infrared is a natural extension of the visual and near-infrared capability provided by CCD's. Intense usage of the KPNO FTS (Hall et al. 1979) over the last decade has generated considerable interest in and experience with 1-5μm spectra among the NOAO user community. The 2-5 μm infrared contains transitions of light molecules, especially CO and H_2, that are of great astrophysical importance. The CO fundamental and first overtone bands are unique probes of cool regions. The 2-4 μm region also contains bands from essentially all the light diatomic molecules of astrophysical interest (CN, CH, OH, SiO, HCl, HF, NH, C_2), and contains bands from several polyatomic molecules (most notably HCN, C_2H_2, and CH_4. The broad region of telluric absorption between 5 and 8 μm naturally limits the 1-5 μm region. The last factor is that in the 2.3-5 μm region the instrument need only be cooled to LN_2 temperature. At longer wavelengths cooling to LHe temperature becomes necessary and the dewar technology is much more complex.

The NOAO cryogenic echelle is intended for the f/8 Ritchey-Chrétien focus of both the KPNO 2.1 m and 4 m telescopes. A prime focus instrument was removed from consideration because of space limitations and cryogen handling problems in the prime focus cage. A prime focus instrument also would be restricted to the 4 meter and S/N calculations indicated that many projects could best be done at the 2.1 meter. Prime focus does have the unique advantage of only one warm reflection. Placing the instrument at f ratio foci larger than f/8 was removed from consideration because of the resulting increase in the instrument's size. One arc second at f/8 on the 2.1 m is 78 μm, easily matched to the detector pixel size. Background subtraction will be done by moving the telescope to alternately place the star at different positions along the slit.

The FWHM seeing disk on Kitt Peak typically is about 1 arc second. The corresponding (predicted) values at 2.2 and 5 μm are 0.89 and 0.76 arcsec. For a Gaussian image profile, a slit width equal to the FWHM will accept 77% of the source flux. Clearly slit widths of 0.5 arcsec or less are highly inefficient.

4.2. Grating Size, Slit Width, and Resolution

The grating size is set by two constraints, the theoretical resolution R, for an echelle

$$R = \frac{2W}{\lambda} \sin\theta_B \cos\theta \tag{1}$$

and the slit width S (in seconds of arc)

$$S = 206265 \frac{Wm\lambda}{aDR} \tag{2}$$

(Bingham 1979). The symbols used above have the following meanings: λ is the wavelength, R is the resolving power, $\lambda/\Delta\lambda$, m is the order number, θ_B is the blaze angle, a is the groove spacing, D is the diameter of the telescope primary, and W is the width of the used area of the grating in the plane of the grating. θ is defined so that the angle of incidence $\alpha = \theta_B + \theta$, and at the peak of the blaze the angle of diffraction $\beta = \theta_B - \theta$. Hence the grating equation may be expressed

$$m\lambda = 2a \sin\theta_B \cos\theta \tag{3}$$

Equation (2) is best restated by substituting this form of the grating equation,

$$S = 412530 \frac{W \sin\theta_B \cos\theta}{DR} \tag{4}$$

From equation (1), the grating size required to provide a resolution of at least 50000 at 4.6 μm, assuming standard values of $\theta_B = 63.4°$ and $\theta = 6°$ (Schroeder, 1970), has a length of $\simeq$ 12.9 cm. However, the grating size also controls the slit width (equation 4). The following discussion of slit width is for the 2.1 meter telescope. For R=50000, W=12.9 cm, and D=2.1 m the slit width is 0.45 arc second. This is unacceptably small, especially if higher resolutions are to be used at shorter wavelengths. Note that S and W are directly proportional, so 1 arc second corresponds to W=28.7 cm. Echelle gratings are available commercially in sizes up to 8 x 16 inches with 5 x 10 inches being in common use. A 5 x 10 inch grating will give a

0.87 arc second slit for R=50000 and a 0.43 arc second slit for R=100000. A 8 × 16 inch grating provides a gain of 1.6 in slit width, i.e. 1.4 arc second for R=50000.

A 8 × 16 inch grating appears to be the optimal choice because of the large slit width that it provides. However, size of the grating also is limited by the need to keep the instrument size and weight under control. Use of a 5 × 10 inch grating could be a desirable option to limit instrument size and weight if R≤100000 is emphasized. A third alternative is to use a 5 × 10 inch grating and over fill it, providing a larger slit size (Diego and Walker 1985). This is done on the 4 meter echelle and results in only a small amount of lost light. In the thermal IR it would improve cold baffling as well. With any choice of echelle size the grating can be shaped to match the collimated beam size. This may save space in the dewar depending on the mechanical design.

4.3. Order Separation

In wavenumbers, the length of an order (free spectral range) of an echelle is given by

$$\frac{1}{2a \sin \theta_B \cos \theta} \qquad (5)$$

(Loewen 1970). For a 31 line mm^{-1}, 63.4° echelle, the free spectral range in Littrow is 175 cm^{-1}. In the 1-5 μm infrared using this echelle grating, the order number ranges from 12 to 60. Thus some scheme is needed to separate the order we wish to observe from the other 47$^+$ orders visible to the detector.

The basic order separation device for any spectrograph is a filter. If a narrow filter is selected one order may be observed or if a broader filter is selected several orders may be observed by using a cross dispersing element. Echelles in the visual when used with a two dimensional detector (CCD, photographic plate) often use a cross disperser.

To examine the usefulness of a cross disperser in the infrared we will consider a best case: the lowest resolution at which the cryogenic echelle will be used, $R \simeq 20000$, i.e. 0.2 cm^{-1} at 4300 cm^{-1}, combined with the an array of 256 × 256 elements, the largest array we foresee. This combination would sample about 25 cm^{-1} (at 2 pixels per resolution element) out of the free spectral range of 175 cm^{-1}. Thus it is possible to detect only about 15% of the spectrum in each order. From the example above, we conclude that: cross dispersion gains little additional information in the infrared because less than 20% of each order may be sampled (array improvements foreseeable in the next 10+ years are covered by this statement!). To make cross dispersion useful the free spectral range needs to be $\simeq$ 50 cm^{-1}. This would be possible if gratings could be made at ≤10 lines mm^{-1}. Rulings this coarse are not possible now or in the foreseeable future. (Note that at 5000 Å the 175 cm^{-1} free spectral range of a 31 line mm^{-1} echelle grating corresponds to $\sim$ 40 Å. At 0.1 Å resolution, the entire free spectral range will fit on 800 pixels so cross dispersion is very desirable in the visual.)

Several other problems with cross dispersion in the infrared are readily apparent. In the thermal infrared the 175 cm^{-1} free spectral range is large enough that different orders will have large differences in background radiation making optimization of integration time impossible for more than one order at a time. Tokunaga (private communication) has investigated cross dispersion in detail and notes that 3 different cross dispersing gratings would be required to provide optimum performance across the 1-5 μm range. Some of these gratings have unobtainable groove spacings. The possibility of fringing is not removed by using a cross disperser since the cross disperser also obeys the grating equation and must have its other orders blocked.

To summarize, cross dispersion does not seem appropriate in the infrared because it adds little information, greatly increases the complexity and size of the instrument, and does not improve throughput or eliminate the possibility of fringing.

The most obvious choice for a filter to pass the individual orders of an infrared echelle is a circular variable filter (CVF). CVF's are available with bandpasses from 1 to 10%. The 175 cm^{-1} free spectral range provided by a 31 line mm^{-1} echelle is 1.9% at 1.1 μm and 8.5% at 5 μm. A 31 line mm^{-1} echelle appears well matched to the range of bandpasses available with a CVF. A 23 line mm^{-1} echelle has a free spectral range of 1.4% at 1.1 μm and could be blocked with a CVF but pushes harder on the narrow bandpass end of CVF technology.

An alternative to a CVF would be a refractive order separator. This option has the advantages of no possible fringing and potentially high throughput. The disadvantage is that refraction is not linear with wavelength making such an order separator optically and mechanically complex. This is especially true since the cryogenic echelle will operate over a factor of 5 in wavelength.

The following table summarizes the conclusions and recommendations based on this design study.[2]

- Resolutions of 100000, 50000 and 25000.

- 8 × 16 inch echelle grating (preferred), 5 × 10 inch (alternate).

- Slit width of 1.4 arc seconds for resolution of 50000 and 0.7 arc seconds for resolution 100000 on the 2.1 meter telescope for 4.6 μm.

- Cross dispersion is not useful at wavelengths much longward of 1 μm with foreseeable echelle grating technology. Improvements in the technology are not expected on a time scale relevant to this project.

- Order separation with CVF's.

- A 31 line mm^{-1} echelle is well matched to a CVF order separator.

We wish to acknowledge very fruitful discussions with Alan Tokunaga and Don Hall on IR echelle spectrometers, and with Richard Joyce on the InSb detector arrays. STR thanks the University of Paris and the Meudon Observatory for their hospitality when this review was under preparation.

REFERENCES

Beckers, J.M., et al., 1987, in *Interferometric Imaging in Astronomy*, J. W. Goad, ed., (National Optical Astronomy Observatories, Tucson), 171.

Bernat, A.P., 1981, *Ap. J.*, **246**, 184.

Bingham, R. G., 1979, *Q. J. R. Astr. Soc.*, **20**, 395.

---

[2]These parameters and other aspects of the cryogenic echelle are under review at NOAO at the time of this writing.

Black, J.H., and Wilner, S.P., 1984, *Ap. J.*, **279**, 673.

Diego, F. and Walker, D. D., 1985, *M. N. R. A. S.*, **217**, 347.

Fowler, A.M., Gillett, F.C., Gregory, B., Joyce, R.R., Probst, R.G., and Smith, R., 1987b, (private communication).

Fowler, A.M., Probst, R.G., Britt, J.P., Joyce, R.R., and Gillett, F.C., 1987a, *Optical Eng.*, **26**, 232.

Goad, J.W., and Gallagher, J.S., 1985, *Ap. J.*, **297**, 98.

Hall, D. N. B., Ridgway, S., Bell, E. A., and Yarborough, J. M., 1979, *Proc. S. P. I. E.*, **172**, 121.

Hinkle, K.H., 1978, *Ap. J.*, **220**, 210.

Loewen, E. G. 1970, *Diffraction Grating Handbook*, (Bausch and Lomb, Rochester, New York).

Merkle, F., 1986, in *ESO's Very Large Telescope*, (ESO, Garching), 443.

Orias, G., Hoffman, A.W., and Casselman, M.F., 1986, *Proc. S. P. I. E.*, **627**, 408.

Ridgway, S.T., and Brault, J.W., 1984, *Ann. Rev. Astron. Astrophys.*, **22**, 291.

Ridgway, S.T., 1987, in *Observational Astrophysics with High Precision Data*, 27th Liége International Astrophysical Colloquium, in press.

Schroeder, D.J., 1970, *P.A.S.P.*, **82**, 1253.

Scoville, N., Kleinmann, S. G., Hall, D. N. B., and Ridgway, S. T., 1983, *Ap. J.*, **275**, 201.

DISCUSSION

LINSKY Will the proposed NOAO cryogenic échelle spectrometer operate at 10 μm and if so what will be the sensitivity gain over the 4 m FTS spectrometer.

RIDGWAY We plan to restrict the spectral range to $\lambda \leq 5$ μm. Operation at 10 μm would require cooling to much lower temperature, and exchangeable detectors and gratings. A 10 μm spectrometer can be simpler if constructed separately from the 1 - 5 μm unit. The gain over an FTS should be considerable, on the order of the square root of the ratio of the FTS bandwidth to the dispersive spectral resolution. Gains of 2 - 3 magnitudes should be easily obtained.

ANDERSEN Did your comparison of performance figures between the spectrograph and the 4m FTS include the wavelength range included in a single observation ?

RIDGWAY No.

SIGNAL–TO–NOISE RATIO AND ASTRONOMICAL
FOURIER TRANSFORM SPECTROSCOPY

J.P. Maillard
Institut d'Astrophysique de Paris
98bis Bld Arago, 75014 Paris (France)

ABSTRACT. The multiplex properties of the Fourier Transform Spectrometer (FTS) can be considered as disadvantageous with modern detectors and large telescopes, the dominant noise source being no longer in most applications the detector noise. Nevertheless, a FTS offers a gain in information and other instrumental features remain: flexibility in choosing resolving power up to very high values, large throughput, essential in high–resolution spectroscopy with large telescopes, metrologic accuracy, automatic substraction of parasitic background. The signal–to–noise ratio in spectra can also be improved: by limiting the bandwidth with cold filters or even cold dispersers, by matching the instrument to low background foreoptics and high–image quality telescopes. The association with array detectors provides the solution for the FTS to regain its full multiplex advantage.

I. INTRODUCTION

FTS has gained the rank of high–performance spectroscopic tool for twenty years now. This instrument has largely contributed to open the new field of *infrared spectroscopy* by giving access to the high–resolution study of new transitions of atoms and molecules in laboratory, and to the infrared spectrum of many astronomical sources. Since that time, FTS have remained without any serious competitors in infrared laboratory spectroscopy — except in the domain of ultra high–resolution where laser techniques are dominating. The situation is different in astronomy. FTS did not become an universal instrument in the observatories. The energy from the astronomical sources is weak, the observing time limited, which put many objects out of reach of deep spectroscopic investigations. When attempted, the signal–to–noise ratio (SNR) which can be reached becomes a fundamental consideration and determines the choice of the instrumentation. However, the harsh telescope environment for any instrument imposes very often the choice and leads to prefer simpler devices, even if they are of limited capabilities. Consequently, few FTS are in operation behind ground–based telescopes. Nevertheless, an extraordinary quantity of results, which could not have been obtained by any other methods, were recorded in solar, stellar and planetary spectroscopy. The last review to date showing the variety of scientific applications was made by Ridgway and Brault (1984). An update review by de Bergh (1987) is limited to the solar system only.

The present symposium gives the opportunity to focus on the SNR aspect of the method, to the light of modern detectors developments and then to examine its future.

II. THE MULTIPLEX GAIN

FT spectroscopy belongs to the class of the multiplex methods. Thirty years ago, P. Felgett pointed out that a dramatic gain in SNR, which was called the *multiplex gain*, could be obtained using such a method instead of sequential techniques, only known way at that time, and for many years, of entering in the era of the photoelectric detectors. Progress in that field have modified this prophetic view. A reexamination is made in Table I through the comparisons of SNR between the three classical classes of spectroscopic systems : multiplex, multichannel, sequential.

TABLE I

per spectral element	multiplex	multichannel	sequential
integration time	T	T	T/M
flux	$wd\sigma T$	$wd\sigma T$	$wd\sigma T/M$
detector noise (I)	NEP $\sqrt{T}$	NEP $\sqrt{T}$	NEP $\sqrt{T/M}$
source noise (II)	$\sqrt{w\Delta\sigma T}$	$\sqrt{w\Delta\sigma T}$	$\sqrt{w\Delta\sigma T/M}$
backgr. noise (III)	$\sqrt{B\Delta\sigma T}$	$\sqrt{B\Delta\sigma T}$	$\sqrt{B\Delta\sigma T/M}$
SNR (I)	$wd\sigma\sqrt{T}/\text{NEP}$	$wd\sigma\sqrt{T}/\text{NEP}$	$wd\sigma\sqrt{T}/\text{NEP}\times M^{-1/2}$
SNR (II)	$\sqrt{wd\sigma T}\times M^{-1/2}$	$\sqrt{wd\sigma T}$	$\sqrt{wd\sigma T}\times M^{-1/2}$
SNR (III)	$w\sqrt{d\sigma T/B}\times M^{-1/2}$	$w\sqrt{d\sigma T/B}$	$w\sqrt{d\sigma T/B}\times M^{-1/2}$

w energy density of the spectrum, B of the background radiation, both per cm^{-1} ; $d\sigma$ limit of resolution and $\Delta\sigma$ spectral bandwidth in cm^{-1} ; M number of spectral elements $\Delta\sigma/d\sigma$. Luminosity and properties of each detector element are supposed to be identical between the three systems.

A practical example of each class is: a FTS, a grating spectrograph + Reticon array, a grating spectrograph working in scanner. The noise sources which are encountered in any observations give respectively condition I, II, III, when one of them becomes the dominant noise origin: (I) detector noise, (II) source photon noise, (III) thermal background noise. The table is made under the simple following assumptions: instruments have same luminosity, cover the same spectral range $\Delta\sigma$, work at the same resolution $d\sigma$, in the same observing time. From this table, it can be concluded that a multiplex device has no advantages under condition I by comparison with a multichannel device, and becomes definitively slow with respect the same system under condition II and III. With the progress on photoelectric detectors, lowering detector noise to fundamental limits, and the general use of large telescopes, increasing available energy from the sources, condition I becomes less and less valid. The FTS is loosing its multiplex gain by a factor $\sqrt{M}$, with the event of large–size detector arrays, where M is the number of spectral elements related to the number of pixels by the sampling factor. In other words, a grating spectrometer equipped with a low readout–noise CCD, or an infrared array, has a fundamental gain in SNR with respect a FTS. But before to draw an abrupt conclusion, all the instrumental parameters which determine the SNR in real instruments: luminosity, resolution, spectral range, have to be carefully compared, which can radically change the conclusions. A FTS can remain an uncomparable tool for high–SNR spectroscopy.

III. INSTRUMENTAL CHARACTERISTICS OF A FTS

The factors which determine the signal–to–noise ratio in FT spectrum are reviewed to show the advantages of the method they contain:

1. Resolving power

The four currently working astronomical FTS (Davis et al 1980, Maillard and Michel 1982, Brault 1979, Hall et al 1979) have a maximum path difference respectively of 50, 60, 100, 140 cm. The limit of resolution in a spectrum depending only on the maximum path difference reached in the corresponding interferogram, low (few hundred) to very high (10^5 or more) resolution can be continously selected. The resolving power can be tailored to the scientific purpose, providing the source is bright enough for the observing time available.

It is on the side of high resolution that a FTS is difficult to beat. In many cases, comparison of SNR as done in the previous paragraph, is without meaning because the grating spectrometer able to reach resolving powers of 10^5 or more does not exist. This situation is particularly true in the infrared. Such a resolution corresponds to scanning over a path difference of 10 cm only at 2 μm and 25 cm at 5 μm, which is much less than the limits offered by existing FTS. Similar resolutions would require gratings, if working at theoretical resolution, 50 cm wide at least.

The lack of resolution can be compensated by the association of a Fabry–Perot with a moderate resolution grating spectrometer, which is working as a monochromator (Wade 1983), or with a circular variable filter (Tanaka et al 1985). These systems are able to reach 3×10^4, (at the best), but suffer of a limited spectral coverage and a lack of versatility in changing of spectral range. Several etalons have to be used to cover for instance the $3-5$ μm region. It is better adapted to the study of the same line profile (as Bα or Bγ) in different objects (Persson et al 1984).

2. Spectral range

Change of spectral range with a FTS is easy. Equipped with InSb detectors, any domain limited by a filter, within the range 0.9–5.5 μm can be selected. Two beamsplitters are enough to optimize the optical efficiency over such an extended range. On the CFH instrument (Maillard and Michel 1982), three beamsplitters (an additional one for the visible) are permanently installed in the instrument, mechanically mounted in such a way that the interchanging does not require any further adjustment. A choice of broad and narrow band filters within the range of sensitivity of the detectors, mounted on a 10–filter wheel, determines the width of the spectrum.

3. Throughput

From the classical relationship between the resolution R of an interferometric spectrometer and the solid angle of the entering beam, an equation can be deduced when the interferometer is matched to the throughput of a telescope of diameter D_T, for a source of angular diameter :

$$R = \frac{8}{\alpha^2}(\frac{D_I}{D_T})^2$$

The instrumental parameter which enters in the equation D_I is the diameter of the parallel beam in the instrument. Numerical applications of this equation shows that for D_I =20mm, an extended source of 10" in diameter can be observed up to a resolving power of 10^5. With a 8-m telescope a beam size of 45 mm is enough to accept the same field at the same resolution. Lower resolutions allow bigger fields, usable for galactic nuclei, planetary nebulea, comets etc...In stellar applications, field-of-view does not seem to be so important. However, a spectrometer must be able to accept the whole seeing disk to reach the maximum luminosity. With modern 4-m class telescopes the luminosity of grating spectrographs becomes seeing-dependent for resolutions higher than 10^4. In the ESO coude echelle spectrograph for instance, a slit width of 1" has to be used for a resolution of 8×10^4. With this slit width, the transmission is equal to 65 % with a 1" (FWHM) seeing and falls to 35 % with 2" (Diego 1985). The luminosity drops dramatically for higher resolution.

4. Substraction of parasitic background

All the modern FTS systems are based on a design which employs cat's-eye retroreflectors. Two equal entrance apertures on the sky are matched on the two output detectors, allowing an automatic substraction of the parasitic sky background. The latter can be of thermal origin, above 2.5μm, or parasitic daylight. This property is part of the photometric quality of FT spectra, particularly essential in infrared. That makes possible also to work in some cases in presence of daylight, which would be completly impossible with grating spectrographs. However, the energy of the source within the filter bandwidth must be brighter than the contribution of the sky in order to get the same SNR than in dark time. By respecting this condition, the multiplex "disavantage" offers the advantage of longer observing time.

5. Metrologic accuracy

In spectra obtained from a FTS, all the line positions are automatically calibrated to an absolute accuracy which can be equal to the accuracy to which the reference line of the instrument is known (generally a stabilized laser, $\Delta\lambda/\lambda \simeq 5\ 10^{-9}$). This property does not contribute to the SNR of a spectrum but takes advantage of high SNR. The accuracy of relative frequency determination is SNR limited. This unique quality, which is fully utilized in high-resolution, high-SNR laboratory spectroscopy, has also many advantages in astronomical applications. The identification of lines, the detection of new molecular species through several lines covering different quantum number, thanks to the large spectral coverage, can be made without ambiguity in crowdy spectra, providing accurate frequencies are also known in laboratory. This property has made the success of FT spectra. Many examples can be find in the reviews cited above. Radial velocities can be also easily measured and small lineshifts of $\simeq$100m/s detected (Maillard and Nadeau 1988).

In conclusion, this type of instrument is particularly suitable for all astronomical projects which benefit the most of all these characteristics: high resolution, large spectral range, accurate frequency calibration, eventually non-stellar angular diameter. The solar photosphere, the cool stellar atmospheres, the circumstellar envelopes, the dense molecular clouds, the planetary atmospheres, all objects with numerous, narrow lines are the fields where the FTS brings an essential contribution. Also, on faint emission-line objects, for which the multiplex

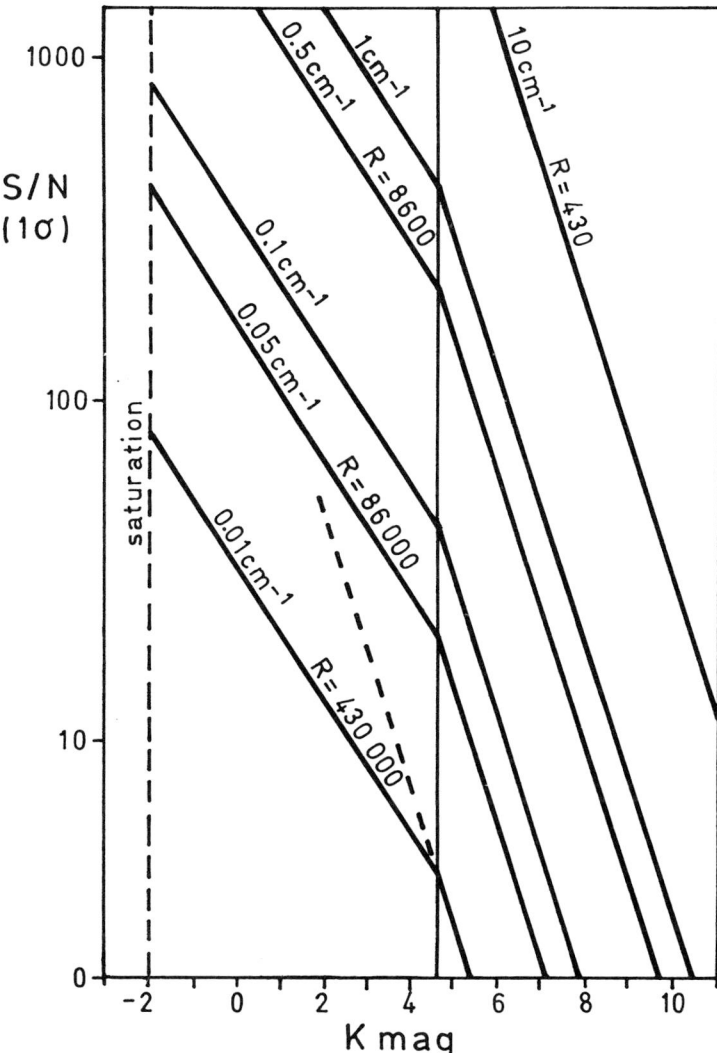

Fig. 1 *Signal-to-noise ratio (1 σ) versus K magnitude for different resolution and 1 hour integration, with the CFH-FTS and a narrow band cold filter (peak 2.32 μm, 0.11 μm BWHM). The faster slopes correspond to detector noise limited conditions. Extrapolating those lines indicates by which factor SNR is deteriorated when photon source noise limited. For sources fainter than K = 4.6 (vertical line) the full gain multiplex is recovered. At a limit of resolution of .1 cm^{-1} such spectra contain more that 2000 spectral elements.*

gain remains almost intact, a FTS offers the advantage of a complete survey over an extraordinary large spectral range.

IV. IMPROVEMENT OF SNR WITH A FTS

Optimization is possible by analyzing the factors which command the SNR in a FT spectrum. All the actions will tend to maximazise the signal and lower the photon noise to approach the detector noise limit.

— telescope size: as a FTS can accept a large throughput, it will be always more efficient behind a large telescope. Under condition I, SNR goes like D^2 and like D under condition II and III. For high-resolution grating spectrograph, it varies only as $\approx D$ and $\approx \sqrt{D}$ for comparable noise conditions.

— telescope plus FTS transmission: careful design with a minimum of optical pieces and selected beamsplitters can produce an instrument with a 70 % optical efficiency against 30 % for the best grating spectrograph. A high-resolution FTS can be compact enough to be installed at the cassegrain focus (Davis et al 1982, Maillard and Michel 1982), providing a gain of a factor ≈ 1.5 in luminosity with respect a coudé focus. If in addition, the infrared focus is used, as for the CFH-FTS, a significant gain is obtained in the thermal infrared by the low emissivity and the low temperature at this focus. Table II gives the final gain in SNR in comparison with a coudé system.

TABLE II

Focus	sky	Transmission telescope	FTS	Temperature
coudé	0.80	0.60	0.60	283 K
IR cass	0.90	0.89	0.60	273 K

λ µm	3.0	3.8	4.0	5.0	10.0
G	2.8	2.6	2.5	2.4	2.2

Gain (G) in signal-to-noise ratio in the thermal infrared obtained by installing the FTS at the IR cassegrain focus of the CFH Telescope with respect a similar instrument in coudé, in a low elevation site. Emissivities and temperatures are lowered.

— spectral range: limiting the recorded spectral range to the useful range is an absolute rule to obtain the best SNR. A modern FTS has to be equipped with a large choice of narrow band filters. With the CFH-FTS and a narrow band filter in the CO 2-0 band (peak 2.32 µm BWHM 0.11 µm), the system becomes detector noise limited for sources fainter than K=4.6 .In this range of brightness, the FTS benefits totally of the multiplex gain and there is no reason to use narrower filters.

— throughput: under condition III, an improvement is obtained also by limiting the cold aperture on the sky. SNR varies as α. Against a widespread opinion, high-image quality in infrared is required, even with a FTS. With the CFH telescope, known for good image quality, diaphragms of 2.5" are currently used in the FTS, which insured a 87 % transmission for 1" seeing (Diego 1985). In case a better seeing occurs (0.8"), 1.5" can be used, making an extra gain of a factor ≈ 1.7.

— instrument thermal background: the background radiation is the sum of three contributions: the sky, the telescope and the instrument. The respective importances of the three terms for the CFH-FTS are: 0.053, 0.066, 0.4, (from the parameters of Table II) which makes the instrument the dominant source of thermal background. Cooling the instrument to -30°C would produce a reduction of a factor ~ 2.5 of the level of background radiation at 4.8μm.

- association with array detectors: the development of array detectors can be a new chance for astronomical FTS.

In presence of thermal background, if a cold grating spectrograph is installed on one of the FTS outputs, it disperses the interferogram on the array detector. Each pixel can approach non-background limited conditions. A large spectral coverage, with all the other characteristics of a FTS, can be obtained, depending on the number of pixels. This option is particularly attractive in the 10μm window, at the maximum of thermal background. A group of Goddard Space Flight Center has already built a single helium-cooled post-disperser (Jennings et al 1986), to use on FTS facilities, on which a 20-element Si: As BIB array detector will be tested. A 7 cm^{-1} spectral range at 10μm will be covered near noise detector limited conditions.

In the visible and the near infrared, in presence of source photon noise, a similar device is the only way of reducing enough the flux received by each individual detector. Such a FTS system could combine the optimum SNR with the spectral coverage, regaining the full multiplex gain of a FTS.

Another use of array detectors associated with a FTS can be considered to perform 2D-spectroscopy on extended object. Using the throughput advantage and the stigmatic image in the plane of the detectors, interferograms from each pixel are recorded in parallel, corresponding to each point of the object. The amount of data to process has prevented developments in this direction. A project of FTS imager has been submitted anyway in view of the Space Telescope Program (Hall 1986).

IV CONCLUSION

Under observing conditions for which a FTS is not detector noise limited, it does not reach the ultimate SNR which would be possible to obtain with an ideal instrument, able to keep the detectors in these noise conditions. The lack of any other alternatives for scientific programs requiring in particular large spectral coverage and high resolution, imposes to accept this limitation. On the other hand, when efforts are made, in particular in the thermal infrared, significant gains in SNR can be obtained, which have been reviewed. Beyond all the improvements already tested with single detectors, a decisive step can be reached with a future generation of FTS systems associating array detectors. Then, all the characteristics of a FTS in resolution, spectral coverage could be exploited from the visible to the thermal infrared. The mixing of very different observing projects in spectral range and resolution, which is already possible during a telescope run, because any time is lost due to Aextensive change, could be completly achieved. With all the others, this factor should be kept in mind for the future instrumentation of coming very large telescopes.

References:

de Bergh, C., *6th International Conference on Fourier Transform Spectroscopy*, Vienna 24 – 28 Aug 1987; proceedings to be published in Micro Chemica Acta (Springer Verlag Pub. Co.).

Brault, J.W., 1979, Ossni. Mem. Oss. Astofis. Arcetri, 106, 33.

Davis, D.S., Larson, H.P., Williams, M., Michel, G., Connes, P., 1980, Applied Optics, 19, 4138.

Diego, F., 1985, Pub. A.S.P., 97, 1209.
Hall, D.N.B., Ridgway, S.T., Bell, E.A., Yarborough, J.M., 1979, SPIE, 172, Instrumentation in Astronomy III, p.121.
Hall, D.N.B., 1986, *Imaging Michelson Spectrometer for Space Telescope*, U. of Hawaii Institut for Astronomy, Jet Propulsion Laboratory.
Jennings, D.E., Deming, D., Wiedemann, G.R., Kedy, J.J., 1986, Ap. J. (Letters), 310, L39
Maillard, J.P. and Michel, G., 1982, in *Instrumentation for Astronomy with Large Optical Telescopes*, IAU Colloquium No 67, C. M. Humphries (ed) p. 213, D . Reidel Pub. Co.
Maillard, J.P. and Nadeau, D., 1988, in The Impact of very high S/N Spectroscopy on Stellar Physics, IAU Colloquium No 132, Meudon 29 June – 3 July 1987.
Persson, S.E., Geballe, T.R., McGregor, P.J., Edwards, S., Lonsdale, C.J., 1983, Ap. J., 286, 289.
Ridgway, S.T. and Brault, J.W., 1984, Ann. Rev. Astrophys., 22, 291.
Tanaka, M., Yamashita, T., Satos, S., Okuda, H., 1985, Pub. A.S.P., 97, 1209.
Wade, R., 1983, SPIE, 445, 47.

Discussion

<u>P.E Nissen</u>: One advantage of the FTS, which you did not mention, is its clean spectral profile and negligible amount of scattered light. This makes the FTS an interesting instrument in connection with accurate determinations of profiles of spectral lines at a resolution R 200,000, to be used in e. g. the study of stellar atmospheric velocity fields.

<u>J.P. Maillard</u>: That's right. However, in this talk I wanted to focus on the instrumental characteristics of a FTS which directly enter in the determination of the S/N. These parameters have to be selected carefully to balance between the best S/N to get and the advantages they offer. A clear example is the spectral range which has to be limited but gives access to many lines to analyze. The clean spectral profile is a supplementary quality but in the example you give a FTS would be required first for its capabilities of very high resolution.

<u>J.E. Beckman</u>: Could you please give us the result of your determination of D:H ratio on Mars?

<u>J.P. Maillard</u>: This ratio was measured with the CFH-FTS in Jan. 87 from the detection of the molecule HDO, at 3.7 μm, for the first time on Mars. An enhancement of this ratio by a factor 6 with respect to the terrestrial value was deduced, giving a new clue to the past history of water on Mars. I gave this example to illustrate the increased power of a FTS installed in a dry infrared site like Mauna Kea, at the focus of a low background telescope like CFHT. The large throughput, to accept the planetary disk, and the high resolution of the instrument were fully exploited.

PRECISE RADIAL VELOCITIES OF F,G,K DWARFS

Bruce Campbell
Physics Department
University of Victoria
Victoria, B.C., V8W 2Y2, Canada

Gordon A. H. Walker and Stephenson Yang
Geophysics and Astronomy Department
University of British Columbia
Vancouver, B.C., V6T 1W5, Canada

ABSTRACT. We have monitored *changes* in the radial velocities of 24 bright F, G and K dwarf stars (known spectroscopic binaries excluded) for the past six years at CFHT by imposing the absorption lines of HF gas in the spectra to act as wavelength fiducials. The average external error in the Δ(velocities) which are based on some 16 stellar lines is 13 m/s corresponds to 0.6 micron in the spectrum or 0.04 of a diode spacing per line. Reductions are complete for 16 stars. There is no evidence for brown dwarf companions in the sample. Two previously unknown spectroscopic binaries were found, and seven stars show indications of significant, long-term, low-level velocity variations which could be interpreted as purturbations by companions of a few Jupiter masses with periods greater than 12 years except for γ Cep, which may have a period of 2.7 years, and ε Eri. Observing time has been guaranteed for at least two more years at CFHT.

1. THE PRECISION

The absorption lines imposed in a stellar spectrum by passing starlight through a captive gas can act as wavelength fiducials which are independent of the large systematic errors associated with conventional calibration techniques. We first proposed and demonstrated the feasibility of using HF as the absorbing gas in 1979 (Campbell and Walker 1979). The wide dynamic range of the Reticon diode array allows us to take single spectra with a signal to noise in excess of 1000 whichis sufficient to measure the positions of sharp lines with a resolution of the order of 0.01 of a diode spacing using the differencing technique developed by Fahlman and Glaspey (1973). In the coudé spectrograph of the CFHT at 4.8 A/mm this corresponds to about 10 m/s at λ 8700 (the wavelength of the (3,0) band of HF used in our technique).

If the technique is free from long term systematic errors, a precision of 10 m/s would be sufficient to detect the slow oscillations in radial velocity of solar or later type stars with Jupiter mass companions when viewed in their orbital plane. We began such an 'unseen' companion search in the winter of 1981 at CFHT and, apart from a gap of six months in 1982, the program continues to be given six pairs of night per annum. Details of the data reduction and the program can be found in Campbell, Walker, Pritchet, Long (1986).

The variation of the instrumental line profile from run to run is the principal instrumental effect which limits the long term accuracy of the technique. The variations are almost certainly due to the 4-grating mosaic in the spectrograph (Richardson, Brealey, Dancey 1971). The line profile is a composite from each of the four gratings. Not only does each grating gives a different image quality and dispersion but alignment is never exact nor completely stable which is serious because of the need to correct the HF line profiles for blends with weak stellar lines. The HF lines are about twice the width of the stellar lines and the instrumental profile. Although the effects of asymmetry in the instrumental profile are modelled in the data analysis there are still residual off-sets of the order of 7 m/s between runs. In consequence we apply run to run corrections based on the average off-set of the Δ(velocities) using measurements for all of the stars observed in a run. In practice it is usually possible to observe 16 or more program stars during a clear run.

This paper appears in full in the proceedings of IAU Colloquium No 99: *"Bioastronomy - The Next Steps"*.

REFERENCES

Campbell, B., Walker, G.A.H., 1979, *Pub. A.S.P.*, **91**, 540.
Campbell, B., Walker, G.A.H., Pritchet, C., Long, B., 1986, *Astrophysics of Brown Dwarfs*, R.S. Harrington and S.P. Maran, eds. (Cambridge University Press), p.37.
Fahlman, G.G., Glaspey, J.W., 1973, *Astronomical Observations with Television-Type Sensors*, J.W. Glaspey and G.A.H. Walker, eds. (Vancouver: British Columbia), p. 347.
Richardson, E.H., Brealey, G.A., Dancey, R., 1971, *Publ. Dom. Ap. Obs.*, **14**, 1.

DISCUSSION

MARCY Epsilon Eri is your best candidate for having a planetary companion. However, that star is well-known for being one of the most magnetically active. Have you made any efforts to see if the line profiles are changing with time, perhaps due to the Zeeman effect?

WALKER I can't rule out such an effect. The unstable line profile means that our information about subtle changes in profile is not very secure. There is no correlation with chromospheric activity.

LIBBRECHT Can you tell me what are the frequency shifty of the HF lines with temperature and pressure in your cell, and are those shifts a problem?

WALKER The cell is maintained at 100° C and the HF pressure by an ice water bath. We've measured the line shifts at large deviations in temperature and pressure, and we make small corrections for these effects. The corrections are small compared to the signal we're reporting, so they're not a problem.

ANDERSEN It may be a bit disappointing that, in Commission 30, we are perhaps most interested in your non-detections in connection with the establishment of the new system of IAU Radial Velocity Standards. I should be most interested to hear about the independent work by the Arizona group, too.

WALKER We hope to submit our data for publication later this year.

A Search for Radial Velocity Oscillations in Procyon

Ken G. Libbrecht
Solar Astronomy 264-33
California Institute of Technology
Pasadena, CA 91125

ABSTRACT. I describe here a new technique for precision radial velocity measurements, using an iodine absorption cell in front of the slit of a high resolution spectrograph. A CCD in a continuous readout mode records a timeseries of stellar spectra with a duty cycle of unity. In January 1987, observations were obtained on two clear nights with bad seeing. No oscillations were detected in the data, but the technique shows promise for significantly better results under better conditions.

Introduction. Acoustic oscillations in the sun have been observed, without spatial resolution, both as radial velocity oscillations and intensity oscillations in integrated sunlight, the latter from spacecraft measurements. A power spectrum of the solar radial velocity signal shows a series of sharp peaks uniformly spaced in frequency with a separation of 70 μHz. The oscillation power is a maximum at around 3 mHz (5-minute period) with an amplitude per mode of roughly 10 cm/sec, and the ratio of intensity to velocity oscillations is

$$\frac{(\Delta F/F)^2}{v^2(\text{cm}^2/\text{sec}^2)} \approx 1.5 \times 10^{-14}.$$

Solar oscillation frequencies are currently being used to learn about the internal structure of the sun, while the mode amplitudes provide information about the convective excitation process. It is likely that observations on other stars will be as informative as on the sun, if oscillations can be detected for some number of stars.

Since the oscillation amplitudes are very small, detection is a formidable challenge, and it is not yet certain if solar-type oscillations have been observed on stars other than Ap oscillators. Perhaps the best detection is that by Gelly, Grec, and Fossat (1986, see also Fossat, these procedings) on Procyon, using a sodium resonance filter. They measured oscillations with periods near 10 minutes, velocity amplitudes of 70 cm/sec per mode, and a mode spacing of 40 μHz. A resonance filter offers the advantage of high stability, but the disadvantage of very limited spectral coverage, namely the Na D lines. If oscillations on fainter stars are to be measured, then it is likely that radial velocity observations using many lines will be necessary.

Observations. I have been investigating a new technique for precision radial velocity measurements, using an iodine absorption cell in front of the slit of a high resolution spectrograph. Several dozen sharp iodine lines around 5350 Å, added to the stellar spectrum, serve as a stable wavelength reference spectrum, and there is no inherent restriction on the number of stellar lines that can be used. In order to obtain an adequate duty cycle for a bright star such as Procyon, the CCD detector was read out slowly and continuously, giving a timeseries of spectra with a duty cycle of unity.

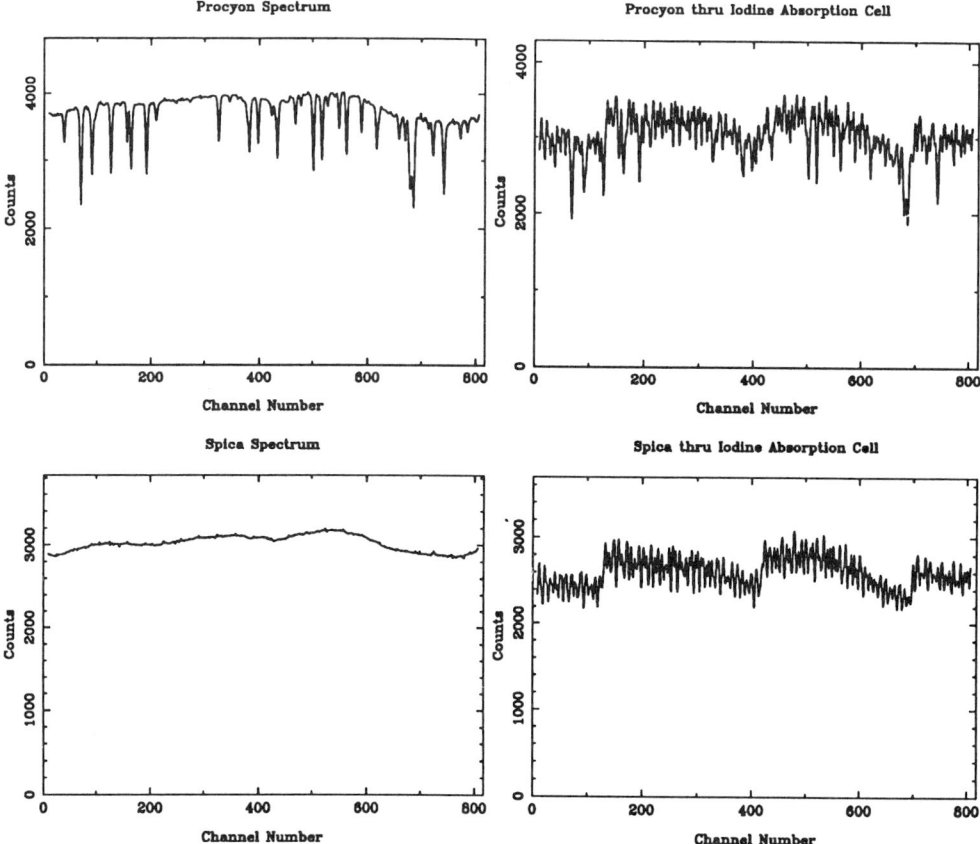

Figure 1. Sample stellar spectra, taken with the 5-meter Hale telescope coude spectrograph. A TI 800 × 800 CCD was used, covering a region of the spectrum 50 Å wide near 5350 Å.

Figure 1 shows a sample of the data. At the beginning of each evening's observations a reference Procyon spectrum was taken, followed by a spectrum of Spica which served as a flat-field, and a spectrum of Spica through the iodine cell. The rest of the night was used collecting a timeseries of spectra of Procyon through the iodine cell. Due to bad weather only two useable nights of observations were obtained, with 3 arcsecond seeing. In order to keep the highest spectral resolution, a fairly narrow 0.25 arcsec slit was used, and measurement showed that the bad seeing caused a factor of 10 in light to be lost at the slit.

In a one-minute integration a velocity accuracy of approximately 15 m/sec was achieved with this data. After subtracting the known shifts due to earth rotation, slow drifts in the measured

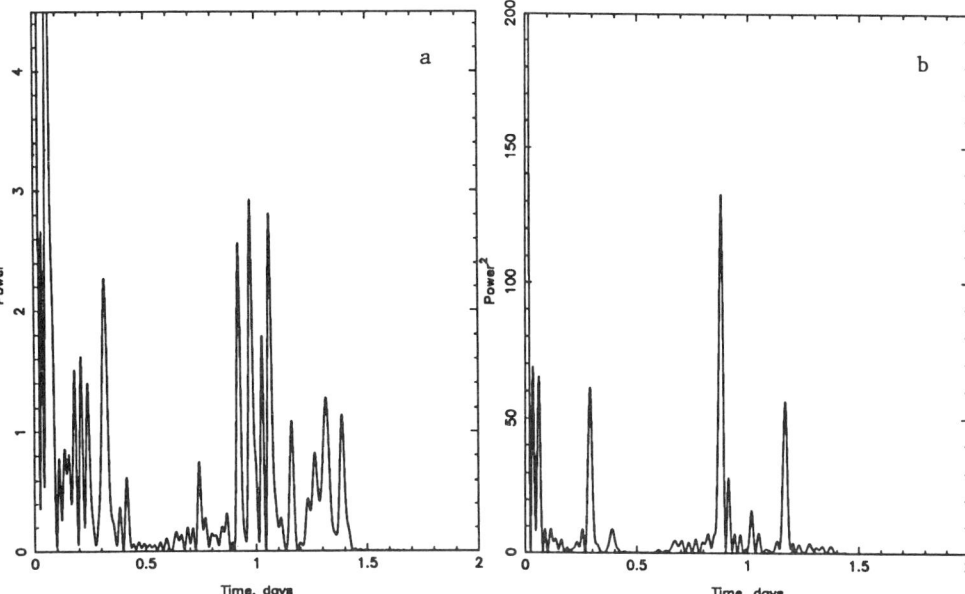

Figure 2. (a) Power spectrum of the velocity power spectrum, where the power outside the region 1.17 to 1.65 mHz was set to zero. The process follows the analysis described by Gelly, Grec, and Fossat, and produces an autocorrelation function. (b) The same analysis applied to the same data, but with an oscillation of 3 m/sec/mode added to the raw data, with a mode spacing of 40 μHz.

velocity were observed at the level of nearly 100 m/sec, which were found to be correlated with fluctuations in the density of iodine in the cell (and thus presumably with the iodine temperature, since the cell heater was not in a servo loop). The stability of the system was adequate for measuring acoustic oscillations, and with better temperature regulation the long-term stability could be greatly improved.

Using the known count/photon ratio of the CCD, and the measured Procyon spectrum in Figure 1, the measured velocity noise was found to be comparable to that expected from simple photon counting noise. However it must be noted that one photon was counted only for every 800 which struck the primary mirror. This poor efficiency was the result of an overall spectrograph efficiency of approximately 2 percent, the factor of 10 loss at the slit mentioned above, and losses from absorption in the iodine cell. Thus the realization of the photon noise limit was due in part to the poor photon counting efficiency.

The timeseries of radial velocity measurments was Fourier transformed, producing a power spectrum which was not distinguishable from random noise. To search for an oscillation spectrum with modes that are uniformly spaced in frequency, the analysis described by Gelly, Grec, and Fossat (1986) was performed. Basically the power spectrum was itself Fourier tranformed over the frequency range 1.17-1.65 mHz, forming an autocorrelation of the data. If a solar-type spectrum were present in the data, then a set of uniformly spaced peaks should appear in the autocorrelation function, along with nulls for the daytime gaps. The results are shown in Figure 2, both for the data and for the data with an artifical oscillation signal added to the measured velocity timeseries. From this we see that no oscillations are present in these data, and that an upper limit of 1 m/sec/mode is indicated. These data are consistent with the French detection of 70 cm/sec/mode.

The Future. Although the present data are not good enough to detect acoustic oscillations in Procyon, I believe the measurement technique described is worthy of further attention. First of all, with better seeing and a good image slicer or scrambler, most of the slit losses could be avoided, giving a factor of 3 increase in the velocity S/N. Although the spectrograph efficiency seems poor, it is unlikely to be greatly improved in the future, since even the very carefully designed Hamilton spectrograph boasts an efficiency of 4 percent (see Vogt, these proceedings). The biggest gain will come by replacing the 50 Å wavelength coverage here by several thousand Angstroms, giving an additional improvement of a factor of 5-10. With these improvements it is likely that oscillations could be measured on many stars.

Although existing echelle spectrographs could provide the extended wavelength coverage desired, it is worth noting that the duty cycle on a bright star is quite low with current machines (say, 5 seconds to expose, followed by 45 seconds to read out), and so unless fast-readout CCDs are pressed into service it will be difficult to achieve these gains.

Reference

Gelly, B., Grec, G., and Fossat, E. 1986, *Astron. Astrophys.* **164**, 383.

S. Vogt: I think that the thing which is limiting you at the moment is your relatively low dispersion. Assuming typical wavelength accuracies of 0.01 pixels, your limit should be $\sim$ 40 m/sec. It's primarily a question of dispersion, not S/N.

K. Libbrecht: Nothing fundamental limits you to 0.01 pixel. For instance, most of my work on solar oscillations is done with 0.25 Å resolution, and we measure down to 1 mm/sec.

G. Walker: Once one tries to measure displacements of less than 10 m/sec there are seeing effects, microseismic effects, and drifts, all which can become important. I agree that high dispersion is called for, combined with a vacuum spectrograph.

K. Libbrecht: The iodine reference spectrum reduces most of these problems, especially seeing across the slit. We may see small systematic errors with higher S/N, but for now all our high frequency errors are explained by simple photon noise. That's the problem we need to really worry about when looking for these short-period oscillations.

D. Baade: Is the precision of you measurements conceivably limited by Spica being a double-lined spectroscopic binary and also a nonradial oscillator?

K. Libbrecht: I don't think that's a problem.

E. Fossat: I would like to remind you, or should I say sell you, of the project of Pierre Connes, of an absolute accelerometer taking advantage of the whole stellar spectrum. Unfortunately, Connes himself has not given a price for this project. I don't know who can pay for it.

HIGH S/N SPECTROSCOPY OF PRE-MAIN SEQUENCE STARS I

I. Appenzeller
Landessternwarte, Königstuhl
D 6900 Heidelberg
Federal Republic of Germany

ABSTRACT. The spectra of PMS stars usually are superpositions of contributions from a relatively normal photosphere, an often strongly enhanced chromosphere, and of circumstellar matter related to stellar winds, accreation flows, jets, and cool circumstellar disks. Only high S/N data allows a reliable separation of these different contributions. Because of the high optical depths of PMS chromospheres, the PMS photospheres often can be observed in the very weak spectral lines only, which are undectable on low S/N spectrograms. Finally, magnetic fields are assumed to play a particularly important role for the appearance and evolution of PMS objects. Their spectroscopic measurement requires very high S/N data.

1. INTRODUCTION

PMS stars are very young stars where significant nuclear burning has not yet started. The low mass PMS stars are classified as T Tauri or Post-T Tauri stars, and their basic properties have been reviewed recently e.g. by Kuhi (1983), Bertout (1984), and Appenzeller (1985). The higher mass PMS stars are called Herbig-Ae-Be stars and have been described e.g. by Herbig (1960) and Finkenzeller and Mundt (1984).

Most known PMS stars have been identified by their characteristic emission lines using very low S/N objective prism spectrograms. Most qualitative properties of these objects were derived from low S/N photographic or image tube spectrograms. On the other hand, for a detailed quantitative derivation of the physical structure of PMS objects high S/N is perhaps more important than for any other class of stars. There are three effects, which make high S/N a prerequisite of success in quantitative investigations of PMS stars: Firstly, PMS stars generally show "composite" spectra containing contributions of a hydrostatic photosphere,

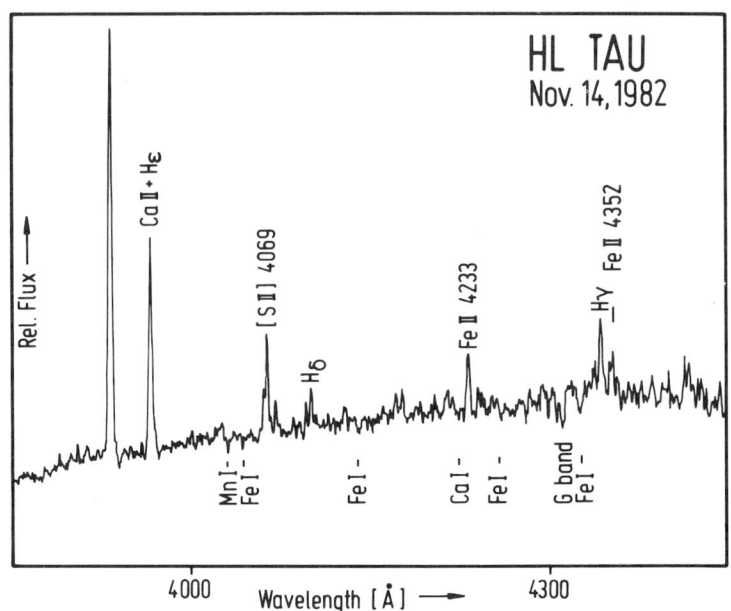

Fig. 1. The spectrum of the young PMS star HL Tau, showing contributions of a late type photosphere (Fe I absorption lines, G-band), a dense chromosphere (producing e.g. the Ca II and He I emission), and of circumstellar gas ([SII], Balmer lines, broad metallic lines).

contributions by an (often strongly enhanced) chromosphere, and contributions of circumstellar matter related to stellar winds, accretion flows, jets, and cool dusty circumstellar disks (cf. Fig. 1). Quantitative studies of PMS stars require a reliable separation and subtraction of the different contributions, which is possible only with high S/N linear spectrograms. Secondly, the PMS stars (and in particular the T Tauri stars) often show chromospheres of exceptionally high optical depths. Hence, in these stars the "strong" (i.e. high optical depth) lines are formed near the temperature minimum or in the chromosphere. As a result the normally strong photospheric lines tend to become very weak or occur in emission (cf. e.g. Cram 1979, Finkenzeller and Basri 1987). In the particularly interesting "extreme" T Tauri stars only the very weak spectral lines can be detected as photospheric absorption features (Appenzeller et al. 1986). Finally, as pointed out in the reviews listed above, we have indirect but reliable evidence for the presence of

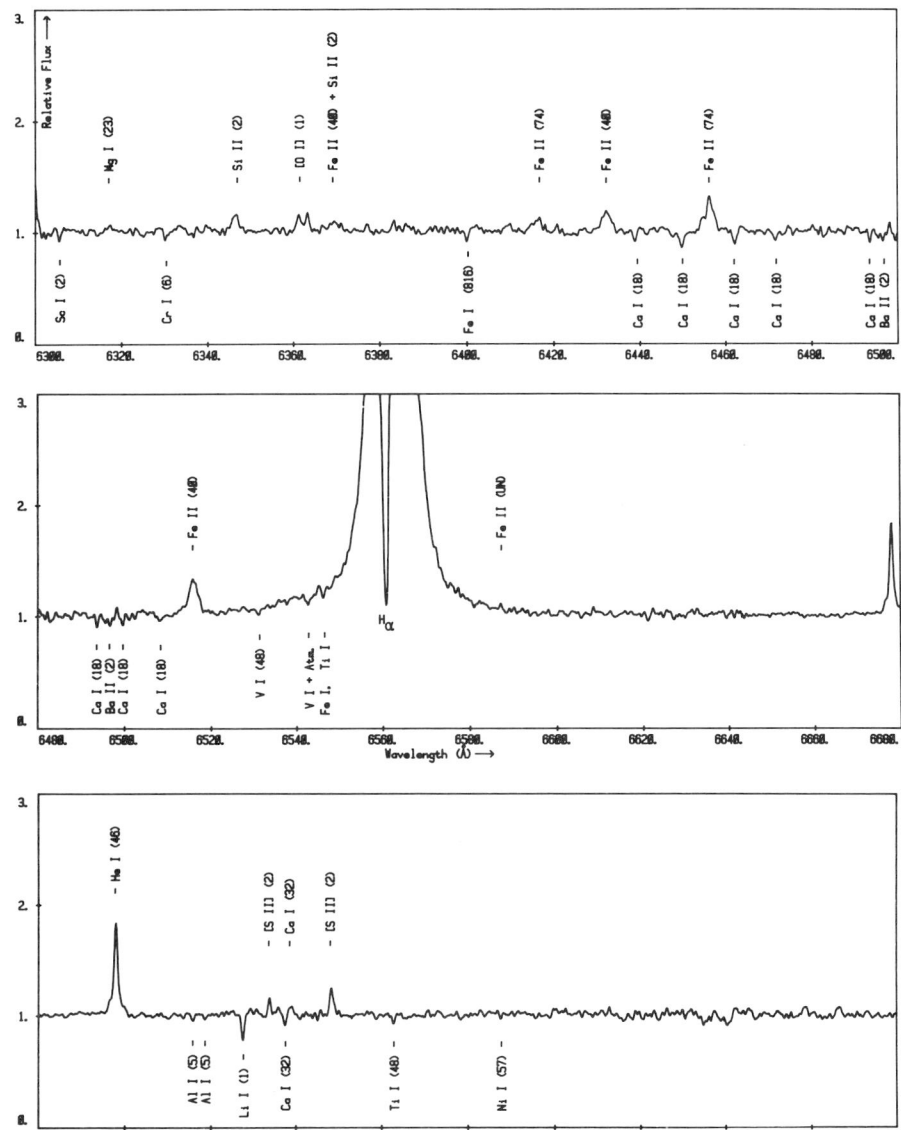

Fig. 2. Section of a high S/N spectrogram of the extreme T Tauri star S CrA.

extended magnetic fields on the surface of PMS stars. Since the evolution of contracting stars depends on their surface properties, these fields probably have a profound influence on the structure and evolutionary time scale of these stars (Appenzeller 1985). Thus, measurements of these fields are

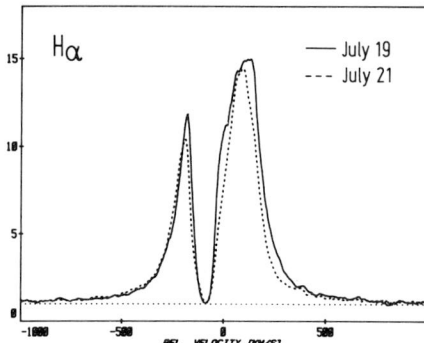

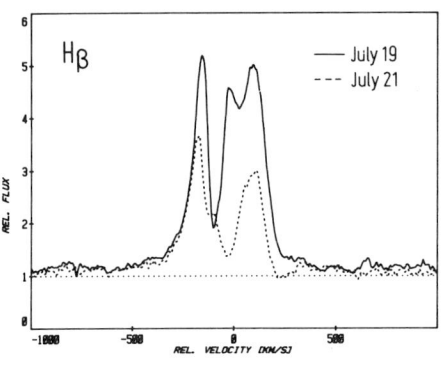

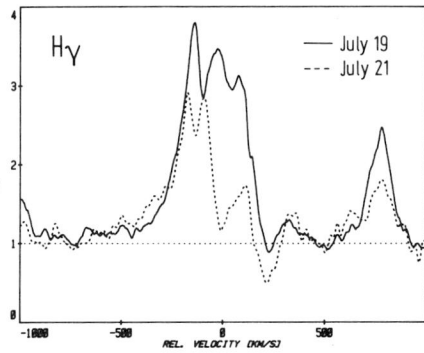

Fig. 3. Balmer line profile variations.

of great importance for a better understanding of PMS stellar evolution. As described by Saar, Marcy, Basri and others elsewhere in this volume, these measurements require very high S/N values. So far only upper limits or very marginal detections have been reported for PMS stars.

2. LIMITATIONS

In the case of massive stars the core hydrogen burning starts already during the IR or "protostellar" evolutionary phase. Hence all PMS stars are of relatively low mass and moderate to low luminosity. Furthermore, there are no star forming regions in the immediate solar vicinity. Therefore, PMS stars are generally rather faint objects. Typically the known PMS stars (as listed e.g. in the catalog of Cohen and Kuhi, 1979) have apparent visual magnitudes between 10^m and 20^m with a median of about 15^m. Following Matthews and Sandage (1963) the photon flux F_N of a star of visual magnitude m_v can be approximated by:

$$F_N = 10^{3-.4 m_v} \text{ ph. s}^{-1} \text{cm}^{-2} \text{Å}^{-1}.$$

Assuming that photon noise (of the signal) is the only source of noise, S/N = 100 requires 10^4 photons per resolution element. Using a 3.6-m telescope and assuming a total efficiency (atmosphere + telescope + spectrograph + detector) of .03 (which is not untypical for modern high resolution instruments with CCD detectors) the integration time for a PMS star of $m_v = 15$ becomes 1.8 hours for a spectral resolution of $R = \lambda/\Delta\lambda = 10^4$ and 18 hours for $R = 10^5$. Obviously, high S/N spectroscopy of PMS stars is time-consuming and feasible only for the

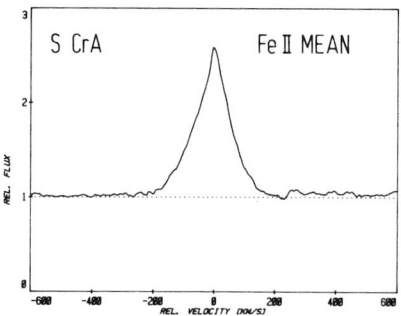 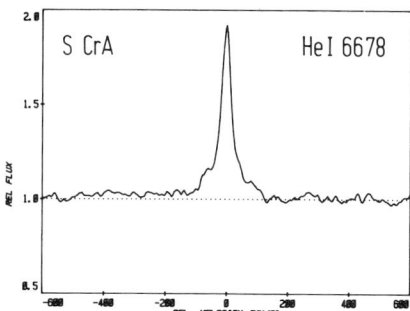

Fig. 4. Mean Fe II (average of four unblended lines) and He I emission line profiles of S CrA.

brighter members of this class of objects.

3. EXAMPLES

A few of the Herbig-Ae-Be stars are bright enough for high S/N observations, even with conventional Coude spectrographs. Using modern solid state detectors, for these bright objects S/N values of several 10^2 can be reached (cf. Catala elsewhere in this volume). For the fainter T Tauri stars high S/N spectroscopy became possible only with the introduction of sensitive photoelectric detectors. Spectroscopy with one-dimensional photon-counting detectors resulted in S/N values up to about 60 (see e.g. Mundt 1984). But high resolution observations of T Tauri stars of really high S/N (>100) became feasible only with CCD detectors at efficient Echelle spectrographs. Examples of these new instruments are the ESO CASPEC and the Lick HAMILTON spectrometers. Below, a few examples obtained with the CASPEC spectrograph are presented. Exciting results obtained more recently with the new HAMILTON spectrograph are given in the following paper by Dr. Basri.

Figure 2 shows (as an example of high S/N spectroscopy of an extreme T Tauri star) a section of the "spectral atlas" of S CrA (Appenzeller et al. 1986). Note that even in this extreme object a photospheric absorption spectrum is visible on high S/N spectrograms. However, in accordance with the theory, the photospheric spectrum is restricted to weak lines only. Also visible in Fig. 2 are the conspicuous differences in the profiles of emission lines formed in different volumes of the S CrA system.

Another important application of high S/N spectroscopy of T Tauri stars is illustrated by Fig. 3: Qualitatively the complex and rapidly variable structure of the Balmer emission lines of PMS stars has been known since years from lower S/N data. But only the new high S/N spectrograms allow a reliable identification of the many different components which form these profiles. Obviously the gas producing these lines is highly nonhomogeneous and follows a highly nonstationary flow pattern.

Examples of metallic and He I emission-line profiles are given in Fig. 4. These profiles are less variable than the Balmer lines. The narrow core of the He I line is only slightly broader than the photospheric absorption lines, confirming earlier suggestions that the (relatively high-excitation) He lines originate mainly in essentially static chromospheric layers. In contrast, from their width it seems clear that the Fe II lines must be formed in the moving gaseous envelope of the T Tauri star. The almost triangular shape of these lines obviously reflects the velocity field of the envelope. A simple curve of growth analysis of the Fe II lines of S CrA shows that these lines are optically thin, making a modelling of their profiles relatively easy. An exact match of the Fe II profile of Fig. 4 appears not possible with simple assumptions on the velocity field and line source function. However, a qualitative approximation of the observed profile is obtained with the assumption that the gas which produces the variable inverse P Cyg profiles at the higher (n>5) Balmer lines (cf. Fig. 3) is also responsible for the Fe II emission. In order to produce the inverse P Cyg profiles this gas must be falling inward towards the photosphere with a line-of-sight velocity component of about the free fall velocity. Assuming for simplicity a spherical and stationary free fall, we have for the velocity $v \sim r^{-1/2}$ and for the density $\rho \sim r^{-3/2}$, where r is the distance to the stellar center. Assuming furthermore that the line emissivity is proportional to ρ^2 we obtain for the line profile (as plotted in Fig. 4):

$$\frac{dI}{dv} = \frac{dI}{dr}\frac{dr}{dv} \sim r^2 \, \rho^2 \, r^{-3/2} \sim r^{1/2} \sim v^{-1} \qquad (2)$$

The resulting hyperbolic profile approximates the Fe II lines reasonably well, if we take into account that Fe II is expected to be the dominant Fe ion in a limited part of the envelope $r^{min} < r < r^{max}$ only.

In spite of the rough qualitative success of the simple spherical model for some details, a consistent description of the T Tauri spectra clearly requires more complex nonspherical configurations. Work on such improved models of PMS stars is in progress in various places. Thus, there is hope that the accurate observational data provided by modern high S/N spectroscopy of PMS stars will soon be matched by equally accurate theoretical model predictions.

REFERENCES

Appenzeller, I., 1985, Physica Scripta T II, 76
Appenzeller, I., Jankovics, I., Jetter, R., 1986, Astron. Astrophys. Suppl. Ser. 64, 65
Bertout, C., 1984, Rep. Prog. Phys. 47, 111
Cohen, M., Kuhi, L. V., 1979, Ap. J. Suppl. 41, 743
Cram, L. E., 1979, Ap. J. 234, 949
Finkenzeller, U., Basri, G., 1987, Ap. J. (in press)
Finkenzeller, U., Mundt, R., 1984, Astron. Astrophys. Suppl. Ser. 55, 109
Herbig, G. H., 1960, Ap. J. Suppl. 4, 337
Kuhi, L. V., 1983, Rev. Mexicana Astron. Astrof. 7, 127
Matthews, T. A., Sandage, A. R., 1963, Ap. J. 138, 30
Mundt, R., 1984, Ap. J. 280, 749

DISCUSSION

CAYREL DE STROBEL Has the G-band in T Tauri stars been studied also in very faint ($\simeq$ 19 mag) T Tauri objects ?

APPENZELLER It can be detected on low resolution spectrograms if the emission lines are weak.

DESHPANDE Do you observe any Doppler shift signatures on these lines ?

APPENZELLER The apparent velocities of the metallic lines vary with the ionization potential. The Balmer line components show complex time variations of their velocity shifts.

ROTATION OF PRE-MAIN SEQUENCE STARS FROM HIGH S/N SPECTROSCOPY

J. Bouvier
Institut d'Astrophysique
98 bis bd Arago
F-75014 Paris

ABSTRACT. Until 1980, only a handful of low-mass, active pre-main sequence (pms) stars had known rotation velocities ($vsini$) /1/. Since then, increasingly sensitive detectors coupled to large telescopes led to high-resolution (a few 10^4) spectroscopic studies of these faint stars (m_v=10-13), with S/N ratio of the order of 100. The measurement of $vsini$ for large samples of pms stars that resulted brought new insights on various pressing questions related to stellar formation and early stellar evolution : how do the rotation rates of pms stars compare with those expected from models of stellar formation ? how does the stellar angular momentum change during pms evolution ? is pms activity linked with rotation as would be expected if activity were triggered by magnetic processes ?

1. STELLAR ROTATION OF LOW-MASS YOUNG STARS

Vogel and Kuhi (1981) /2/ obtained mostly $vsini$ upper limits of 25-35 km.s^{-1} for a sample of pms stars they investigated spectroscopically. That low-mass pms stars were rotating far below their break-up velocity ($\simeq$ 300 km.s^{-1}) led them to conclude that most of the angular momentum presumably contained in the inital interstellar gas cloud had been removed before the star became visible at optical wavelengths.

Bouvier et al. (1986) /3/ measured the rotation rates of 28 low-mass pms stars. While confirming previous results of Vogel and Kuhi, the low detection threshold (a few km.s^{-1}) of CORAVEL, a cross-correlating spectrograph, allowed us to derive the $vsini$ distribution of the sample, with $vsini$ ranging from 8 to 73 km.s^{-1}, including no upper limits. Activity-rotation relations were then searched for in pms stars (see /4,5/).

Independently, Hartmann et al. (1986) /6/ applied cross-correlation analysis on high-quality spectrograms of 50 pms stars to measure $vsini$ with a detection threshold of 10 km.s^{-1}, leaving undertermined the rotation rates of 20% of their sample.

At the same time, the rotation rates of low-mass stars in several young clusters were obtained by Stauffer et al. (1984,1985) /7,8/, with a detection threshold of 10 km.s^{-1}.

These studies offer an observational sampling of the rotation of low-mass stars at various ages that constrains models such as Endal and Sofia's (1981) /9/ which describes the rotational evolution of a 1 M$_\odot$ star during the pms phase and up to the Sun's age. Figure 1 shows solutions of this model compared to the observational constraints. Starting from 10^7 years, the surface rotation first increases as the result of stellar contraction and because a radiative core develops in the initially completely convective star. Then, assuming magnetic braking acts to remove angular momentum from the outer convective zone, the surface velocity starts decreasing as the star approaches the main-sequence. Model calculations of the transfer of angular momentum inside the star predict that, as the star reaches the main-sequence, the outer convective zone uncouples from the rapidly rotating radiative core and is abruptly slowed down. Although the observed trends (pms spin-up and ms spin-down) are accounted for by the model, quantitative disagreements remain. Rapid rotators in the α Persei cluster suggest that the pms spin-up is larger than that predicted by the model. Also, the early main-sequence braking seems to act on a time-scale much shorter than the model predicts. Several assumptions of the model may be responsible for these disagreements. New models which investigate various braking laws for stars of various masses are being currently developed by the Sofia group at Yale University and may eventually provide a better account of the observations. Ultimately, such models should enlight how angular momentum is lost during early stellar evolution (winds, magnetic fields) and will also be helpful to estimate the amount to which internal rotation may influence the depletion of light elements as the star evolves.

The observed widening of the $vsini$ distribution from the pms stars to the α Persei cluster (see Fig.1) remains puzzling. Although several suggestions have been made to account for the existence of slow-rotators in the young clusters (age spread /7/, spread of initial angular momenta /3/, rotation-independent braking law /13/) further high resolution, high S/N spectroscopic studies are needed to settle the true $vsini$ distribution of rotators having $vsini$ upper limits of 10 km.s^{-1}. Pertinent observations are planned for 50

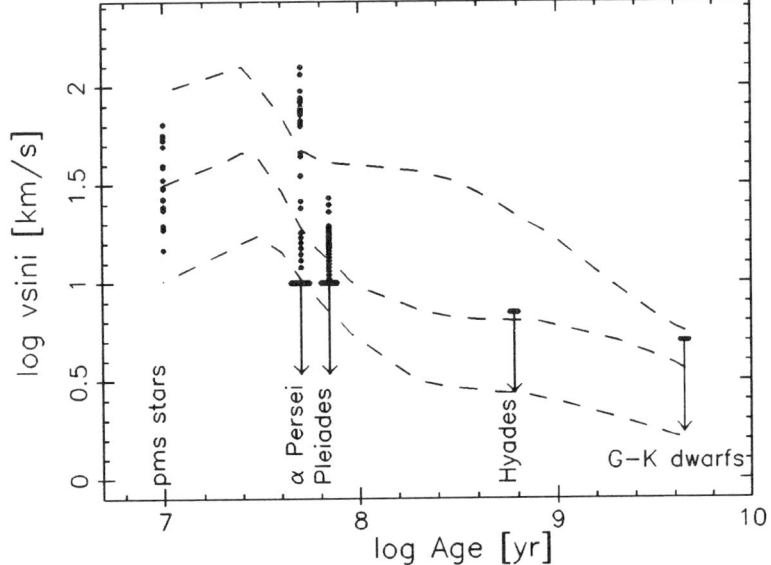

Figure 1: The distribution of surface rotational velocities of low-mass stars as a function of age. References: pms stars ($M \leq 1.2 M_\odot$) /3/, α Persei and Pleiades G0-K2 stars /10/, Hyades G0-K2 stars /11/, G-K dwarfs /12/. The $vsini$'s of pms stars ($\simeq 10^6$ years) have been projected to 10^7 years assuming solid-body rotation (dots). Vertical arrows denote upper limits on $vsini$: in α Persei and in Pleiades clusters, approximately half of the stars in the considered mass range have $vsini \leq 10$ km.s^{-1}. For these clusters, the $vsini$'s have been grouped in 10 km.s^{-1}-wide bins (dots). The dashed curves show solutions of Endal and Sofia's model for different amounts of initial rotation (see text).

pms, α Persei and Pleiades slow rotators. Also, Lithium abundances of the slow and rapid rotators in the young clusters should be measured to explore a possible age spread.

2. PROSPECTS

2.1 Doppler Imaging

The surface of several active pms stars are believed to be covered with spots /5,14,15/. Up to now, the spots properties were derived, with large uncertainties, from photometric observations. From now on, the new CCD camera coupled with the CES at ESO 1.4m telescope provides high enough S/N (about 100) at a spectral resolution of 7.10^4 to start Doppler imaging the surface of the brightest active pms stars. Long-term observations dedicated to the study of spots (growth and decay, activity cycles, latitudinal migration → surface differential rotation) should bring strong constraints on stellar dynamo models.

2.2 Accretion disks

Recent observational and theoretical studies have strengthened the belief that young stellar objects have circumstellar disks (e.g. /16/ and references therein). The inner parts of Keplerian disks, which are thought to radiate mostly in the near-IR range (see /17/), are expected to rotate at a few hundred $km.s^{-1}$, i.e., much more rapidly than the stellar photosphere. Forthcoming observations of pms disk-candidates with IRSPEC, a newly available near-IR spectrograph at the 3.6m ESO telescope, will aim at testing this expectation.

References

1. Herbig, G. H., 1957, *Astrophys. J.*, **125**, 612
2. Vogel, S. N., Kuhi, L. V., 1981, *Astrophys. J.*, **245**, 960
3. Bouvier, J., Bertout, C., Benz, W., Mayor, M., 1986, *Astron. Astrophys.*, **165**, 100
4. Bouvier, J., 1987, in *Molecular Clouds and Protostars*, Collège de France, Paris, ed: T. Montmerle and C. Bertout, CEA Saclay: 1987, in press.
5. Bouvier, J., 1987, Thesis, Paris
6. Hartmann, L., Hewett, R., Stahler, S., Mathieu, R. D., 1986, *Astrophys. J.*, **309**, 275
7. Stauffer, J. R., Hartmann, L., Soderblom, D. R., Burnham, N., 1984, *Astrophys. J.*, **280**, 202
8. Stauffer, J. R., Hartmann, L., Burnham, N., 1985, *Astrophys. J.*, **289**, 247
9. Endal, A. S., Sofia, S., 1981, *Astrophys. J.*, **243**, 625
10. Stauffer, J. R., Hartmann, L., 1986, *Publ. Astron. Soc. Pac.*, **98**, 1233
11. Benz, W., Mayor, M., Mermilliod, J. C., 1984, *Astron. Astrophys.*, **138**, 93
12. Gray, D. F., 1982, *Astrophys. J.*, **261**, 259
13. Stauffer, J. R., Hartmann, L., 1987, *Astrophys. J.*, **318**, 337
14. Bouvier, J., Bertout, C., Bouchet, P., 1986, *Astron. Astrophys.*, **158**, 149
15. Vrba, F. J., Rydgren, A. E., Chugainov, P. F., Shakovskaya, N. I., Zak, D. S., 1986, *Astrophys. J.*, **306**, 199
16. Bertout, C., 1986, *Proc. of I.A.U. Symp. no.122, " Circumstellar Matter"*, Heidelberg, June 1986, ed. I. Appenzeller and C. Jordan, Reidel: 1987
17. Bertout, C., Basri, G., Bouvier, J., 1987, *Astrophys. J.*, submitted

DISCUSSION

   BASRI   The PMS stars inhabit an unfortunate regime of rotations. They are mostly too slow for good Doppler imaging (10 – 25 km/s) and mostly too fast for measurement of magnetic line broadening (also 10 – 25 km/s). It will therefore be difficult to pursue either of these techniques with them ; though certainly worthwhile if successful.

   BOUVIER   There are a couple of stars rotating fast enough 70 – 100 km/s to use for Doppler Imaging, and they are also both fortunately bright.

   MOROSSI   How did you determine the effective temperatures of your T Tauri stars ? What about spectral classification uncertainties ?

   BOUVIER   Effective temperatures are deduced from the spectral type according to Cohen and Kuhi's (1979) Sp-Teff scale. Uncertainties on the spectral type are of the order of 1 or 2 subclasses.

High Quality Echelle Observations of T Tauri Stars

G. Basri
Astronomy Department
University of California
Berkeley, California USA

ABSTRACT. This is a very brief review of the high resolution line profile work that has been done on very young stars. The spectral anamolies peculiar to these stars are mentioned, with some discussion of what may give rise to them. The Hα line is discussed most extensively, as the most work has been done with it. While progress has been made in understanding the general nature of T Tauri spectra, there are very large gaps in our current understanding of the emission lines from these stars.

The T Tauri stars were noticed as a class because of their spectral peculiarities. While the primary identifying characteristic is the presence of Hα emission, as their spectra are studied in greater detail the number of puzzling features increases. We now understand them to be pre-main sequence stars which have recently emerged from their enveloping star-forming molecular cloud cores. Until recently it was rather difficult to study these stars with both high resolution and signal-to-noise, because the brightest of them is tenth magnitude and most are fainter than twelfth magnitude. The problem is particularly difficult in the blue, since most of them are K-M stars and the reddening is often substantial. I report on results obtained primarily at the MMT, the CASPEC at ESO, and the Hamilton spectrometer at the Lick Observatory. A number of very good reviews of T Tauri stars have appeared in the last few years; perhaps the most relevant to the topic at hand is that by Bertout (1984).

There are several peculiarities apparent in T Tauri spectra compared with their main sequence counterparts. The photospheric lines are sometimes "veiled", meaning their equivalent widths are reduced as though an extra source of continuum light were present. This effect generally increases at shorter wavelengths. A closer look also reveals that the reductions tend to be greater for lines which are stronger in standard stars of the same spectral type (note further that many diagnostics of spectral type itself can be disturbed in the T Tauri stars). The latter type of line weakening is called differential veiling. In the more extreme T Tauri stars the absorption line spectrum can be completely absent; the emission line spectrum becomes increasingly strong and includes the Balmer lines, Ca II lines, Fe I lines and other strong lines like Na D and Mg B and He I. The profiles of these lines become increasingly broad, and exhibit asymmetries diagnostic of mass flows. In some lines, sharp or broad high velocity absorption components are seen (usually blueshifted). The widths of the lines can acheive several hundred km/s. Finally, there can be great variability of both the line profiles and line strengths, on time scales from hours to years. None of the above effects can be properly studied without high quality spectra (meaning velocity resolution of at least 10000 and S/N of at least 30:1).

A few atlases of line profiles have been published. Mundt and Giampapa (1982) address the question of profile variability, Hartmann (1982) produced a catalog of line profiles relevant to the envelope, Mundt (1984) also studies primarily the absorption line components arising from outflowing shells in Hα and Na D but with higher quality, Ulrich and Knapp (1985) display the appearance of several diagnostic lines from a number of stars, and Appenzeller, Jankovics, and Jetter (1986) published the entire spectrum of two extreme T Tauri stars including many line identifications. It is now possible to have spectra which contain essentially all the information available in the optical bandpass obtained at the same time. These are simultaneously useful for a number of fruitful lines of investigation, and provide a large increase in the observational contstraints under which any theoretical understanding must proceed.

Closest examination of complete spectra has been made by Finkenzeller and Basri (1987) who study 7 moderate T Tauri stars, provide absolute calibration of the spectra, and make comparisons with spectral standards. This data can be used for a direct confrontation between physical models and observations throughout the entire region of spectral formation Their analysis of the absorption line spectrum makes it clear that the phenomenon of "differential veiling" is the predicted effect of a deep chromosphere on the line source functions. The stronger line cores are formed closest to the temperature minimum and lower chromosphere where the temperatures are higher than for an inactive main sequence star of the same spectral type. In fact, division of a T Tauri spectrum by an appropriate standard is the software equivalent of a solar eclipse and yields a rather similar psuedo-emission spectrum. In conjunction with observations of X-ray flaring and dark spots on T Tauri stars, it is clear that the stars themselves have fairly normal deep photospheres overlain with strong stellar magnetic activity.

Of particular interest is the study of the breadth of the Ca II K line compared with Hα. In all but one of the stars in their sample, the K line was narrow and symmetric as should be expected from a low-lying stellar chromosphere, while even when the Hα line was weak it still had much greater breadth. In one case, a broader asymmetric component was evident in the wings of the K line (and also in the infrared triplet lines of Ca II), but the chromospheric central component was also still in evidence (BP Tau is another example of this). This suggests that the Hα line is primarily a feature of what I have referred to as the "envelope" (meaning simply that it is not from the stellar surface), and that as the envelope becomes a stronger emitter it starts to cover over the surface line profiles for lines of increasingly smaller optical depths. In the most inactive T Tauri stars, even the Hα line eventually comes to resemble the Hα lines of the RS CVn stars: active post-main sequence subgiants which are thought to be primarily chromospheric. In the more active T Tauri stars the envelope increasingly becomes the primary source of the emission lines, making them broader, stronger, and less symmetric.

The importance of detailed profile studies is illustrated by the remarkable behavior of SU Aur (G2IV). Its Hα line is relatively weak and appears in the Hartmann (1982) atlas as rather anamolous, with a suggestion of emission cut by several absorption components. In Fig. 1a are shown the profiles from Oct. 1986 and April 1987, which are rather different from it, but like each other. The abscissa is a relative velocity scale in km/s. Here we see a simple flat-topped profile interrupted by a single central absorption component. Such a profile would traditionally indicate spherical optically thin constant velocity outflow under a cool stationary shell. Less than a month later, the profiles in Fig. 1b were obtained. Now the central absorption has become an emission feature flanked on either side by small absorption features, and a major absorption feature is seen at the blue edge of the emission on one night, but disappears the next night. In Fig. 1c (Dec. 1986) something similar to the original 1982 profile has appeared. The flanking absorption components have moved out and deepened, and the central emission is being eaten away by a newly emerging absoption component. There appears to be reasonable evidence for a cycling of the profile components, which might serve as the key to unlocking a rather complex structure.

SU Aur is a rather atypical case, but now that repeated high quality observations are becoming available, it is clear that probably most stars show some interesting variability in their Hα profiles. Two other examples are given in Fig. 2. CO Ori is another G star, with the almost triangular shape also seen in GW Ori. At least that was true in Oct. 1986; by Dec. 1986 the triangular wings had noticeably weakened and a low, broad very blueshifted emission component had appeared. The strong X-ray emitter ROX-29 shows significant changes in both the blue and red wings on two successive nights. My impression is that the weaker Hα profiles show more rapid variations, but this remains to be confirmed by a much more extensive set of observations. A very strong and broad Hα line (as in DF Tau or RW Aur) seems to retain its basic structure over time, although variations in the absorption components and the peaks are certainly seen.

I believe that interpretation of these profiles in terms of the usual P Cygni explanation is dangerous at best. Firstly, one wouldn't expect the rapid variability of the emission wings if they really arose from a large region around the whole star. Secondly the general appearance of the lines has been successfully interpreted by a variety of models (Bertout 1984) so the P Cygni paradigm is not required. Thirdly, it has been shown that a variety of possibly relevant radiative transfer effects which are not accounted for in the simple picture can markedly change the expected appearance of the line. The great width of the line is something of a puzzle, since velocities at the edges of the emission line are not usually seen in absorption components. The simultaneous presence of red and blueshifted components is also an indication that a simple picture is not sufficient.

As an example of an alternative interpretation which must be considered, it has lately become increasingly apparent that many (and perhaps all but the "naked") T Tauri stars are still surrounded by disks of gas and dust. The evidence is accumulating (Bertout, Basri, and Bouvier 1988) that a number of these are still active accretion disks, with an active boundary layer where the accreting material meets the slowly rotating star. While very little is known about the details of the boundary layer structure, there is evidence that the Balmer continuum emission and perhaps most of the line emission could arise in such a layer.

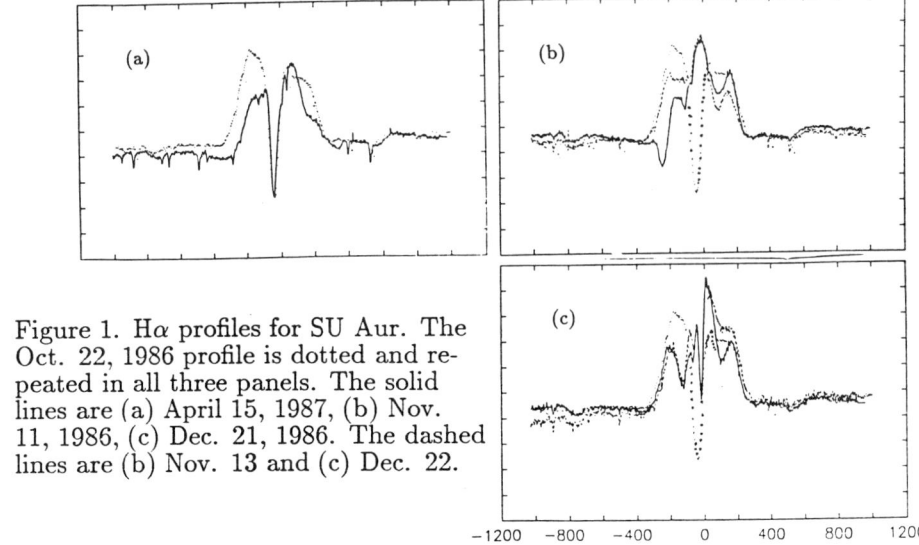

Figure 1. Hα profiles for SU Aur. The Oct. 22, 1986 profile is dotted and repeated in all three panels. The solid lines are (a) April 15, 1987, (b) Nov. 11, 1986, (c) Dec. 21, 1986. The dashed lines are (b) Nov. 13 and (c) Dec. 22.

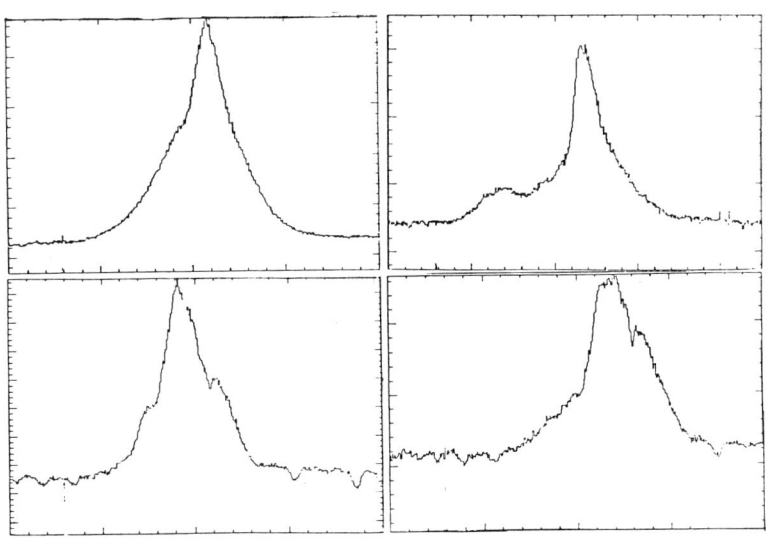

Figure 2. Hα profiles for CO Ori (upper panels) on Oct. 22, and Dec. 21, 1986; and for ROX-29 (lower panels) on June 9 and 10, 1987.

This might be expected to yield a symmetric component of emission with about the observed breadths. Bipolar outflows could give rise to absorption components, whose appearance would depend somewhat on the viewing angle of the system. The geometry would therefore have at best axial symmetry, and shadowing by the disk becomes an important ingredient. This would have the virtue of explaining why the redward emission from $H\alpha$ is usually seen, while it is never seen in the forbidden line emission which arises over hundreds of AU from the star. Redward absorption components could arise from material flowing from the disk to the star. One would expect cyclical variations if the system has a non-axial component. The weaker, flatter profiles might be from stars which have largely lost their disks and are winding down from the strong mass loss phase.

At the moment, the above is almost completely speculative. Clearly there needs to be much more detailed observational and theoretical work before an understanding of line emission from T Tauri stars is acheived. Our underlying physical conception of what we are looking at must be solidified before a detailed quantitative analysis will be really useful. The most crucial observations will be from repeated high quality echelle spectra, which provide full velocity information from many lines of different optical depths and formation conditions at the same times. A new threshold in research on young stars has been opened up by the modern capability for high resolution and signal-to-noise stellar spectroscopy of faint objects.

References:

Appenzeller, I., Jancovics, I., Jetter, R., 1986, Astron. Ap. Supp. **64**, 65.
Bertout, C., 1984, Rep. Prog. Phys. **47**, 111.
Bertout, C., Basri, G., Bouvier, J., 1988, Ap. J. (submitted)
Edwards, S., Cabrit, S., Strom, S.E., Strom, K.M., 1987, Ap. J. (in press)
Finkenzeller, U. and Basri. G., 1987, Ap. J. **318**, 823.
Hartmann, L., 1982, Ap. J. Supp. **48**, 109.
Mundt, R., 1984, Ap. J. **280**, 749.
Mundt, R. and Giampapa, M.S., 1982, Ap. J. **256**, 156.
Rydgren, A.E., Strom, S.E., Strom, K.M., 1976, Ap. J. Supp. **30**, 307.
Ulrich, R. and Knapp, G., 1985, preprint.

DISCUSSION

GRAY One might look for different rotation rates in chromospheric and photospheric lines.

BASRI I think that the low chromospheric diagnostics, (narrow symmetric emission or filling in) are not formed significantly above the photosphere, and so will have the same v sin i. The broader components are probably formed farther out, so they may well have a different rotation.

APPENZELLER How good are homogeneous chromosphere models in view of the spotty structure of T Tauri photospheres indicated by the rotational variations?

BASRI They are even worse than homogeneous chromospheric models for active main sequence stars. There really isn't spatial information in the observed narrow symmetric lines. The best hope is through studies of rotational modulation as mentioned by Dr. Bouvier.

HENRICHS How would you identify the characteristic timescales of the spectral variations you observe ? Are they comparable to flow timescales or rotational timescales?

BASRI There are all timescales observed, from hours to months or years. These should not be thought of as arising from a common mechanism ; some are rotational, some due to changes in the flow, some due to changes in activity or in the geometry of regions, etc.

PRADERIE When you speak of a deep chromosphere, how deep means deep? Don't you have to take curvature effects in the computation of your non-LTE line spectrum?

BASRI The "deep chromosphere" means the part directly above the photosphere and so a plane parallel approximation is sufficient for this. It is at mass column density of $\simeq 1$ g/cm^2. The filled in absorption lines and narrow symmetric emission lines should be formed there. Any broader or asymmetric components are likely formed farther out where geometry should be accounted for.

LINSKY Please comment on the spectral resolution needed to study the emission lines and the narrow circumstellar absorption features?

BASRI It depends somewhat on what you want to learn. The work done with Finkezeller had R = 15000 which is sufficient for studying the broad components and for separating chromospheric and envelope contributions. To study narrow circumstellar features one prefers R > 30000. To look at magnetic fields requires R > 100000.

SEARCH FOR ROTATIONAL MODULATION IN PRE-MAIN SEQUENCE
HERBIG Ae/Be STARS*

C. Catala, J. Czarny, P. Felenbok
Observatoire de Paris, Section de Meudon

1 - Introduction: The Herbig Ae/Be stars (Herbig, 1960; Finkenzeller and Mundt, 1984) are widely believed to be intermediate mass (2-5 $M_\odot$) pre-main sequence stars. In the past few years, a big effort has been made to model their outer layers, and it has been shown that they possess stellar winds and extended chromospheres (Catala et al., 1984; Catala et al., 1986a; Catala and Kunasz, 1987).
    Periodic variations have been reported in the Ca II K line and in the Mg II h and k lines for one of these stars, AB Aur (Praderie et al., 1986; Catala et al., 1986b). A 32 hour period has been found for the Ca II K variations, whereas the period is of 45 hours for the Mg II h and k variations. These variations have been interpreted as a rotational modulation. A qualitative model of the wind has been built, in which fast and slow streams alternatively cross the line of sight. In this model, the radial differential rotation of the star's extended atmosphere explains the period difference mentioned above : the Ca II K line is formed in a compact region at the base of the wind, where the structures due to the inhomogeneous wind are stable, and is therefore modulated with the star's rotation period ; on the other hand, the blue edge of the Mg II resonance lines is formed further away in the wind, in a region where the streams have merged, and is therefore modulated with the rotation period of the extended atmosphere in this region.
    The stream structure discovered in the wind of AB Aur is obviously attached to the stellar photosphere, and therefore related to azimuthal inhomogeneities at the star's surface, like for instance the alternation of regions of closed and open magnetic loops. The observed rotational modulation of lines formed in the wind of AB Aur constitutes evidence for the existence of activity in this star.
    The question that must be addressed now is whether the other Herbig Ae/Be stars show the same rotational modulation phenomenon.Since the rotation periods of these stars are typically of the order of 30-40 hours, this observational programme requires ground-based multi-site

(*) communication presented by F. Praderie

or space-based observations. The present paper describes two ground-based multi-site campaigns that have been carried out in 1986.

2 - <u>Observations and discussion</u> :

2-1 - <u>HD250550</u>: The Hα line of this star (m_v=9.1, A0) shows long term variations (Catala et al., 1986a), and a vsini of 110 km s^{-1} was estimated by Finkenzeller (1985).

We observed this star in February 1986, with the Coudé spectrograph of the 3.6 m Canada-France-Hawaii telescope (CFHT), the Cassegrain spectrograph of the 2.5 m "Isaak Newton" telescope (Canary Islands), and the fiber-fed "Isis" spectrograph (see Felenbok and Guérin, these proceedings) on the 1.93 m telescope at Observatoire de Haute-Provence (OHP). The aim was to obtain a series of Ca II K line profiles with the best time coverage possible, in order to find short term variations and to determine an eventual period for these variations. Unfortunately, the weather was not good enough to reach this goal. However, the few spectra we obtained are sufficient to show that the Ca II K line is spectacularly variable on a very short time scale. In particular, two spectra obtained at CFHT within two hours exhibit strong variations in the Ca II K line profile (see Fig.1).

We therefore conclude that HD 250550 is an excellent candidate for a search for rotational modulation in the Ca II K line.

2-2 - <u>BD+46°3471</u>: The vsini of this Herbig star (m_v=10.1, A4) is estimated to about 150 km s^{-1} (Finkenzeller, 1985).

A multi-site observation campaign has been carried out on the Hα line of this star in July 1986, involving the "Hamilton" Coudé spectrograph on the 120 inch telescope at Lick Observatory, and the fiber-fed "Isis" spectrograph on the 1.93 m OHP telescope. The Ca II K line was not chosen, because it was impossible to obtain a reasonable S/N ratio in this line for this magnitude at OHP. Due to the small longitude difference between the two sites, the time coverage was not optimal, but still sufficient to detect an eventual rotational modulation. We obtained 27 spectra at Lick observatory and 16 at OHP. The total duration of the campaign was 5 days.

We detected variations in the Hα line, which appears purely in emission. Fig. 2 shows the two most different spectra of the series. We looked for a period in the variations of the blue wing (where they are the most intense), using the technique described in Catala et al. (1986b), but with no success. This indicates that the observed variations are not simply related to the star's rotation.

This result can be interpreted in several ways. First, it is possible that BD+46°3471 is not active, in which case the activity phenomenon would not be general among Herbig Ae/Be stars. We would still have to explain the origin of the detected variability. Another possibility is that the activity of Herbig Ae/Be stars is cyclic, and we have observed BD+46°3471 in a quiet phase. Finally, the Hα line can be formed far from the stellar photosphere (Catala and Kunasz, 1987), and considering a model similar to the one built far the case of AB Aur (Praderie et al., 1986; Catala et al., 1986b), it is possible that no structure related to the rotation is present in its region of formation, even if the star is active.

New observations, in particular in the Ca II K line, should provide answers to these questions.

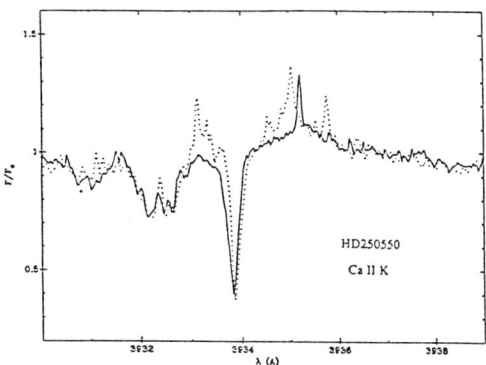
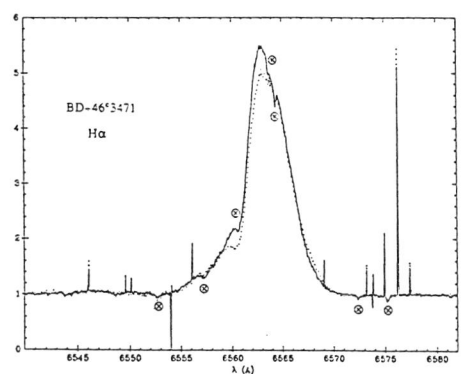

Fig. 1: Two spectra of the Ca II line obtained at CFHT. The dotted spectrum has been observed 140 minutes after the other one. The spectacular short-term variability of this line is obvious. The blue edge of the main absorption component varies by 10 km.s^{-1}, while the spectral resolution is 7 km.s^{-1}.

Fig. 2: These 2 Hα profiles obtained during the Lick-OHP campaign shows the most important variation in the observed series. Telluric absorption lines are indicated by crossed circles. The other sharp features are due to blind columns of the CCD. Although the variability of the Hα profile is obvious (9% at the top of the emission component, 16% in the blue wing), no periodicity was found in the variations.

3 - <u>Conclusion</u>: The two multi-site observation campaigns described in this paper have brought some new insight into the problem of the activity of Herbig Ae/Be stars. Apart from AB Aur, which clearly shows rotational modulation of its wind, we now have a second excellent candidate for this kind of phenomenon, HD250550, and an "ambiguous" case, BD+46°3471.

Let us mention here that a third possible case of rotational modulation has been detected with IUE : preliminary results of an IUE observation campaign on HD163296 show a possible 32 hr period in the Mg II h and k line variations. We have not presented these results here, because they are not based on high S/N observations.

The problem of the origin of this activity of the Herbig Ae/Be stars is still open. By analogy with the solar wind, one might think that the magnetic field is responsible for the wind structuration. However, in these stars where the presence of a magnetic field has not been established, other mechanisms could be at work.

Acknowledgements: The multi-site campaigns described above require a lot of efforts and involve many people. Many thanks are due to P.S. Thé, H.R.E. Tjin-A-Djié, A. Talavera, M. Dreux and J. Guérin, who took an active part in the observation of HD250550, and to U. Finkenzeller, J. Guérin, M. Dreux and T. Fauconnier who participated in the observation of BD+46°3471. Finally, we thank F. Praderie who originated the study of activity in Herbig Ae/Be stars, and who presented this paper at the conference.

Références:
Catala, C., Kunasz, P., Praderie, F.: 1984, *Astron. Astrophys.* 134,402
Catala, C., Czarny, J., Felenbok, P., Praderie, F.: 1986a, *Astron. Astrophys.* 154,103
Catala, C., Felenbok, P., Czarny, J., Talavera, A., Boesgaard, A.M.: 1986b, *Astrophys. J.* 308,791
Catala, C., Kunasz, P.B.: 1987, *Astron. Astrophys.* 174,158
Finkenzeller, U., Mundt, R.: 1984, *Astron. Astrophys. Suppl.* 55,109
Finkenzeller, U.: 1985, *Astron. Astrophys.* 151,340
Herbig, G.H.: 1960, *Astrophys. J. Suppl.* 4,337
Praderie, F., Simon, T., Catala, C., Boesgaard, A.M.: 1986, *Astrophys. J.* 303,311.

DISCUSSION

BAADE You mentioned possible evidence from IUE spectra for rotational modulation of line profiles in HD 163296. Otmar Stahl and I just observed several optical lines of the same star at high spectral and time resolution as well as low noise. The profile variability we are seeing in all lines selected appears to be of the same type as in OB stars where it is caused by nonradial pulsations. If this possibility is borne out by a more thorough analysis, it may
a) shed some light on the interaction between different types of variabilities seen in pre-main sequence stars, and
b) help to identify the (as yet unknown) driving mechanism of NRPs in OB stars. The latter potential derives from the fact that HD 163296 is classified as A0, i.e. has a Teff where NRPs have not so far been seen in MS or slightly evoled stars. This might support the view that there is a core excitation mechanism and it could, e.g., already be active in HD 163296 because the star's core is farther evolved than its envelope is.

PRADERIE I would not be surprised that such a star shows non radial pulsation. We face a problem with the hydrodynamics of the interior, whether we want to give account of non radial pulsation or of the non-axisymmetric wind, modulated by rotation, that we detect. I would like to know the period of your non radial pulsation.

High S/N simultaneous optical and IR spectrophotometric

observations of Herbig Ae/Be stars

A.Vittone(1),C.Rossi(2),E.Covino(1),F.Giovannelli(3)

1-Osservatorio Astronomico di Capodimonte,Napoli- I.
2-Istituto Astronomico,Università "La Sapienza" Roma- I.
3-Istituto di Astrofisica Spaziale,C.N.R. Frascati Italia

ABSTRACT- We present a preliminary report on simultaneous optical and infrared observations on a sample of Herbig Ae/Be stars carried out at the European Southern Observatory in La Silla-Chile in March 1984 and in March-April 1985.
In this paper Z CMa, V380 Ori, R Mon and HD53367 are analyzed A strong variability in both energy distribution and Hα line profile is detected in Z CMa, while for HD53367 only He I and Hα profile's variations are revealed; V380 Ori and R Mon show non significant variability.

1. INTRODUCTION

The Herbig Be star Z CMa is localized in a large expanding nebular region and it is the most interesting object of an association of very young stars called CMa R1.According to Finkenzeller et al(1984) Z CMa is a double star system. The light curve is characterized by irregular fluctuations, but from a comparison between the present behaviour and the old one,it seems that the star is now on a relatively quiescent phase,(Covino et al,1984). The spectral type is peculiar:the absorption spectrum looks like that of a moderately rotating F star superimposed on a late B spectrum (Strom et al,1972,Covino et al,1984).
V380 Ori is the illuminating star of NGC 1999; it is an irregular variable of T Tauri type of early spectral class (B8-A2 e). In the optical range the spectrum is characterized mainly by the very strong emission lines of H,FeII,TiII and CaII; In the near infrared the emission lines of OI,CaII and HeI are the most important features.
R Mon is the nucleus of the nebula NGC 2261 and according to Herbig (1960) it is not a star, but resembles a small coma; probably the star is deeply embedded in a dust nebula and it is not directly observable. The spectral class

is unknown and the emission spectrum reveals a similarity with the one of V380 Ori.

HD53367 is probably a spectroscopic binary which might explain its emission line spectrum (Finkenzeller et al 1984). The IR colours indicate that the object may be a normal Be star (accidentally) associated with a dark cloud material.

2. OBSERVATIONS

2.1-Optical Medium dispersion spectra (114 Å/mm) in the range 4700-7000 A were obtained with the Boller and Chivens spectrograph mounted at the Cassegrain focus of the ESO 1.52m telescope; the Image Dissector Scanner was used with a double 8" aperture deker. Standard ESO reduction has been made. Rms values of the noise have been computed on many line free region of the spectra and the relative signal to noise ratios at the continuum level have been computed; typical values range from 50 to 150, while for the most relevant emission features the S/N ratios are ranging between 100 and 300. In Figures 1,2 and 3 the spectra of Z CMa, V380 Ori and R Mon are reported.

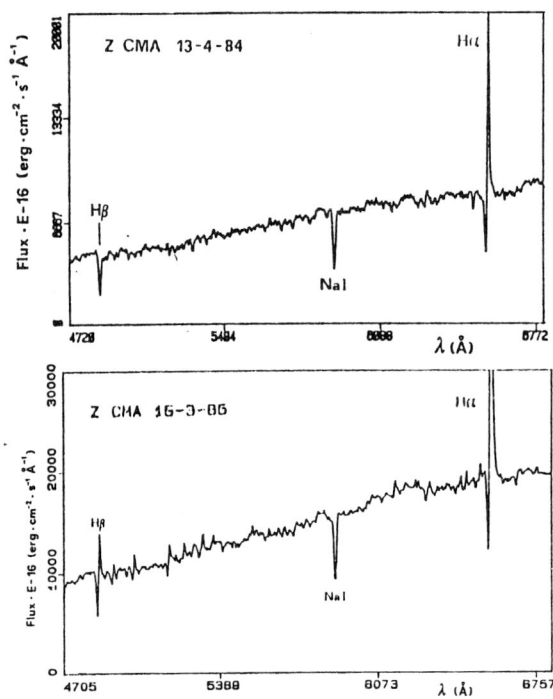

Fig 1. a-up b-low 1.52m ESO + IDS

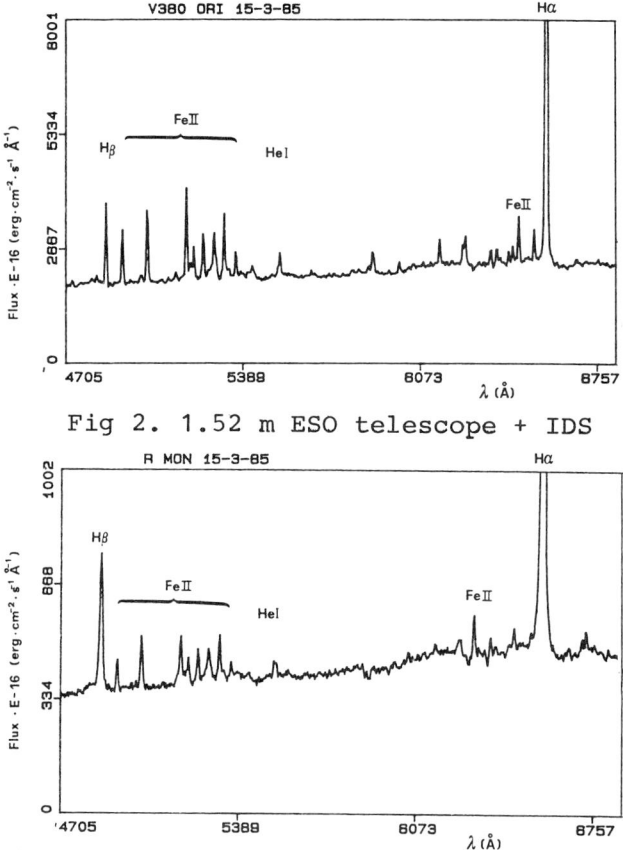

Fig 2. 1.52 m ESO telescope + IDS

Fig 3. 1.52 m ESO telescope + IDS

High dispersion line profiles have been taken with the Coudé Echelle Spectrograph fed by the 1.4 CAT. the resolving power was 80000. In figure 4 the Hα profile of Z CMa is shown, while in figure 5 the Hα and HeI 5876 line profiles are plotted.

We emphasize the importance of the accuracy during the reduction procedure; as an example the S/N ratio of the Z Cma spectrum of fig.4b at the continuum level was measured for the raw data and after the complete reduction (background subtraction and flat field division). The S/N value improuved from the original 41.5 to the final 170.

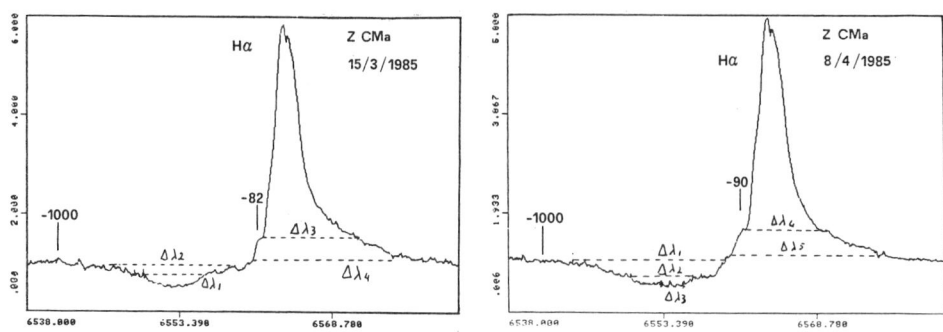

Fig 4. a-left b-right   ESO 1.4m   CAT + CES

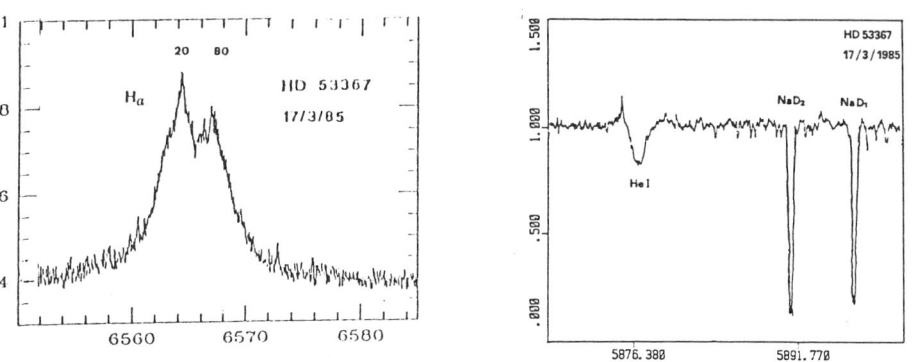

Fig 5. ESO 1.4m CAT + CES

2.2-Infrared Low dispersion spectrophotometry with $\lambda/\Delta\lambda=100$ has been carried out at the ESO 1m telescope. Three different Circular Variable Filters were used to cover the range 1.4-5.3μ with an InSb detector. In figures from 6 to 9 the energy distributions are shown for Z Cma, V380 Ori, R Mon and HD53367 respectively.

Table 1

velocity fields referring to the Z CMa CAT observations (given in Km/sec)

15/3/1985	Δv1	Δv2	Δv3	Δv4	
	283	598	461	813	
8/4/1985	Δv1	ΔV2	ΔV3	Δv4	Δv5
	663	274	151	342	681

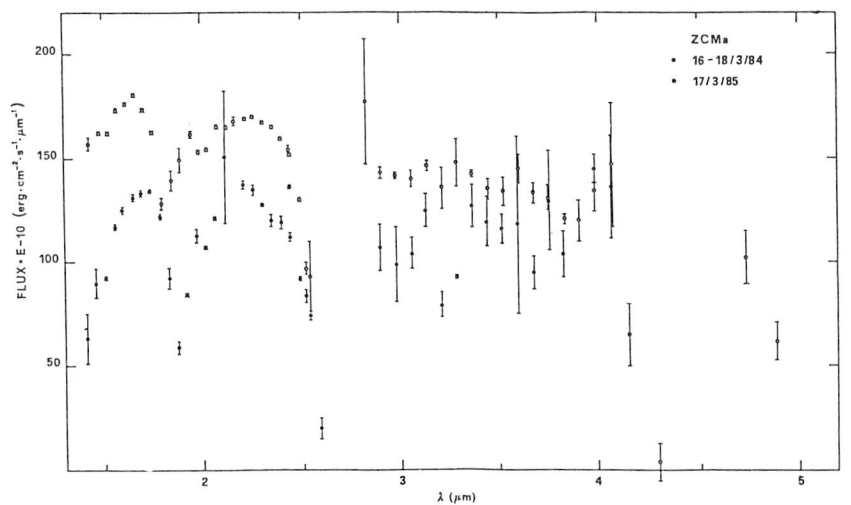

Fig 6. ESO 1m telescope + In b detector

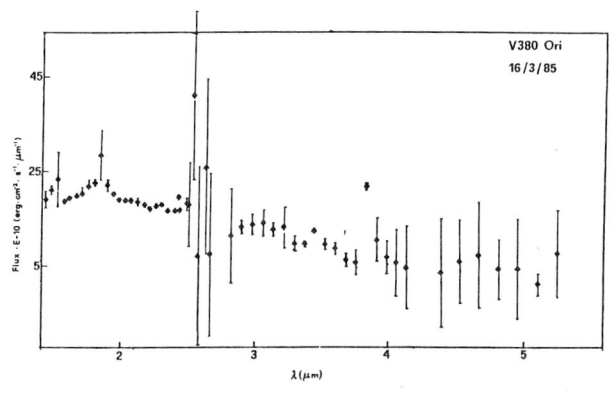

Fig 7.

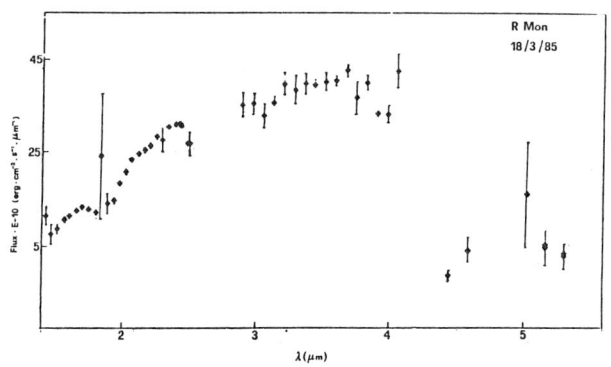

Fig 8.

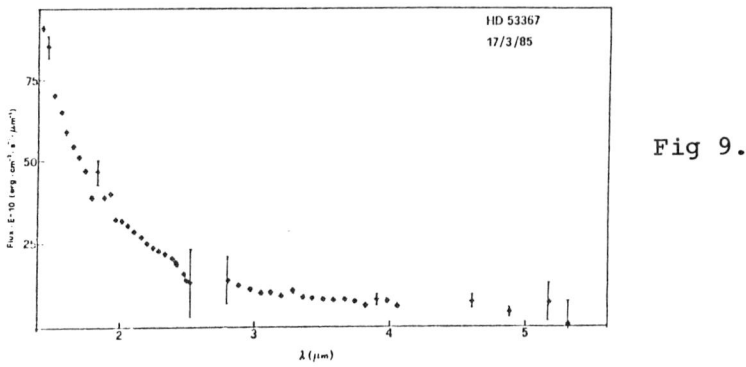

Fig 9.

3. RESULTS

Referring to Z CMa a strong variability is detected in the whole spectral range. In the 1984 IDS spectrum (fig 1a) the Hβ and FeII lines are seen only in absorption and the spectrum is very similar to the one observed in 1983 when the star had the same V magnitude (V=9.3). In 1985 (V=8.8) the object presents very strong P Cygni profiles in the Balmer and FeII(42),FeII(49) lines (see fig 1b). The absorption components are fairly broad and blue shifted of 300Km/sec. Other FeII lines are present only in emission. The high dispersion observations revealed changes in the Hα line profile, as can be seen from fig 4. In particular while the emission component does not shows any significant change in the time scale of 1 month, the shallow and dish like absorption is variyng both in central wavelength and velocity fields of the multiple ejected shell system (see table 1). Infrared spectrophotometry indicate a brightening on long time scale as shown in fig 6, while only minor variations are detected in the two 1985 observations.

A strong variability has been detected in the high resolution spectrum of HD53367. Comparing our data with those by Finkenzeller et al,1984 , we note tha in 1985 the object shows a double peaked Hα and a weaker blue shifted emission peak with a stronger unshifetd absorption component as can be seen in fig 5. In contrast with the other stars analyzed HD53367 has a very weak infrared excess (see fig 9) which suggests that this star is likely to be a normal star with strong emission line activity.

A complete analysis of the data will be given in a forthcoming paper

REFERENCES

Covino E.,Terranegra L.,Vittone A.,Russo G.:1984,Astron.J. 89,1868

Finkenzeller U.,Mundt R.:1984,Astron.Astrophys.Suppl.55,109

Herbig G.H.:1960,Astrophys.J.Suppl. 4,337

Strom S.E.Strom K.M.,Yost J.,Carrasco L.,Grasdalen G.:1972, Astrophys.J.173,353

IMPROVED NON-LTE MODEL ATMOSPHERES FOR SUBLUMINOUS O-STARS

U. Heber, K. Hunger, T. Rauch and K. Werner
Institut für Theoretische Physik und Sternwarte
Olshausenstr. 40
D2300 Kiel 1
Federal Republic of Germany

ABSTRACT. Improved non-LTE model atmospheres designed for the analysis of very hot subluminous O stars are presented. The calculations are based on the new method of the accelerated lambda iteration (ALI) which proves capable of treating up to 100 levels of various ions. Presented here are improved calculations for (i) pure hydrogen model atmospheres including Stark broadening, (ii) for hydrogen- and helium-composed model atmospheres and (iii) first preliminary models which include in addition a detailed carbon model atom. These models remove an apparent mismatch of Balmer line profiles noted previously and fit high S/N, high-resolution hydrogen and helium spectra obtained with the ESO-Cassegrain echelle spectrograph very well.

1. INTRODUCTION

The sublumious O stars are amongst the hottest known pre-white dwarfs, with effective temperatures ranging from about 40000 K to 100000 K. Their evolutionary status, however, is still not well understood. In the absence of reliable distance determinations for these stars, the best way to discuss their properties is to find their positions in the (g, T_{eff})-diagram by means of model atmosphere analyses. In view of the very high effective temperatures it is clear that deviations from LTE inevitably occur and detailed non-LTE calculations are required.

In conventional NLTE model atmosphere calculations, hydrostatic equilibrium and planeparallel geometry is assumed. The sdO atmospheres meet these assumptions very closely, since their surface gravities are high ($4.5 \lesssim \log g \lesssim 6$) and mass loss rates are small ($\lesssim 10^{-9}$ M⊙/yr, Hamann et al., 1981). In some cases the atmospheres are so calm that diffusion (gravitational settling) occurs, which means that the mass outflow must be greatly suppressed (Heber, 1986) in order to maintain downward diffusion. Different from massive O stars (see Bohannan et al., these proceedings), wind blanketing can be neglected for the subluminous O stars. Therefore, the sdO stars are ideal test objects for conventional NLTE model atmospheres.

2. NLTE MODEL ATMOSPHERES

The classical method of computing NLTE model atmospheres is the complete linearization approach of Auer and Mihalas (1969), which has been applied frequently to calculate hydrogen- and helium-composed model atmospheres for sdO stars (e.g. Kudritzki and Simon, 1978). The complete linearization method accounts for the coupling of all variables amongst all depth points in the model atmosphere. Due to numerical limitations the method allows no more than 100 variables to be treated. The most urgent point is the large number of transfer equations needed to describe the radiation field adequately. Therefore, only a small number of levels can be allowed to depart from LTE. Typical applications use 9 (hydrogen and helium) levels, 6 line transitions and 65 frequency points. The question arises whether such crude models can reproduce high S/N observations (ESO-CASPEC) which became available recently.

Such a comparison immediately reveals a mismatch for the Balmer line cores (see Fig. 1): the observed cores are somewhat deeper than predicted by theory. According to our experience the discrepancy in the Balmer line cores is a general phenomenon which is always observed for sdO stars whenever high S/N spectra are obtained (Heber and Kudritzki, 1986; Herrero, 1987). Obviously, there is a need for more elaborate model atoms.

The first attempt to construct such models was made by Anderson (1985), who reduced the set of unknown variables by dividing the frequency spectrum into several blocks which are characterized by their total energy density of radiation. The detailed distributions within the blocks are updated afterwards by lambda iteration.

We followed a different approach using the so-called accelerated lambda iteration (ALI). This method was first applied to line formation calculations by Werner and Husfeld (1985) and Herrero (1987). Werner (1986) demonstrated that the ALI method is a powerful tool to construct model atmospheres in NLTE. (A detailed description of the method is given by Werner, 1987a,b.) The basic aim is to remove the radiation field variables altogether from the set of linearized equations which have to be solved simultaneously. Simply separating the treatment of the radiative transfer from the constraint equations (i.e the equations of statistical equilibrium, of hydrostatic and radiative equilibrium, of number and charge conservation), however, results in a lambda iteration which is known to converge very slowly at large optical depths. Hence the radiative transfer has to be solved simultaneously with the constraint equations. Following an idea of Scharmer (1981) the radiative transfer can be simplified by using approximate lambda operators and can be iterated to obtain the exact solution. In this way the radiation field variables can be removed from the set of linearized equations if the approximate lambda operator chosen is either local or one-directional. This approach results in an iteration scheme that consists of two nested iteration cycles. In the inner cycle, the linearized constraint equations are solved by Newton-Raphson iteration. In the outer iteration cycle ("Scharmer"-iteration), the radiative transfer is updated. It turned out that it is not necessary to include

the equation of hydrostatic equilibrium in the Newton-Raphson iteration
which makes the method more flexible since it can be started from the
inner boundary as well as from the outer one.

The great advantage of the ALI method is that it is no longer
limited by the number of frequency points (NF) as is the complete linearization method. For the ALI method the computing time scales as
$c_1 NF + c_2 (NL+3)^3$ (NL is the number of levels), and the limiting quantity
is NL which must be lower than about 100 due to numerical limitations
(matrix inversion in the Newton-Raphson iteration).

In the following sections we shall discuss recent applications of
the ALI method.

3. IMPROVED HYDROGEN AND HELIUM COMPOSED NLTE MODEL ATMOSPHERES

As a first application of the ALI method, improved NLTE models for pure
hydrogen atmospheres were calculated (Rauch, 1987) allowing for up to
15 NLTE levels, up to 105 line transitions and up to 1845 frequencies.
An internal accuracy of 1% in the line profiles as well as in the temperature structure is achieved if ten levels and all corresponding line
transitions are considered. (The absorption coefficient is approximated
by a Doppler profile. Including Stark-broadening in the statistical
equilibrium calculations has a marginal effect only.) The outermost
layers of the atmospheres are found to be heated considerably by the
additional line transitions. The new models predict deeper line cores
than the previous ones (up to 10% deeper for Lyman lines, 20% for
Balmer lines and 40% for Paschen lines), while the line wings remain
unchanged.

In the second step helium was added (Werner, 1987b) and hydrogen-
and helium- composed models constructed. In these calculations, 23 NLTE
levels (the lowest 10 levels of H I and He II, respectively, the lowest
level of He I and the levels of H II and He III) and 72 line
transitions are considered. (The overlap of H I and He II lines is
accounted for.) 478 frequency points are needed to describe the
radiation field adequately. These models have already been used for the
analysis of high-resolution, high-S/N spectra of very hot sdO stars
(see Heber, Hunger and Werner, these proceedings).

To give an example for the improvements achieved by the ALI
models, we compare theoretical H_γ-line profiles to a high-resolution
spectrum of the helium-deficient sdO LB 3459 obtained with the ESO-
CASPEC (see Fig. 1). Low-resolution spectra of LB 3459 have already
been analyzed for the atmospheric parameters (Kudritzki et al., 1982),
which yielded T_{eff} = 40000 K, log g = 5.3, n_{He}/n_H = 0.003 (by number).
As can be seen from Fig. 1, the ALI model fits the entire H_γ profile
very well. Further examples for improved line-profile fits from high-
resolution, high-S/N visual spectra are given by Herrero (1987). We can
therefore conclude that the Balmer line discrepancy mentioned above is
completely removed when the improved ALI models are used.

4. METAL LINE BLANKETED NLTE MODEL ATMOSPHERES

Of course, the next step for improving the model atmospheres is to include metal ions in the calculations. Recently, Werner (1987b) added the important element carbon represented by 12 NLTE levels (10 levels for C IV and the ground-state levels of C III and C V, respectively). First model calculations were carried out for T_{eff} = 75000 K, log g = 5 and solar helium and carbon abundances. Fig. 2 shows the temperature structure for models with and without carbon. As can be seen, considerable cooling occurs around log m $\approx$ -3 when carbon is included. This can be traced back to the desaturation of the carbon resonance lines in these layers. The emergent H- and He-line profiles, however, remain almost unchanged because they are formed deeper inside the atmosphere.

The latter result is important since it gives us some confidence that simple H- and He-composed NLTE atmospheres (used up to now in all analyses) might be sufficiently accurate for the analysis of the visual spectra, provided the metal abundances are low (say of solar type). (Additional metals, of course, have to be included to check this.) However, in some hot pre-white dwarfs some metals are known to be strongly enriched. For the (pulsating) hydrogen-deficient PG 1159 stars (GW Vir stars), for instance, it has been conjectured that C and O dominate their envelopes (Starrfield et al., 1984). The ALI method is ideally suited to the construction of carbon and oxygen rich models, which shall be used to analyse high S/N CCD spectra of PG 1159 stars obtained recently with the 3.5m telescope at the DASZ (Calar Alto, Spain).

5. REFERENCES:

Anderson, L.S: 1985, in "Progress in Stellar Line Formation Theory", J.F. Beckman, L. Crivellari (eds.), Reidel, p.225
Auer, L.H., Mihalas, D.: 1969, Astrophys. J. **158**, 641
Hamann, W.-R., Gruschinske, J., Kudritzki, R.P., Simon, K.P.: 1981, Astron. Astrophys. **104**, 249
Heber, U.: 1986, Astron. Astrophys. **155**, 33
Heber, U., Kudritzki, R.P.: 1986, Astron. Astrophys. **169**, 244
Herrero, A.: 1987, Astron. Astrophys. **171**, 189
Kudritzki, R.P., Simon, K.P.: 1978, Astron. Astrophys. **70**, 653
Kudritzki, R.P., Simon, K.P., Lynas-Gray, A.E., Hill, P.W.: 1982, Astron. Astrophys. **106**, 254
Rauch, T.: 1987, diploma thesis, Kiel
Scharmer, G.: 1981, Astrophys. J. **249**, 720
Starrfield, S.G., Cox, A.N., Kidman, R.B., Pesnell, W.D.: 1984, Astrophys. J. **281**, 800
Werner, K.: 1986, Astron. Astrophys. **161**, 127
Werner, K.: 1987a, in "Numerical Methods in Radiative Transfer", W. Kalkofen (ed.), Cambridge University Press, in press
Werner, K.: 1987b, Ph.D. thesis, Kiel
Werner, K., Husfeld, D.: 1985, Astron. Astrophys. **148**, 417

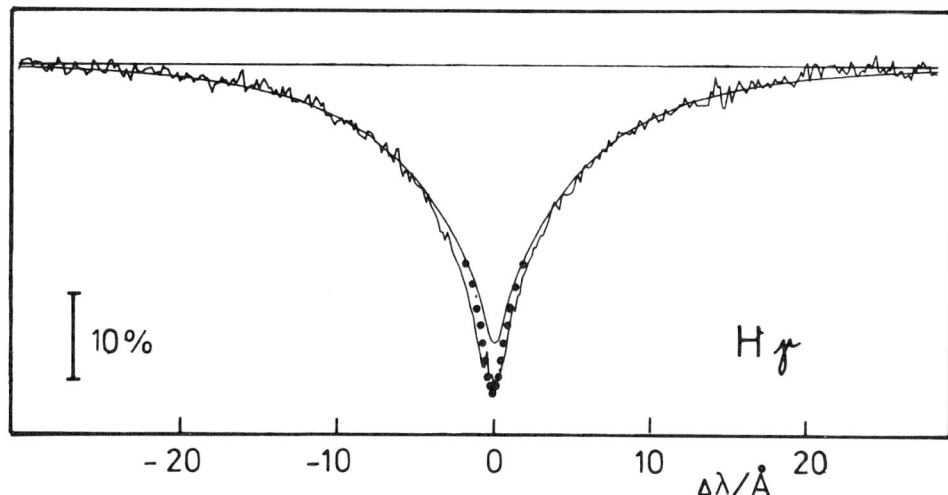

Figure 1. Comparison of a CASPEC spectrum of LB 3459 to theoretical Hγ line profiles. Fully drawn: models calculated using the complete linearization method. Dotted: improved models calculated using the ALI method. The model parameters are T_{eff} = 40000 K, log g = 5.3, no helium.

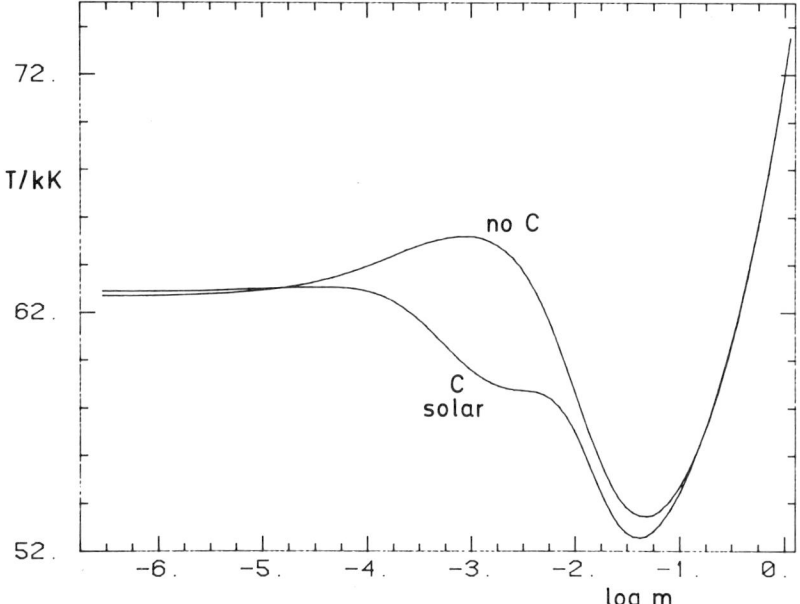

Figure 2. Temperature structure of two ALI models (T_{eff} = 75000 K, log g = 5, solar helium abundance), computed without any carbon and for a solar carbon abundance.

DISCUSSION

LINSKY   What signal-to-noise is required for your spectra ?

HEBER   The S/N ratio required for the analysis depends on the individual case, especially whether He I lines are observed or not. In the case of ROB 162, S/N ≈ 25 is sufficient since He I 4471 and 4713 are observed, which give an accurate Teff. If He I lines are not observed, higher S/N is required since the atmospheric parameters (Teff, log g and He abundance) have to be determined from He II and Balmer lines alone.

MAGAIN   Don't you have problems to define the continuum for such broad lines on Echelle spectra ?

HEBER   Yes, indeed. The rectification of CASPEC data is tedious and time consuming. The best results were obtained when suitable standard stars were used to define the Echelle blaze functions. In this case, Helium-B-giants are well suited because they display only very weak Balmer lines in their spectra (Wλ < 200 mÅ).

GRAY   Could you give us some indication of the resolution of the observations you use to test your theory and how well the H-line cores are actually resolved in the regions critical to your comparison.

HEBER   The ESO-CASPEC spectra have a resolution of 0.25 A and the Balmer line cores are well resolved. The line profiles are well defined in all parts (wings and cores). Hence their is no need for higher resolution.

BOHANNAN   What helium abundances do you get for these stars ?

HEBER   The range in helium abundances is very large. In some cases helium is depleted (by up to a factor 100). In other cases we cannot find a trace of hydrogen and the atmospheres seem to consists almost entirely of helium.

A SEARCH FOR CORONAL EMISSION LINES FROM EARLY-TYPE STARS*

D. Baade and L.B. Lucy
European Southern Observatory
Karl-Schwarzschild-Str. 2
D-8046 Garching, Fed. Rep. Germany

1. INTRODUCTION

For more than ten years now, a controversial issue in studies of stellar winds has concerned the existence or not of a coronal zone ($T \sim 10^6$ K) at the base of the cool winds ($T \sim T_{eff}$) of early-type stars. The latest revival of interest in this possibility is due to Wolfire et al. (1985) who showed that Waldron's (1984) recombination stellar wind (RSW) version of the hot corona – cool wind model (Hearn 1975; Cassinelli et al. 1978) yields models for ς Puppis (O4 If) that are consistent with both *IRAS* and *Einstein* IPC data, thus refuting an earlier claim (Lamers et al. 1984) to have excluded the existence of a coronal zone.

2. ZETA PUPPIS

Stimulated by this work of Wolfire et al. (1985), we have repeated with greater sensitivity the search for coronal line emission previously attempted at the Washburn Observatory (Nordsieck et al. 1981). In our initial investigation, we concentrated on ς Puppis because the narrow range of permissable coronal parameters (Wolfire et al. 1985) allowed a decisive experiment to be carried out for this prototypical and much-studied example of an early-type star with a high velocity, massive wind.

With the coronal parameters that Wolfire et al. (1985) obtain by fitting the Einstein IPC data for ς Puppis, the strongest coronal line should be that of [Fe XIV] at λ 5303 Å, the expected equivalent width being ~5 mÅ. In the event, no signal was detected, even though the excellent, high S/N data obtained with the ESO CES+Reticon system would have allowed a statistically significant detection down to ~1 mÅ, the exact limit depending on the assumed kinematic properties of the coronal zone – see Baade and Lucy (1987). Accordingly, our investigation of Puppis excludes a coronal zone with the parameters determined by Wolfire et al. (1985). Moreover, given that our detection limit is substantially lower than the predicted signal, it now seems highly unlikely that a coronal structure can be found that accounts for the observed X-ray emission and is at the same time consistent with our limits on [Fe XIV] λ 5303 Å emission.

Because full details of our investigation of ς Puppis have now been published (Baade and Lucy 1987), further discussion here is not necessary except perhaps to emphasize that the coronal line strength calculations used to interpret our data were *not* carried out in the low density limit. Thus, in addition to radiative de-excitations, we included also those due

---

*Based on observations obtained at the European Southern Observatory, La Silla, Chile

to electron and proton collision (Mason 1975). We also allowed for the weakening of the emission line due to the formation of an underlying absorption line.

3. OTHER STARS

Even though coronae are often said not to be expected for early-type stars because of the absence of a surface convection zone, violent motions observed in their reversing layers (*e.g.* Slettebak 1956) imply that a flux of mechanical energy is certainly available for dissipative heating. Moreover, a long-standing puzzle concerning the *Copernicus* UV spectrum of τ Sco can perhaps be resolved by postulating a base coronal zone. The observation is of inward moving cool gas (Lamers and Rogerson 1978), a surprising result since cool gas can be, and is in fact seen to be accelerated away from the star by radiative forces. However, if there is a coronal zone, the transition to a cool, radiatively-driven flow probably occurs at subsonic velocities (Baade and Lucy 1987) in which case the transition will be subject to the Rayleigh-Taylor instability. One can then imagine that this results in the formation of blobs of cool gas that are able to fall through the coronal gas because they are optically thick in the UV resonance lines. The constraint that the observed red-shifted absorption is non-saturated is met by the small expected covering factor of the stellar disk by the blobs.

Because the existence of base coronal zones in some early-type stars is thus not implausible, we have been motivated to continue our search for coronal line emission. Thus far data has been acquired but not yet analyzed for the stars τ Sco (B0 V), θ Car (B0 Vp), ϵ Ori (B0 Iae), κ Ori (B0.5 Iae), β Cru (B0.5 III), β Cen (B1 III), α Lup (B1.5 III), ϵ CMa (B2 III), and β Ori (B8 Iae).

Preliminary inspection indicates that the low detection limits achieved for ζ Puppis probably will not be reached: in the cooler stars the problem with line blending is quite severe, and the spectra require a more careful treatment of instrumental effects.

REFERENCES

Baade, D., Lucy, L.B.: 1987, *Astron. Astrophys.* **178**, 213.
Hearn, A.G.: 1975, *Astron. Astrophys.* **40**, 277.
Lamers, H.J.G.L.M., Rogerson, J.B.: 1978, *Astron. Astrophys.* **66**, 417.
Lamers, H.J.G.L.M., Waters, L.B.F.M., Wesselius, P.R.: 1984, *Astron. Astrophys.* **134**, L17.
Mason, H.E.: 1975, *Mon. Not. R. ast. Soc.* **170**, 651.
Nordsieck, K.H., Cassinelli, J.P., Anderson, C.M.: 1981, *Astrophys. J.* **248**, 678.
Slettebak, A.: 1956, *Astrophys. J.* **124**, 173.
Waldron, W.L.: 1984, *Astrophys. J.* **282**, 256.
Wolfire, M.G., Waldron, W.L., Cassinelli, J.P.: 1985, *Astron. Astrophys.* **142**, L25.

DISCUSSION

DOAZAN You tried to detect the [Fe XIV] 5303 Å line for diagnosing the existence of a hot corona ($\simeq 10^6$K) at the base of the wind of ζPup. But what about the density at which such a line may be formed ?
In the solar corona, this forbidden line is formed at a density of $\simeq 10^8$ cm^{-3} and Te $\simeq 10^6$K. In ζPup, the wind is much more massive (dM/dt $\simeq 10^{-6}$ M$_\odot$ yr^{-1}), i.e. about 10^8 times higher than in the Sun. The generally quoted value for Ne <u>in the wind</u> of ζPup is $\simeq 10^{11}$ cm^{-3} (see for ex. Hamann's ζPup model). Therefore, at the base of the wind, just after the photosphere, the density would be <u>at least</u> $\simeq 10^{12}$ cm^{-3}. Do you expect this [Fe XIV] forbidden line to be formed at such densities ?

BAADE Mason (1975) has tabulated the emission rates for [Fe XIV] λ 5303 in dependence on both electron density, Ne, and electron temperature, Te. It can be seen from his work that increasing Ne from 10^7 cm^{-3} to 10^{12} cm^{-3} at constant coronal Te actually increases the emission rate by a factor of $10^6 - 10^7$. Note that our predicted equivalent widths for the [FeXIV] λ5303 line are not obtained from the emissivity in the solar corona but by solving the equations of statistical equilibrium and radiative transfer for the various coronal models considered for ζPup. Thus, following Mason (1975), we adopt a 12-level model atom for FeXIV and include the effects of proton collisions within the ground configuration.

APPENZELLER How well is the critical density of [Fe XIV] λ 5303 really known and on what work is the number which was just quoted based on ?

DOAZAN I do not quote any particular work. I only quote what is known for the Sun, where the [Fe XIV] coronal line is formed at a density of $\simeq 10^8$ cm^{-3} and Te $\simeq 10^6$K (see for ex. Mihalas book on Stellar Atmospheres).

Effective Temperatures and Gravities for O-type Stars Determined from High Precision Line Profiles and Wind-Blanketed Model Atmospheres

B. Bohannan, S.A. Voels, D.C. Abbott, and D.G. Hummer
Joint Institute for Laboratory Astrophysics
Astrophysical, Planetary and Atmospheric Sciences
University of Colorado, Boulder, Colorado, U.S.A.

I. Introduction

Analysis of line profiles obtained with astronomical instrumentation capable of high signal-to-noise ratio spectroscopy have contributed significant new precision to the determination of the basic stellar parameters of hot, luminous, mass-loosing stars. Accurate measurement of such stellar properties as effective temperature and helium abundance for stars of spectral type O and early B is important not only to the physics of these stars but also to the environment in which they are located. The overall goals of the work we summarize here are to refine the spectral-type $vs.$ temperature calibration for the most massive stars and to determine helium abundances for stars that are loosing mass at a rate sufficiently high to affect their evolution. Details of our procedures are described in the analysis of ζ Puppis by Bohannan et al. (1986).

For stars hotter than roughly 35,000K, accurate values of effective temperature cannot be determined from the observed continuum flux distribution (Hummer et al. 1988); instead, effective temperature, surface gravity and helium abundance must be simultaneously measured through an analysis of high-precision line profiles using model atmospheres that contain the best available physics. The theoretical atmosphere used in the analysis we describe here assumes non-local thermodynamic equilibrium and includes the effect of heating of the photosphere by radiation backscattered by the stellar wind, the wind-blanketing effect (Hummer 1982). A description of the wind-blanketed model atmospheres is given by Abbott and Hummer (1985).

To achieve results of sufficient accuracy that will contribute significantly to such areas as the study of the initial mass function and the observational resolution of various theories of stellar structure and evolution, our analysis requires observations of high photometric precision. For stars with effective temperature on the order of 40,000K, the wings of Hγ must be determined with an accuracy of 0.2% to obtain gravities with an error of 0.1 dex; to measure temperatures with an error of 5% requires that the core-wing region of neutral and ionized helium lines must be observed at the 1% level.

2. The Observations

Photometric accuracy at the levels dictated by our goals are not possible with photographic techniques, but can be obtained with currently available electronic detectors capable of linear response over a wide dynamic range. For a number of O and early B-type stars with mass loss rates well-determined from radio measures, we have observed essentially all of the hydrogen and helium lines in the wavelength region 3900Å to 7100Å with the Kitt Peak National Observatory coudé-feed telescope and coudé spectrograph equipped for most of our runs with a TI 3-phase 800x800-pixel CCD. The spectral resolution is 0.30Å in the blue and 0.45Å in the red.

The photometric precision of our line profiles is 0.5% or better. We have found that, because of calibration of flat-fielding effects, the photometric precision of the observed line profiles is less than that predicted from photon statistics. Apparently the CCD in the coudé spectrograph is not identically illuminated by the starlight as by the flat-field lamp used to remove pixel-to-pixel variations. Our present technique is to trail the star along the slit; in this way we spread the spectrum over 10 to 40 columns of the CCD in an attempt to average residual fringing effects. We achieve with less observing time a higher signal-to-noise ratio spectrum with the trailing strategy than was possible by adding together spectra taken at different grating settings. However, there is still some residual fringing remaining which we attribute to fringing effects that are strongly wavelength dependent and which lie perpendicular to the dispersion, effects that do not average out when the spectrum is trailed. Additionally, there is some uncertainty in the profiles associated with fitting the continuum to the very broad absorption line profiles found in early-type stars.

3. The Model Atmospheres

The wind-blanketed model atmospheres used in our analysis are those of Mihalas (1972) with a modified upper boundary which allows for back-scattering of radiation by an outflowing stellar wind. The photospheric model is based upon assumptions of radiative and hydrostatic equilibrium, plane-parallel geometry, non-local thermodynamic statistical equilibrium and contains only hydrogen and helium. The stellar wind model includes radiation from the photospheric model and multiscattering by overlapping lines and by electrons. Some 10,000 lines between 200Å and 10,000Å are included. We use the mass loss rate set by the radio flux and the terminal velocity as measured from ultraviolet resonance lines.

In general, we find a good fit by the model line profiles to the observations (for examples, see Bohannan *et al.* 1986 and 1988). However, we do not model well those lines formed high in the atmosphere where the atmosphere may be spherically extended and possibly not static.

4. The Black-Art of Line Profile Fitting

To achieve the most accurate results possible, one cannot use equivalent widths of just a few lines to determine basic stellar parameters. Rather, effective temperature, gravity and helium abundance must be set simultaneously to as many lines as feasible. In our work, we include a large number of transitions to overcome the effects of unresolved blends, of observational errors and of modeling deficiencies. For a starting point in temperature, we use HeI $\lambda 4471$Å and HeII $\lambda 4542$Å profiles; for a starting gravity, Hγ $\lambda 4340$Å. We then vary the temperature and gravity until we get a good fit to all of the hydrogen and helium transitions observed, usually about 10 lines in all.

Through a plot of depth of formation, we identify those transitions formed high in the atmosphere which may not produce good fits. In general, we find that strong lines of the dominate ionization stage of helium are not well modeled by our code. For example, lines like HeII $\lambda 5411$Å and HeII $\lambda 4542$Å in ζ Puppis (O4If) and HeI $\lambda 5876$Å and HeI $\lambda 5411$Å in α Cam (O9.5Ia), transitions formed high in the respective atmospheres, are not fit with the same precision as other transitions of helium. Note that this discrepancy affects the two lines, HeII λ4542Å and HeI λ4471Å, that are the fundamental temperature criterion in defining the spectral classification of O-type stars. Since we are able to fit lines formed deep near the continuum, we do not feel that this deficiency seriously affects the accuracy of the stellar parameters we measure.

To determine the helium fraction, we use a self-consistency test on the effective temperature and helium fraction using equivalent widths of all of the neutral helium and unblended ionized helium lines. We iterate with a set of temperature, gravity and helium abundance until we get a consistent fit to all of the lines formed deep in the atmosphere.

5. Discussion

The table below summarizes our results to date. The temperature scale for evolved early O-type stars has been reduced some 20% over that of Conti (1973). For spectral type O4If, allowance for an appropriate gravity made possible by high precision line profiles has changed the temperature by -8000K, inclusion in the model atmosphere of the effect of wind-blanketing accounted for a change of -4500K, and the use of high precision line profiles and the analysis of many lines of helium produced a change of +4000K. We estimate the errors of our stellar parameters as follows: in effective temperature, ±1500K at O4 and ±1000K at O9.5; in gravity, ±0.1 dex at O4 and ±0.05 dex at O9.5; in helium fraction, [Y], ±0.03.

Adopted and Derived Stellar Parameters of Some Early-type Stars
From High Precision Line Profiles and Wind-Blanketed Model Atmospheres

Star	Sp. Type	log L/L$_O$	log M /M$_O$	T$_{eff}$(°K)	log g	[Y]
ζ Pup	O4If	6.00	-5.3	42,000	3.5	0.17
9 Sgr	O4V	6.11	-5.4	46,000	3.9	0.10
α Cam	O9.5Ia	5.78	-5.4	30,000	2.85	0.20
ζ Ori	O9.5Ib	5.77	-5.8	31,000	3.2	0.10
δ Ori	O9.5II	5.60	-6.3	33,000	3.4	0.10

The high helium abundance observed for ζ Puppis is consistent with the evolved nature of this star which has lost some 40 M$_O$ of material from its original mass of 70-90 M$_O$. A similar conclusion has been reached for α Cam. There is a hint from photographically derived equivalent widths of a large number of O-type stars that this result may hold true for many, if not all, evolved O-type stars.

References:

Abbott, D. C. and Hummer, D. G. 1985, *Ap. J.* **294**, 286.
Bohannan, B., Abbott, D. C., Voels, S. A., Hummer, D. G. 1986, *Ap. J.* **308**, 728.
Bohannan, B., Voels, S. A., Abbott, D. C., Hummer, D. G. 1988, submitted to *Ap. J.*
Conti, P. S. 1973, *Ap. J.* **179**, 161.
Hummer, D. G. 1982, *Ap. J.* **257**, 724.
Hummer, D. G., Voels, S. A., Abbott, D. C., Bohannan, B. 1988, submitted to *Ap. J.*
Mihalas, D. 1972, NCAR TN/STR-76, *Non-LTE Model Atmospheres of B and O Stars*.

DISCUSSION

EBBETS You mentioned that you did not see variability in αCam. I remember observing Hα variability several years ago, and having HeI λ6678 on the same spectra. This was an absorption line, and it did show night to night variations in profile.

BOHANNAN Hα and other "wind" lines are certainly variable in line strength. We have tried to concentrate on lines formed primarily in the photosphere. For example, we have multiple observations of Hγ in αCam that appear to be constant in profile. We would like to believe that any bump and wiggles seen in line profiles have only a small effect on the temperatures and gravities that we derive.

EDVARDSSON Just how sensitive are your surface gravities to, say, a 2000 K error in the effective temperature of the star.

BOHANNAN We solve for temperature, gravity and helium abundance in an iterative fashion. The starting gravity comes from Hγ – the starting temperature from neutral and ionized helium lines. We then adjust the temperature and gravity until we get a best fit to <u>all</u> of the lines. The interaction between temperature and gravity depends strongly upon what spectral type star is under consideration. But there will not be a systematic error in gravity caused by an uncertainty in temperature because different lines have different sensitivities to temperature and to gravity. For αCam we have a sensitivity of 1000 K in effective temperature and 0.05 dex in gravity.

GRATTON What do you mean by Y ?

BOHANNAN [Y] here is the number fraction of helium.

GRIFFIN You said that co-adding spectra in order to increase S/N "did not work". Could you please say why not ?

BOHANNAN Adding together CCD spectra taken at the same grating setting does not remove, rather strengthens, the effects of residual fringing. Adding together spectra at slightly different grating settings has the potential of achieving the highest S/N but is inefficient of telescope time as it requires independent flat-field exposures at each grating setting. Our present technique is to trail the star over 10 – 40 lines on the CCD. However, some low level ($\approx 0.2\%$) residual fringing remains after extraction because fringing is a wavelength dependent phenomenon and any residual fringing present will tend to lie perpendicular to the dispersion.

VARIABILITY OF SOME Be STARS ON HIGH-RESOLUTION, HIGH S/N SPECTRA

A.M. Hubert, H. Hubert, B. Dagostinoz AND M. Floquet
DASGAL et UA 337
Observatoire de Paris, Section d'Astrophysique de Meudon
F-92195, Meudon Cedex, France

ABSTRACT. Rapid variability in Be stars could be understood by non radial pulsations or by rotation of an inhomogeous surface brightness distribution...
The structure and the variability of the Hα and of the HeI λ6678 lines have been investigated with an optical fiber spectrograph and a CCD camera. The signal to noise ratio, measured in the continuum, is between 300 and 500.
Weak changes in the Hα emission line profile of γ Cas have been detected on time-scale of hours and days. This line has an asymmetric profile exhibiting only one blue-shifted maximum while the HeI λ6678 has a double-peak emission, superimposed to the photospheric contribution, with a violet to red peak ratio V/R>1.
The Hα emission line of φ Per exhibits a complex structure with significant changes in its core, from night to night and on a short time scale <1hr. The HeI λ6678 presents a blue-shifted asymmetric emission (red-winged) superimposed to the photospheric contribution.
Furthermore the HeI photospheric line λ6678 of the B6 star o And has presented notable variations in its profile during the 2 observational campaigns, which do not seem correlated to the photometric period of 1.57 day.

1. INTRODUCTION

Rapid variability in Be stars (several hours, 1 or 2 days) could be induced by non radial pulsations (g or r modes), rotational modulation of superficial inhomogeneities, or in the case of 1-2 day period, by contact binary systems. Short-time scale spectroscopic variability of some Be stars has been investigated with the 1.93 m telescope of the Haute Provence Observatory, equipped with the optical fiber spectrograph ISIS (Felenbok et al., 1986) and a CCD camera. High S/N (300 to 500) spectra have been obtained for 2 resolutions (0.18 and 0.70 Å). Flat-field spectra and wavelength calibration (thorium-argon) were made each night of observations. All data reductions have been performed with a VAX 8600 computer.

2. RESULTS

Our preliminary results concern γ Cas, φ Per and o And. The 2 early Be stars γ Cas and φ Per are well-known to exhibit variable mass loss, and o And is a B6III star which exhibits temporary emission and/or shell features (episodic mass loss phenomena).

2.1 γ Cas

The Hα emission line fluctuations of γ Cas have been searched on 21 spectra taken on 5 consecutive nights (1986, September 19-24), at the resolution 0.18 Å.

1 During several nights we have found very weak variations (over about 3 hours interval) at the center of the Hα emission line (1% in relative flux Fλ/Fc), at the limit of the detectability of a CCD camera. Such variations could be due to non radial pulsations.

2 Faint night-to-night variations (about 5% in relative flux Fλ/Fc) have been only detected over the 3 last nights of observations. These results are in agreement with those of Chalabaev and Maillard (1983).

The HeI line λ6678 Å (resolution 0.70 Å) has presented, in September 1986, a double asymmetric V>R emission superimposed to the broad photospheric absorption. So at the epoch of observations the inner layers of the cool disk where this double emission line is formed, were not only dominated by the rotational velocity law, but also by an additional inwards motion.

2.2 φ Per

8 Hα emission line profiles at the resolution 0.18 and 0.70 Å have been obtained in 1986, September 20-24. A complex structure (several peaks) with significant changes in its core has been noted, fig 1a and b.

a) On September 22, a notable variation of the central depth and of the equivalent width (6%) has been seen on 2 consecutive spectra obtained at UT=3h 30mn and UT=4h 05mn, fig 1a. Stellar surface activity could explain such rapid fluctuations in intensity.

b) During 4 consecutive nights, a gradual enhancement of emission has been observed with an inversion of the most prominent peaks, fig 1b. Such variations could result from a temperature and density increase in the cool envelope by travelling shock waves induced by non monotonic winds.

At the same epoch, the HeI λ6678 has exhibited an asymmetric rather narrow emission at the center of the broad photospheric line

a) some disturbance occured in the wings during the night September 22, as rapid changes in Hα was noted

b) the narrow emission component is formed in external layers of the cool envelope where the rotational velocity is low. As in γ Cas, HeI had a profile characteristic of infalling circumstellar layers.

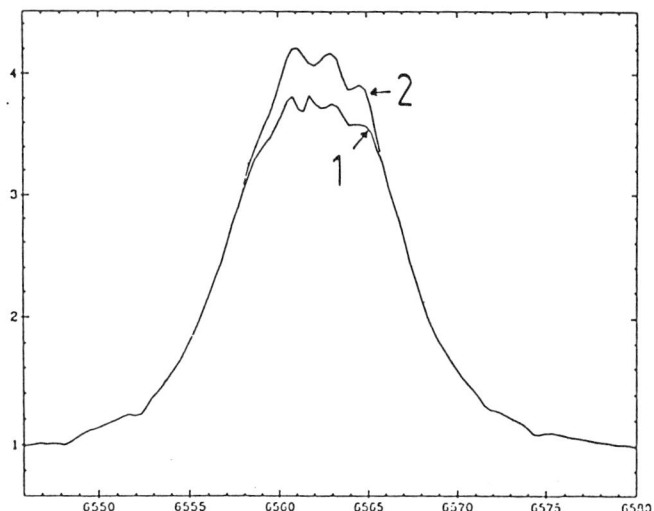

Fig.1a : Variation of the Hα line of φ Per during the night of 1986, September 22 (JD2446695); 1: UT=3h 30m, 2: UT=4h 05m.

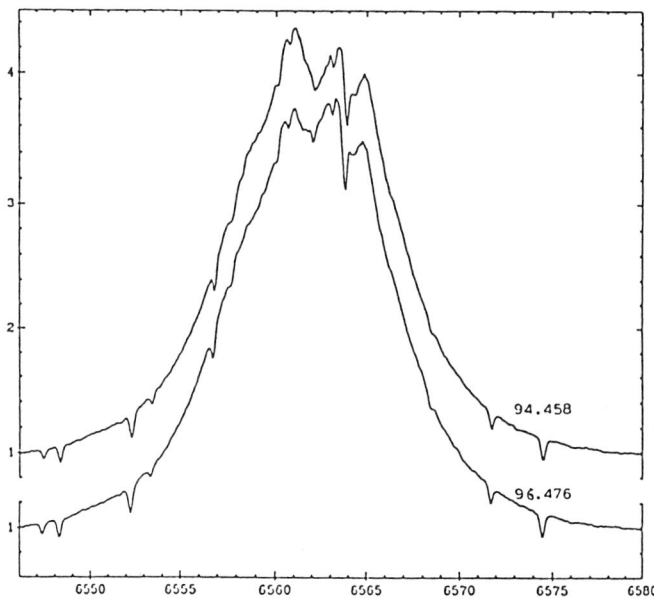

Fig.1b : Hα line profiles of φ Per. Julian dates are indicated on each profile (JD2446600 +). Resolution is R=0.18 Å.

2.3 o And

Several photometric periods have been proposed for o And, the most probable value given by Harmanec (1984) is 1.57 day, but the light curve is variable in shape and amplitude.

We have analysed 33 spectra of the HeI photospheric line λ6678 Å obtained in 1986, August 14-22, and September 19-23 fig 2 with a resolution 0.70 Å and an exposure time 20mn.

  a) no variation over a time scale ≤2 hours has been detected
  b) changes in line profile have been observed from night-to-night: temporary presence of asymmetry
  c) asymmetry is more pronounced in August than in September
  d) radial velocity of centroïd is quite different in August and in September. It is a result in favour of duplicity of this star (P of several days or tens of days)
  e) line profile variations do not follow the photometric period of 1.57d but would probably have a longer time scale.

In conclusion no-short time scale periodicity has been found on the HeI line λ6678 of o And, but mid-term variability in favour of duplicity has been deduced from our spectra.

Acknowledgements

We are indebted to Drs. P. Felenbok, D. Gillet and J. Guérin for their valuable help.

3. REFERENCES

Chalabaev, A., Maillard, J.P.: 1983, Astron. Astrophys. 127, 279
Felenbok, P., Guérin, J., Czarny, J.: 1986, Colloque ESO/OHP sur les CCD
Harmanec, P.,: 1984, IAU Inf. Bull. Var. Stars N°2506.

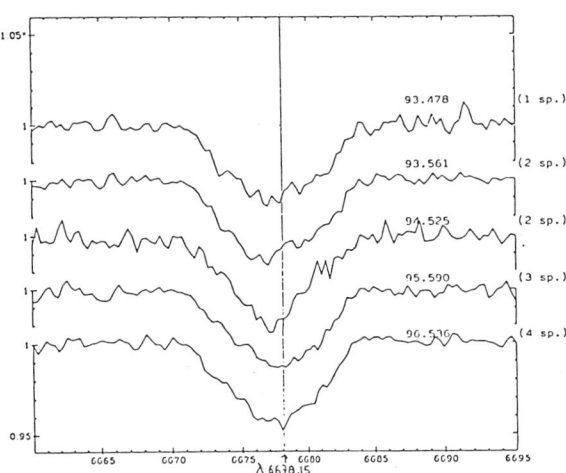

Fig.2 : Variability of the HeI line λ6678.15Å of o And. Number of average spectra and Julian day of observations (2446600 +) are indicated on each line profile. S/N is between 300 and 500.

A MODELLING OF LINE TRANSFER IN SUPERGIANT ENVELOPES

Krzysztof Gesicki
N.Copernicus Astronomical Center
Laboratory for Astrophysics
ul.Chopina 12/18
87-100 Torun, Poland

ABSTRACT. A computer code that solves the transfer equation in moving atmosphere is prepared for analysis of variability in circumstellar spectral lines. This model gives a theoretical line shape, which consists of a symmetric photospheric line and superposed on it a P-Cyg type component originated in the envelope. Comparison between calculated and observed line profiles will be used to analyze 6 A/mm spectra of F-K supergiants.

1. INTRODUCTION

It is well known that some of the high luminous F-K supergiants are surrounded by circumstellar envelopes. Circumstellar shells of Rho Cas and HR8752 were investigated by Sargent (1961), Smolinski and Climenhaga (1985), Lambert et al. (1981). One of indicators of circumstellar matter is the observed splitting in low-excitation-potential spectral lines. The interpretation of this feature is that a single and symmetric line formed in the photosphere changes its shape in the process of scattering while passing through the envelope. As the envelope is spherical, extended and expanding, the photospheric profile is affected by an absorption component shifted towards shorter wavelengths as well as by an emission component shifted towards longer wavelengths. The resulting profile is complicated and its analysis requires model computations. Such analysis was made for late type supergiants by Sanner (1976), Bernat (1977) and by Hagen (1978).

2. THE CHOSEN METHOD AND THE RESULTING CODE

In my analysis of 6A/mm spectra of supergiants I have concentrated at first on Rho Cas (F8Ia$^+$), which shows interesting changes in the shell line profiles (Fig.1). Having decided to study the variability on the basis of a model computations I have written a computer code. There are many ways of handling line transfer in circumstellar shells, among known authors are Mihalas, Kunasz, Hummer, recently Sanner, Bertout.

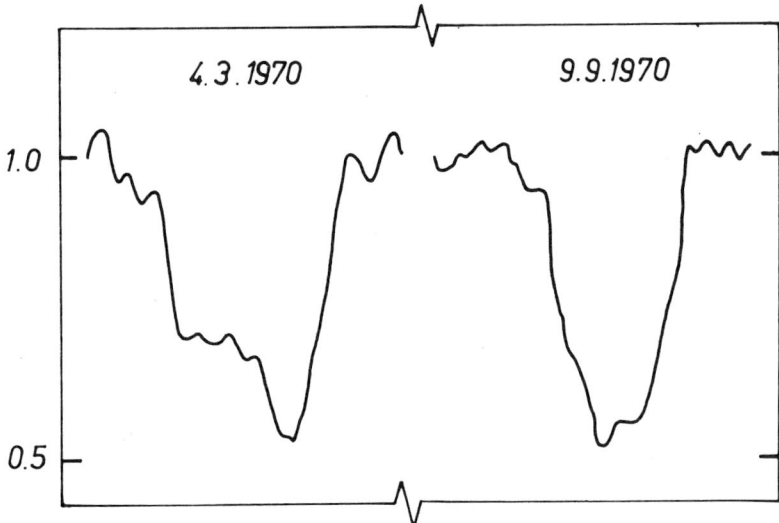

Figure 1. An example of line Ba II 5854 A in supergiant Rho Cas. The variability of the shortwave component, which is of circumstellar origin, is readily visible.

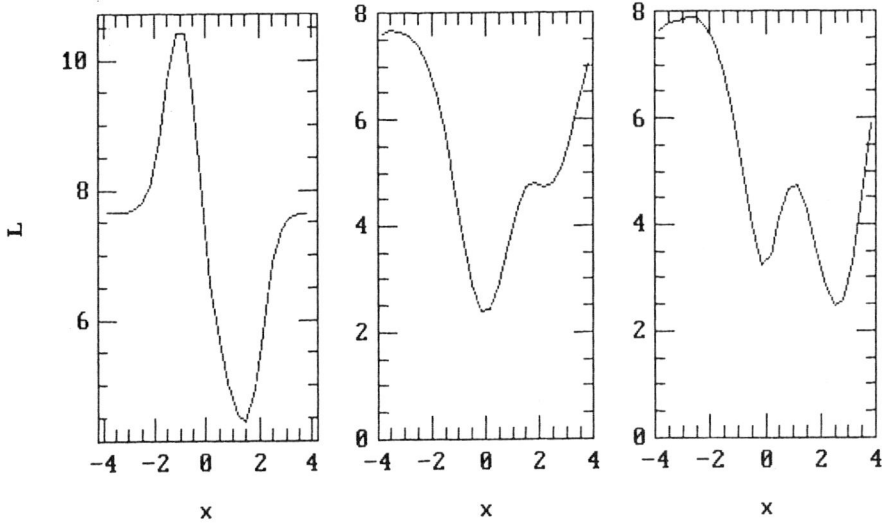

Figure 2. Luminosity profiles produced with the code. The horizontal axis is the frequency x in Doppler units. First from left is the profile obtained with constant stellar continuum, then follow two profiles with underlying absorption line and with different mean opacity T of the expanding envelope, T = 1 and 3 respectively.

I have chosen a method of Kunasz and Hummer (1974) since it has already been successfully used to perform similar analysis (Bernat, Hagen). The transfer equation is written in the observer's frame along a set of rays parallel to the direction to the distant observer. Intersections of these rays with the set of radii which divide the envelope into concentric shells, constitute the grid used to numerical solution. Spherical symmetry and the application of Feautrier variables result in an efficient code that does not require large computer memory. This method allows for an arbitrary density and velocity profile across the envelope. I have changed the boundary conditions so that one boundary is in the middle of the envelope. It enables to put the star surface inside the envelope. This version allows for an arbitrary shape of the underlying photospheric continuum which can be in the form of an absorption line.

The code is written in FORTRAN77 and runs on an IBM PC/XT computer. Some examples of line profiles obtained with it are shown in Fig.2. They represent the preliminary results. However a comparison between Fig.1 and Fig.2 shows that the model produces results similar to observations. The analysis of variability of supergiant envelopes requires a construction of a whole atlas of theoretical profiles to compare with real spectra. This work is in progress.

ACKNOWLEDGEMENTS

The computations were partially performed on an IBM XT compatible computer, on loan from Princeton University.

REFERENCES

Bernat, A.P., *Astrophys.J.*, 213, 756, (1977)
Hagen, W., *Astrophys.J.Suppl.*, 38, 1, (1978)
Kunasz, P.B., Hummer, D.G., *Mon.Not.R.Astr.Soc.*, 166, 57, (1974)
Lambert, D.L., Hinkle, K.H., Hall D.N.B., *Astrophys.J.*, 248, 638, (1981)
Sanner, F., *Astrophys.J.Suppl.*, 32, 115, (1976)
Sargent, W.L.W., *Astrophys.J.*, 134, 142, (1961)
Smolinski, J., Climenhaga, J.L., *Proc.IAU Symp.No.116*, (1985)

DISCUSSION

PRADERIE Is there an observational or theoretical evidence for a finite optical thickness T in these circumstellar envelopes ?

GESICKI Presence of line splitting in low-excitation potential spectral lines and the absence of the splitting in other lines is one of the indicators of circumstellar shell. Calculations, such as presented by me, will allow to estimate the optical thickness. Earlier estimations based on the curve of growth method cannot be accurate.

NLTE-EFFECTS IN PROFILES OF PASCHEN AND BRACKETT LINES OF AN A0IV STAR γ GEM

A.A. Chalabaev[1], J. Borsenberger[2], J.P. Maillard[3], and F. Praderie[2]

[1] CNRS, Observatoire de Haute-Provence, F-04870, Saint-Michel-l'Observatoire, France
[2] Observatoire de Paris, Section de Meudon, F-92195, Meudon, France
[3] CNRS, Institut d'Astrophysyque, 98bis bd Arago, F-75014, Paris, France

Abstract   Profiles of the hydrogen Paβ, Paγ and Brγ lines of an A0IV star γ Gem were obtained at a high resolution to investigate at what degree the hydrogen IR lines can be useful for fine tests of the NLTE model atmospheres. All three lines show a deep and narrow absorption core in a good qualitative agreement with the predictions of the available NLTE models for B stars, thus proving that the NLTE effects are important for A stars as well. New NLTE models were computed to extend the grid down to 9500 °K and to include 8 discrete levels of H atom into the energy balance. However, further theoretical work is necessary to compare the observed profiles and the theory.

1. Introduction.   About 20 years ago it has been realized that the departures from LTE in atmospheres of O and B main sequence stars change dramatically the temperature profile near the stellar surface (Strom and Kalkofen, 1966; Auer and Mihalas, 1969a). The subsequently built NLTE model atmospheres for the hot stars were tested by using observations obtained in the visible (Strom and Kalkofen, 1967; Auer and Mihalas, 1970). Auer and Mihalas (1969b) also suggested to undertake spectroscopy in the near IR. Namely, they pointed out that the profile of the hydrogen Brα 4.05 μm line should be very sensitive to the departures from LTE and have a narrow emission core in the spectra of B stars. However, high resolution spectroscopy of this line is difficult because of the strong thermal background at these wavelengths. In the present study we wish to adress two questions. First, whether a high resolution spectroscopy of other hydrogen lines in the near IR ($\lambda\lambda$ 1-2.5 μm), where the thermal background is negligible, could be useful in testing model atmospheres, and second, if the NLTE effects are still important for A stars. We selected γ Gem (A0IV, T_{eff} = 9500°K) which is a slow rotator, v·sin i = 7 km·s^{-1} (Boesgaard and Praderie, 1981)

2. Observations.   The Fourier transform spectrometer at the 3.6 m CFH telescope on Mauna Kea (Hawaii) was used to obtain the spectrum of the star from 1 to 2.5 μm at the resolution of 1 cm^{-1}. The instrument was described by Maillard and Michel (1982). The hydrogen lines of interest in this spectral range are Paβ 1.2818 μm, Paγ 1.0938 μm and Brγ 2.1655 μm. Their observed profiles are displayed in Fig. 1.

3. Discussion and model atmosphere computations.   All three lines show broad wings and a deep and narrow absorption core which is particularly conspicious in the profile of Paβ. Such a core was predicted by Auer and Mihalas (1970) and is due to departures from LTE. However, the lowest effective temperature they considered in

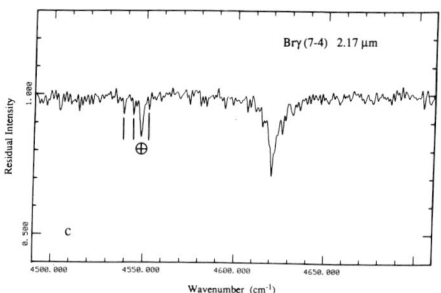
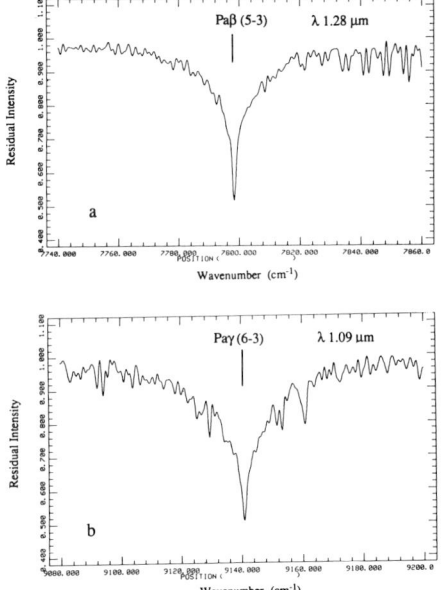

Fig. 1 *The observed profiles of the hydrogen lines in the near IR spectrum of γGem (A0IV): a - Paβ; b - Paγ; c - Brγ.*

their NLTE computations was 12500 °K, therefore the agreement with their results is only qualitative. The grid of the NLTE model atmospheres had to be extended to temperatures lower than 12500 °K. At these temperatures, the ionisation front of hydrogen is situated in the atmosphere, which leads to instability of iterations. Further, the statistical equilibrium equations in the models of Auer and Mihalas include the first 5 discrete levels of the H atom while to have a correct source function of Brγ (transition 4-7) one must consider at least 8 discrete levels. As the starting point we used the code described by Mihalas *et al.* (1975). It makes use of the complete linearisation method. The code was modified to include 8 discrete levels (in total, 21 transitions were considered; the Lyman transitions are assumed to be in the detailed balance). A great care was taken to handle instability problems due to lower T_{eff}. The code underwent additional modifications to be used on VAX mini- and micro-computers. A grid of NLTE model atmospheres was thus obtained for log g = 4 and for T_{eff} ranging from 9500°K to 15000°K with the step of 250°K. For each model, the profiles of temperature in the atmosphere, of the hydrogen ionisation and of populations of the first 8 levels of H atom were computed in a *self-consistent* manner. It was found that including 8 levels in the energy balance leads to the outward temperature rise up to about *1000 °K more* than in the models where only 5 discrete levels were considered (Borsenberger and Gros, 1978) . As an example, the temperature profile obtained for T_{eff}=9500°K is displayed in Fig. 2. (There, we also indicated the depths in the atmosphere where the hydrogen lines of interest reach the unity optical depth. One can see that the Brackett and Paschen lines probe the region of the temperature minimum). Another result to be mentioned is the emission core predicted in the Brα 4.05 μm line at T_{eff} as low as 9500°K. A more detailed description of the model computations and the results will be given elsewhere (Borsenberger, in preparation).

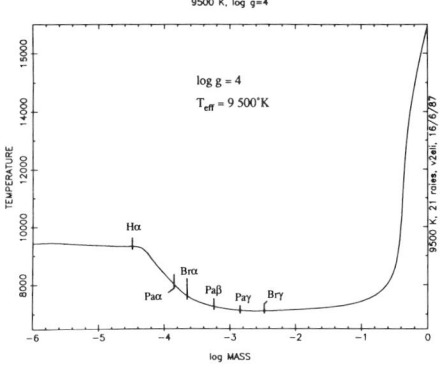

Fig. 2 The temperature profile of the NLTE model atmosphere at $T_{eff}=9500$ °K. The energy balance accounts for 8 discrete levels of hydrogen and the continuum. We also indicated the depths where different hydrogen lines become optically thick.

So far, only a progress report can be done on the computations of the hydrogen IR line profiles. The source functions are NLTE and frequency independent. The optical depths and the source functions are fully consistent with the models. The line broadening includes the thermal one and the Stark broadening. The quasi-static approximation was used for ions (Edmonds *et al.*, 1967). The electron broadening was treated quasistatically in the wings and using the impact approximation in the line center (Hoang-Binh,1982). The resulting computed profiles turned out to have too narrow wings and do not match the observed profiles. After analysis, we noted that for a large part of the higher series H line profiles neither the impact approximation nor the quasi-static one can be used. Unfortunately, laboratory profiles of those lines or theoretical profiles computed in the unified or quantum theory are lacking. Further work on improving the line broadening computations is necessary before we could produce a comparison between the observed profiles and the theory.

4. Conclusions. The reported study shows that the profiles of the hydrogen lines in the 1-2.5 μm spectral range, arising deeper in the atmosphere than the Balmer lines, are useful as a fine test of the model atmospheres. In A-type stars, the IR lines probe the region of the temperature minimum. The observed profiles demonstrate that the NLTE effects are important for T_{eff} as low as 9500 °K. A new grid of plane-parallel NLTE model atmospheres was computed from 9500°K to 15000 °K. The energy balance takes into account 8 discrete levels of H atom and the continuum level. The outward temperature rise was found to be up to 1000°K more than in the models where only 5 discrete levels and the continuum one were considered. The models predict an emission core in the Brα 4.05 μm line for T_{eff} as low as 9500 °K. Further work is in progress to obtain theoretical profiles of the IR lines.

References:
Auer, L.H., and Mihalas, D., 1969a, Ap.J., **156**, 157
Auer, L.H., and Mihalas, D.,1969b, Ap.J., **156**, L151
Auer , L.H., and Mihalas, D., 1970, Ap.J., **160**, 223
Boesgaard, A.M., and Praderie, F., 1981, Ap.J., **245**, 219
Borsenberger, J., and Gros, M., 1978, As.Ap.Suppl., **31**, 291
Edmonds F.N., Schüster, H., Wells, P.C., 1967, Mem .R.A.S., **71**, 271
Hoang-Binh, D., 1982, As.Ap., **112**, L3
Maillard, J.-P., and Michel, G., 1982, in "Instrum. for Astronomy with Large Opt. Teles.", ed. C.M.Humphreys)
Mihalas, D., Heasley, J.N., and Auer, L.H., 1975, NCAR Techn. Note 104)
Strom, S., and Kalkofen, W., 1966, Ap.J., **144**, 76
Strom, S., and Kalkofen, W., 1967, Ap.J., **149**, 191

DISCUSSION

        NISSEN How long integration times did you need to obtain these FTS spectra for γGem ?

        CHALABAEV About 30 minutes. This is a bright star and the passband is very large so that one has a "multiplex disadvantage" with respect to a conventional spectrograph + mosaic detector. However, the comparison is not so straightforward. The total number of spectral elements was 8000 (although only a part of them can be considered as useful). How many settings one needs to obtain the same spectrum by a conventional spectrograph ? Further, there is no need in calibration exposures (flat-fields, comparison spectra, etc.), so that the telescope time is used efficiently.

(See also review by J.P. Mailllard in this volume).

BALMER EMISSION PROFILES IN RADIALLY PULSATING STARS : THE CASE OF THE DOUBLE Hα EMISSION

D. Gillet
Observatoire de Haute-Provence
04870 Saint Michel l'Observatoire
France

ABSTRACT. We present new high resolution profiles of Hα emissions in radially pulsating stars (RR Lyrae, W Virginis, RV Tauri, Mira stars and Classical Cepheids). Depending on phase, these profiles can show an apparent absorption within the emission close to the Hα laboratory wavelength. The origin of this feature is discussed in the framework : (a) of a composite of a single emission and a deep photospheric absorption (RR Lyrae, W Virginis and RV Tauri stars); (b) of a true absorption caused by cool hydrogen above the shock wave producing the emission; (c) of a geometric model in which the absorption is a lack of emission. Recent observations (Hα profiles of Mira stars S Car and T Cen), which show that the absorption dip is below the continuum level, are in favour of the true absorption model.

1. INTRODUCTION

The existence of shock waves in the atmospheres of radially pulsating stars is a well-known phenomenon. They are characterized by a discontinuity in the radial velocity curve due to double absorption lines, by hydrogen and metallic emissions and sometimes by the presence of fluorescence lines. Hydrogen emissions in Classical Cepheids (pulsating stars of population I) is weak and shock waves have been suggested only in a few cases (for instance in β Doradus by Hutchinson et al., 1975). In Cepheids of population II, only RV Tauri and W Vir stars show both helium and hydrogen emission (Preston, 1962; Abt, 1954 and Wallerstein, 1959 respectively) while RR Lyrae stars have only weak hydrogen emission (Preston, 1964). Finally, Mira stars (both populations I and II) present during a large fraction of their periods strong Balmer emission lines (Joy, 1947), but helium emission is never observed.

In all these stars, atmospheric motions are the consequence of radial pulsations although the cause of pulsation can be different (Cox, 1974). Thus, we must expect to observe some spectral similarities between these different kinds of stars; namely double absorption lines. Are there also similarities in observed emission profiles ? This is certainly affected by the disparity of stellar parameters (gravity, $T_{eff.}$, atmospheric composition,...) between these stars. The aim of this paper is to show that the double Hα emission must be observed in all classes of radially pulsating stars and to suspect a similar origin (§ 2). The different explanations for the emission are then discussed in the case of Mira stars (§ 3) for which a large number of profiles are already available.

2. DOUBLE Hα EMISSION IN RADIALLY PULSATING STARS

The double Hα emission is only observed around or after the luminosity maximum. It is characterized by an absorption close to the Hα laboratory wavelength within an apparent broad emission. In the case of cool stars (Mira or RV Tauri stars), other smaller absorptions at different wavelengths can also be observed (see fig. 1b).

Fig. 1a shows the double Hα emisson observed in the spectrum of the RR Lyrae star X Ari at the Canada-France-Hawaii (CFH) Coudé Spectrograph equipped with a 1,872-diode Reticon as detector. The resolution is 0.2 Å ($R \equiv \lambda/\Delta\lambda = 33{,}000$) and the signal-to-noise ratio 60. The phase of this observation is $\varphi = 0.93$, i.e. just before the luminosity maximum. A comparison with the profile (dashed line) without emission ($\varphi = 0.86$, i.e. 1 hour earlier) shows conspicuous double emission components. The blueshifted contribution is stronger than the redshifted one. These observations suggest that the double emission is only apparent and is the consequence of a composite between a single large emission and a strong photospheric absorption. This is consistent with an absorption dip, smaller at phase 0.93 than at phase 0.86.

The Hα profile of the hot Mira star S Car (population II) in fig.1b is more complex because other absorptions are also visible within the emission. The resolution is close to 0.11 Å ($R = 60{,}000$) and the signal-to-noise ratio at the continuum level is approximately 200. The Coudé Echelle Spectrometer (CES) of the European Southern Observatory (ESO) equipped with a 640 x 1,024 RCA-CCD detector has been used. The AAVSO predicted phase is 0.79, i.e. before the luminosity maximum. Since a large number of absorptions is visible, it is not easy to give an accurate position of the continuum level. Nevertheless, in contrast with X Ari, the central absorption is not obviously below the continuum. Its origin will be discussed in the next section.

The following profile (fig. 1c) was obtained with the same spectrometer at the same resolution. The double Hα emission profile of the relatively faint star (V ~ 10.0, corresponding to phase 0.21) W Vir (the prototype of W Virginis stars) is very easily visible. As with S Car, the blueshifted emission component is stronger than the redshifted one and the central absorption is above the continuum level.

Finally, the last profile (fig. 1d) concerns the bright RV Tauri star R Sct at its second luminosity maximum. The CES equipped with a 1,872-diode Reticon was used ($R = 60{,}000$ and S/N ~ 300). This profile is similar to that of X Ari. The central absorption is clearly below the continuum level. Unfortunately, it is not possible, as in X Ari, to observe the normal absorption profile without emission but we must accept that the main contribution to the central absorption is due to the photospheric absorption.

All these stars are of population II and at the moment of their observation their spectral types were respectively A8 II (X Ari), K5e III (S Car), F0 Ib (W Vir) and G0eIa (R Sct). In Mira stars, the Hα absorption is never observed from luminosity maximum to luminosity minimum (for S Car see profiles in Gillet et al. 1985). The strong TiO band-head and TiO lines of the γ-system in the Hα region affect considerably the spectral visibility (Hänni, 1987). The lack of high resolution Hα profiles of W Vir does not allow one to appreciate the intensity of the photospheric absorption although it is easily visible at phase 0.395 within the Hδ profile (Abt, 1954). Thus, quantitative estimates of the photospheric contributions would be useful.

Classical Cepheids show large distortions of the Hα profile (for instance, Wallerstein, 1972) while visible emissions remain weak. In other words, the intensity of the photospheric absorption compared to that of emission always dominates. Thus, no manifest double emission profiles are observed.

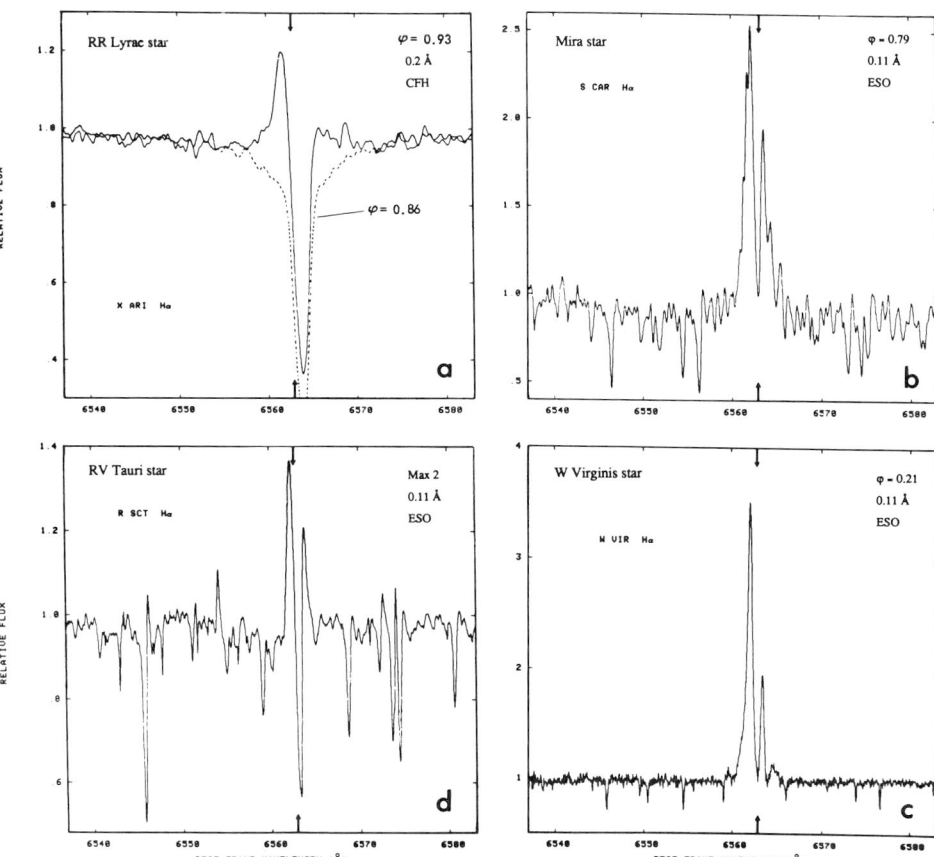

Fig. 1 High resolution profiles of Hα lines in four different radially pulsating stars. The resolution, the observatory and the phase φ (at luminosity maximum φ ≡ 0.0) are given. The position of the Hα laboratory line is marked by an arrow on the stellar rest frame wavelength axis. All the profiles show a double emission structure which is not always present.

3. THE CASE OF MIRA STARS

Because of their very low effective temperature (< 3000 K), the visibility of the double Hα emission profile is unusual. Up to now, approximately twenty stars with this kind of profile have been reported (Gillet, 1987). In S-type and very hot (early M2) M-type Mira stars the double profile is quite visible from luminosity maximum, while it appears only around phase 0.4 in typical (i.e. cool) M-type Mira stars. Untill now, no double emission has been observed in C-type Mira stars. Thus, this suggests that TiO absorption is not the cause of the observed double structure although the R1a (~ 6562.8 Å) TiO band-head of the γ-system at the Hα laboratory wavelength (6562.817 Å) can contribute in some cases to this structure (Hänni, 1987).

Two kinds of explanations have been proposed. The first, due to Bidelman and Ratcliffe (1954), claims that the absorption is a true absorption caused by cool hydrogen above the shock wave. The problem is to show how a cool gas (T < 2000 K) can absorb Hα-photons. Wallerstein and Cox (1984) in their discussion on W Virginis stars, suggest that Lα-photons produced in the shock wake would be absorbed in the umperturbed atmosphere above the shock. This Lα scattering with the help of the metastability of the 2S state, would provide the overpopulation of the n = 2 level, inducing the observed Hα absorption.

The second explanation has first been proposed by Willson (1976) and later refined by Gillet et al. (1983,1985). It assumes that the shock is far from the photosphere. Thus, we can see emission from the far side of the star (the receding part of the shock) giving the redshifted component while the advancing part of the shock produces the blueshifted component. This geometrical interpretation shows that the central absorption could be only a lack of emission. The ratio of the thickness of the de-excitation zone behind the shock front to the photospheric radius is very much smaller than 10^{-4}. The profile resulting from this narrow and radially moving shell has been investigated by Wagenblast et al. (1983) and Bertout and Magnan (1987). Due to the photospheric occultation, the width of the redshifted contribution may be smaller than the blueshifted one as observed. The larger column density for photons coming from the receding part of the shock explains the weaker intensity of the redshifted component. As shown by Wagenblast et al. (1983), the absorption dip is not necessarily at the Hα laboratory wavelength.

In the last model, the central absorption cannot be below the continuum except if the photospheric absorption is visible, but this does not seem to be the case in Mira stars. All double emissions previously reported have their absorption dip above the continuum (see Gillet, 1987). Figure 2 presents two recent observations which prove that the absorption dip can be below the continuum. As phase advances, the intensity of the central absorption decreases and becomes higher than the continuum level (fig. 3). All these observations have been done with the CES at ESO La Silla (R = 60,000 and S/N ~ 200-400). Since T Cen is a very hot (up to K0 at luminosity maximum) M-type Mira star of population I, its true continuum level is close to a relative flux of unity. This is not the case for S Car (fig. 1b) which was observed at phase 0.68. Indeed, with a spectral type close to M5, molecular absorptions are very concentrated and the true continuum level may be around 1.1-1.2 in relative flux (Piccirillo et al., 1981).

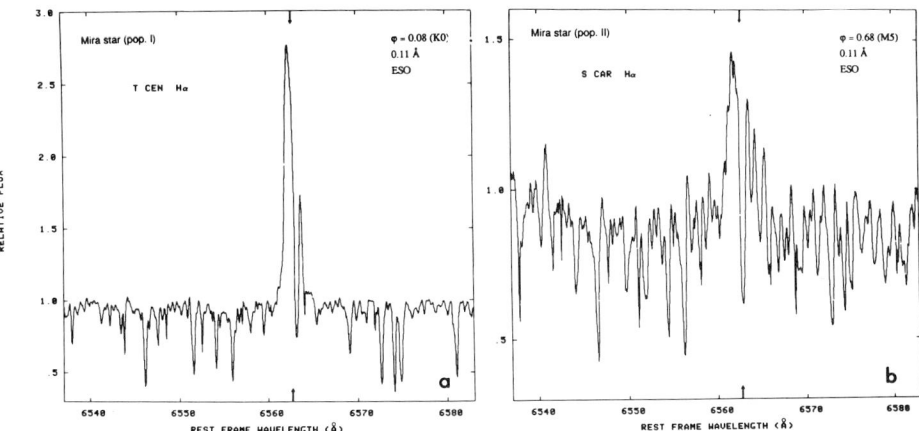

Fig. 2 Double emission structure of two Mira stars. The striking feature is that the central absorption dip is under the stellar continuum level, in contrast to previous observations (Gillet, 1987).

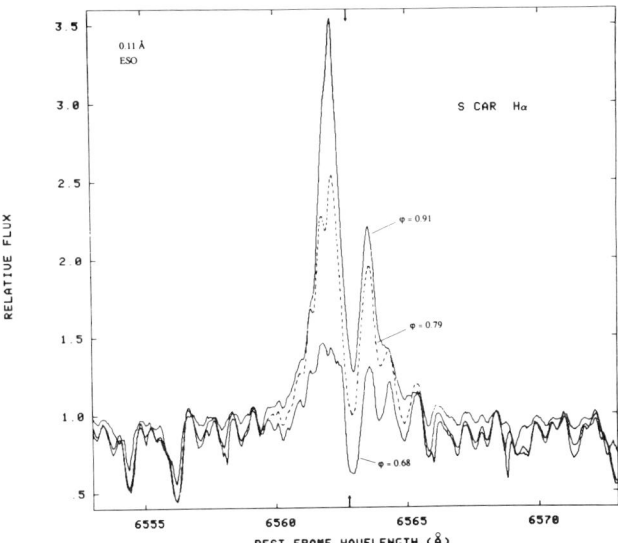

Fig. 3 Evolution in phase of the double Hα emission of S Car. At $\varphi = 0.91$ (luminosity maximum $\varphi \equiv 0.0$), the central absorption dip is above the continuum, in contrast to that at phase 0.68.

4. CONCLUSION

The high resolution Hα-profiles of the Mira stars T Cen and S Car show that the central absorption can be below the stellar continuum. Since the Hα photospheric absorption is not observed in these kinds of stars, this is not in favour of the geometric model. Thus, the central absorption dip could be explained by the self-absorption of the cool hydrogen above the shock wave.

In RR Lyrae and RV Tauri stars the contribution of the photospheric absorption to the double emission profile is certainly significant. The same conclusion applies to Classical Cepheids and perhaps also to W Virginis stars but further observations would be useful. The case of Mira stars seems different because the evaluation of the photospheric contribution, which is never observed, is difficult to estimate. It would be also useful to consider if the hydrogen self-absorption can contribute to double emission profiles in RR Lyrae, RV Tauri, W Virginis and Classical Cepheid variables.

Acknowledgements : The author is particularly grateful to Drs. D. Baade, R. Crowe, and P. Magain for their contribution to observations presented in this paper and to Dr. R. Crowe who supervised the english.

REFERENCES

Abt, H.A. : 1954, Astrophys. J. Suppl. **1**, 63
Bertout, C., Magnan, C. : 1987, Astron. Astrophys. in press
Bidelman, W.P., Ratcliffe, A.E. : 1954, Publ. Astron. Soc. Pacific **66**, 255
Cox, J.P. : 1974, Rep. Prog. Phys. **37**, 563
Gillet, D., Maurice, E., Baade, D. : 1983, Astron. Astrophys. **128**, 384
Gillet, D., Maurice, E., Bouchet, P., Ferlet, R. : 1985, Astron. Astrophys. **148**, 155
Gillet, D. : 1987, Astron. Astrophys. in press
Hänni, L. : 1987, private communication
Hutchinson, J.L., Hill, S.J., Lillie, C.F. : 1975, Astron. J. **80**, 1044
Joy, A.H. : 1947, Astrophys. J. **106**, 288
Piccirillo, J., Bernat, A.P., Johnson, H.R. : 1981, Astrophys. J. **246**, 246
Preston, G.W. : 1962, Astrophys. J. **136**, 866
Preston, G.W. : 1964, Ann. Rev. Astron. Astrophys. 2, 23
Wagenblast, R., Bertout, C., Bastian, U. : 1983, Astron. Astrophys. **120**, 6
Wallerstein, G. : 1959, Astrophys. J. **130**, 560
Wallerstein, G. : 1972, Publ. Astron. Soc. Pacific **84**, 656
Wallersein, G. Cox, A.N. : 1984, Publ. Astron. Soc. Pacific **96**, 677
Willson, L.A. : 1976, Astrophys. J. **205**, 172

SOME CONSTRAINTS ON CHROMOSPHERIC MODELLING FOR SOLAR-TYPE STARS WITH HIGH S:N SPECTRA

B.H. Foing[1], F. Castelli[2], G. Vladilo[2], J. Beckman[3]
1. LPSD/IAS Verrièrres, France.
2. Osservatorio Astronomico di Trieste, Trieste, Italy.
3. Instituto de Astrofisica de Canarias, Tenerife, Spain.

ABSTRACT. High signal to noise spectra are required in chromospheric modelling because chromospheric emission lines are formed in a boundary layer under conditions of NLTE and in non-hydrostatic equilibrium, as well as in multiple magnetically-controlled streams, with horizontal structure on several scales, and vertical velocity fields. To obtain useable estimates of energy dissipation with height we have obtained sequences of spectra from F8 to K5 and for stars of different activity levels. These comprise CaII H and K, the IR triplet, and MgII h and k obtained with the ESO CAT+CES or IUE. We outline the constraints such observations place on models and indicate theoretical and observational difficulties.

1. SEMI-EMPIRICAL MODELS

Solar physicists have produced semi-empirical model chromospheres (Vernazza et al., 1981) either from EUV continua, obtained using Skylab, or by modelling strong lines, CaII H and K from the ground, or MgII h and k and Lyα from space (cf. OSO8, Lemaire et al., 1981). The model of Vernazza et al (1981) is given in Fig. 1 showing heights of formation of features in a plane parallel, horizontally homogeneous model, with hydrostatic equilibrium and uniform microturbulence. Such models allow one to calculate net radiative losses in key lines: Ly α, H α, CaII and MgII as functions of height, and show, <u>grosso modo</u>, which lines dominate the energy balance at each height.

2. INHOMOGENEITIES

Ca spectroheliograms, or EUV images, as in Fig. 2, show the inhomogeneity of the solar chromosphere on large scales for the plages or the network, and on small scales for the properties of elementary flux tubes. Multicomponent models are needed, of the quiet sun in the almost field-free supergranular cells, as well as of the network and of plages. There is growing evidence that inside flux tubes the temperature minimum is higher, and the chief cooling agent is H^-, whereas outside flux tubes the temperature minimum is lower, and the chief coolant is CO, which dissociates much above 4000 K.

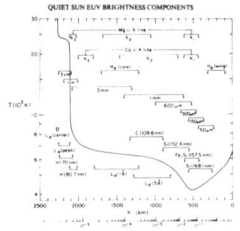
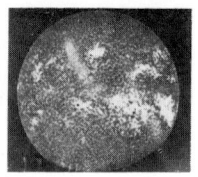

Fig1: Schematic one-stream model of temperature v. height in the chromosphere (Vernazza et al., 1981) indicating spectral features used to probe heights.

Fig. 2: EUV spectroheliogram, shown here to illustrate plages (white areas) network (light mottling) and cell (dark mottling) components, and the need for multi-stream models.

If such considerations are partially understood on the sun, it is correspondingly more difficult on stars without angular resolution. Among the resolved solar phenomena anticipated in other chromospheres are systematic downflows in the network, and time varying responses to shocks and oscillations which can introduce profile asymmetries. Network downflows have now been observed in MgII lines from dwarfs and giants (Vladilo et al., 1987).

3. OBSERVATIONS.

These have been of CaII H and K, and IR triplet lines at $\lambda/\Delta\lambda \sim 10^5$ with S:N >200 for bright solar-like quiescent and active dwarfs, using the ESO CAT+CES combination and MgII h and k with $\lambda/\Delta\lambda \sim 2\times10$ using IUE for some of these objects. For IUE a set of two-dimensionally extracted (Franco et al., 1984) absolutely calibrated spectra of stars strongly expected to be free of Local Interstellar Absorption were co-added to produce spectra of exceptionally high (for IUE) S:N, up to 100 in the continuum and 30 in the h1 and k1 minima. An example is shown in Fig. 3 for the quiescent G5 dwarf δ Pavonis. All spectra have been flux-calibrated, to compare with height-dependent predictions. In the case of the Ca H line, an extensive modelling programme has helped us to establish fluxes to ±20% in a heavily line-blanketed part of the spectrum.

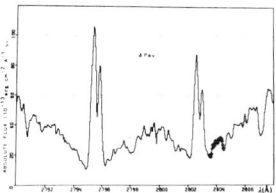

Fig. 3: MgII h and k emission from quiescent G5V star δ Pavonis. In this one of the highest S:N spectra from IUE yet published (S:N ≃ 30 at h1 and k1, S:N ≃ 100 in continuum) red-shifted h3 and k3 are clearly seen.

4. ACTIVE AND QUIET CHROMOSPHERES

With the aim of separating "active" spectra from the spectra of the surrounding cooler medium, we compare and subtract normalized spectra of quiescent stars from active stars of the same luminosity and spectral type. In Fig. 4 we show an example of such a comparison for Hα and the CaII triplet λ8542 line for the active star ε Eri (K2V) compared with the quiescent object O_2 Eri (K1V). The differential spectra are indicators of the incremental emission of an "active" stream. Careful comparison with solar analogs will help yield an "atlas" set" of ideal active and quiet profiles which can be modelled independently using monotonically varying models, and the classical tools of radiative line transfer.

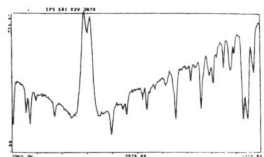 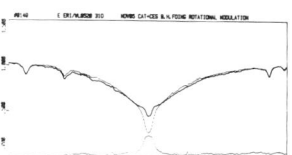

Fig. 4a: CaII H emission from the chromosphere of the active dwarf ε Eri (using CAT+CES). Note strong emission and red-shifted h3 (downflow in network)

Fig 4b: CaII IR triplet line at λ 8542 A for ε Eri (K2V) compared with same feature in quiescent star O Eri of similar spectral type (K1V). The different spectrum illustrates the net chromospheric flux from ε Eri.

5. VELOCITY FIELDS

The comparison of the MgII h and k profiles of the "quiescent" star δ Pav in Fig. 3, and the CaII H profile of the active star ε Eri in Fig. 4 shows the common feature of red-shifted h3, k3 and H3. The presumption is that these features all sample downflow and hence all originate in the network. High S:N spectra, coupled with good absolute velocity calibration can supply true vertical velocity fields with reference to the photospheric rest frame and hence give dynamically valid model parameters (see Vladilo et al., 1987).

6. SYNTHESIS

Although we are still some way from chromospheric models which are anything other than schematic, high S:N, high resolution line spectra are beginning to show us when, and in what admixture, classical (NLTE, PRD) line transfer can be used to probe chromospheres and to what extent further refinements in these techniques are of no value when confronting real chromospheres because of boundary value constraints.

References
Beckman, J.E., Crivellari, L., Foing, B.: 1984, ESO Messenger **38**, 24.
Franco, M.L. et al.: 1984, Astron. Astrophys. Suppl. Ser. **58**, 693.
Lemaire, P., Gouttebroze, P., Vial, J.E., Artener, G.E.: 1981, A & A **103**, 160.
Vernazza, E., Avrett, E.H., Loeser, R.: 1981, Ap. J. Suppl. **45**, 635.
Vladilo, G. et al.: 1987, Astron. Astrophys (In Press).

A METHOD FOR CALIBRATING, IN ABSOLUTE FLUX UNITS, CA II H PROFILES OF
LATE TYPE STARS OBSERVED AT ESO

F. Castelli(1), P. Gouttebroze(2), J. Beckman(3),
L. Crivellari(1), B. Foing(2)

(1) Osservatorio Astronomico di Trieste
    Via Tiepolo 11-I34131 Trieste-Italy
(2) Laboratoire de Physique Stellaire et Planetaire
    BP 10,F-91376 Verriers-le-Buisson-France
(3) Instituto de Astrophysica de Canarias
    La Laguna-Tenerife-Spain

ABSTRACT. In this paper we have applied to the Sun a method for calibrating, in absolute flux units, Ca II H profiles of late-type stars. After comparing, in the region 3948-3882 A, an LTE synthetic spectrum with the data of the solar flux Atlas by Kurucz et al.(1984), we have defined the wavelength ranges where observations agree with computations, based on specific radiative equilibrium models and collisional broadening parameters. By fitting in these regions the spectrum of the moon observed at ESO with the corresponding synthetic spectrum, we derived a calibration factor that enables us to calibrate, in absolute flux units, the whole observed range.

1. INTRODUCTION

In this paper we present the method that we will adopt for calibrating, in absolute flux units (10^6 erg sec^{-1} cm^{-2} A^{-1}), a set of Ca II H profiles of stars of spectral type F2-K5 observed at ESO with high resolution. The stars are listed in Table I; the observations and the data reduction are described in Crivellari et al.(1987).
    For late-type stars it is practically impossible to define a true continuum level in the region below 4000 A; therefore, only profiles that are absolutely calibrated can be compared with each other and with the computed ones, and can be used for deriving chromospheric radiative losses.
    The problem of the calibration of the Ca II H region observed at ESO is by no means trivial, because the most widely used methods require either observations in the whole range 3925-3975 A (Linsky et al.,1979) or observations at 3950 A (Duncan,1981; Catalano,1979), where we suppose there is a pseudo-continuum level. The useful range of most of the observations in Table I is 3954-3985 A.
    We therefore decided to apply the calibration procedure proposed by

Ayres et al.(1976). The observed wing profiles are calibrated with the absolute flux wing profiles computed with a radiative equilibrium (RE) model photosphere. This method is independent of the observed range and can at the same time give an estimate of the chromospheric radiative losses in the core of the line.

Because the procedure is model dependent, before applying it to the observed stars we tested it on solar data.

Table I: The observed stars.

HR	HD				Teff(K)	logg	N(H)	
77	1581	ζ Tuc	F9	V	5832	4.82	1	Q
88	1835	δ Cet	G2	V	5814	4.59	2	A
98	2151	β Hyd	G2	IV	5747	4.45	3	Q
509	10700	τ Cet	G8	V	4975	4.52	2	Q
591	12311	α Hyd	F0	V			1	Q
1084	22049	ε Eri	K2	V	4998	4.80	4	A
5459	128620	α Cen A	G2	V	5770	4.50	6	Q
5460	128621	α Cen B	K1	V	5300	4.54	7	Q
5544	131156	ζ Boo A	G8	V			4	A
5568	131977		K4	V			1	A
5897	141891	β Tr A	F2	III			1	Q
6094	147513		G2	V			3	A
6098	147584	ζ Tra	F9	V	6054	4.66	2	A
6102	147675	γ Aps	G8/K0	III	5050	3.10	1	A
6752	165341	70 Oph A	K0	V			4	A
7665	190248	δ Pav	G6/8	IV	5563	3.81	4	Q
7703	191408		K3	V	4903	4.54	2	Q
7776	193495	β Cap	F8+A0	V	4876	4.54	1	Q
8387	209100	ε Ind	K4/K5	V	4590	4.57	6	A
		Moon	G2	V	4770	4.44	2	

Q or A indicates whether the star is quiescent or active. N(H) indicates how many observations of the CaII H profile have been made for each star.

2. THE DATA

The profile to be calibrated is the average of two observations of the solar light reflected by the moon. The spectra were obtained with CES plus the 1.4 m CAT of ESO. The resolution is $\lambda/\Delta\lambda = 8\ 10^4$. The two observations have been performed on 1/02/1985 and have an exposure time of 300 sec and 600 sec respectively; the useful range is 3954.592-3985.175 A.

3. THE CALIBRATION METHOD

Ayres et al.(1976) calibrated the observed K lines of αCen A and αCen B by fitting the far wing profiles ($\Delta\lambda > 5A$) with computed profiles based on radiative equilibrium (RE) models. The wing intensities for a particular RE model were calculated on the basis of an LTE partial coherent scattering formalism which considers a five-level representation of Ca II (Ayres,1975).

With ATLAS8 code (Kurucz,1986) we computed a RE line-blanketed solar model with parametrs Te=5770 K, log g=4.44, opacity distribution functions with 2 Km/s and mixing length to scale height ratio l/H=1. We want to stress that we have not considered specific empirical or semi-empirical solar models, but rather a theoretical model, because the calibration method should be generalized to the stars observed.

With the RE model we computed Ca II H profiles both in NLTE with the hypotheses of partial (PR) and complete redistribution (CR) and in LTE. We found that with the RE model adopted, without any increase in temperature in the upper layers, the Ca II H profiles computed in PR, CR and LTE do not show any remarkable difference. We conclude that with RE models we can use LTE to compute the photospheric CA II H profile that will be used to calibrate the observations.

4. TO WHAT EXTENT THE COMPUTED PROFILE IS RELIABLE

To test if the so-computed profile well represents the observed flux, we compared, in the region 3948-3982 A, the computed spectrum with the absolutely calibrated solar spectrum derived from the Solar Flux Atlas by Kurucz et al.(1984). The synthetic spectrum was computed only with the most important lines, namely Ca II K, Al I 3961.52, Ca II H and $H\epsilon$. The result is that we cannot reproduce the whole extent of the Ca II H wings with the same Van der Waals parameter. Figure 1 and Figure 2 show the comparison between the profile of the Solar Atlas and the profiles computed with:

$$\gamma vw = 1.7 \ 10^{-8} \ NH(T/5000)^{0.3} \quad \text{(Ayres,1975)}$$

and

$$\gamma vw = 1.45 \ 10^{-8} \ [NH+0.42NHe+0.85H2](T/10000)^{0.3}$$
(Kurucz and Avrett,1981)

where $1.45 \ 10^{-8}$ is the Van der Waals parameter for pure hydrogen γ_H/NH computed with the tables of Deriddier and Van Rensbergen (1981). The radiative damping and the Stark broadening parameters for Ca II lines are $\gamma rad=1.5 \ 10^8 sec^{-1}$ and $\gamma s=3 \ 10^{-6} Ne$ respectively (Shine and Linsky,1974). We adopted the Ca abundance $\log\epsilon=-5.67$ (Lambert and Warner,1968) and a microturbulence $\xi=1Km/s$. The use of the Bell et al.(1976) solar model gives results analogous to those of Figure 1 and Figure 2. The conclusion is that to obtain a reliable absolute calibration with the theoretical RE models available we should fit the

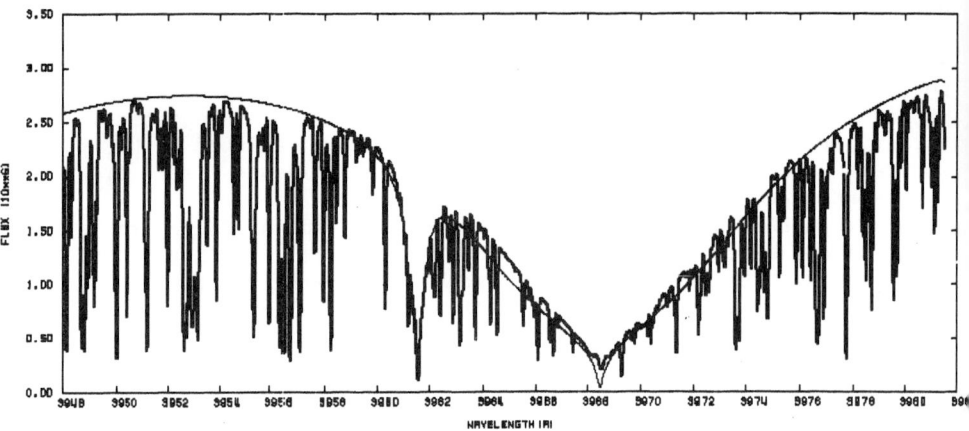

Figure 1. Comparison between the observed Ca II H profile from Kurucz et al.(1984) Solar Flux Atlas (thick line) and the profile computed with a RE Kurucz's model and γ vw=1.7 10^{-8}NH(T/5000)$^{0.3}$ The units of the ordinates are: 10^6 erg sec^{-1}cm^{-2}Å^{-1}

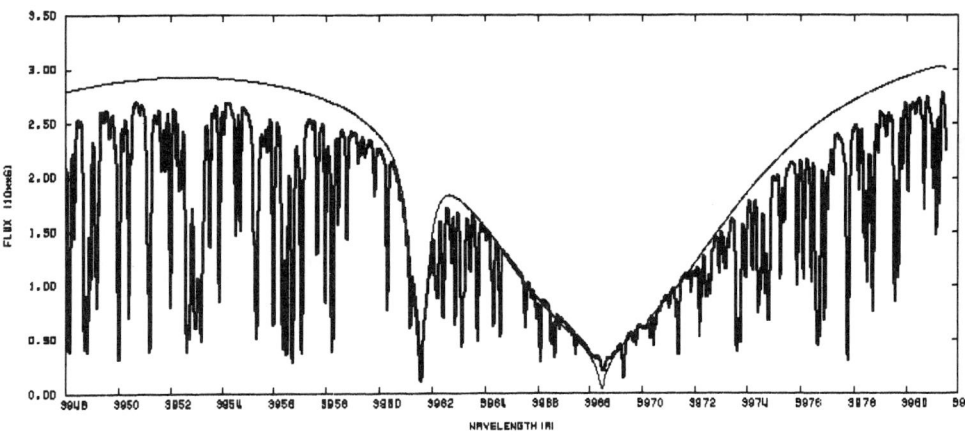

Figure 2. The same as Figure 1, but with
γ vw=1.456^{-8}[NH+0.42NHe+0.85NH2](T/10000)$^{0.3}$

observed to the calculated spectrum only in appropriate wavelength ranges and with appropriate γ_{vw} parameters. Table II lists the useful wavelengths ranges and Van der Waals parameters for calibration purposes to be used with Kurucz's models (K models) or with Bell et al. models (BG models).

Table II: Useful wavelength ranges and Van der Waals parameters for calibration purposes.

Model	Van der Waals parameter γ_{vw}	wavelength ranges	
K	$1.70\ 10^{-8} NH(T/5000)^{0.3}$	3948-3958	3976-3982
K	$1.45\ 10^{-8}[NH+0.42NHe+0.85NH_2](T/10000)^{0.3}$	3962-3967	3970-3974
BG	$1.45\ 10^{-8}[NH+0.42NHe+0.85NH_2](T/10000)^{0.3}$	3948-3958	3975-3982
BG	$1.00\ 10^{-8}[NH+0.42NHe+0.85NH_2](T/10000)^{0.3}$	3962-3967	3970-3973

---

5. THE ABSOLUTE CALIBRATION OF THE MOON SPECTRUM

To calibrate the ESO spectrum of the moon we have to fit the observed spectrum with the computed one and the closer the computed spectrum is to the observed one, the better the fit will be. Therefore, we computed a synthetic spectrum in the region 3948-3982 A with all the lines with residual flux in the line center $F_\lambda/Fc \leq 0.99$ predicted by either K or BG models. According to the model adopted and Van der Waals parameter chosen, we made the fit in the wavelength regions given in Table II.

If $Yc(\lambda)$ and $Yo(\lambda)$ are the calculated and the observed flux points and if $Yc(\lambda)=C\ Yo(\lambda)$, with C a multiplicative factor, the fitting procedure consists in deriving that C which minimizes the quantity

$$\chi^2 = \sum_{i=1}^{N}[Yc(\lambda)-CYo(\lambda)]^2$$

where N is the total number of points. We will call C the calibration factor.

As an example, if we adopt the K model and the Van der Waals parameter

$$\gamma_{vw}=1.45\ 10^{-8}[NH+0.42NHe+0.85NH_2](T/10000)^{0.3}$$

the fit has to be made in the regions 3962-3967 A, and 3970-3974 A, according to Table II. In this case the resulting calibration factor is $C=11.578\ 10^6$.

We obtain the whole Ca II H profile of the moon spectrum calibrated in absolute flux units by multiplying all the observed points by C.

Figure 3 compares the calibrated moon spectrum with the spectrum of the Solar Flux Atlas(Kurucz et al.,1984). Both spectra are in absolute flux units. The agreement is quite satisfactory. We obtained similar results for the other cases of Table II.

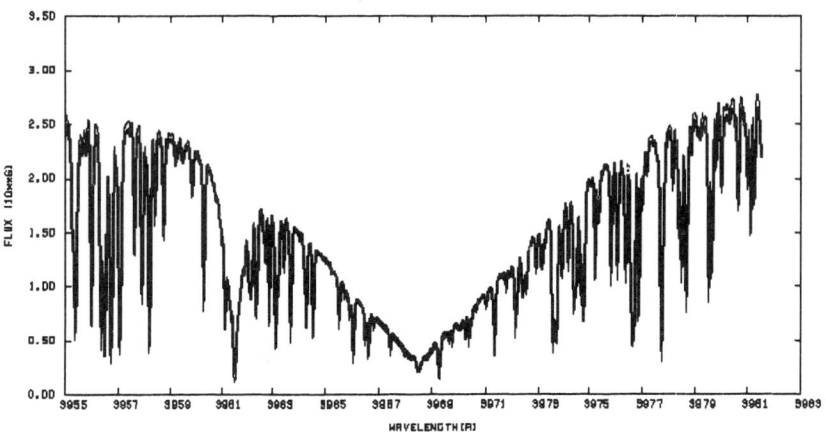

Figure 3. Comparison of the calibrated moon spectrum (thick line) with the profile in absolute flux units of the Solar Flux Atlas by Kurucz et al. (1984) (thin line). The units of the ordinates are 10^6 erg sec^{-1} cm^{-2} A^{-1}.

Table III: Solar H1 indices obtained from different sources.

	$\mathcal{F}$(H1)	$\mathcal{F}'$(H1)	
1. Moon	$3.9\ 10^5$	$2.0\ 10^5$	This paper
2. Moon	$3.7\ 10^5$	$2.1\ 10^5$	Linsky et al.(1979)
3. Mean Sun	$3.8\ 10^5$	$2.2\ 10^5$	Linsky et al.(1979)
4. Sky	$4.6\ 10^5$	$3.0\ 10^5$	Linsky et al.(1979)
5. Sun	$3.8\ 10^5$	$1.9\ 10^5$	Beckers et al.(1976) Atlas
6. Sun	$4.6\ 10^5$	$2.8\ 10^5$	Kurucz et al. (1984) Atlas

The values in rows 5 and 6 have been measured directly from the Atlases.

6. THE CHROMOSPHERIC RADIATIVE LOSSES

Table III compares the solar H1 indices obtained from our calibrated data with those of other sources in the literature. For the definitions of $\mathcal{J}$ (H1) and $\mathcal{J}'$ (H1) see Linsky et al.(1979).

Our value agrees with those of Linsky et al.(1979) and with that derived from the Beckers et al.(1976) Solar Atlas. The values from the Kurucz et al.(1984) Solar Atlas and from the sky data of Linsky et al.(1979) are higher.

7. CONCLUSIONS

From the comparison of data already calibrated in absolute flux units (Kurucz et al. Solar Flux Atlas, 1984) with an LTE synthetic spectrum in the region 3948-3982 A we have seen that the adopted RE theoretical models for the photosphere are inadequate to reproduce the wings of the profile in the whole region. Nevertheless, in some ranges the computed flux matches the observed one, but, for a given Ca abundance, the regions differ with a different choice of the Van der Waals parameter. We can obtain a satisfactory calibration of the ESO data if we fit the observed spectrum to the computed one in the wavelength regions appropriate to the adopted model.

For stars different from the Sun, the major source of uncertainty for the proposed method would be, in our opinion, the values of the elemental abundances relative to the model parameters, Te, log g and microturbulence. In fact the existing quantitative analyses for the stars of Table I are mostly limited to curve of growth methods performed several years ago. For some stars, abundances for only a few elements are available. For a worthwhile re-analysis of the photospheres of these stars, which is a necessary premise for a better understanding of the chromospheric properties, we need both obervations in large visual ranges with high resolution and a high S/N ratio, and more realistic blanketed models of cool stars, for instance with increased opacities.

In the meantime, we estimate that our analysis is the best we can do with the RE model method and with the available models and data. We have seen that it is valuable for the Sun. As a next step, we shall apply it to the stars of Table I.

References

Ayres,T.R.:1975,Astrophys.J. 201, 799.
Ayres,T.R.,Linsky,J.L.,Rodgers,A.W.,and Kurucz,R.L.:
    1976,Astrophys.J. 210, 199.
Beckers,J.M.,Bridges,C.A.,and Gilliam,L.B.:1976,
    'A High Resolution Spectral Atlas of the Solar
    Irradiance From 380 to 700 Nanometers',
    Sacramento Peak Obs. Project No. 7649.
Bell,R.A.,Eriksson,K.,Gustafsson,B,and Nordlund,A.:
    1976, Astron. Astrophys. Suppl. 23,37.

Catalano,S.:1979,*Astron.Astrophys.* **80**, 317.
Crivellari,L.,Beckman,J.E.,Foing,B.H.,and Vladilo,G.: 1987,*Astron.Astrophys.* **174**, 127.
Deridder,G.,and Van Rensbergen,W.:1976,*Astron.Astrophys. Suppl.* **23**, 147.
Duncan,D.K.:1981,*Astrophys.J.* **248**, 651.
Lambert,D.L., and Warner,B.: 1968,*Mon.Not.R.astr.Soc.* **140**, 197.
Linsky,J.L.,Worden,S.P.,McClintock,W.,and Robertson,R.M.: 1979,*Astrophys. J. Suppl.* **41**, 47.
Kurucz,R.L.:1986,private comunication.
Kurucz,R.L.,and Avrett,E.: 1981, *SAO Sp.Rep.* 391.
Kurucz,R.L.,Furenlid,I.,Brault,J.,Testerman,L.: 1984,'Solar Flux Atlas from 296 to 1300 nm', National Solar Observatory, Sunspot, New Mexico.
Shine,R.A.,and Linsky,J.L. 1974, *Solar Physics* **39**,49.

DISCUSSION

SNEDEN   Did you use more than a few lines in your final synthetic spectra ?

CASTELLI   Yes in my final synthetic spectrum I used about 380 lines for the region 3948-3982Å. These are all the lines with a residual flux in the line center $F\lambda/Fc \leq 0.99$ predicted by the model.

SNEDEN   How would you propose to add more lines (not yet identified) into your calculations?

CASTELLI   In the CaII region there are only few unidentified lines. To add them in my calculations I have to wait that someone identifies them and provides the relative gf values.

GUSTAFSSON   We cannot feel quite sure that the completeness of recent published line-lists is sufficient for this purpose. Which line-list did you use?

CASTELLI   I used an extended and corrected version of the Kurucz-Peytreman line list. I continually update this list with the $\log gf$ values from the literature.

BAADE   I believed I noted that, in all spectra you showed, the extreme blue wings (:blue ends of the spectra) were more depressed than the red wings are. CES spectra are known to suffer some vignetting at their blue end. Are you sure that this defect has been well compensated by your rectification procedure?

LINSKY   Could you comment on how the scattered light at the CaII H and K lines is determined and substracted from the CES spectra? This is an important point for determining the absolute flux of the emission cores of the CaII lines because these features lie in the cores of very deep photospheric absorption lines.

CRIVELLARI   Effects due to scattered light and vignettings are certainly present in CES spectra. As everybody, we are well aware of the problem and did our own measurements to estimate the percentage of the contamination of the signal. However we wish to stress that:

i- In our opinion the main source of contamination is the remanence effect in the RETICON (see Crivalleri and Foing, this proceedings).

ii- We think that, at this stage of the calibration work, the actual correction of the observed profiles is quite accurate for our purposes. One should also note that the comparison in the present paper is between a computed synthetized profile and the average spectrum of only two high quality observations of the solar light reflected by the moon (high S/N, low parasitic effect).

SPECTROSCOPY OF COOL STARS FROM *IUE* DATA

P. G. Judge
Department of Theoretical Physics,
University of Oxford,
1 Keble Road,
Oxford OX1 3NP,
England.

ABSTRACT.
After nine years of operation the *IUE* satellite continues to provide valuable spectra of cool stars from 1200 to 3100 Å. The impact of these spectra has been greatest in studies of the outer regions of the atmospheres, above the photospheres, allowing the general properties of stellar chromospheres, transition regions and winds to be established. After outlining these properties, I focus on studies based on high signal-to-noise echelle spectra ($\lambda/\Delta\lambda$ ~1.2×10^4) of single stars, showing how high quality emission line profiles have been used to derive constraints on the outer atmospheric structure, which in turn have been used to examine models of heating and mass loss.

1. INTRODUCTION

Prior to the launch of the *International Ultraviolet Explorer* (*IUE*) satellite in 1978, only a few observations of cool stars had been made in the ultraviolet (UV) region below 3000 Å (see e.g. the review by Linsky, 1980). Since then, *IUE* has revolutionized our knowledge of the outer atmospheres of cool stars, largely through the study of spectral lines, and now we possess a quite detailed knowledge of stellar chromospheres, transition regions (TR's), coronae and winds in the cool half of the HR diagram.

The aim of the present review is to highlight advances in this field using *IUE* data with high signal-to-noise (S/N) ratios. Despite the relatively low quality of *IUE* data compared with modern ground-based spectroscopy (the maximum count/pixel with *IUE* is 255), data are of *sufficient* quality in the essentially unexplored and important UV spectral region that great strides have been made in studies of cool stars and other fields of astronomy (see the remarkable book edited by Kondo *et al.* (1987)).

2. AN OVERVIEW OF COOL STAR SPECTRA IN THE ULTRAVIOLET

I give here a brief description of the general characteristics of the spectra of cool stars across the HR diagram. More comprehensive discussions can be found in the reviews of Jordan & Linsky (1987) and Dupree (1986). Figure 1 (from Dupree, 1986) shows spectra of three stars representing the different basic types of outer atmospheres identified in part from *IUE* observations.

(i) "coronal" stars include all observed dwarfs cooler than ~ F0 and lower gravity stars as late as K0 III. The UV spectra appear similar to those of regions of various levels of activity in the Sun showing both chromospheric and TR emission lines (up to temperatures of T_e ~2×10^5 K from N V $\lambda 1240$). X-ray observations with the *EINSTEIN* satellite (Ayres *et al.*, 1981) revealed the presence of coronal emission at the levels expected when scaled from a Solar-like outer atmosphere.

(ii) "non-coronal" stars are cooler and more luminous than spectral type K0 III. *IUE* spectra revealed relatively strong chromospheric emission lines (especially of neutral species), an absence of detectable plasma above T_e ~ 2×10^4 K, and asymmetric profiles with enhanced red wings indicating significant velocity gradients, probably associated with a massive wind, in optically thick chromospheric lines (e.g. Mg II k). No X-rays have yet been detected from *single* non-coronal stars: upper limits of surface fluxes are much lower than in "coronal" stars.

(iii) "hybrid" stars are mostly luminosity class II stars which lie in the "non-coronal" region of the HR diagram. *IUE* spectra show evidence both for 10^5 K plasma *and* massive winds of higher terminal velocity than their non-coronal counterparts.

The regions in the HR diagram where the different atmospheric types are found are shown in Figure 2 (from Mullan & Stencel, 1982).

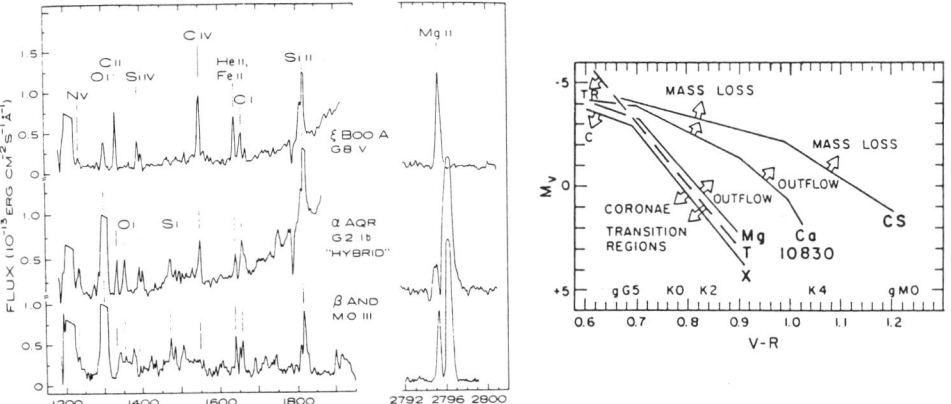

Figure 1: Spectra of 3 cool stars illustrating the characteristic features of "coronal", "hybrid" and "non-coronal" stars. (From Dupree 1986).

Figure 2: An HR diagram showing the regions where different types of outer atmospheric structure are found according to various spectral signatures. "Coronal" stars lie to the left of the lines markes "T" and "X" (from Mullan & Stencel, 1982).

3. EXAMPLES OF "HIGH" S/N STUDIES OF THE OUTER ATMOSPHERES OF COOL STARS.

3.1 "Coronal" Stars

Although we can study the Sun's outer atmosphere in much more detail than any other coronal star, spectra of large numbers of such stars have been obtained with *IUE* which have established important correlations of e.g. "activity" with fundamental stellar parameters such as rotation and age which are crucial to the understanding of the mechanisms responsible for heating chromospheres and coronae (e.g. Jordan & Linsky, 1987). In addition, detailed studies of high-S/N spectra of the brightest individual stars have, by various semi-empirical techniques applied earlier to the Sun, yielded models for the upper chromospheres and TRs of stars of widely differing gravities. Jordan et al. (1987), for example, applied the technique of emission measure analysis (based on the integrated fluxes of emission lines) with additional constraints from observed line widths and ratios to derive models for 5 G-K dwarf stars. These models span conditions of different regions found in the Sun, and Jordan et al. concluded that the processes determining the structure of the outer atmosphere are basically the same as for the Sun, with the same associated problems (Jordan, 1980).

Lower-gravity "coronal" stars have also been studied using similar techniques (e.g. Brown et al., 1984) and using model- atmosphere approaches (e.g. Eriksson et al., 1983). Similar results to the solar case are found but the lower gravities lead to lower pressures, as expected, and larger scale heights. An important discovery was made by Ayres et al. (1983): in high quality spectra of the primary of Capella (G6 III), lines formed at TR temperatures ($\gtrsim 5 \times 10^4$ K) are significantly *red-shifted* (by up to 20 km s^{-1}) relative to the photosphere. Ayres et al. also found similar shifts between TR and chromospheric lines in the 4 other coronal stars examined, and concluded that the shifts probably arise from definite *downflows* analogous to those observed in the solar network boundaries. Such large-scale flows must be accounted for in the energy and momentum requirements of the TR/ chromosphere.

3.2 "Non-Coronal" Stars

Owing to significant differences between UV spectra of "coronal" and "non-coronal" stars, modelling of "non-coronal" atmospheres is at a more basic stage, since solar techniques mentioned above cannot be directly applied. Following early work on rocket spectra by Haisch et al (1977) prior to *IUE*, advances were initially made in line identifications and excitation mechanisms (see the discussion by Jordan & Judge, 1984). This work was done in the light of chromospheric density estimates from a valuable diagnostic (line ratios within the C II] λ2325 multiplet) by Stencel et al. (1981), which (when modified by up-dated atomic data of Lennon et al., 1985) showed that electron densities are typically $\sim 10^9$ cm^{-3} for late K giants, substantially smaller than in the solar chromosphere where $N_e \sim 10^{11}$ cm^{-3} (Vernazza, Avrett & Loeser, 1981). Figure 3 shows the C II] lines observed in α Tau (K5 III) and β Gru (M5 III). Such low chromospheric densities allow radiative processes, such as fluorescence, line-locking, photoionization and line broadening due to multiple scattering to become crucial in forming the observed spectra (Jordan & Judge, 1984; Judge, 1986a). Nevertheless, detailed empirical studies based on emission measures, linewidth measurements, opacity sensitive line ratios and density sensitive lines have been succesfully applied to three stars (α Boo (K2 III), α Tau (K5 III) and β Gru (M5 III)) to derive useful constraints on the chromospheric emitting regions (Judge, 1986a,b).

Detailed profile modelling of even well-understood lines (e.g. Mg II k) under the conditions derived by Judge (1986a,b) is complicated by partial redistribution effects, Doppler diffusion (Basri, 1980) and geometry in spherically expanding atmospheres (Drake & Linsky, 1983; Drake, 1985). Drake (1985) has computed Mg II *k* line profiles, taken these problems into account, and has derived models for the expanding chromosphere of α Boo (Figure 4) which also satisfy radio (f-f) constraints from the VLA, yielding the first realistic mass loss rate for a typical (single) non-coronal giant.

Profile modelling work is currently in progress (Judge, Avrett & Loeser, 1988) using all available data for α Boo in an attempt to unify the various techniques and diagnostics available.

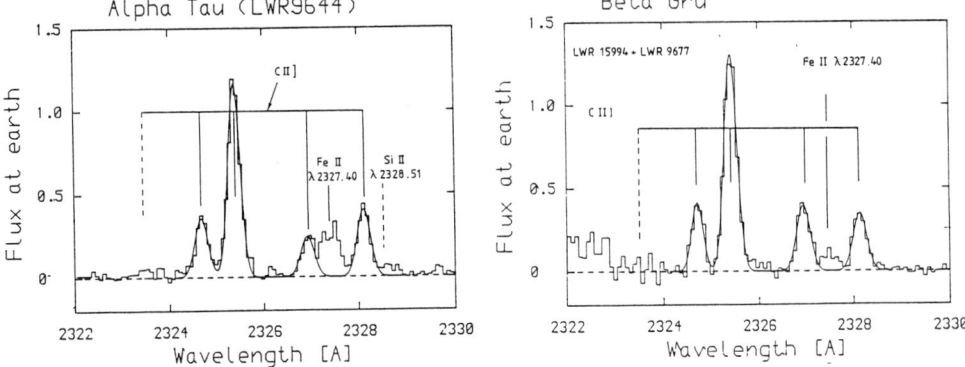

Figure 3: Example of line profiles from IUE echelle spectra. The figure shows observed lines of the density-sensitive C II] multiplet (histograms) and least-square Gaussian fits (solid lines). The fits yield a mean electron densities of 10^9 and 5×10^8 cm^{-3} for the emitting regions of α Tau and β Gru, respectively (From Judge, 1986b).

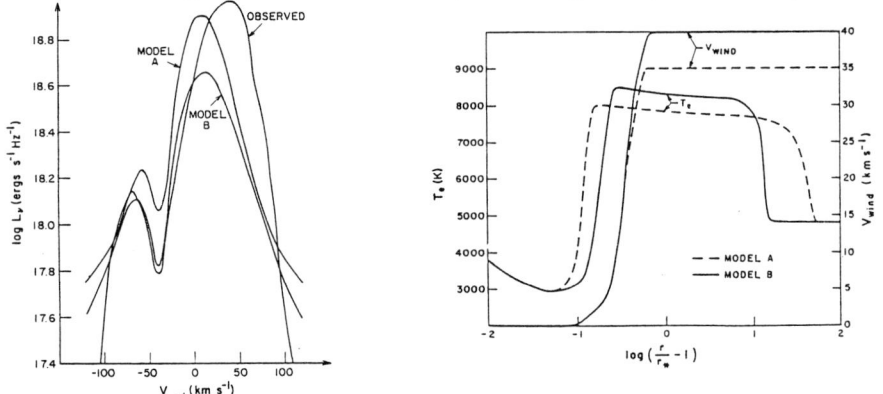

Figure 4: Observed and computed Mg II k line profiles and wind models from Drake (1985). A mass loss rate of 2×10^{-10} M$_\odot$ yr^{-1} was derived from these profiles.

Perhaps the best constraints from *IUE* on the outer atmospheres of non-coronal stars have been obtained by studying *absorption* lines observed when the chromosphere and wind of the primary component of e.g. a ζ Aur system (K– supergiant) eclipses the secondary (a B dwarf). As reviewed by Reimers (1987) radially-dependent chromospheric and wind parameters for the primary have been derived yielding accurate mass loss rates (factor ~3 uncertainties). This method is, however, restricted to a small region of the HR diagram containing suitable systems. Nevertheless valuable constraints on theoretical models have been derived using these techniques.

Comparisons of theoretical and observed line profiles have been made only for the case of an Alfven wave driven wind model of α Ori (M2 Iab) (Hartmann & Avrett, 1984). Although able to account for global properties of the spectra (e.g. the radio (f-f) emission), the model's velocity field differs from that inferred from the

IUE profiles. Work is currently in progress for acoustic- shock driven wind models (Cuntz *et al.*, 1987), another mechanism currently in favour proposed to account for "non-coronal" chromospheres and winds.

3.3 "Hybrid" Stars

Since the (dozen or so) known hybrids have spectral features common to both "coronal" and "non-coronal" stars, techniques applied to these have been extended to the hybrids (e.g. Hartmann *et al.*, 1985). Hartmann *et al.* found that the UV line fluxes are profiles in α TrA (K4 II) could be accounted for in a hydrostatic wave-supported atmosphere which could perhaps extend into an Alfven-wave driven wind. However, Mendoza (1984) finds that the wind models which give the best fit to line fluxes predict differential expansion velocities which are too large at $T_e \leqslant 6 \times 10^4$ K. Also, S/N ratios are lower for the visually fainter hybrids than for the brightest "coronal" or "non-coronal" stars and higher S/N observations are necessary for further progress to be made (see below).

4. FUTURE UV SPECTROSCOPY OF COOL STARS

IUE has provided the first insight into the outer atmospheric structure of many stars in the cool half of the HR diagram. The *IUE* spectra indicate that a similar leap in our understanding will occur when the *HRS* on Hubble Space Telescope (*HST*) is in operation. Some important improvements over *IUE* will be: (i) the resolution of previously unresolved lines formed in the chromospheres, winds and TRs of cool stars (see e.g. the partially resolved profiles in Fig. 3). This will enable much deeper comparisons of theory and observation to be made concerning the heating and momentum deposition in cool star outer atmospheres; (ii) the higher S/N will reveal much higher quality emission line profiles for lines already observed with *IUE*. It will allow the detection of weaker absorption features (e.g. wind components) and emission components (e.g. density- sensitive quadrupole lines of C III], faint emission components in the wind), if present, and the ability to study variations at the level of a few percent.

REFERENCES.

Ayres T.R., Linsky J.L., Vaiana G.S., Golub L. & Rosner R.,1981. Astrophys. J. 250,293.
Ayres T.R., Stencel R.E., Linsky J.L., Simon T., Jordan C., Brown A. & Engvold O., 1983. Astrophys. J. 274, 801.
Basri G.S., 1980. Astrophys. J. 242, 1133.
Brown A., Jordan C., Stencel R.E., Linsky J.L. & Ayres T.R., 1984. Astrophys. J. 283, 731.
Cuntz M., Hartmann L. & Ulmschneider P., 1987. In *"Circumstellar Matter"*, IAU Symp. 122, *325*. (Eds. Appenzeller I. and Jordan C.). Reidel (Dordrecht).
Drake S.A. & Linsky J.L., 1983. Astrophys. J. 273, 299.
Drake S.A., 1985. In *'Progress in Stellar Spectral Line Formation Theory'*,(Eds. Beckman & Crivellari), p.351. Reidel (Dordrecht).
Dupree A.K., 1986. Ann.Rev. Astron. Astrophys. 24, 377.
Eriksson K., Linsky J.L, & Simon T., 1983. Astrophys. J. 272, 665.
Haisch B.M., Linsky J.L., Weinstein A. & Shine R.A., (1977). Astrophys. J. 214, 785.

Hartmann L., Jordan C., Brown A. & Dupree A.K., 1985.   Astrophys. J. **296**, 576.
Hartmann L. & Avrett E.H., 1984. Astrophys. J. **284**, 238.
Jordan C., 1980. Astron. Astrophys. **36**, 355.
Jordan C., Ayres T.R., Brown A., Linsky J.L. & Simon T., 1987.   Mon. Not. R. astr. Soc. **225**, 903.
Jordan C. & Judge P.G., 1984.   Physica Scripta T**8**, 43.
Jordan C. & Linsky J.L., 1987.   In *"Exploring the Universe with the IUE Satellite"*, p. 259. Eds. Kondo *et al.*, Reidel (Dordrecht).
Judge P.G., 1986a.   Mon. Not. R. astr. Soc. **221**, 119.
Judge P.G., 1986b.   Mon. Not. R. astr. Soc. **223**, 239.
Kondo Y. *et al.*, (eds.), 1987. *"Exploring the Universe with the IUE Satellite"*, Reidel (Dordrecht).
Lennon D.J., Dufton P.L., Hibbert A. & Kingston A.E., 1985.   Astrophys. J. **294**, 200.
Linsky, J.L., 1980.   Ann. Rev. Astron. Astrophys. **18**, 439.
Mendoza B.M., 1984.   D. Phil. Thesis, University of Oxford.
Mullan D.J. & Stencel R.E., 1982. In *"Advances in UV Astronomy: Four Years of IUE Research"*, NASA CP-2238, p. 235.
Reimers D., 1987. In *"Circumstellar Matter"*, IAU Symp. **122**, 307. (Eds. Appenzeller I. and Jordan C.).   Reidel (Dordrecht).
Stencel R.E., Linsky J.L, Brown A., Jordan C., Carpenter K.G., Wing R.F. & Czyzak S., 1981. Mon. Not. R. astr. Soc. **196**, 47P.
Vernazza J.E., Avrett E.H. & Loeser R., 1981.   Astrophys. J. Suppl. **45**, 635.

DISCUSSION

CRIVELLARI   It is well known that the Mg II h and k interstellar component is strong enough (even within a few parsecs from the sun) to significantly alter the profile in the emission cores. We have evidence that (Vladilo et al., 1987, in press : Astron. Astrophys.) in specific cases the local interstellar medium (LISM) contamination (the precise wavelength of the LISM component is derived by Crutcher's LISM flow velocity) can reverse the expected V/R asymmetry of the emission core.

JUDGE   I agree that the LISM is an important factor which should be taken into account when examining Mg II profiles. Space Telescope should, with its factor 10 gain in resolution over IUE, enable the "wind" and LISM components to be separated better.

EBBETS   You mentioned several times your anticipation that observations to be made with the Space Telescope High Resolution Spectrograph will make important new contributions to your field. Could you identify what you consider to be the most important HRS performance parameters for your work, and what types and precisions of calibrations will be required.

JUDGE   The most important single improvement over IUE will be the increased resolution ($\simeq$3km/s) which will allow, for the first time, the profiles of optically thin/thick lines to be examined, yielding tight constraints on energy and momentum deposition. Accurate (<0.5 km/s) radial velocities could be very useful. Absolute photometric calibration better than $\simeq$10% would be adequate, but it would be nice to measure variability in line fluxes at a level $\simeq$1%.

SHOCK PHENOMENA IN β CEPHEI STARS

Richard Crowe
Canada-France-Hawaii Telescope Corporation, U.S.A.

Denis Gillet
Observatoire de Haute-Provence, France

ABSTRACT. In this paper, we present new observations of β Cephei stars made with the CFHT coudé Reticon. Spectral sequences of the extreme star BW Vulpeculae show large velocity and line profile changes in Hα and in the C II doublet ($\lambda\lambda 6578, 6582$) over 0.04 of a cycle, during three different cycles on three different nights. In fact, there are two such phases of line doubling where the velocities are nearly discontinuous before and after the so-called velocity "stillstand" (near maximum light). Also, the line doubling phases coincide with pronounced peaks in the Hα core residual intensity. The most straightforward interpretation of these observations is that there are two shock waves, separated by about 50 minutes, which generate double absorption components. For ν Eridani and γ Pegasi, there are no discernible changes in the Hα profiles, despite rather regular velocity variations. This argues against the shock-wave interpretation for "classical" β Cephei stars.

1. INTRODUCTION AND OBSERVATIONS

The importance of rapid and complex line profile variations in the short-period variable β Cephei stars has long been recognized. Recent studies of these objects with new detectors include those by Goldberg, Walker and Odgers (1976) and Young, Furenlid and Snowden (1981) of BW Vulpeculae; by Smith and McCall (1978) and Le Contel and Morel (1982) of γ Pegasi; and by Smith (1983) of ν Eridani. The study reported by Young *et al.* concerned the behaviour of Hα and C II $\lambda\lambda 6578, 6582$, while the other studies concentrated on Si III $\lambda 4567$. The extreme star BW Vulpeculae is of particular interest because it is observed to have the largest variability of light, radial velocity and line profiles among members of the β Cephei class.

It has been suggested that all β Cephei stars have moving shells which arise from atmospheric shocks associated with non-linear radial pulsation (Smith 1983). The original hypothesis was made by Odgers (1955), based on observations of BW Vul. The presence of a shock wave is deduced from the appearance of an extended "shell" feature to

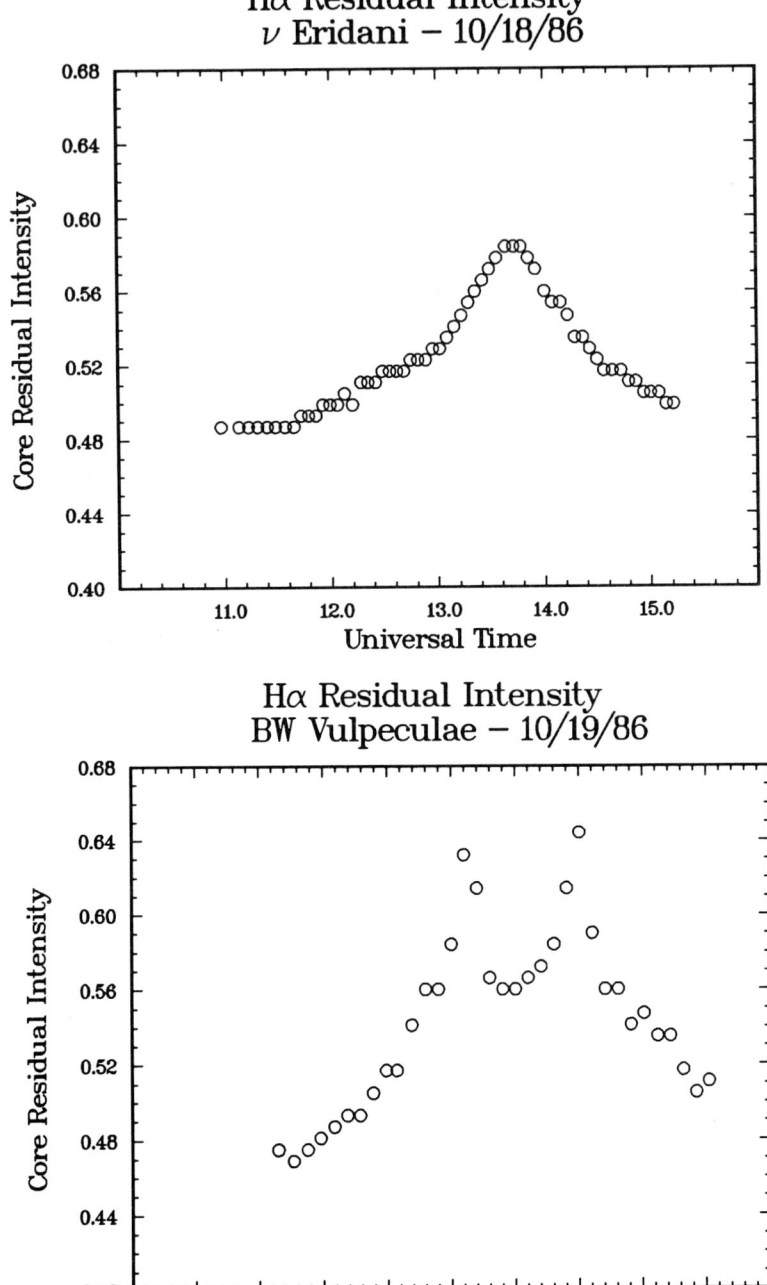

the red or blue of the photospheric profile, which appears and
disappears on a time scale of less than 0.1 of a period. In the
spectra of BW Vul, line doubling and a prominent "stillstand" on the
descending branch of the radial velocity curve can be seen; these
effects are attributed to the movement of an ejected shell above the
deeper photospheric layer. Thus, the velocity stillstand apparently
corresponds to a brief period of rest for the lower photosphere
before the upper layer is once more impulsively ejected. It has been
shown that the line profiles can be generated by strong shock waves
associated with radial pulsation (Campos and Smith 1980). However,
some investigators (*cf.* Young *et al.* 1981; LeContel and Morel 1982)
have expressed concern that an interpretation of the line profile
variations based solely on Doppler displacements of the spectral wing
features is naive, because of radiative transfer effects which should
be taken into account. Nonetheless, radial velocity variations of 50
km sec^{-1} or more, such as are found in BW Vul and ν Eri imply that
shock velocities should be present. Propagating shocks in stellar
atmospheres usually produce line doubling and emission in the Balmer
lines, such as in Mira variables, RR Lyrae stars, W Virginis stars
and RV Tauri stars (Gillet 1988). At least, we would expect the
residual intensity of the Hα core to be filled in by emission during
shock phases. Therefore, we have obtained more time-resolved spectra
of BW Vulpeculae and a few classical β Cephei stars at Hα.

The spectroscopic observations were made with the CFHT f/8.2
coudé spectrograph and 1872-element Reticon on the nights of
October 17-21, 1986. The 830 groove mm^{-1} grating was used to deliver
a linear reciprocal dispersion of 4.83 Å mm^{-1}. Each Reticon pixel
accepted 72 mÅ, and the entire array comprised a bandpass of 135 Å.
We followed BW Vul through three cycles on three different nights.
On U.T. October 18, we obtained 13 spectra over 0.70 cycle with a
time resolution of 12 minutes. The signal-to-noise ratios varied
from 250-160. During that same night, we observed ν Eri over 1.02
cycles, obtaining 59 spectra exposed 4 minutes apart to get S/N
ratios varying from 560-380. The following night, we obtained a
series of 34 spectra of BW Vul over 0.70 cycle with a time resolution
of 6 minutes. This time, the signal-to-noise ratios varied from
150-75. Finally, on U.T. October 21, we obtained 32 spectra of
BW Vul over 0.66 cycle, again with a time resolution of 6 minutes.
Since the star was at a slightly higher air mass, the achieved
signal-to-noise was lower, varying from 100 to 40. In addition,
γ Peg was observed over 0.30 cycle, during which time we obtained 12
spectra spaced 6 minutes apart, with signal-to-noise ratios between
650 and 400. Heliocentric radial velocities in the stellar rest
frame were derived for Hα and for both lines in the CII doublet
$\lambda\lambda 6578, 6582$ of each spectrum. The largest internal velocity errors
at Hα were of order ± 4.5 km sec^{-1}. However, the mean errors were
smaller, corresponding to about 3 km sec^{-1} in BW Vul and about
1 km sec^{-1} in ν Eri and in γ Peg.

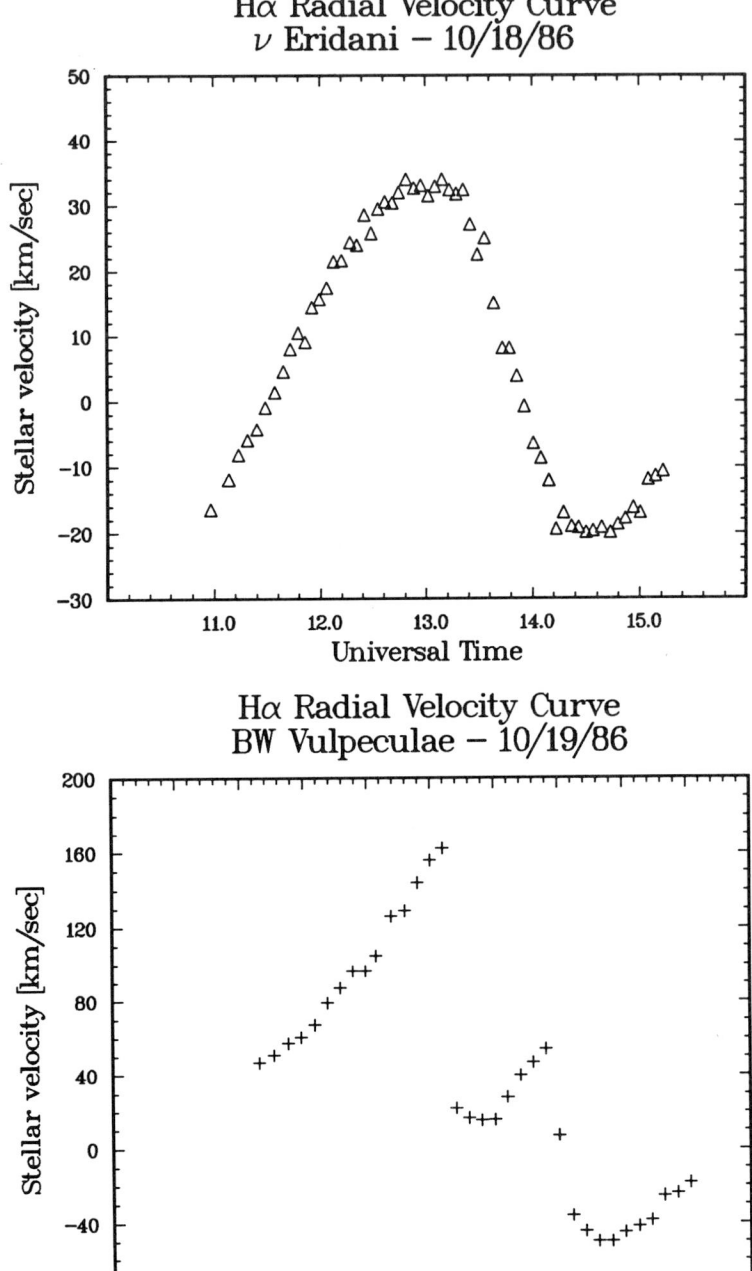

2. SUMMARY OF RESULTS

From these new high-resolution spectra, we conclude that the line profile changes and radial velocity curves for BW Vulpeculae are consistent with a double-shock model. The shell components are clearly visible, and there is line doubling and asymmetry in the Hα profiles before and after the velocity "stillstand" which is associated with maximum light. The similarity of the profiles, during two phases of nearly discontinuous velocity change 50 minutes apart, leads us to the hypothesis that the same shock-wave mechanism produces both phases of line doubling. The velocity variations through the stillstand phases are consistent with rapid upward acceleration, deceleration and infall toward maximum compression, implying that the "stillstand" is actually a recovery from a shock acceleration. The core residual intensity of Hα exhibits pronounced peaks during the phases of largest velocity variation, implying that there is filling in by emission at these phases. The most straightforward interpretation of these observations is that there are two shock waves, separated by about 50 minutes, which generate photoionizing precursors and double absorption components.

The radial velocity curve of ν Eridani is skewed, implying some kind of impulsive atmospheric motion; however, the velocity variations are not discontinuous at any time through the cycle, suggesting that the perturbations are not sufficient to generate shock waves. Also, the Hα core residual intensity peaks at phase 0.13, relative to maximum radial velocity, which is near maximum compression; this is consistent with an increase in the intensity of the source function as a result of atmospheric contraction. Thus, although a shock wave model for ν Eri cannot be ruled out, we find the evidence for it unconvincing. In γ Peg we see no evidence for shock activity, since there is no significant change in the Hα or CII profiles during the moving-shell phases. It seems that only the largest-amplitude pulsators (*e.g.*, BW Vul) show clear evidence for shock pulsation. In fact, since BW Vul is the only β Cephei star to display two velocity discontinuities, we surmise that the double-shock mechanism only applies to BW Vul.

3. REFERENCES

Campos, A.J. and Smith, M.A., 1980, *Astrophys. J.* **238**, 250.
Gillet, D., 1988, in these Proceedings.
Goldberg, B.A., Walker, G.A.H. and Odgers, G.J., 1976, *Astron. J.* **81**, 433.
Le Contel, J.-M. and Morel, P.-J., 1982, *Astron. Astrophys.* **107**, 406.
Odgers, G.J., 1955, *Publ. Dom. Astrophys. Obs. Victoria* **10**, No. 9, 215.
Smith, M.A., 1983, *Astrophys. J.* **265**, 338.
Smith, M.A. and McCall, M.L., 1978, *Astrophys. J.* **221**, 861.
Young, A., Furenlid, I. and Snowden, M.S., 1981, *Astrophys. J.* **245**, 998.

DISCUSSION

FURENLID I'd like to make two comments.
First, it is hard to discover shocks in Hα in hot stars, where hydrogen is almost completely ionized. Second, the two episodes of highest residual flux may have their origin in emission from pressure induced recombination.

CROWE We were discussing these points before the session, and I do not necessarily disagree with the view expressed by Dr. Purenlid. (We need to find an explanation for why "classical" β Cephei stars show little or no evidence for shock waves at Hα. I will be looking forward to seeing Dr. Purenlid's et al. recent work in print).

WHITE DWARFS AS OBSERVED AT HIGH SIGNAL TO NOISE

Jesse L. Greenstein

Palomar Observatory and Department of Astronomy
California Institute of Technology, Pasadena, CA 91125, USA

ABSTRACT. The goal is largely historical, 30 years of instrumental progress in a difficult new field, faint white dwarfs, and some results. High signal-to-noise spectrophotometry at 40–160 Å resolution revealed the separation between hydrogen- and helium-rich atmospheres, and provided a temperature scale from models. The white-dwarf color-luminosity relation proved narrow. Their simple spectra made brute-force averaging possible to 14th magnitude at good photographic resolution. Features as shallow as 5 % and 200 Å wide included C_2, but in magnetic stars some strong absorptions remain unidentified. Metals are deficient, gravitational diffusion setting the surface composition. The Palomar double CCD spectrograph can now give $S/N \approx 100$ to 17^m. Some polarized white dwarfs have Zeeman triplets in magnetic fields near 20 megagauss. In one, Zeeman components are shifted up to 2000 Å at 300 megagauss. Rotation is small in all white dwarfs, angular momentum mostly lost. Non-LTE cores of $H\alpha, H\beta$ exist and permit improved gravitational redshifts. An evolutionary phenomenon is progressive steepening of the Balmer decrement below 7000 K, cool atmospheres being helium-dominated.

1. INTRODUCTION; EARLY HISTORY OF SPECTROPHOTOMETRY

I am indeed honored that this Symposium of the IAU is to be dedicated to me, and especially regret that I cannot be physically present. It is even more exciting to see how many new workers and revolutionary topics fill the program. Stellar spectroscopy has gathered new strength from technology, area detectors with high quantum efficiency and massive data handling. For theorists model atmospheres seem no longer tedious and atoms are better understood. While new fields are glamorous (e.g. astroseismology and stellar granulation), others from the 1950's are exciting, like lithium abundances, compositions of globular clusters, heavy-element abundances in population II, the C, N, O ratios and the chemically-peculiar stars. The Symposium deals with classical and novel problems in the mainstream of astrophysics strengthened by the new possibility of observations with high signal-to-noise ratio, (hereafter S/N). My paper will emphasize the history of observational data about white dwarf spectra, with current opinions on their physical state. I illustrate mainly white dwarfs with strong magnetic fields; many viewgraphs of typical spectra shown at the Symposium are omitted for economy of space. A few are shown at the end.

The concept of high S/N seems implausible to those whose experience began with photographic stellar spectroscopy which could not have high S/N. The invaluable *Utrecht Solar Atlas*, had good S/N, obtained from high-resolution spectrographs with high-contrast emulsions. Poor S/N was especially characteristic of spectra of the white dwarfs, (hereafter WD), mostly fainter than 14th mag. My own first use of high S/N was in the attempt by Greenstein and Richardson (1951) to observe both ^7Li and ^6Li in the Sun. Its depletion to 1 % of the terrrestrial value lead us to the hypothesis of its destruction by nuclear reactions after mixing to about two million degrees.

For stars, the Schmidt cameras and Babcock gratings of the Mount Wilson coudé spectrograph gave widened photographic spectra at up to 1.0 Å mm^{-1}. Greenstein (1956a) compares a photographic coudé spectrum of the daylight sky, processed through a linearizing microphotometer, with the *Utrecht Solar Atlas*. Both have low noise; the lower resolution of the coudé makes weak lines shallow but still shows some lines of Rowland intensity -1. Oke pioneered spectral scanning at Mount Wilson and later at Palomar; one of his first instruments was a two-channel, high-resolution scanner at the coudé; its output (Oke and Greenstein 1961) gave high S/N on bright stars, slowed by the need to scan.

In that paper van Maanen 2 (EG 5, 0046+05) a cool 13th mag white dwarf is also illustrated. From 4000 Å to the ultraviolet, strong, broadened and blended $FeI, CaII, MgI$ are seen. The features have low contrast, many too broad to be easily "visible" by inspection. Tracings of 5 Palomar coudé spectra were measured objectively at constant wavelength intervals without smoothing plate grain and then averaged. The resolution is 1.8 Å ; noise is reduced by brute-force averaging. Reductions were tedious but features 5 % deep were found. Weidemann (1960) analyzed these spectra to show that metals were deficient by factors of $\approx 10^{-4}$. The helium opacity is low; the low metal abundance was the first evidence of how abnormal was the composition of WD's. We now believe that gravitational diffusion separates metals from the helium. With the IUE the star was found to be 19th mag in the 2800 Å line of $MgII$.

Brute-force averaging of photographic spectra is particularly effective in giving a "collective" high S/N, to detect features spanning many resolution elements. Greenstein and Matthews (1957) used this averaging procedure on spectra of faint WD's; Figure 1 illustrates Wolf 219 (EG 24, 0341+18), with a broad features 10 % deep, 200 Å wide on three spectra. The 4670 Å feature was identified as C_2 with extreme pressure broadening; a band 3% deep at 4355 Å is also C_2. Later the (0,0) band at $\lambda 5165$ was found. The atmosphere is dominantly helium and the ultraviolet resonance lines of CI were found with IUE in carbon-rich WD's, (type DQ). Using the Kitt Peak Image-Intensified Dissector Scanner and helium-dominated models, Wegner and Yackovich (1984) studied abundances in DQ's from C_2 . They found the C/He ratio to be $10^{-2} - 10^{-7}$ with H, N, O in low abundance over a range of T_{eff} where carbon is mixed to the surface from the degenerate core.

Figure 2 here shows the extraordinary object Grw +70°8247 (EG 129, 1900+70), with mean profiles from five spectra (Greenstein 1956b, Greenstein and Matthews 1957). The quality is comparable to the less noisy CCD spectrum illustrated in Greenstein, Henry and O'Connell (1985) at 12 Å resolution and 20000 counts per pixel. Only $\lambda 4135$ was "visible" in Minkowski's original discovery in 1938. This resisted identification; in 1984 I identified $\lambda 1347$ on the IUE spectrum as the Lyman-α Zeeman component, 1s0 - 2p-1. Then 10 Zeeman components of Hα, Hβ and Hγ were successfully identified, from quantum-mechanical computations of hydrogen in strong magnetic fields. Figure 3 of λ4135 shows the fit to 2s0 – 4f0 of Hβ (Greenstein, Henry and O'Connell 1985; Angel, Liebert and Stockman 1985).

The 1957 mean of five spectra of HZ 29 showed doubled lines of $He\,I$ averaging 5% in depth and never exceeding 10 %. This star is a helium-rich mass-transfer variable, AM CVn (EG 91, 1232+37); it probably consists of two white dwarfs that will ultimately merge, and according to theorists, produce a supernova of type I. The line profiles did not move or change. A similar interacting pair of helium white dwarfs is G61-29, showing broad emission lines of $He\,I$.

I mention these objects because the spectra have not been substantially surpassed in published recent papers using electro-optical sensors. Some numbers and a critique concerning photographic spectrophotometry of faint stars may be useful. A typical WD spectrum at the 200-inch prime-focus, widened to 0.3 mm at 180 Å mm^{-1}, took 60 minutes at 15th mag. What is the noise in such spectra? The projected slit included ≈ 20 developed plate grains. The number of photons required by a single plate grain was ≈ 50, so the photon noise (from the required 1000 photons) is not serious. Plate grains are essentially opaque and their summed response is non-linear; the small number of grains per resolution element is the noise on the microphotometer tracing. For most purposes, the resolution of ≈ 5 Å was higher than needed, because the WD's stark- or pressure-broadened lines are wider. Plates are more uniform in response than are pixels on a CCD, so flat-fielding is unnecessary. In summary, careful photographic spectrophotometry could detect subtle WD features to a depth of about 3 %. None of these objects needed sky subtraction, being photon-starved by the inefficiency of the plates, rather than sky-limited. What we had was "detectability" rather than high S/N. Other wavelength regions are more contaminated by airglow emission than the blue, where photographic spectrophotometry was limited to 15th mag.

A serious problem encountered in pioneering high-dispersion coudé studies of normal stars are worth recall. We were forced to a much effort on the scales of line intensities, which varied between observatories and depended on dispersion. IAU Commissions became interested and involved in an international attempt to provide standard equivalent widths and profiles (Wright et al. 1964) resulted. The optical quality of the older coudé spectrographs was excellent and the photographic calibration careful; scattered light and grating ghosts were investigated. I wonder whether similar attention is paid now to such questions? Solid-state detectors and especially echelle spectrographs in new spectrometers might usefully be intercompared by observations of standard stars.

2. THE WHITE DWARFS; PROGRESS IN CAPABILITY

Given the limitations on S/N it is fortunate that the apparent simplicity of most WD spectra is real; problems arising from a background of weak, highly pressure-broadened, invisible lines have not surfaced. The important result of much analysis is that a dichotomy of composition exists. The more numerous group is hydrogen-dominated (the so-called DA types), the other is helium-dominated (the non-DA's). In the DA's the abundance ratio is $H/He \geq 10^3$; in the non-DA's, $He/H \geq 10^4$. The DA's have essentially no metals, while the non-DA's sometimes have carbon and metals, but always in abundances far below their solar values; mixed H + He compositions are rare. In DA's the Balmer lines coalesce beyond H8; pressure broadening destroys the higher states depressing the ionization potential. The Balmer continuum is strong; hydrogen lines and continuum apparently disappear at $T \leq 7000$ K (see §III). Among the non-DA's, He II is seen rarely (defining the DO type) at high temperatures; over a limited range of temperatures, He I is strong, broad and shows forbidden components in DB's. Carbon is present in cooler stars, called type DQ; metals appear weakly in a few WD's, mostly below 12000 K, called DZ's. Both carbon and metals are confined to the non-DA's and result from the low opacity of helium as well as from genuine abundance anomalies at the surface.

Given the relative simplicity of the WD spectra the multichannel spectrophotometer (the MCSP) designed and built by J. B. Oke proved nearly ideal for surveys. The high quantum efficiency of 31 photomultipliers provided simultaneous observation in two apertures, sky, and sky + object, with high accuracy in short times. To illustrate the sensitivity of the MCSP, a DA star GD 356 (Gr 329, 1653+53) has $m_V = 15.3$; it gave $\approx 2 \times 10^4$ detected photons in 360 seconds at 5400 Å with 40 Å resolution ($3100 \leq \lambda \leq 5700$) and 80 Å resolution ($5700 \leq \lambda \leq 10000$). This visual flux corresponds to 2.7 milliJanskies; I discuss this star in §III. By stepping the grating to five wavelengths and switching apertures, the entire spectrum is covered, with redundancy. We obtain m_ν on the Oke AB scale, and $log\, f_\nu$, at 124 wavelengths, in 60 minutes, most with photon-counting statistics of ±0.01 mag. Data are better suited to model-atmosphere fitting than are the broadband photoelectric magnitudes, because of the nearly monochromatic fluxes.

Colors observed with the MCSP for over 300 WD's with $\approx 1-2$ % errors are in Greenstein (1984), on the absolute AB79 scale of Oke and Gunn (1983). The correlation of color with luminosity for WD's with modern parallax is excellent (Greenstein 1985). The intrinsic (or cosmic) scatter appears to have a σ near ±0.25 mag; the resulting spread in radius is only 12 %, less if the color-T_{eff} relation is not unique. Weidemann and collaborators at Kiel have fitted model atmospheres to fluxes observed in both hydrogen and helium WD's. The absolute calibration of AB is central to the temperature scale; spectroscopic gravities average $log\, g = +8.0$ (Koester, Schulz and Weidemann 1979), agreeing with those from parallaxes and temperatures. Weidemann and his group specialized in obtaining spectroscopic masses. Predicted and observed gravitational redshifts agree if the core of the WD has $\mu_{el} = 2.0$, i.e. is helium or carbon and the typical mass is near 0.6 M $_\odot$(see §5). The MCSP and hydrogen models define the limited range of temperatures in which rapid light variation of the ZZ Ceti type occurs. These complex oscillations occur in DA's, with T_{eff} confined to within ±500 K of 11200 K (Greenstein 1982), in agreement with the theory of non-radial modes of hydrogen envelopes.

The wide wavelength range covered by the MCSP at good S/N favors detection of broad features, success depending on the smoothness of the heterochromatic calibration. The MCSP revealed broad "waves" in otherwise continuous spectra of WD's with circular and linear polarization. Many of these features remain unidentified. The cool degenerate G240-72 (Gr 372 1748+70) has a dip 2400 Å wide, reaching nearly 20 % and centered on 5260 Å (Greenstein 1974). GD 229 (Gr 333, 2010+31) is shown in Figure 4. It is hot and has still unidentified strong dips at wavelengths 4185, 5280, 7090 and 8000 Å which are correlated with the wavelength dependence of circular polarization. Greenstein and Boksenberg (1978) counted 5000 photons per pixel at the coudé with the IPCS to show that the 4185 Å feature had ≈ 300 Å FWHP, with details of structure unchanged from four years earlier with the MCSP. It may have cyclotron absorption. Liebert et al. (1976) found Zeeman components of $H\,I$, $He\,I$ in Feige 7 (Gr 267, 0041-10) shifting as it rotates with in a 2.2 hour period.

A fascinating problem remains unresolved for the coolest degenerates.

What explains a spectrum without any lines? In early work, I called WD's with no lines deeper than 5 % type DC. Helium opacity is so low in a hydrogen-free atmosphere that unsaturated resonance lines should be visible, if even only a trace of a metal exists. Can lines be pressure-broadened into invisibility by the gas pressure, which exceeds 100 atmospheres? Can all bound states except the ground state be destroyed? Laboratory gases at such pressures have extraordinary spectra with broad asymmetric "waves", sometimes located near resonance lines. Quasi-molecules and polymers are formed. While I doubt that the apparent lack of lines in cool white-dwarfs is caused only by pressure broadening, the observational evidence is still marginal. The only attempt was a study (Greenstein 1979) of the smoothness of the mean fluxes in nine faint, red DC degenerates with $M_V \geq 15$. The residuals from the mean flux and its quadratic representation were evaluated from 8000 to 4200 Å. They had the remarkably small scatter of $2\sigma \leq 0.02$ mag, except from 4000 to 3500 Å where they are rough and depressed by 0.05 mag. Unfortunately the calibration of sdG's in this region is uncertain at that level, because of metallic lines. The search for "invisible" lines remains an important problem in absolute calibration for high S/N detectors.

3. HIGHER RESOLUTION AND S/N; COMPOSITION; ZEEMAN EFFECT

That higher S/N would be important for study of physical processes in WD's was demonstrated by Gary Wegner, with Kitt Peak National Observatory spectrometers. He showed that many apparently line-free, so-called DC's actually had weak lines, sometimes hydrogen in the cool DA's, or C_2 bands in the DQ's. Wegner and Yackovich (1982) and Wegner (1983) observed DC's at 17 Å resolution; of 46 WD's originally called DC, 18 remained line free, 4 had Hα, 8 had HeI, 9 had C_2, 1 had CI, and 6 remained uncertain. The improved S/N revealed helium-rich atmospheres, and weak hydrogen, given the extra effort. Wegner's results broadened our ideas about the carbon-rich WD's. Wegner and Yackovich (1984) found the H/He ratio low and the C/He ratio to range over 10^5 critically dependent on temperature. Convection occuring over a limited range of T_{eff} dredges carbon up from the degenerate core.

I had only one chance to use the Oke-Gunn double spectrograph at the cassegrain focus of the 200-inch, a genuinely high S/N instrument for moderate resolution on stars. It is provided with two excellent CCD's of high quantum efficiency, flat fielding and sky subtraction. Saturation is near 30000 detected photons per pixel; short exposures reach 17th mag. I observed 140 white dwarfs in eight nights. In 1982, after my retirement, I used it with low-dispersion gratings to cover the entire WD spectrum; lines of equivalent width below 1 Å are just detectable.

From these CCD spectra a major new result (Greenstein 1986) was that the fraction of WD's whose atmosphere contains hydrogen lines falls rapidly at $T_{eff} \leq 6000$ K. The Balmer decrement changes; 80 % of WD's with $T_{eff} \geq 7000$ K show $W(H\alpha)/W(H\beta) \approx 0.8$, the normal ratio. Their atmospheres are dominantly hydrogen. This fraction drops below 20 % among cooler WD's, which are dominantly helium. Hα finally disappears at 4500 K (at $M_V \geq +15$) but remains surprisingly sharp. Figure 5 shows LHS 5023 (Gr 462, 0102+21A), which is among the coolest known degenerates. The Figure also graphically illustrates the power of the CCD spectrograph and the virtue of even moderately high S/N. The star is near 18th magnitude, the observing time was one hour, the line is 4 % deep and visible in at most 4–6 pixels. This type of Hα is not present in all cool DC's and may be enhanced by resonance broadening or non-LTE effects, stronger than expected from pure hydrogen models. Greenstein (1986) illustrates a mean profile formed to improve S/N in three other faint red degenerates near 4500 K; Hα is 3 % deep and visible in several pixels, but no significant absorption is visible at Hβ. In contrast, Hβ, (measured in a second-order grating spectrum at better resolution) decreases as predicted by the same models. High S/N on faint red degenerates is an important and still difficult task for large telescopes. The observed ratio of $W(H\alpha)/W(H\beta)$ increases to above 2 at 6000 K, as Hβ fades into invisibility faster than Hα. The most probable explanation comes from computations of atmospheres containing mixtures of helium and hydrogen. An abnormal Balmer decrement in WD's was first explained by Wehrse (1977) before it was observed. Liebert and Wehrse (1983) computed models with different He/H abundance ratios, for a star of 7400 K, much hotter than those I have observed. As gas pressure is increased by increasing the He/H ratio both Balmer lines strengthen. Even with 1 % hydrogen Hα remains as strong as in pure hydrogen, but Hβ, formed at large optical depth, broadens into invisibility. Calculations of models for cool stars, by Wehrse and by the Montreal group (Fontaine, Wesemael and collaborators) were reported at the IAU Colloquium No. 95 on *Faint Blue Stars* in Tucson, June 1987. The deduction that the coolest degenerates have helium-dominated atmospheres is quit important in interpreting the balance between convection, gravitational diffusion and possible accretion.

4. HIGH S/N IN SOME MAGNETIC STARS

The DA star GD 356 was used in §II as a quantitative illustration of MCSP sensitivity. Because it had an apparently "rough" MCSP spectrum, it was re-observed at Hα and Hβ with the Oke-Gunn double CCD spectrograph, at 6 Å and 4 Å per pixel. The resolution is degraded to about 2 pixels by the large apertures but the $S/N \approx 1\%$. Greenstein and McCarthy (1985) found the Hα structure to be a Zeeman emission triplet which is fitted by a pole-on magnetic dipole field of 20 megagauss; the Hβ emission is weaker and distorted by the quadratic Zeeman effect. The absorption Zeeman triplet was found in the polarized, cool degenerate G 99-47 (Gr 290, 0553+05). Neither a dipole magnetic field of 15 (pole-on) or 27 megagauss (equator-on) fits the observations as well as a tangled field of 14 megagauss with a gaussian distribution of the scalar field, with $\sigma \approx 1.4$ megagauss. It may seem to be a non-physical model; no polarization has been observed in GD 356, so a tangled magnetic field may have some merit.

In the strongly magnetic star Grw +70°8247, the CCD spectra show details of the λ4135 line (Greenstein, Henry and O'Connell 1985). This 2s0 - 4f0 component of Hβ is not forbidden in a strong magnetic field. Various values of the dipole magnetic field, with assumed tilts were tried; the noise level is low, and the fit is fairly good with a nearly pole-on field of 250 – 300 megagauss, as illustrated in Figure 3. The strongest line in the red is 2s0 - 3p0 of Hα at 5910 Å , best fitted with 300 megagauss field. A strong Hα component exists at 8500 Å which should persist at even higher magnetic field.

5. ROTATION. GRAVITATIONAL REDSHIFT

The photon yield for WD's was improved over the photographic plate by the image-intensifier tube; it permitted Greenstein and Trimble (1972) to see sharp absorption cores in Hα and Hβ. The sharp core occurs over a limited range of T_{eff}. With resolution of 1.1 Å our estimate was that the rotation $v \sin i \leq 30$ km s^{-1}. Greenstein and Peterson (1973) confirmed the existence of the sharp, deep cores and advanced the non-LTE explanation, as did Shipman. Hydrogen atoms at small τ_{cont} and therefor at low gas pressure see a depressed radiation field ±50 Å from the line center, at a low boundary temperature. Low rotation means that about 99 % of the angular momentum has been lost as the red giant lost 80 % of its mass. The typical main-sequence parent is a late ≈ 4 M$_\odot$) B star, rotating at 60 km s^{-1}. When it shrinks to the WD radius of 0.01 R $_\odot$ it should rotate at 10^4 km s^{-1}, were its angular momentum conserved. Pilachowski and Milkey (1984, 1987) with much improved resolution at the Kitt Peak echelle spectrograph beautifully confirmed the slow rotation of DA's. In their 1987 paper, they have good data for 15 stars in the most favorable temperature range for non-LTE cores. They strengthen the limit to rotation (which averages ≈ 30 km s^{-1}); rotation "is indistinguishable from zero in 10 stars". The highest $v \sin i = 60 \pm 10$ km s^{-1}. Most of the specific angular momentum of the parent star has been lost during the AGB evolution. Cyclic variation in circular polarization showed the magnetic WD's to be even slower rotators than the non-magnetic WD's. I had observed a sharp core in 40 Eri B (EG 33 0413-07) with the Boksenberg IPCS system (0.8 Å per pixel) at Palomar (Greenstein et al. 1977), for which Pilachowski and Milkey (1987) find rotation of 19 ± 5 km s^{-1}. The non-LTE cores have a steep Balmer decrement, as expected from transition probabilities. The Hβ core is easily seen, but no core is seen at Hγ even with 2500 counts per pixel. Pressure broadening, which strengthens high series members and flattens the decrement, does not operate on the cores.

The sharp core permit velocity measures at relatively high accuracy in a few of the brighter DA's; it should not be pressure-shifted and is objectively measureable. Lines are 6000 km s^{-1} wide; if the apparent centroid of this wide line is used photographic measurement is a delicate art. Trimble and Greenstein (1972) obtained a statistical value of the Einstein gravitational redshift. Velocities corrected for solar motion were averaged to derive the K-term; a few WD's in resolved binaries and clusters gave redshifts. Our mean gravitational redshift was 51 km s^{-1}. Wegner (1974) measured redshifts in southern WD's and found the statistical value of 40-50 km s^{-1}. Currently, some resolved binaries (DA and dM) give individual redshifts on electronic spectra, and therefor reliable M/R determinations. At higher dispersion these may have different sensitivity to pressure shifts; they indicate smaller mean redshift (near 25 km s^{-1}) and therefor lower masses. Astrometric masses are rare for white dwarfs, e.g. 0.4 and 1.1 M $_\odot$ for 40 Eri B and Sirius B. Astrophysical masses use model atmospheres, temperatures and surface gravity and give typical mass near 0.6 M $_\odot$. Photographic redshifts suggested about 0.7 M $_\odot$; these new, unpublished measures in binaries suggest 0.5 to 0.6 M $_\odot$. Improved techniques with such different methods should converge to establish the M/R relation soon.

In conclusion, I again must express my gratitude to those who organized this IAU Symposium and provided to all of us another encounter with modern spectroscopy. In addition to personal friendships we share the good fortune of work in an important and exciting field.

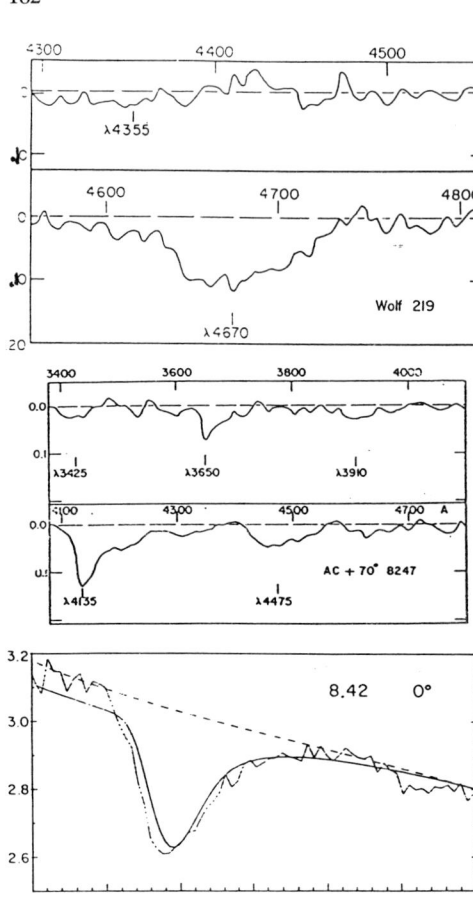

Figure 1. Mean (2) coudé photographic spectra (1957) illustrating the pressure-broadened C_2 (1957) in a white dwarf Wolf 219; absorption in magnitudes.

Figure 2. Mean (5) coudé photographic spectra of the polarized, magnetic white dwarf Grw+70°8247. Absorptions, unidentified until 1984, are Zeeman components of hydrogen.

Figure 3. (Light data points). Single high S/N Palomar CCD spectrum of $Grw + 70°8247$ at the $\lambda 4135$ feature, a component of $H\beta$. Ticks on wavelength scale are 10 Å, flux in milliJanskies. (Solid). Predicted 2s0-4f0 transition of $H\beta$ in a pole-on magnetic field, 260 megagauss.

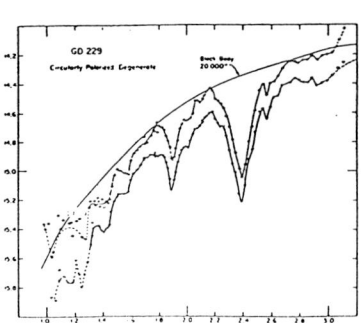

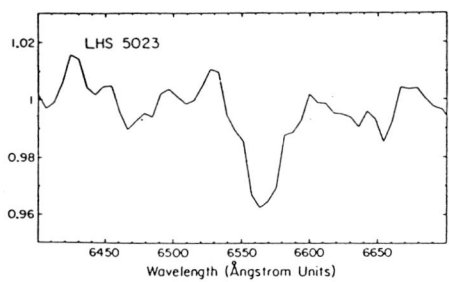

Figure 4. Two multichannel spectra (vertically offset), mag vs. $1/\lambda$ in μm^{-1} of GD 229, a polarized hot white dwarf with still unidentified features.

Figure 5. $H\alpha$ at moderate noise in an 18th mag red degenerate LHS 5023 (probably low H/He); Palomar CCD spectrograph, 1 hour.

REFERENCES

Angel, J. R. P., Liebert, J. and Stockman, H. S. 1985, *Astrophys. J.*, **292**, 260.
Greenstein, J. L. 1956a, *Vistas in Astronomy*, ed. A. Beer, (Pergamon: London), **2**, 1299.
Greenstein, J. L. 1956b, *Proc. 3rd Berkeley Symp. Math. Statistics and Probability*, ed. J. Neyman, (Univ. of Calif. Press: Berkeley), **3**, 11.
Greenstein, J. L. 1974, *Astrophys. J. (Letters)*, **194**, L51.
Greenstein, J. L. 1979, *Astrophys. J.*, **227**, 244.
Greenstein, J. L. 1982, *Astrophys. J.*, **258**, 661.
Greenstein, J. L. 1984, *Astrophys. J.*, **276**, 602.
Greenstein, J. L. 1985, *Pub. Astron. Soc. Pacific*, **97**, 827.
Greenstein, J. L. 1986, *Astrophys. J.*, **304**, 334.
Greenstein, J. L. and Boksenberg, A. 1978, *Mon. Notices Roy. Astron. Soc.*, **185**, 823.
Greenstein, J. L. 1985, Henry, R. J. W. and O'Connell, R. F. 1985, *Astrophys. J. (Letters)*, **289**, L25.
Greenstein, J. L. and Matthews, M. S. 1957, *Astrophys. J.*, **126**, 14.
Greenstein, J. L. and McCarthy, J. K. 1985, *Astrophys. J.*, **289**, 732.
Greenstein, J. L. and Peterson, D. M. 1973, *Astron. Astrophy.*, **25**, 29.
Greenstein, J. L. and Richardson, R. 1951, *Astrophys. J.*, **113**, 536.
Greenstein, J. L. and Trimble, V. 1972, *Astrophys. J. (Letters)*, **175**, L1.
Koester, D., Schulz, H. and Weidemann, V. 1979, *Astron. Astrophys.*, **65**, 262.
Liebert, J., Angel, J. R. P., Stockman, H. S., Spinrad, H. and Beaver, E. A. 1976, *Astrophys. J.*, **214**, 457.
Liebert, J. and Wehrse, R. 1983, *Astron. Astrophys.*, **122**, 297.
Oke, J. B. and Greenstein 1961, *Astrophys. J.*, **133**, 349.
Oke, J. B. and Gunn, J. E. 1983, *Astrophys. J.*, **266**, 713.
Pilachowski, C. A. and Milkey, R. W. 1984, *Pub. Astron. Soc. Pacific*, **96**, 821.
Pilachowski, C. A. and Milkey, R. W. 1987 *Pub. Astron. Soc. Pacific*, in press.
Trimble, V. and Greenstein, J. L. 1972, *Astrophys. J.*, **177**, 441.
Weidemann, V. 1960, *Astrophys. J.*, **131**, 638.
Wegner, G. 1983, *Astron. J.*, **88**, 1034.
Wegner, G. 1974, *Mon. Notices Royal Astron. Soc.*, **166**, 271.
Wegner, G. 1978, *Mon. Notices Royal Astron. Soc.*, **182**, 111.
Wegner, G. and Yackovich, F. H. 1982, *Astron. J.*, **87**, 155.
Wegner, G. and Yackovich, F. H. 1984, *Astrophys. J.*, **284**, 257.
Wehrse, R. 1977, *Mem. Soc. Astr. Italiana*, **48**, 13.
Wright, K. O., Lee, E. K., Jacobson, T. V. and Greenstein, J. L. 1964, *Pub. Dom. Astrophys. Obs.*, **12**, 173.

THE BUYING POWER OF HIGH SIGNAL-TO-NOISE RATIOS IN SPECTROSCOPY

David F. Gray
Department of Astronomy
University of Western Ontario
London, Ontario, Canada N6A 3K7

ABSTRACT. High S/N is a good first step toward accurate profiles.

1. INTRODUCTION

I developed an interest in high S/N spectroscopy (S/N is used here for signal-to-noise ratio) in the early 1970's, and I built a photoelectric line scanner consisting of a photomultiplier behind an exit slit placed in the focal plane of a 13 meter focal-length camera. The slit and photomultiplier were marched repeatedly across the spectral line with about a minute of integration at each position. I can still remember spending a whole night measuring about 30 points across Na D lines of a fourth magnitude star and attaining the stupendous signal-to-noise ratio of 50! I'm not sure what gave us the faith to continue, except that we knew there was a great deal of untapped information in those line profiles. We can now measure the same star at about 2000 wavelength points with five times higher spectral resolution, and in one hour surpass a S/N of several hundred. With some effort, S/N ~ 1000 can apparently be achieved. I say apparently because the meaning of S/N varies with the application and because there is more to dependable line profile measurement than a low noise level, as I shall point out below. But lowering the noise level is a good start, allowing us to buy our way into whole new research endeavors.

2. INTERPLAY OF S/N AND RESOLUTION

In most line profile work, spectral resolution and S/N are related. Typical absorption line profiles are unimodal, and so differences in shape appear as "higher order" effects, which means that the differences appear toward higher Fourier frequencies. Figure 1 illustrates this. Both high S/N and high spectral resolution are needed to see the high frequency portions of these transforms.

The cool-faint portion of the HR diagram is populated by stars having the narrowest spectral lines. For K dwarfs, we need resolution of at least 2.5 km/s, but as the noise is pushed down, and the Fourier

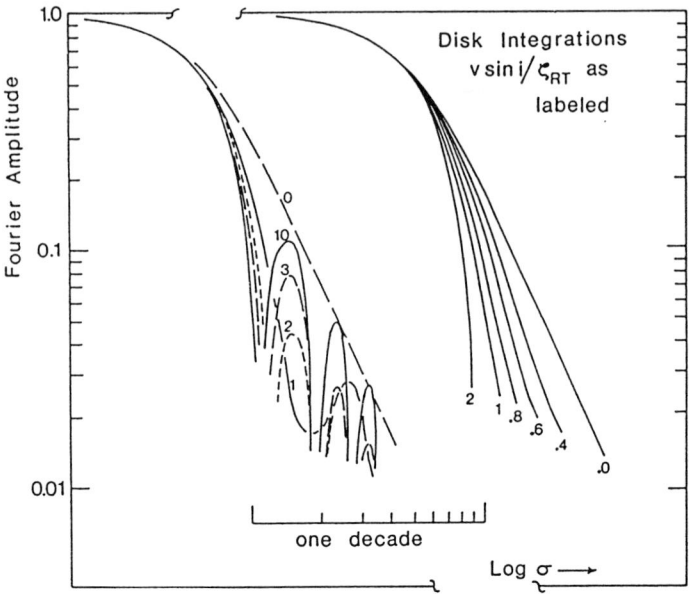

Fig. 1. Model disk integrations were used to combine the Doppler-shift distributions of rotation and radial-tangential macroturbulence. Their Fourier transforms are shown for various ratios of rotation to macroturbulence dispersion, ζ_{RT}. The differences are seen to be larger at higher Fourier frequencies where the amplitudes are smaller and (in the observations) closer to the noise level.

transforms are followed to ever higher frequencies, we may need even S/N higher resolution. Naturally we can expect exposure times to go up as we strive for higher S/N and higher resolution. With silicon-diode-type detectors, S/N improves linearly with exposure time initially because detector noise dominates, but eventually we cross into the photon-noise-dominated domain.

The cost of higher resolution depends on whether or not a good image slicer is used. Without a slicer, we pay at the entrance slit and at the detector, so the exposure increases as the square of the resolution. With a slicer, the exposure increases with the first power of the resolution. In both cases we loose field in proportion to the increase in resolution.

3. THE ROLE OF S/N IN LINE BROADENING ANALYSIS

The key to separating macroturbulence from rotation in F, G, and K stars is to be able to place the observations in a grid like the one shown in Figure 1. Unless we clearly see the transform of the stellar lines above the noise, we will be unsuccessful. In the case where rotation dominates ($v \sin i/\zeta_{RT} > 2$), the information on the size of the macroturbulence dispersion, ζ_{RT}, is contained almost exclusively in the sidelobe height (left side of Fig. 1). Most G and K stars have v sin i/ ζ_{RT} < 1; then the right side of Figure 1 is relevant.

Fourier analysis has been applied to a significant number of stars in the F, G, and K portion of the HR diagram. A summary of the results for macroturbulence is given in Gray and Toner (1987). Simply stated, the macroturbulence dispersion increases monotonically with luminosity and effective temperature. It turns out that uncertainty in the luminosity classification is a significant source of scatter in this diagram, and it may be that the size of ζ_{RT} is a more sensitive measure of luminosity than the standard classification methods.

Some recent observations of σ Dra (K0 V) are shown in Figure 2. I have combined a number of separate exposures to force the noise level down. Each successive step toward the right in the diagram includes more data, and as the S/N pushes past ~ 500, one begins to see the sidelobe emerge. This sidelobe arises from the onset of saturation in the line profile and so it is sensitive to desaturation mechanisms such as microturbulence.

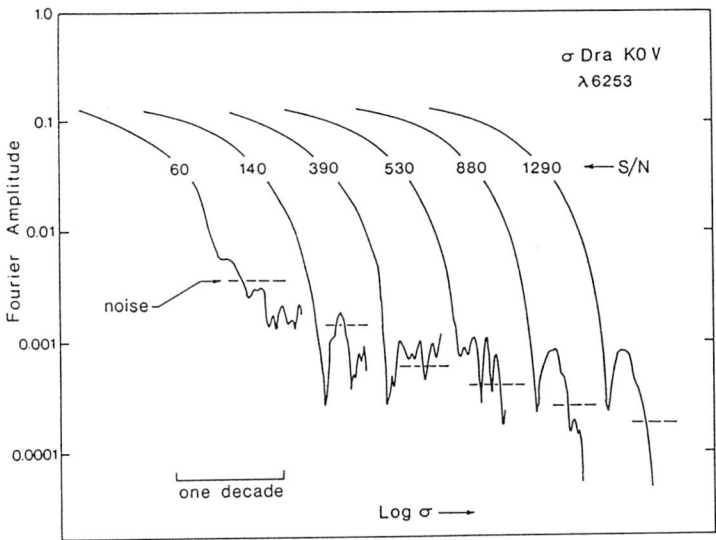

Fig. 2. The Fourier transform of Fe I λ6252.57 is shown for various S/N values in the continuum of the spectrum. A weak sidelobe appears for the highest S/N cases toward the right.

Currently attainable S/N is already sufficient to show up weak points in the kinematic modeling. In some instances, ζ_{RT} seems to be a function of depth (Gray 1982b), and this is not included in the usual modeling. Further, real stellar line profiles are rarely if ever symmetric, but the line shape analyses usually ignore this important aspect for sake of tractability. But let us turn our attention to it now.

4. THE ROLE OF S/N IN STELLAR GRANULATION MEASUREMENTS

The signature of stellar granulation is the asymmetry seen in the spectral lines. Most commonly this is displayed using the line bisector (see Dravins et al. 1981, Gray 1982a). In a typical solar bisector, the rising hot granules produce a blue shift in the main portion of the line because they dominate the light. The cooler, less bright, falling material produces a positive tail on the overall velocity distribution, and this is seen as a depression of the red wing of the profile which in turn produces a redward bend in the top of the bisector. In the sun, the granulation contrast has largely disappeared at the higher layers where the line cores are formed, and so the blue shift does not occur in the lower part of the bisectors of strong lines. The net effect is C-shaped bisectors. Stars hotter and more luminous than the sun show blue shifts even in the line cores, so we know that granulation penetrates higher in their photospheres than for the solar case.

Elementary considerations tell us that our ability to measure bisectors depends on the S/N in the profile coupled with the slope of the profile. A given photometric error translates into a larger velocity error in the wings and core of the line (Gray 1984). Midway down a strong line, the slope might be ~ 10% per km/sec, leading to expected bisector errors ~ 70 m/sec for S/N of 100. Since a typical bisector has a span of only ~ 100 m/sec, we clearly need S/N in excess of 100.

Interestingly, the importance of spectral resolution on the bisectors is very different from that of profile-shape analysis. Instead of a continuous interplay between S/N and resolution, here we have a semi-binary situation: the exact resolving power is not very important as long as it is in excess of about 100,000. This comes about because the bisectors themselves do not have high Fourier components, making really high resolution unnecessary, while resolving power < 50000 results in smearing over a significant fraction of the profile, wiping out the asymmetry of the line. A number of observational experiments along this line have been done by Dravins (1987).

Even with the modest S/N of a few hundred, we have been able to show that granulation is stronger and penetrates higher into the stellar photosphere as the luminosity and/or the effective temperature increases. One uncertain part of this overall behavior was for the K dwarfs, where there was some indication of deviation from a monotonic trend (Gray 1982b). In Figures 3, I show some more intensive studies on dwarfs. The main improvement here over former studies is in pushing the S/N from ~ 100-300 to ~ 1000. We see in the figure the C shape so typical of solar bisectors, and we see a very small velocity span, ~ 100 m/s, also typical of the solar case. Dravins (1987) has found similar results for α Cen A (G2 V) and B (K1 V). So it now appears that there is indeed a general decrease in the vigor of granulation toward the lower right portion of the HR diagram, consistent with the behavior of macroturbulence dispersion.

Toward hotter stars, an interesting reversal of the classical cool-star line asymmetry occurs. The behavior is rather striking in Ib supergiants (Gray and Toner 1986). Some important atmospheric changes are taking place as a star evolves across the HR diagram.

One interesting need for ultra-high S/N in line asymmetry work is with the "rotation effect." Simple numerical simulations (Gray and Toner 1985) predicted bisector displacements enhanced by rotation, and I have attempted to use this effect to pin down actual granule rise velocities (Gray 1986a). Previously we had thought that absolute velocity measurements for bisectors were impossible because of the arbitrary radial velocity of the star. But the Doppler-shift distribution, broadened by

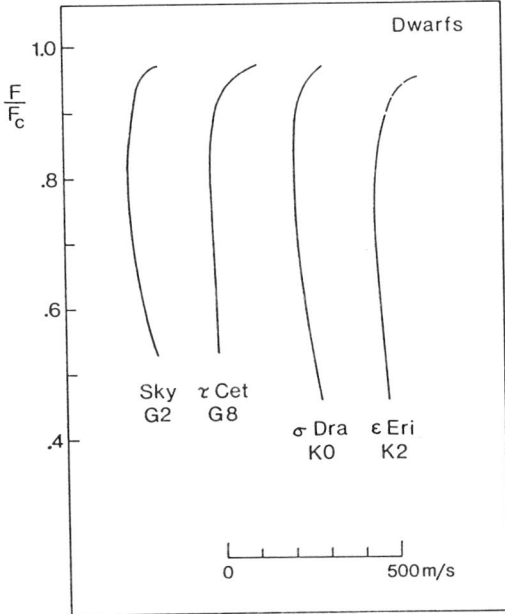

Fig. 3. Cool dwarfs do have C-shaped bisectors, as illustrated here. These bisectors are means of several lines and many exposures. Notice that the characteristic velocity span is ~ 100 m/s.

rotation, supplies the needed zero-velocity position. Rise velocities of 1.5 km/s seems to be typical of granules. The reason ultra-high S/N is of value here is because profiles broadened by rotation are very shallow--have small slopes which magnify the photometric errors into large bisector noise. So although the rotation effect may prove to be a useful tool for granulation studies, we will have to work diligently on improving the S/N if we hope to exploit it.

5. BEYOND HIGH S/N

High S/N is just the first step on a much more comprehensive buying plan. It has given us buying power to study turbulence, rotation, granulation, magnetic fields, starspots...., but it will not be long before we will have to face up to other limitations to accurate spectroscopic measurements. Among these are:

1) <u>line blending</u> - it varies markedly with spectral type and with spectral region, but always becomes more serious as S/N increases because then the more numerous weak blends make a difference instead of being buried in the noise.

2) <u>instrumental profile</u> - the only way uncertainty here could be completely eliminated is by making the instrumental profile very narrow compared to the stellar lines. This is feasible only for the sun. For other stars, we are photon starved and cannot afford the luxury of such a narrow instrumental profile. Problems arise because it is usually necessary to measure the instrumental profile at a grating setting different from that used when measuring the star; the polarization properties of the instrumental profile from image slicers, mirrors, filters, grating, and detector are probably significant but usually ignored.

3) <u>scattered light</u> - a difficult contamination to measure since it can vary with grating orientation, wavelength, and techniques of order sorting.

4) <u>diurnal rotation of the earth</u> - the projection of the earth's rotational velocity vector onto the direction of the star can vary enough during a long exposure to broaden the spectral lines by up to several tenths of a km/s (Gray 1986b).

5) <u>detector stability and geometry</u> - mechanical and thermal drifting can occur. At some level the irregularities of the pixel size and spacing introduce errors.

6) <u>detector electrical flaws</u> - modern detectors can have latent images, incomplete charge transfer, spurious clocking periodicities, etc.

7) <u>nonlinear response</u> - no detector is perfectly linear. Many show zero-point problems, or saturation effects, or integration errors. Interference fringing can also introduce a modulated response.

8) <u>nonuniform illumination of optics</u> - collimation errors and differences in illumination, especially of in-beam obstructions, for starlight, flat-field lamps, and instrumental profile lamps can cause distortions and various field errors.

Alas, the S/N can be very high and we still may not have data suitable for comparison with theory. Nevertheless, high S/N has given us the buying power to open up the "new" spectroscopy.

REFERENCES

Dravins, D., Lindegren, L., and Nordlund, A. 1981 Astron. Ap. 96, 345.
Dravins, D. 1987 Astron. Ap. 172, 211.
Gray, D.F. 1982a Ap.J. 255, 200.
Gray, D.F. 1982b Ap.J. 262, 682.
Gray, D.F. 1984 P.A.S.P. 95, 252.
Gray, D.F. 1986a P.A.S.P. 98, 319.
Gray, D.F. 1986b in I.A.U. Symp. 118, "Instrumentation and Research Programmes for Small Telescopes", (Reidel: Dordrecht), p. 401.
Gray, D.F., and Toner, C.G. 1985 P.A.S.P. 98, 499.
Gray, D.F., and Toner, C.G. 1986 P.A.S.P. 97, 543.
Gray, D.F., and Toner, C.G. 1987 Ap.J. Nov. 1st issue.

DISCUSSION

SODERBLOM Can you tell us the name of the "friendly dwarf" please ?

GRAY Eventually, yes. But I am not at liberty to do so quite yet because a PhD thesis is involved.

PRADERIE You report a variation of 7% in equivalent width. What is then the accuracy on the measurement of W ?

GRAY Well, the apparent precision can be better than 0.25%.

BOHANNAN One thing that has become clearer to me during this symposium is that the buying power of reducing observations has to be considered when designing instrumentation and planning observing programs.

GRAY Sure.

EBBETS Your illustration of the interplay between signal to noise ratio and resolution might suggest that for a given resolution there is a maximum S/N that is worth obtaining. The higher Fourier frequencies are attenuated by the instrumental broadening. Isn't it true though that higher S/N allows you to detect and measure weaker and weaker lines, you just can't distinguish their shapes from that of the instrumental profile ?

GRAY You are quite right in the sense that lower and lower noise will not let us see any Fourier amplitudes that have been severely filtered by the instrumental profile transform. Then higher spectral resolution is the answer to such a situation.

DOPPLER IMAGING OF VARIABLE EARLY-TYPE STARS*

Dietrich Baade
Space Telescope-European Coordinating Facility
European Southern Observatory
Karl-Schwarzschild-Str. 2
D-8046 Garching, W. Germany

ABSTRACT. Line profile-variability (LPV) is very wide-spread among early-type stars. With the exception of inhomogeneous surface abundance distributions associated with magnetic fields, the LPV of *bona fide* non-magnetic stars is consistent with (often only with) nonradial pulsation (NRP). Peculiar surface chemistry and NRP may even be mutually exclusive, and there are other indications that in spite of the ubiquity of NRPs the distribution of their characterizing parameters in the HRD is far from being uniform. This may be important for uncovering the driving mechanism of NRPs in early-type stars.

1. INTRODUCTION

In ideal stars the rotational Doppler effect only broadens the spectral lines whereas on many real line profiles the fingerprints of surface inhomogeneities are imposed as various irregularities and distortions. From a single observation one may only conclude that there are variations with stellar longitude. A series of spectra not only complements the information to the full stellar circumference but may also reveal dependences on latitude. However, extracting this information and reconstructing a stellar image requires an inversion of the observations, and it is a matter of definition if this favour to the observer is called Doppler imaging or whether that term is reserved for the solution of the inversion problem.

If the inhomogeneities are merely of a radiometric nature the correspondance between the location of a feature on the rotating star and its propagation in the observed spectra is relatively unique, and the inversion reduces largely to a geometrical-mathematical problem (see the papers by Vogt and by Hatzes in these proceedings). However, if the inhomogeneities are kinematic ones due to an atmospheric velocity field, the position of a feature in the line profile depends on the location of the absorbing (or emitting) gas on the stellar surface *and* its velocity. The solution of the inversion problem is, then, so little constrained by the observations alone that the usage of a physical model is indispensible. Intermediate between these two extremes are Zeeman broadened lines of strongly magnetic stars.

2. DOPPLER MAPPING THE HERTZSPRUNG-RUSSEL DIAGRAM

The spatial resolution of Doppler images is the higher, the faster the star rotates, so that

*Based in part on observations obtained at the European Southern Observatory, La Silla, Chile

this technique is especially applicable to early–type stars. However, with increasing broadening spectral lines become shallower. With regard to the subject of this symposium it is therefore interesting to note that the first detections of line profile variability (LPV) were made in narrow–lined early–type stars several decades ago whereas efficient low–noise detectors permitted a systematic exploration of the upper left corner of the HRD only rather recently. Additions of major groups of early–type stars to the LPV domain have been:
- β Cephei stars (Henroteau 1918),
- narrow–lined B stars (Petrie and Pearce 1962, Smith 1977),
- Be stars (Baade 1979, Walker et al. 1979, Vogt and Penrod 1983),
- medium– (Petrie 1958, Smith 1985) and broad–lined (Smith and Penrod 1984) B stars,
- broad–lined ($\geq$ 100 km/s) OB supergiants (Baade and Ferlet 1984, Baade 1987),
- helium–variable stars (Bolton et al. 1987; Bohlender and Landstreet, these proc.),
- δ Scuti stars (Yang and Walker 1987),
- Ap stars (Hatzes, these procedings; Landstreet, these proceedings)
- Herbig Ae stars (Baade and Stahl 1988, in preparation).

LPV has been detected in spectral types as early as O4 (Baade 1987b), and for the normal members of any of the groups listed exceptions from LPV have not so far been reported. The emerging ubiquity of LPV may, therefore, eventually make non–detections not less significant results and give them their own diagnostic value. Known LPV voids or very low amplitude regions in the HRD include:
- the zone between about B7 and the blue edge ($\sim$ A2) of the δ Scuti instability strip (Smith and Penrod 1984, Baade 1986) with the possibly important exception of the Herbig A0e star HD 163296 (Baade and Stahl 1988, in preparation).
- narrow–lined ($v \sin i \leq$ 100 km/s) OB supergiants (Baade 1987b).
- some chemically peculiar stars (Smith and Stern 1979; Baade unpublished).

Occasionally, also Be stars may display no detectable LPV; but as they are then probably merely in a 'dormant' phase (see Baade, these proceedings), the phenomenon is different.

3. CLASSIFYING THE PHENOMENA

The necessary pre–classification (see Introduction) of the LPV is the easiest for stars with measured magnetic fields because the solution of the inversion problem is more constrained. Unless the star is also pulsating, there is just one period which along with information about the geometry of the magnetic field is usually known from photometry or polarimetry. Lines with the least magnetic sensitivity can then be used to infer the abundance geometry of various ions. From the analysis of lines with different Landè g factors and the comparison with the magnetic field measurements, further refinements are possible as detailed by Bohlender and Landstreet, by Hatzes, and by Landstreet in these proceedings.

Without magnetic field, the distinction between radiometric and kinematic inhomogeneities can only be based on line profiles and hopefully also on simultaneous photometry (for a clearer separation of velocity and temperature effects on the line profiles). The criteria for the discrimination between corotating surface inhomogeneities and nonradial pulsation have been summarized many times in recent years (Vogt and Penrod 1983, Smith 1986, 1987; Percy 1987; Baade 1986, 1987a). Most compelling in favour of NRPs are multiperiodicity and phase velocities that differ drastically from the equatorial rotation velocity. In fact, spectroscopy has not so far detected any non–magnetic early–type star with spots.

However, the interpretation in terms of NRPs requires a model (see Introduction), and there is always the danger that the analysis merely 'confirms' the assumptions it is based upon. This is all the more so since in principle the eigenfunctions of rotating nonradially pulsating stars are the superposition of an infinite number of spherical harmonics which form a complete set and therefore can reproduce any observation if only the amplitudes can

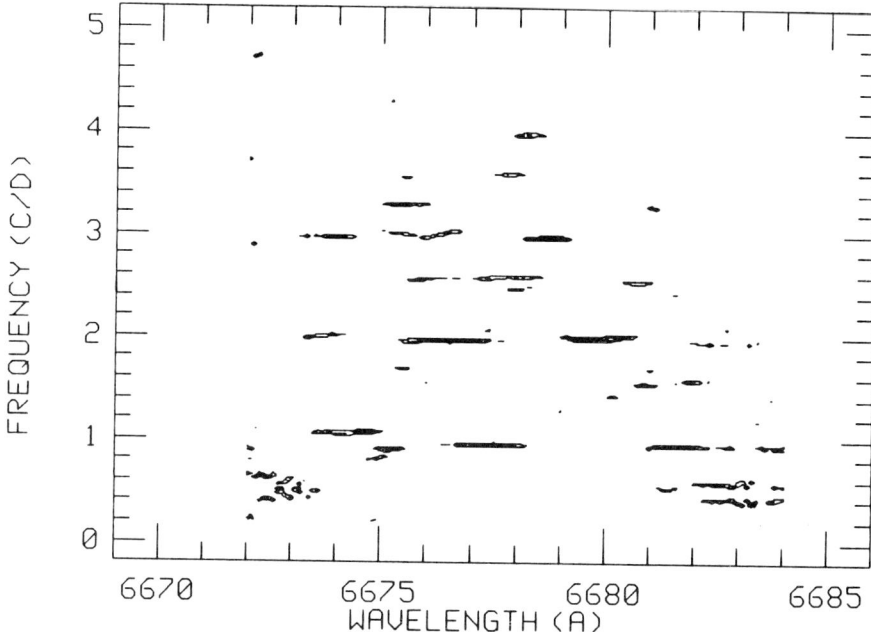

Figure 1: Composite power spectrum for 122 He I λ 6678 Å profiles obtained of μ Cen (B2 IV–Ve) over 9 nights in 1987 April (Baade, these proceedings). Individual power spectra have been calculated for time series of 0.1 Å wide bins, deconvolved with the window spectrum and the CLEAN algorithm, and combined according to wavelength. Power contours are shown for 0.1, 0.2, 0.3, 0.4, and 0.5 in arbitrary units; step size in frequency is 0.02 c/d.

be suitably chosen. Fortunately, recent work by Lee and Saio (1986) suggests considerable constraints on the number of terms with significant amplitudes. But to minimize the unavoidable model dependence, any possibility to determine free parameters beforehand should be exploited. This notably concerns the number of modes and their periods.

Unfortunately, photometry with its conveniently Fourier–analyzable scalar data in practice finds it difficult to detect multiple periods because the amplitudes are so low (see, e.g., Smith et al. 1987). Spectroscopy appears in a better position because a fit of a single line profile would even fix the *phase*. However, with multiply periodic stars this turns into a shortcoming when, after the gross LPV has been accounted for, profile fitting has to decide whether to add more pulsation modes or whether to adjust the ephemeris of the modes identified already. The former alternative will in practice be difficult to choose because errors in the periods, phases, or amplitudes of the modes included may render any additional modes undeterminable from the residuals. The second one may imply curious results such as that the observer 'knows' the pulsation ephemeris better than the star does.

Balona (1986) has proposed to use the first four moments of the line profiles to obtain model–independent periods. However, sacrificing the high spectral resolution appears a doubtful strategy. A method that without a model uses all observations simultaneously at all wavelengths has recently been applied by Gies and Kullavanijaya (1987) to ε Per (B0.7 III) and by Baade (these proceedings) to μ Cen (B2 IV–Ve). For each resolution element in wavelength of a given spectral line, time series are constructed from the flux measurements in all spectra. Each time series can then be independently analysed for periodicities by conventional techniques. An example where the individual power spectra obtained have been joined together according to their respective wavelengths is shown in Fig. 1. Not unexpectedly, this technique finds more periods than other analyses of the same stars had before. But the confirmation, within the errors, of the previously determined periods also gives credit to the earlier results. Gies and Kullavanijaya additionally offer independent evidence that the mode orders, m, had been correctly identified by Smith et al. (1987). It remains to be seen if inclusion of the additional modes can alleviate some of the former oddities such as rapidly varying O–C residuals of the pulsation phase.

4. DOPPLER IMAGING CIRCUMSTELLAR MATTER

Doppler imaging is also possible of moving circumstellar structures. In σ Ori E, corotating matter above the intersection of magnetic and rotational equator produces complex variations of emission and absorption components in Hα (Bolton et al. 1987). In Be stars, double–peaked emission lines arise from a rotating envelope and their V/R ratio often varies with the period of the stellar low–order NRP mode. (Baade 1979, Smith and Penrod 1984, Baade 1987a). The pulsation therefore locally leads to either an enhancement of the circumstellar matter density or a change of the ionization balance caused by the variable radiation field associated particularly with a vertical component of the NRP velocity field. Similar V/R variations are also seen in broad–lined OB supergiants (Baade 1987b).

Non–saturated lines formed in the extremely fast winds of early–type stars also provide maps which, however, so far have proven to be difficult to read. Especially the narrow components (Henrichs 1987) are not understood because they are often nearly stationary for many wind flow times. But others have also been observed to drift red–to–blue across the wind profiles so that they could be due to density enhancements being accelerated in the ambient wind. Similar, complex variability has recently been detected in optical absorption lines of several WNL+abs stars (Stahl, Vreux, Magain and Baade 1988, in preparation).

5. SUMMARY AND CONCLUSIONS

While early–type stars do not find it easy to escape NRP, the contrary is true of theoretical models. Covering a factor of ~4 in T_{eff}, any surface driving mechanism would be little constrained by the observations. Not at last for that reason, Osaki (1987) and Lee and Saio (1987) consider the coupling between oscillatory convection of the core and a gravity mode of the envelope. In fact, the periods of low–order modes are often too long for p–modes (Smith 1986). The range of super–periods, i.e. the time needed for one complete revolution (in the observer's frame) of the pulsation pattern about the star, is smaller than expected on the basis of the range of probable surface rotation periods (Smith and Penrod 1984, Baade 1986). This may provide additional circumstantial evidence of the core dominating over the envelope and/or the rotation acting as an active mode filter.

The distribution in the HRD of the most prominent pulsation modes is clearly non-uniform. For instance, low–order modes seem to be the distinguishing characteristic of Be stars vs. Bn stars. A more detailed mapping of the HRD may therefore eventually provide some insight into the structure and evolutionary state of various early–type stars. Particularly interesting are deviations from the general pattern. E.g., the ^{3}He rich star 3 Cen A (Smith and Stern 1979) and the mild OBN star θ Car (B0 Vp; Baade, unpublished) show no indication of NRP related LPV. An exclusion mechanism between unusual surface abundances and NRP is therefore possible (but θ Car is also a close binary) and should be investigated in more detail. The Herbig A0e star HD 163296 shows LPV with all the symptoms known of nonradially pulsating stars. If this preliminary classification is borne out by a more detailed study (Baade and Stahl, in preparation), not only the first nonradial pulsator among pre-main sequence stars would have been discovered, but also the first one in the apparent NRP void between B7 and A2 (Smith and Penrod 1984, Baade 1986). This combination triggers the speculation if the core of this star has already reached the evolutionary stage where pulsation à la Osaki (1987) and Lee and Saio (1987) is sustained while the surface temperature is still lagging behind.

Acknowledgements: I thank Drs. Myron Smith, Alex Fullerton, and John Percy and Drs. Doug Gies and Anchana Kullavanijaya for sending me preprints of their respective papers on ϵ Per. Doug Gies and Myron Smith also provided useful comments on the manuscript.

REFERENCES

Baade, D.: 1979, *The Messenger* (ESO) No. **19**, p. 4.
Baade, D.: 1986, in *Highlights of Astronomy*, J.-P. Swings (ed.), D. Reidel, Vol. **7**, p. 255.
Baade, D.: 1987a, in Proc. IAU Coll. No. **98** *Physics of Be Stars*, A. Slettebak and T.P. Snow (eds.), Cambridge Univ. Press, Cambridge, p. 361.
Baade, D.: 1987b, in *O, Of and Wolf-Rayet Stars*, P.S. Conti and A.B. Underhill (eds.), NASA/CNRS *Monogr. Ser. on Nontherm. Phenom. in Stell. Atmosph.*, in press.
Baade, D., Ferlet, R.: 1984, *Astron. Astrophys.* **140**, 72.
Balona, L.A.: 1986, *Mon. Not. R. astr. Soc.* **220**, 647.
Bolton, C.T., Fullerton, A.W., Bohlender, D., Landstreet, J.D., Gies, D.R.: in Proc. IAU Coll. No. **98** *Physics of Be Stars*, A. Slettebak and T.P. Snow (eds.), Cambridge Univ. Press, Cambridge, p. 82.
Gies, D.R., Kullavanijaya, A.: 1987, *Astrophys. J.*, submitted.
Henrichs, H.F.: 1987, in *O, Of and Wolf-Rayet Stars*, P.S. Conti and A.B. Underhill (eds.), NASA/CNRS *Monogr. Ser. on Nontherm. Phenom. in Stell. Atmosph.*, in press.
Henroteau, F.: 1918, *Lick Obs. Bull.* **9**, 155.
Lee, U., Saio, H.: 1986, *Mon. Not. R. astr. Soc.* **221**, 365.
Lee, U., Saio, H.: 1987, *Mon. Not. R. astr. Soc.* **224**, 513.
Osaki, Y.: 1987, in *Instabilities in Luminous Early-type Stars*, H. Lamers and C. de Loore (eds.), D. Reidel, Dordrecht.
Percy, J.: 1987, in Proc. IAU Coll. No. **98** *Physics of Be Stars*, A. Slettebak and T.P. Snow (eds.), Cambridge Univ. Press, Cambridge, p. 49.
Petrie, R.M.: 1958, *Mon. Not. R. astr. Soc.* **118**, 80.
Petrie, R.M., Pearce, J.A.: 1962, *Pub. Dominion Astrophys. Obs.* **12**, 1.
Smith, M.A.: 1977, in *Proc. Workshop on Puls. B Stars*, C. Sterken and G.E.V.O.N. (eds.), Obs. Nice, Nice, p. 317.
Smith, M.A.: 1985, *Astrophys. J.* **288**, 266.
Smith, M.A.: 1986, in *Hydrodynamic and Magnetohydrodynamic Problems in the Sun and Stars*, Y. Osaki (ed.), Univ. of Tokyo Press, Tokyo, p. 145.
Smith, M.A.: 1987, in *Pulsation and Mass Loss in Stars*, L.A. Willson and R. Stalio (eds.), Obs. Astr. Trieste, Trieste, in press.
Smith, M.A., Fullerton, A.W., Percy, J.R.: 1987, *Astrophys. J.*, in press.
Smith, M.A., Penrod, G.D.: 1984, in *Relations between Chromospheric-coronal Heating and Mass Loss in Stars*, R. Stalio and J.B. Zirker (eds.), Trieste, p. 394.
Smith, M.A., Stern, S.A.: 1979, *Astron. J.* **84**, 1363.
Vogt, S.S, Penrod, G.D.: 1983, *Astrophys. J.* **275**, 661.
Yang, S., Walker, G.A.H.: 1986, *Publ. Astron. Soc. Pacific* **98**, 1156.
Walker, G.A.H., Yang, S., Fahlmann, G.G.: 1979, *Astrophys. J.* **233**, 199.

DISCUSSION

   PRADERIE   I wonder how you identify the order and degree of the non-radial oscillations, in the case of the stars you study ? What is the systematics ?

   BAADE   If you continuously observe a single line in a single star over about 10 days, you'll find many similar (if not nearly identical) single profiles, however hardly ever identical sequences. This almost certainly means multiperiodicity. So, the first step to bring some system into this apparent chaos is to establish reliable periods without imposing a model early on. This has been done by Gies and Kullavanijaya (1987) for ϵ Per (B0.7 III), and in my next paper I'll show the same type of analysis for μ Cen (B2 IVe). In either case, multiple periods were found. To what extent mode identifications can be obtained depends not at last on whether the eigenfunctions of non-rotating stars are acceptable approximations for the rapidly rotating stars observed whose eigenfunctions are the superposition of in principle infinitely many spherical harmonics. If you believe in such an approximation, the modes can be identified by line profile fitting, analysis of the observed acceleration of the line profile bumps (cf. the work by Gies and Kullavanijaya), etc. The radial overtone, n, can only be derived from the comparison with suitable stellar models.

   BOHANNAN   Do you get a consistent period of the absorption line features in ζPuppis from observing session to observing session ?

   BAADE   For 2 consecutive observing seasons this seems to be the case if my preliminary analysis is correct. However, with a one year gap in the data it most probably won't be posssible to check the phase coherence.

DOPPLER IMAGES OF RAPIDLY ROTATING Ap STARS

Artie P. Hatzes
Lick Observatory, Board of Studies in Astronomy and Astrophysics
University of California, Santa Cruz, CA 95064

ABSTRACT. The magnetic Ap stars are characterized by the presence of large magnetic fields which undergo periodic variations. These magnetic field variations are accompanied by spectral variations caused by the inhomogeneous distribution of elements on the stellar surface. It is believed that the magnetic field plays an important role in determining this distribution. Accurate maps of the surface distribution of elements would provide valuable probes as to the field geometry as well as provide clues to the role of the magnetic fields in the atmospheres of these stars. We have developed a new technique for mapping the local equivalent width on a stellar surface from the observed spectral line variations.

I. THE INVERSE PROBLEM

The technique employed is a modified version of the Doppler imaging technique which incorporates the maximum entropy formalism. Since this technique is described in greater detail in the paper by Vogt in this publication, we will only give a brief outline of the technique and refer the reader to Vogt's contribution.

The inverse problem (mapping local equivalent width on a star from spectral variations) is posed as a matrix equation. The image of the star is divided into n cells which are unwrapped to form an image vector $\mathbf{I}$. Each pixel represents the local equivalent width of the star at the location of that pixel. Observed spectral lines as a function of phase are attached end-to-end to form a data vector $\mathbf{D}$ of m elements. To map from image space to data space requires a transfer matrix $\mathbf{R}$ ($n \times m$ elements) such that $\mathbf{D} = \mathbf{I} \cdot \mathbf{R}$

The elements of transfer matrix $\mathbf{R}$ represent the response of a datum pixel to changes in an image pixel. These elements are merely the specific line intensities computed by an LTE atmosphere divided by the equivalent width of the profile. Information about the star such as rotational velocity, inclination, as well as information about the stellar atmosphere (limb darkening, macroturbulence, $etc.$) are used in construction of the matrix $\mathbf{R}$.

Since the problem is ill-posed, $\mathbf{R}$ cannot be inverted to solve for the image vector $\mathbf{I}$. Instead we approximate the solution by searching in image space until a vector is found which fits the observed data to within the noise level. Since the number of possible image vectors consistent with the data set can be large (non-uniqueness) we impose the additional criterion that the final image is the simplest or smoothest one consistent with the observed data. The entropy of an image, which is defined as the negative of the information content, provides a convenient measure of this smoothness. The image with the least information thus has the maximum entropy. Our technique uses the algorithm developed by Drs. J. Skilling and S.F. Gull to find the image with the maximum entropy (least information) consistent with the data. The reader is referred to the article by Vogt, Penrod, and Hatzes (1987) for extensive tests conducted with the technique.

II. DATA ACQUISITION AND RESULTING IMAGES

Data sets were obtained using the 6347 Å line of Si II in γ^2 Ari and the 4824 Å line of Cr II in ϵ UMa. The data were obtained using the Lick Observatory 3 meter telescope at coudé focus and a TI 800×800 3-phase CCD detector. We have been able to achieve signal-to-noise of 300-500 per pixel and with a resolving power of 60,000 for Cr II and 51,000 for the Si II line.

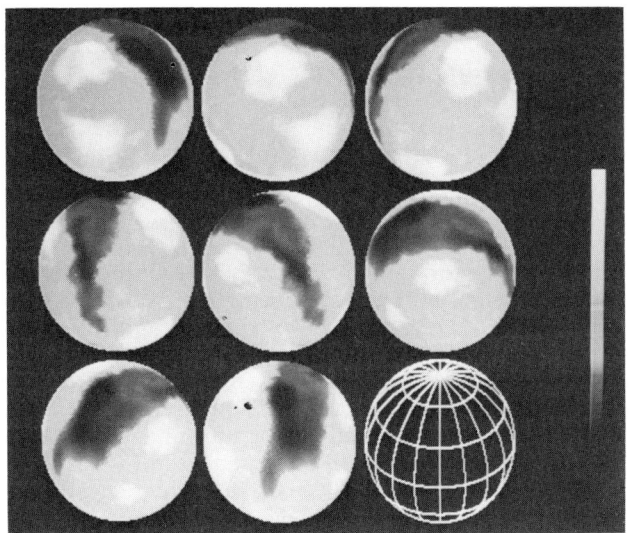

Figure 1a: Local equivalent width map of Cr II for ϵ Uma. Minimum = 70 mÅ (black), Maximum = 250 mÅ (white), Mean = 120Å (grey).

ϵ UMa is an A0pCr star with a well known period of 5.0887 days. For the modeling a $v \sin i$ of 33 km s^{-1} and an inclination of 54° were used. Figure 1a shows the resulting equivalent width map for the 4824 Å line of Cr II. In the image black represents underabundance with respect to the mean while white represents overabundance. Figure 1b shows the resulting line fits to the data as a function of phase derived from the reconstructed map (crosses represent observed data and lines represent the fit). The most prominent feature is the large arc or annulus of depleted abundance which passes by at phase 0.375 and 0.875. The plane in which this arc lies appears to miss the center of the star by about 1/5 of the stellar radius. The equivalent width inside this arc is about 70 $m\text{Å}$ while the mean for the star is about 110 $m\text{Å}$. Also present are three overabundant spots situated in a circle of radius 50° about the point 0° latitude and phase 0.125 The maximum equivalent width in these spots is about 250 $m\text{Å}$.

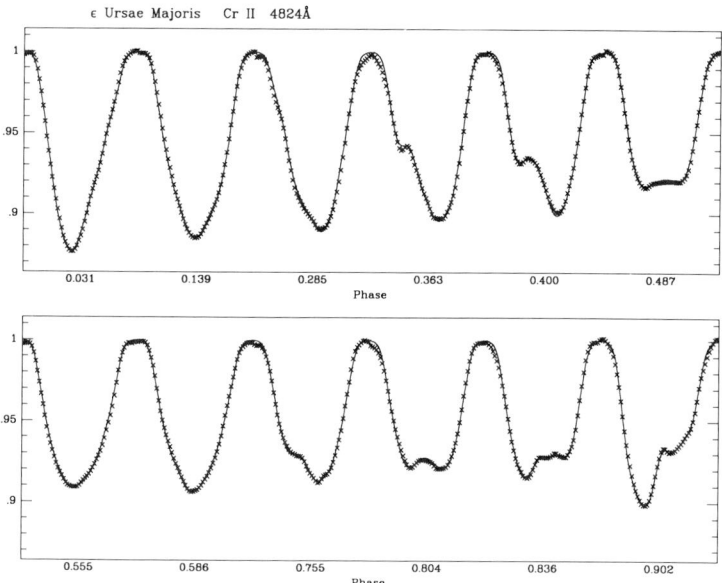

Figure 1b: Observed spectral line as a function of phase (crosses) for Cr II in ϵ UMa and the resulting fits (line) from the equivalent width distribution in Fig. 1

Figure 2 represents the maximum entropy solution to the local equivalent width of silicon in the star γ^2 Ari. Space does not permit showing the line fits to the data. For modeling we used a measured $v\sin i$ of 64 km s^{-1}, an inclination of 55°, and the magnetic period of 1.6093^d. There is a prominent underabundant spot 50° in latitude from the rotation axis. The equivalent width in this spot is at least a factor of 5 less than that of the mean value of 170 $m\text{Å}$ across the stellar surface. This spot is coincident in longitude to the negative magnetic pole as measured by Borra and Landstreet (1980). Surrounding this spot is a non-uniform annulus of overabundance extending in radius from about 30° to 60° and centered on the underabundant spot. The equivalent width in this annulus attains a peak value of 3 times the mean. There is also a much less prominent underabundant spot at lower latitudes and 180° in longitude from the first.

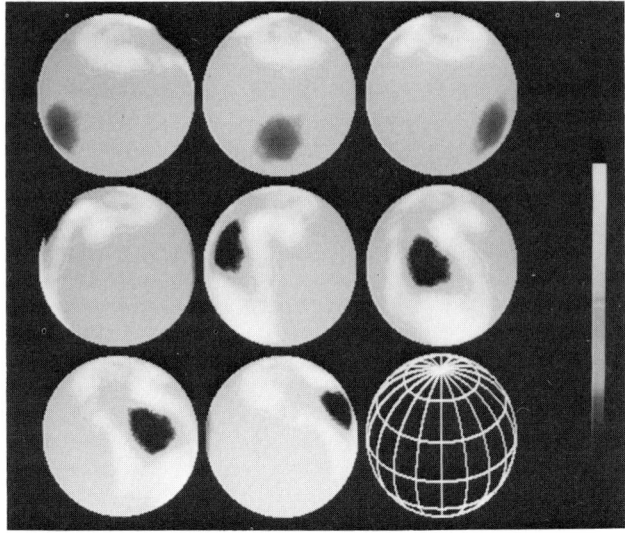

Figure 2: Equivalent width map of Si II 6347Å for γ^2 Ari. Maximum = 400Å (white), minimum = 10Å (black), mean = 170Å (grey).

III. INTERPRETATION OF THE IMAGES

The diffusion mechanism of Michaud (1970) is one explanation for the spectral variations. Work by Vauclair, Hardorp, and Peterson (1979) suggests that silicon should accumulate where fields lines are horizontal. The silicon image of γ^2 Ari seems to be consistent with this prediction. The two underabundant spots represent the two poles (obliquity angle = 50°) while the overabundant annulus must lie near the magnetic equator. This annulus is at too high a magnetic latitude to be accounted for by a pure dipole field. This can be explained by the addition of a quadrupole component or by using a dipole field whose center is displaced from the star's center toward that of the negative pole.

The interpretation of the chromium image of ϵ UMa is uncertain due to the lack of theoretical work on diffusion of chromium in the presence of magnetic fields. It is clear, however, that the great underabundant annulus most likely represents the magnetic equator. If chromium, like silicon, is depleted where the field lines are vertical, then the equivalent width map for this star suggests a quadrupole field. However, an alternate explanation is to invoke horizontal diffusion. Mégessier (1984) argued that for more evolved stars, like ϵ UMa, horizontal diffusion should become important and that the silicon band should migrate towards the magnetic poles as the star evolves. The overabundant spots may represent the enhanced abundance band which was initially at the equator, but which has migrated poleward leaving a depleted band at the equator.

IV. CONCLUSION

In conclusion we believe that this technique applied to Doppler imaging of the magnetic Ap stars represents a powerful tool for deriving accurate local equivalent width maps on these stars. Early indications are that silicon accumulates where the field lines are horizontal and is depleted where the field lines are vertical. These results are consistent with predictions of diffusion theory and may represent the first direct observational evidence that this mechanism is present in the atmospheres of these stars.

REFERENCES

Borra, E.F., and Landstreet J.D. 1980, *Astrophys. J. Supp.*, **42**, 421.

Mégessier, C. 1984, *Astron. Astrophys.*, **138**, 267.

Michaud, G., 1970, *Astrophys. J.*, **160**, 640

Vauclair, S., Hardorp, J., and Peterson, D.M. 1979, *Astrophys. J.*, **227**, 526.

Vogt, S.S., Penrod, G.D. and Hatzes, A.P. 1987, *Astrophys. J.* In press.

DISCUSSION

LANDSTREET:

Remark 1: Latitudes of magnetic poles determined from Balmer line Zeeman analyzer data of Borra and Landstreet are in general determined with rather low precision (errors in inclination of rotation axis, obliquity of magnetic field of $\simeq 20°$). So one can't really say that pole of equivalent width distribution coincides with magnetic pole, only that the two determination are compatible.

Remark 2: Borrar and Landstreet's measurements are not very sensitive to presence of a quadrupole component. A quadrupole/dipole ratio of $\simeq 1$ could easily be present without being obvious in the data. Thus a "quadrupolar" map of equivalent width is not contradicted by observation of a "dipolar" field.

Question: How well is inclination of distribution pole (or rotation axis) to the line of sight determined?

HATZES: The distribution of equivalent width found is almost independent of assumed inclination.

LANDSTREET: Then the inclination is not accurately determined.

HATZES: The inclination (absolute orientation) of the star is not well determined. However, the location of spots and other features (which determines the obliquity angle of the field) with respect to the rotation axis is rather insensitive to the assumed inclination of the star.

LINSKY: Please comment on the S/N and the number of obvservations per rotation period required for your mapping technique to work.

HATZES: That of course depends on the complexity of the spot configurations one is trying to reconstruct. For the Ap stars which exhibit complex distributions, I like to have S/N ≥ 300 and 16-20 phases. For spot configurations with simple shapes good reconstructions can be made with S/N as low as 150-200 and 8 equally spaced observations.

GUSTAFSSON: Two questions: 1 - Do you restrict the "local" value of the equivalent width (apart from $W_\lambda \geq 0$) or is it a completely free parameter? and 2 - Do you limit the geometrical complexity of your solution?

HATZES: 1 - The technique assumes non-negative pixel values (i.e., $W_\lambda \geq 0$), no other restrictions are made on the equivalent width values, and 2 - The only restriction of our solution is that it is the smoothest image consistent with the data set (the maximum entropy formalism).

HIGH TIME RESOLUTION SPECTROSCOPIC OBSERVATIONS OF STELLAR SHOCK WAVES

P.L. Cottrell, W.A. Lawson and S.M. Smith
Mount John University Observatory
Department of Physics
University of Canterbury
New Zealand

ABSTRACT. An absorption line-splitting phenomenon, first reported by Cottrell and Lambert (1982a), has been shown to occur at about maximum light in the semi-regular pulsations of the R Coronae Borealis (RCB) star, RY Sgr (Lawson 1986). This has been interpreted as a shock wave propagating through the photospheric layers (Lawson and Cottrell 1986). We present spectroscopic observations of this star, taken to coincide with this line-splitting event. A sequence obtained during 1986 October revealed that this event extended over about 6 days (out of a period of about 40 days) and began at about the bluest B-V. This colour maximum, which corresponds to maximum photospheric temperatures and minimum radius, leads the V maximum by about 6 days.

1. INTRODUCTION

Shock wave phenomena have been observed in a number of different types of relatively normal composition stars (e.g., W Vir stars : Sanford 1952; RR Lyrae stars : Oke 1966; RV Tauri stars : Abt 1955; Mira stars : Hinkle et al 1984; β Cephei stars : Crowe and Gillet 1988). The shocks are most graphically seen by the discontinuity in the radial velocity curve. However, with their appearance in one RCB star, which as a group have hydrogen deficient carbon rich atmospheres, one can study the propagation of these waves under substantially different atmospheric opacity conditions. This paper presents some preliminary observations. Also we indicate the importance of simultaneous photometric observations and the usefulness of a small University operated observatory, where we can obtain key sequences of data in these semi-regular pulsators.

2. OBSERVATIONS

All the observations were obtained using telescopes at Mount John University Observatory.
The spectra were acquired using an échelle spectrograph and image

tube attached to a 0.6m Boller and Chivens telescope. The observations presented here were recorded on Kodak IIaD plates and traced using a Joyce-Loebl microdensitometer at the University of Canterbury. These plates were only obtained at phases which were known from previous cycles to be the most likely to show splitting of the spectral lines. Later this year, the image tube and photographic plate combination will be replaced by an 1872 element Reticon linear diode array detector, designed and built by one of our graduate students, Phillip MacQueen.

The UBV photometry was undertaken on a 0.6m Optical Craftsmen telescope using a single channel photometer, as part of a larger program to determine the amplitude and period of southern RCB stars.

3. DETAILS OF THE LINE-SPLITTING EVENT IN RY SGR

In Figure 1 we show the B-V, V and radial velocity determinations for the interval, 1986 August - October, while Figure 2 is a sample sequence of spectra which illustrate the development of the line splitting phenomenon. In particular, the latter figure shows the growth of another line (presumably produced by the shock wave) in the blue wing of the stronger lines in the spectrum. This is illustrated here by the two SiII lines and one FeII line. It should be noted that although the two silicon lines are of the same multiplet and excitation potential, they

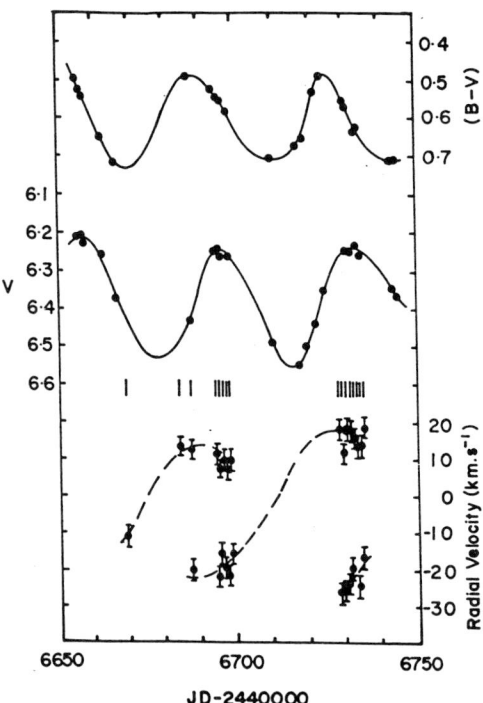

Figure 1. B-V, V and radial velocity curves for the RCB star, RY Sgr. The vertical lines between the V and velocity curves indicate the dates on which spectroscopic observations were obtained. The curves drawn on this figure are 'eye-ball' best fits to the data.

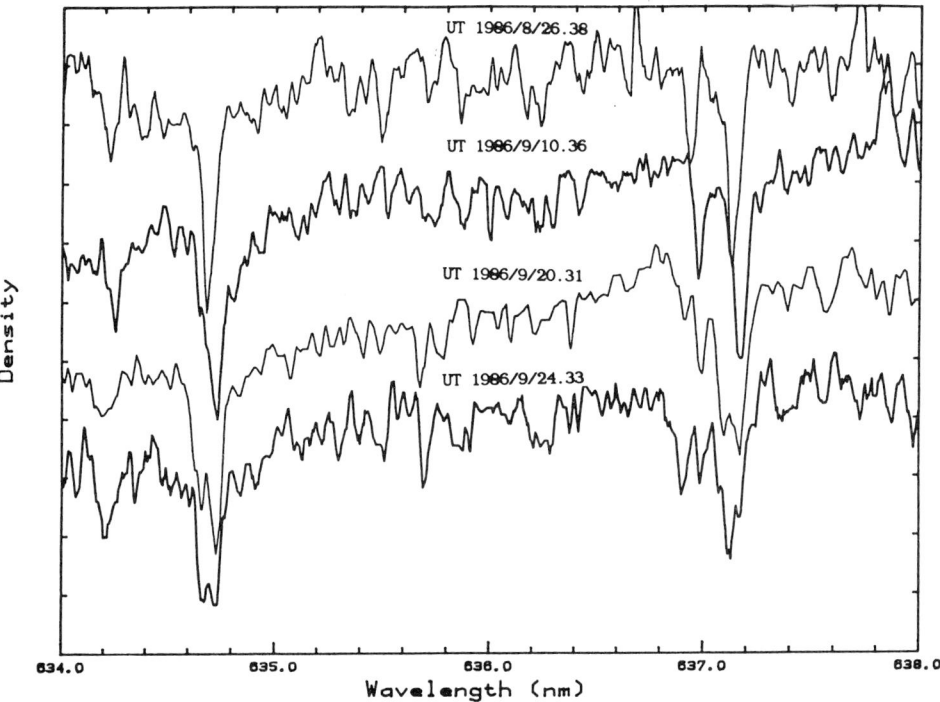

Figure 2. Sample spectra of RY Sgr acquired at the times shown immediately above each plot. The shift of the lines towards longer wavelength is clearly seen between the first and second spectra. The gradual strengthening of the blue (shock) component of the lines is shown in the second to fourth spectra.

show different development during the cycle. The continuous spectral coverage in 1986 October show similar effects in the shock induced features, but any detailed variation from cycle to cycle is yet to be investigated. We expect to do this with the Reticon system, which will enable much **higher signal to noise** and **higher time resolution** data to investigate the propagation of the shock wave and to look for depth dependence effects in the spectral lines. Already we have seen (see Ti II line in Figure 1 of Cottrell and Lambert 1982b) some lines which show emission above the continuum as a consequence of the shock.

The onset of the splitting phase commences near the maximum on the colour curve (minimum B-V, highest surface temperature and minimum radius) as the shock reaches the deepest photospheric layers. This **colour maximum leads** the **visual maximum** by between 4 and 8 days.

4. THE FUTURE

We hope that our continued spectroscopic and photometric coverage of RY Sgr will lead to a better understanding of the propagation of shock waves in stellar material. Our observational program also involves the investigation of many other southern RCB stars, the first phase of which is to obtain good photometric ephemerides of the pulsation of these stars with a view to determining appropriate phases for detailed spectroscopic work.

On the theoretical side, we have recently begun work with Peter Wood (Mount Stromlo and Siding Spring Observatories) to model the pulsations and determine the conditions under which shock waves will be produced, and also whether realistic pulsational amplitudes can be obtained, in these hydrogen deficient, carbon rich stellar envelopes.

One of us (PLC) would like to thank both the IAU and the Council of the University of Canterbury for generous travel support which enabled participation in this symposium. We would also like to thank Pam Kilmartin for obtaining many of the photometric observations and Alan Gilmore for acquiring some of the spectra when teaching commitments kept us away from the mountain.

5. REFERENCES

Abt, H.A. 1955, Ap.J., 122, 72.
Cottrell, P.L., and Lambert, D.L. 1982a, Obs., 102, 149.
Cottrell, P.L., and Lambert, D.L. 1982b, Ap.J., 261, 595.
Crowe, R., and Gillet, D. 1988, this symposium.
Hinkle, K.H., Scharalach, W.W.G., and Hall, D.N.B. 1984, Ap.J.(Suppl.), 56, 1.
Lawson, W.A. 1986, in *IAU Colloquium 87, Hydrogen Deficient Stars and Related Objects*, ed K. Hunger et al. (Dordrecht:Reidel), p. 211.
Lawson, W.A., and Cottrell, P.L. 1986, Obs., 106, 169.
Oke, J.B. 1966, Ap.J., 145, 468.
Sanford, R.F. 1952, Ap.J., 116, 331.

DISCUSSION

CROWE At what luminosity phase does the line splitting occur?

COTTRELL The line splitting begins at minimum B-V (maximum temperature) which can lead the V maximum by between 4 and 8 days. This then gives a phase $\phi = 0.8$ to 0.9 for the line splitting.

CROWE Yes, it seems that the velocity discontinuity usuallly occurs at phase $\phi = 0.9$, with respect to visual light maximum, for the Population II pulsating variables as well as for BW Vulpeculae (and for Miras).

COTTRELL Precise phasing has not been done because of the semi-regular nature of the star and incomplete spectral coverage.

Seismology of the stellar cores

Eric Fossat
Département d'Astrophysique,
Université de Nice, FRANCE.

The tremendous development of helioseismology in the last ten years has included extremely important results obtained through full disk measurements, both in Doppler shifts and broad band photometry. It was then looking very attractive to try to detect the same kind of small amplitude p-modes on other stars.

The amplitude of oscillations to be detected are in the range of 1 to 50 cm/s in Doppler, or 10^{-5} to 10^{-6} magnitude in photometry. Periods range from a few minutes to hours. For this purpose, very specialized instruments are under development or in project. I will describe briefly the main directions of this development.

I will also describe the first (still controversial) results obtained on Procyon, Alpha Centauri and Epsilon Eridani.

At first order, the initial results of asteroseismology will provide two important parameters: First the frequency of the fundamental mode of vibration of the star, which is an almost direct measurement of the stellar radius, and second a small frequency splitting between radial and quadrupolar modes, which measures the sound speed gradient inside the stellar core. This gradient is sensitive to stellar age, by the change of chemical composition implied by hydrogen burning along the main sequence life. These two parameters are among the main ingredients of the theory of evolution and their precise measurement on the sun can help a calibration of their role in the evolutive models.

It can now be predicted that with together Hipparcos and asteroseismology, the precision of the HR diagram will be improved in the near future at a level which would have looked impossible not so long ago.

DISCUSSION

G. CAYREL Have you already estimates the age of α Cen ? You may know that there is a discrepancy between its age as estimated from isochrones and its age as estimated from its Li abundance.

SODERBLOM The age of α Cen is 6 to 8 Gyr (e.g., Flannery and Ayres 1978) and for α Cen A its lithium is twice the solar abundance (Soderblom and Dravins 1984). Although α Cen A is a spectroscopic twin to the Sun, these values are compatible because we know this star's mass, namely 1.1 $M_\odot$. Having a thinner convective zone, a star more massive than the Sun depletes Li at a slower rate.

THEORETICAL CONSTRAINTS FROM ASTEROSEISMOLOGICAL HIGH S/N OBSERVATIONS

W. Däppen
Observatoire de Paris-Meudon
92195 Meudon
France

ABSTRACT. The two firmly established cases of many-mode observations, the Sun and Kurtz's Ap stars, suggest that in stars extremely small pulsation amplitudes are to be expected. If spectroscopic Doppler techniques are used to measure the velocity pattern, then it is obvious that with better resolution and higher S/N ratio of the spectral lines the sensitivity to detect velocity amplitudes increases. Observed pulsation frequencies of any star will put new constraints on the theory of stellar evolution. Besides addressing the two most important issues, namely determination of age and chemical composition, observed periods will also help resolve open questions in the _physics_ of stellar interiors.

1. INTRODUCTION

   Stellar acoustic oscillation frequencies will likely be observed accurately in the near future, in analogy to the well-known solar five-minute oscillation frequencies. Of course we will never expect the wealth of the solar data, which is a result of the high spatial resolution of the Sun. Therefore we will not be able to solve the inverse problem, that is to probe physical quantities as functions of depth. Furthermore, in the case of the Sun, the large number of observed frequencies allows a unique mode identification (see, for instance, Deubner and Gough 1984). This will not be true for the small number of expected stellar frequencies. As has been emphasized by Däppen et al. (1988), asteroseismology should therefore not simply be understood as the stellar analogue of helioseismology (which it is not and cannot be), but rather as a method of testing stellar structure and evolution theory, using _all_ available pulsation data, and not just oscillation frequencies.
   Despite the lack of analogy with the solar case, prospects for asteroseismology are excellent. Given the fundamental importance of the theory of stellar evolution in astrophysics (for instance in the determination of age and chemical composition), it is clear that any observed quantity besides temperature and luminosity can be used to reduce existing uncertainties in the physics of stellar interiors

(e.g. convection, opacity, equation of state, nuclear reaction, etc.).
While this is intuitively clear, one is of course also interested to
know how much observational errors propagate into the answers that
asteroseismology promises to deliver, even if the theory were perfectly
known. Only then one will be able to asses the gain from high S/N, high
resolution spectroscopy.

2. SIMPLE (ASYMPTOTIC) THEORY OF OSCILLATION FREQUENCIES

Assuming a perfectly spherical star (that is with no distortion of
the equilibrium state by rotation, magnetic fields or something else),
the (linear) oscillation modes about the equilibrium can be classified
by the angular degree ℓ of the spherical harmonic associated with the
spatial variation of the surface velocity field, and by the number n of
radial nodes of the velocity pattern inside the star. The radial
number n can of course not be seen; it has to be deduced theoretically.
The simplest theoretical analysis of oscillation frequencies is
asymptotic theory (Tassoul 1980), which - to second order - yields the
following expression for the frequencies $\nu_{n,\ell}$

$$\nu_{n,\ell} = (n + \ell/2 + \sigma)\, \nu_0 + \varepsilon_{n,\ell} \tag{2.1}$$

Here, σ is a constant of order unity, and $\varepsilon_{n,\ell}$ is small compared with
$\nu_{n,\ell}$. At this point it is useful to introduce two definitions pertaining
to the structure in the periodogram of high order pulsators.

2.1. Large and small frequency separations

(i) Large Separation:

$$D_{n,\ell} \equiv \nu_{n+1,\ell} - \nu_{n,\ell} . \tag{2.2}$$

To first order asymptotic theory it is well known that

$$D_{n,\ell}^{-1} \approx \nu_0^{-1} \equiv 2 \int_0^R (1/c)\, dr . \tag{2.3}$$

In simplified stellar models it is easy to show that

$$\nu_0 \propto (g/R)^{\frac{1}{2}} = (GM/R^3)^{\frac{1}{2}} . \tag{2.4}$$

(ii) Small Separation:

$$d_{n,\ell} \equiv (\nu_{n,\ell+1} - \nu_{n,\ell}) - \tfrac{1}{2}(\nu_{n+1,\ell} - \nu_{n,\ell}). \tag{2.5}$$

The small separation serves to cancel the first-order term of (2.1), and thus reveals second-order details, which pertain to the central regions of the star (see below). The ratio between small and large separation is, to second-order asymptotic theory, given by (Tassoul 1980)

$$\frac{d_{n,\ell}}{D_{n,\ell}} \approx \frac{\ell+1}{2\pi^2 \nu_{n,\ell}} \int_0^R \frac{dc}{dr} \frac{dr}{r} \tag{2.6}$$

Since sound speed increases steeply from the surface to the centre of a star, $D_{n,\ell}$ probes more the surface regions and $d_{n,\ell}$ more the central regions.

3. DETERMINATION OF STELLAR AGES FROM SEISMOLOGY

The small separation carries an important signature of stellar age, because as hydrogen is converted into helium in the stellar core, the mean molecular weight μ increases, which causes a decrease of sound speed, thus reducing the small separation. An excellent diagnostic diagram that extracts the information contained in the small and large separation has been invented by Christensen-Dalsgaard (1984) (for a more detailed calculation see Christensen-Dalsgaard 1988). In this diagram, contours of constant stellar mass and age are plotted against the theoretically computed large and small separations. Since the mass and age contours are rather perpendicular than parallel to each other, they reveal the considerable diagnostic potential of these diagrams (hereinafter called JCD diagrams).

Going a step further, Gough (1987) discussed the accuracy of seismological mass and age determination, using JCD diagrams and calculations by Ulrich (1986, 1988). His discussion is purely formal: taking the theoretical model for granted, he computes the uncertainty in the mass and age determination, assuming given errors for the observed stellar parameters (effective temperature, luminosity, heavy-element abundance, large and small frequency separation). Gough's (1987) result is that mass and age determination are so sensitive to the heavy-element abundance that they cannot be carried out in this way, unless other stellar parameters are known by independent means. If, for instance, in the case of a binary system we can determine mass, or if we can estimate it from surface gravity (whose observation obviously profits from high

S/N and high-resolution spectroscopy), then the large separation can reveal the evolutionary information contained in the deviation from the simple relation (2.5) (otherwise the large separation mainly fixes M/R^3). Thus a more accurate age determination could become possible (see Gough 1987).

4. THE PROBLEM OF THE EQUATION OF STATE

An important physical issue to be addressed by solar and stellar oscillations is the equation of state. The principal open problem is the number of excited states of hydrogen and helium in the zones of partial ionization. While for many astrophysical applications simple equation-of-state recipes can be sufficient, it has been shown (Däppen 1987) that for finer helioseismological applications (e.g. helium abundance determination) such simple formalisms are not adequate. In contrast to various other improvements over the simple Saha equation, about which no disagreement exist, there are widely divergent opinions on the internal partition function of bound systems. There has been a recent controversy about the so-called Planck-Larkin partition function (Rouse 1983, Ebeling et al. 1985). The Planck-Larkin partition function essentially limits the number of excited states to those having a binding energy $\geq kT$. Optical hydrogen spectra, however, show more lines than predicted by the Planck-Larkin partition function (Däppen et al. 1987a). Rogers (1986) explains this fact by allowing resonances that are not counted in the partition function but could be seen in optical spectra. Thus the Planck-Larkin controversy cannot be resolved with optical experiments, but thermodynamical properties will have to be known. Stellar models with and without Planck-Larkin partition function will have to be compared. While thermodynamical quantities based on the Planck-Larkin partition function will soon become available (Rogers, private communication), an advanced and very smooth version of a more conventional equation of state has been developed in the framework of an ongoing opacity re-computation (Hummer and Mihalas 1987, Mihalas et al. 1987, Däppen et al. 1987b). If observational constraints on the partition functions can be obtained, helio- and asteroseismology could help answer this question from microphysics.

5. CONCLUSION

It is clear that high S/N and high-resolution spectroscopy will improve the quality of results from asteroseismology. Firstly, the determination of stellar parameters (age, mass, chemical composition), using given theoretical models of stellar evolution, will become more precise. Secondly, the observational data will enable us to develop better physical models for the theory of stellar evolution (equation of state, convection, opacity, nuclear reactions, etc.). Since first (rather crude) observations of stellar oscillation frequencies have already have led to theoretical problems [see, e.g. the case of ε Eridani (Noyes et al. 1984, Guenther and Demarque 1986, Soderblom and

Däppen 1987)], improvements of the observations will be most welcome.

Acknowledgments: I thank D.O. Gough and W. Dziembowski for stimulating discussions.

REFERENCES

Christensen-Dalsgaard, J. 1984, in *Proc. workshop on space research prospects in stellar activity and variability* (eds. A. Mangeney and F. Praderie, Paris Observatory Press) 11-45
Christensen-Dalsgaard, J. 1988, in *Advances in Helio- and Asteroseismology* (IAU Symp. 123, Reidel, Dordrecht)
Däppen, W. 1987, *Strongly coupled plasma physics* (eds. F. Rogers and H. Dewitt, NATO ASI Ser., Plenum, New York)
Däppen, W., Anderson, L.S., Mihalas, D. 1987a *Astrophys J.* 319, 195-206
Däppen, W., Dziembowski, W., Sienkiewicz, R. 1988 in *Advances in Helio- and Asteroseismology* (IAU Symp. 123, Reidel, Dordrecht)
Däppen, W., Mihalas, D., Hummer, D.G., Mihalas, B.W. 1987b *Astrophys. J.* (submitted)
Deubner, F.-L., Gough, D.O. 1984, *Ann. Rev. Astron. Astrophys.* 22, 593-619
Ebeling, W., Kraeft, W.D., Kremp, D., Röpke, G. 1985, *Astrophys. J.* 290, 24
Gough, D.O. 1987, *Nature* 326, 257-259
Guenther, D.B., Demarque, P. 1985, *Astrophys. J.* 301, 207
Hummer, D.G., Mihalas, D. 1987, *Astrophys. J.* (submitted)
Mihalas, D., Däppen, W., Hummer, D.G. 1987 *Astrophys. J.* (submitted)
Noyes R.W., Baliunas, S.L., Belserene, E., Duncan, D.K., Horne, J., Widrow, L. 1984, *Astrophys. J. Letters* 285, L23
Rogers, F. 1986, *Astrophys. J.* 310, 723-728
Rouse, C.A. 1983, *Astrophys. J.* 272, 377
Soderblom, D., Däppen, W. 1987, *Astrophys. J.* (submitted)
Tassoul, M. 1980, *Astrophys. J. Suppl. Ser.*, 43, 469-490
Ulrich, R.K. 1988 in *Advances in Helio- and Asteroseismology* (IAU Symp. 123, Reidel, Dordrecht)
Ulrich, R.K. 1986 *Astrophys. J.* 306, L37-L40

NONRADIAL PULSATIONS AND THE Be PHENOMENON*

Dietrich Baade
Space Telescope-European Coordinating Facility
European Southern Observatory
Karl-Schwarzschild-Str. 2
D-8046 Garching, W. Germany

ABSTRACT. Following a summary of the observations which suggest that the outbursts of classical Be stars are caused by nonradial pulsations, properties, implications and requirements of a model based on this notion are evaluated. A preliminary analysis of new observations of μ Cen is presented which for the first time in a Be star reveals two relatively closely spaced non-commensurate periods. Such a result would render implausible speculations that the variability of Be stars is due to corotating surface features.

1. INTRODUCTION

Two extensive overviews (Percy 1987, Baade 1987a) of the numerous symptoms of the rapid variability of Be stars, their interpretation in terms of nonradial pulsations (NRPs), and the possible causal connection between NRP and mass loss have been given only recently. It suffices, therefore, if here only the main points are recalled:
- Observations of both the mass *loss* (resonance absorption lines extending beyond the escape velocity) as well as of indicators of the mere *presence* of circumstellar matter (Balmer emission, net continuum polarization, narrow 'shell' absorption lines, IR excess) show that the mass loss from many Be stars is highly variable. The mass loss spectrum of these stars probably consists of many minor and much fewer major events, and perhaps does not even include a significant continuous component.
- Because of this variability, rapid rotation alone is not likely the cause of the mass loss as Struve's (1931) model had assumed. This conclusion can be drawn more firmly from a simple analysis of the Bright Star Catalogue where for every Be star there is a Bn star with the same $v \sin i$ but without a record of observed Hα emission.
- High-S/N spectroscopy may have uncovered a more fundamental difference between Bn and Be stars: While both groups, like the overwhelming majority of all other early-type stars with sufficiently broad lines (*cf.* Baade, these proceedings), show a finestructure of pairs of quasi-absorption and -emission components which in time resolved observations move blue-to-red across the photospheric line profiles, only Be stars also show line profile variations on a coarser scale. This latter variability of their absorption lines may therefore be the defining characteristic of the classical emission-line B stars.
- The amplitude of the line profile variability (LPV) is variable with time; what few periods have been determined so far do not give rise to the suspicion that the periods, too, are variable. On a few occasions and in different stars, the amplitude has been observed to decrease with some delay (days, weeks) after a Be outburst.

*Based in part on observations obtained at the European Southern Observatory, La Silla, Chile

- The attribution of the LPV to NRPs, the similarity of the kinetic energy needed for a typical Be outburst and the energy contents of some NRP modes, and the ability of non–axisymmetric NRP modes to transport angular momentum and energy between different radial zones of a star have immediately led to the suggestion that mass loss episodes are *caused* by the pulsation. Because no dependence of the mass loss rate on the pulsation amplitude at the time of the outburst has been found, long–term effects of the pulsation appear to be the most important. It has therefore been conjectured that the long–term effect of low–order NRPs may be a secular change of the atmospheric scale height until a critical combination of scale height, rotation and pulsation is reached where the excess energy is released in an outburst with mass loss.
- Low–order NRP modes also occur in slowly rotating B stars, however without leading to mass loss. But, while the observed periods of the two groups show considerable overlap, the periods in the *corotating* frame are much longer in the Be stars, rendering their pulsations highly non–adiabatic, *i.e.* after each pulsation cycle the atmosphere may be in a slightly different state as required by the above picture of Be outbursts.

The self–consistency of this notion is quite satisfactory, but it is based on observational inferences that need to be well established.

2. CRITICAL CORNER STONES OF THE NRP MODEL

Four points have to be considered: first, the occurrence of large–scale line profile variations in Be stars but *not* in Bn stars, second, the discrimination between NRP and corotating surface features, third, the clear observational documentation of individual mass loss *events*, and, fourth, the unambiguous establishment of a correlation between mass loss events and amplitude changes of the pulsation. The critical details may be summarized as follows:

1.) Two independent samples of 10 Be and 10 Bn stars each (Smith and Penrod 1984, Penrod 1987; Baade 1986) have both shown that significant LPV on scales comparable to the line width is (a) restricted to Be stars and (b) one time or another occurs in all of them. Even though neither data have been published yet, there can be no doubt that the LPV patterns of Be and Bn stars differ dramatically. But a more detailed study is clearly in order to find out, *e.g.*, if the amplitude of large–scale LPV in Bn stars is zero or just at the detection limit and if there are no exceptions to the rule.

2.) Because of the failure of current techniques to detect magnetic fields in Be stars (Barker 1987, see also Vogt and Penrod 1983 on the 'spoke model') there is little dissense that the small–scale LPV is due to high order NRP; but the idea that the large–scale LPV can be explained by the same model meets much more resistance (Harmanec 1984, Balona and Engelbrecht 1986, Balona *et al.* 1987, Sareyan *et al.* 1987, Clarke and McGale 1987). The reason of the latter is rather simple: the observed periods are close to the typical rotation period of Be stars so that – considering only Be stars – the assumption of surface features corotating with the star appears the most straightforward explanation.

However, this situation arises only in Be stars which are among the fastest rotators whereas LPV with the same 1 day timescale is in fact also observed all the way down to slowly rotating stars (Smith 1986). Among slower rotators, an excess of the phase velocities, v_{ph}, over the equatorial rotation velocities, v_{equ}, is implied that increases with decreasing rotation rate and eventually turns into compelling model-independent evidence against corotating surface features and in favor of NRP. Thus, does the near–identity of the two time scales in Be stars justify or even necessitate a second model for the same observational phenomenon? Note that Smith and Penrod (1984) even argue that in the most rapidly rotating B stars v_{equ} not only approaches v_{ph} but actually exceeds it so that in the corotating frame the putative spots or theirlike

would propagate *opposite* to the direction of the rotation.
3.) As stated before, any effect NRPs may have on the mass loss should primarily have the character of a time *integral* because no evidence has been reported that at a given *moment* the mass transfer rate to the Hα emitting envelope and the pulsation amplitude are correlated. Furthermore, Osaki (1986) and Ando (1986) have shown that NRP induced changes of the mass loss rate with time may be relatively slow. The evidence that mass loss from Be stars is (partly) driven by NRP would then be rather circumstantial and depend mainly on the difference in the LPV of Be and Bn stars. However, if tight limits of the order of the star's dynamical time scale can be placed especially on the rise time of a mass loss episode, the need for an explanation other than by rotation, or radiation pressure, or static magnetic fields, *etc.* becomes much more pressing and the notion of the violent release of some extra energy more plausible.

Examples of very short ($\sim$ day(s)) rise and much longer ($\sim$ week(s)) decay times of indicators of circumstellar matter have been given by Baade (1987a). Recent high-quality Hα observations of at least 3 small outbursts within 5 weeks in μ Cen (Baade 1987b, Baade *et al.* 1987) confirm the earlier conclusions and add an important new lower limit on the frequency of such events during active phases of a Be star.
4.) Existing conjectures about a link between NRP and mass loss in Be stars have so far only been qualitative, and nobody has worked out if a decrease in pulsation amplitude after an outburst as reported by Penrod (1986, 1987) is *necessarily* to be expected. If detailed calculations showed this not to be the case, one would again have to resort to more indirect indicators such as item 1 above. The possibility of a positive correlation is therefore certainly very appealing. On the other hand, there may be a rather basic problem with it. If, as has recently been observed (*cf.* point 3), the time between two successive outbursts becomes comparable to the typical time of decline in LPV amplitude, can one safely speak of a correlation? One possible way out is that the size of an event, the subsequent change of the pulsation amplitude, and the time needed by the atmosphere to return to the pre-outburst state are coupled quantities.

In summary, for 3 of the 4 issues discussed the observations either provide or firmly promise the necessary minimum of evidence that in Be stars mass loss events and photospheric variability on a timescale of one day are interconnected. The interpretation of the variability as NRP is more controversial, mostly with single-channel observers and some theorists. The latter are skeptical in part because the observations of so extremely rapidly rotating stars are outside the domain to which theories developed for non-rotating stars may be extrapolated. The former, not being directly confronted with the plethora of phenomena which at high S/N and spectral resolution become observable thanks to the rotational Doppler effect (*cf.* Baade 1987a), occasionally tend to integrate stellar disk-integrated data further into one number, 'the' period, and make it the key to the understanding of Be stars. One reason for this foreshortening are, paradoxical though it may appear, the available analyses of spectroscopic data which have not so far succeeded to detect multiple periods in the LPV which however a) are to be expected from NRPs and b) would eliminate corotating surface features from the model scene. (The ubiquitous presence of LPV with short spatial periods is usually discarded as evidence of multiperiodicity because the phase velocities of long- and short-wavelength patterns are often very similar.)

3. MULTIPERIODICITY

If the periodicity observed in Be stars is to be explained by surface features, these features must form a roughly periodic pattern on the stellar surface. For simplicity and following the same arguments as for NRPs (Baade 1987a), it can be assumed that to first order the pattern defines m identical sectors. If the polar axis of these sectors is inclined with respect to the rotational axis (Harmanec 1984, Clarke and McGale 1987), the corresonding segments of, e.g., a light curve will be unequally long and have unequal amplitudes. Neglecting this possible complication, the relation between the frequencies in the corotating frame and the observer's frame is for m star spots as well as NRP with mode order $\pm m$:

$$f_{obs} = f_{corot} + |m| \times \omega \qquad (1)$$

where ω is the star's rotation rate. The formal difference between star spots and NRPs in this equation is that f_{corot} is zero for spots with a fixed location while for traveling waves it is not. Accordingly, only (two or more) spot patterns would have m–commensurate frequencies (or periods): $f_{obs,m}/f_{obs,n} = m/n$.

If the usual assumption for Be stars is made, that their range in ω is small, Equ. (1) shows that for small m ($m \approx 2$ is appropriate for the large–scale LPVs of most Be stars) the chances to find a star with large (absolute) f_{corot}, i.e., where the distinction between the two models is the least ambiguous, are best for stars with extreme f_{obs}. One reasonable candidate is, therefore, μ Cen with its 0.505 day period (Baade 1984) which is one of the shortest periods known of Be stars (see the list compiled by Percy 1987).

Nearly 20 nights worth of observations have been obtained of this star in three different seasons. Typical nightly examples can be seen in Fig. 2 of Baade (1987a). Although there have been no dramatic changes of the amplitude and all nights look basically similar, there is not a single pair of nights or major sections thereof in which the large–scale LPVs were identical to within many σ of the noise. The residuals are almost certainly not due to small–scale ($m \approx 10$, Baade 1984) LPV, either, so that the beating of at least two large–scale LPV patterns, i.e. the existence of *different* periods, is a strong possibility.

The problem is to measure the presumably not very different periods of some processes which each modulate a series of one–dimensional data strings (line profiles) in a similar fashion. The least model–dependent method that furthermore utilizes the data to their full extent (Baade, these proceedings) is to separately analyse the flux in thin wavelength slices which extend over all observations (spectra) and together cover the entire spectral line observed. Such a study has been carried out for six nights of observations of He I λ 6678 Å in μ Cen obtained in 1987 April. The discrete Fourier transform (DFT) and the phase dispersion minimization method both gave nearly identical results. The wavelength series of DFT power spectra after deconvolution with the window spectrum by the CLEAN algorithm is reproduced in Baade (these proceedings).

The preliminary analysis confirms the 0.505 day period reported earlier (Baade 1984) and reveals a second set of peaks (mostly 1 c/d aliases) arising from a 0.391 day period. The difference of the period ratio from 5:4 appears significant. However, if Equ. 1 were to be simultaneously satisfied with both $m = 4$ and 5 and the observed frequencies, ω would have to be ~0.5 c/d. Assuming 5 $R_\odot$ for the radius of μ Cen (B2 IVe), this would translate into $v_{equ} = 130$ km/s, i.e. less than the observed $v \sin i$ of 155 km/s (Slettebak 1982) although the apparent retrograde propagation (Baade 1984, 1987a) of some features in the line profiles suggests that the star is seen at relatively low inclination (also consistent with the $v \sin i$ value which is low for a Be star).

Thus, μ Cen is almost certainly truely multiply periodic. A third period, either 0.305 day or its 1 c/d alias 0.440 day (the peak of the former is stronger, in the second case all three frequencies are, within the errors, equally spaced), is in fact likely. An unresolved puzzle, however, is the very strong variation of the Fourier power with wavelength (Fig. 1 in Baade, these proceedings). Gies and Kullavanijaya (1987) found a rather similar pattern in their analysis of ϵ Per. Insufficient sampling, especially of long beat periods, clearly is a possible explanation. But the vector character of NRP velocity fields may also play a role. This should be easy to verify from synthetic data.

4. CONCLUSIONS

The Be phenomenon is essentially one of variable mass loss. The timescales observed in indicators of circumstellar matter may be dominated by the time needed to disperse this matter (*e.g.* perhaps as a radiatively driven wind). More characteristic of the initial mass loss mechanism(s) are therefore the rise and repetition times of mass loss events. Recent observations have put fairly low upper limits on both timescales. Current evidence is furthermore consistent with the conclusion that the same process that causes the large-scale LPV of Be stars is involved in their mass loss which is neither otherwise explained nor for spectral types later than $\sim$ B2 paralleled by other stars. The detection of genuine multiperiodicity with significant phase velocities in the corotating frame rules out star spots as an explanation of the LPV. Line profile fitting and analyses of more stars should permit a firm conclusion to be reached that Be stars are low-order nonradial pulsators and that they owe (much of) their mass loss and associated line emission to this property.

Acknowledgement: I thank Dr. Doug Gies for helpful comments on the manuscript.

REFERENCES

Ando, H.: 1986, *Astron. Astrophys.* **163**, 97.
Baade, D.: 1984, *Astron. Astrophys.* **135**, 101.
Baade, D.: 1986, in *Highlights of Astronomy*, J.-P. Swings ed., D. Reidel, Dordrecht, Vol. **7**, p. 255.
Baade, D.: 1987a, in Proc. IAU Coll. No. **98** *Physics of Be Stars*, A. Slettebak and T.P. Snow (eds.), Cambridge Univ. Press, Cambridge, p. 361.
Baade, D.: 1987b, *Be Star Newsletter* No. **15**, 18.
Baade, D., Dachs, J., van de Weygaert, R., Steeman, F.: 1988, to be submitted to *Astron. Astrophys.*
Balona, L.A., Engelbrecht, C.A.: 1986, *Mon. Not. R. Astr. Soc.* **219**, 131.
Balona, L.A., Engelbrecht, C.A., Marang, F.: 1987, *Mon. Not. R. astr. Soc.* **227**, 123.
Barker, P.K.: 1987, in Proc. IAU Coll. No. **98** *Physics of Be Stars*, A. Slettebak and T.P. Snow (eds.), Cambridge Univ. Press, Cambridge, p. 38.
Clarke, D., McGale, P.A.: 1987, in Proc. IAU Coll. No. **98** *Physics of Be Stars*, A. Slettebak and T.P. Snow (eds.), Cambridge Univ. Press, Cambridge, p. 197.
Gies, D.R., Kullavanijaya, A.: 1987, *Astrophys. J.*, submitted.
Harmanec, P.: 1984, *Bull. Astron. Inst. Czechoslov.* **35**, 193.
Osaki, Y.: 1986, *Publ. Astron. Soc. Pacific* **98**, 30.
Penrod, G.D.: 1986, *Publ. Astron. Soc. Pacific* **98**, 35.
Penrod, G.D.: 1987, in Proc. IAU Coll. No. **98** *Physics of Be Stars*, A. Slettebak and T.P. Snow (eds.), Cambridge Univ. Press, Cambridge, p. 463.
Percy, J.: 1987, in Proc. IAU Coll. No. **98** *Physics of Be Stars*, A. Slettebak and T.P. Snow (eds.), Cambridge Univ. Press, Cambridge, p. 49.
Sareyan, J.P., Alvarez, M., Chauville, J., Le Contel, J.M., Michel, R., Ballereau, D.: 1987, in Proc. IAU Coll. No. **98** *Physics of Be Stars*, A. Slettebak and T.P. Snow (eds.), Cambridge Univ. Press, Cambridge, p. 78.
Slettebak, A.: 1982, *Astrophys. J. Suppl.* **50**, 55.
Smith, M.A., Penrod, G.D.: 1984, in *Relations between Chromospheric-coronal Heating and Mass Loss in Stars*, R. Stalio and J.B. Zirker (eds.), Trieste, p. 394.
Smith, M.A.: 1986, in *Hydrodynamic and Magnetohydrodynamic Problems in the Sun and Stars*, Y. Osaki (ed.), Univ. of Tokyo Press, Tokyo, p. 145.
Struve, O.: 1931, *Astrophys. J.* **73**, 94.
Vogt, S.S, Penrod, G.D.: 1983, *Astrophys. J.* **275**, 661.

DISCUSSION

VOGT My graduate student Don Penrod has observed a fairly large sample of both normal and emission-line B stars and finds the high order modes to be almost always present in both groups. However, only those rapid rotations which develop the l=2 mode ever exhibit Be mass loss outbursts. So, both the presence of an l=2 mode and rapid rotation seem to be required for Be outbursts. Hopefully, these results will be published within the forseeable future.

BOHANNAN The outbursts of Be stars are seen in the grossest ways. What signal-to-noise ratio spectra are required to detect the rotationally related line profile variations in the presence of longer period variations ?

BAADE If you want to detect <u>small</u> outbursts <u>spectroscopically</u>, you also need high spectral resolution ($R \approx 20 - 30\,000$) and low noise ($S/N \approx 200$). More efficient, however, appears polarimetry. For the proper mapping of the line profile modulations the S/N should not be much worse than 300. But this number depends on the intrinsic amplitude and the $v \sin i$ because large line broadening reduces the contrast.

ULTRAVIOLET SPECTRAL IMAGING

James E. Neff
Joint Institute for Laboratory Astrophysics
Univ. of Colorado
Boulder, CO 80309-0440, USA

ABSTRACT. I discuss the general problem of determining the spatial structure in the outer atmospheres of active, late-type stars. There are several major differences between the problems of imaging photospheres and of imaging chromospheres. Because of these differences, chromospheric spectral imaging must be based on a direct deconvolution of the observed emission line profiles. I present results based on IUE spectra of AR Lac (=HD 210334).

1 INTRODUCTION

High S/N spectra are often required because one desires to measure a small signal against a bright background. One virtue of ultraviolet spectra of late-type stars is that the photospheric background is faint, allowing the measurement of small relative signals, even with moderate S/N spectra. Stellar chromospheres and transition regions are best studied using ultraviolet emission line spectra.

Doppler imaging (see Vogt 1988) can be used to produce images of stellar photospheres using high-resolution, high S/N, optical absorption line profiles. In order to produce similar images of stellar *chromospheres*, similar procedures should be developed for the analysis of ultraviolet emission line spectra. In this talk, I discuss the general problem of producing images of rapidly rotating stars using observed spectral line profiles ("spectral imaging").

By combining spectral images of the photosphere and the chromosphere with other information about the structure of the corona (derived from radio and x-ray observations), it is possible to develop a three-dimensional picture of stellar atmospheres. With such pictures, the size, location, and brightness of magnetic active regions can be studied as a function of time to determine stellar cycles and to measure differential rotation. It will then be possible to model the physical conditions within each atmospheric component separately.

2 ULTRAVIOLET VERSUS OPTICAL SPECTRAL IMAGING

The goal of spectral imaging is to utilize all of the spatial information that exists in the observed line profiles from rapidly rotating stars. A discrete region of high contrast on the surface of a rapidly rotating star produces an effect on the line profile that moves from blue to red across the line profile as the star rotates. The amplitude of this velocity shift is $v_{rot} \sin i \cos \theta$ (where v_{rot} is the rotational velocity of the star, i is the inclination, and θ is the latitude of the region), if the region is on the stellar surface. When the region is on the central meridian, there is no velocity shift. By measuring the relative velocity of the feature with respect to the center of the line profile as a function of rotational phase, its longitude and latitude can be determined. If the region is spatially extended, the width of the spectral feature as a function of rotational phase can be used to determine the shape of the region on the stellar surface. For various reasons, a spectral image can not be derived by simply identifying the wavelength of "bumps" on the line profiles (e.g. Gondoin 1986).

While all spectral imaging techniques share these same fundamental principles, there are several noteworthy differences between optical (Doppler imaging) techniques (e.g., Hatzes 1988; Jankov 1987; Vogt 1987; Vogt and Penrod 1983; Vogt, Penrod, and Hatzes 1987) and ultraviolet techniques (Neff 1986; Neff 1987a, 1987b; Walter et al. 1987).

First and foremost, the physical properties of stellar *chromospheres* are not known. In fact, the goal of imaging is to determine these properties. The parameters needed to synthesize the intrinsic chromospheric profile are not available. For example, the variation in brightness of a chromospheric active region between the limb and disk center can not be determined until a region is spatially resolved. Available chromospheric models typically represent only a global average, not the physically distinct regions that we expect to be present.

Second, in a given spectral line, most of the emission probably arises from discrete regions. This is emission above a very faint continuum. The more active the star, the brighter the discrete emission. By contrast, photospheric spots are dark regions on a bright background. Further, the photospheric lines become shallower the more rapidly the star rotates.

Because of these differences, chromospheric spectral imaging must be based on a direct *deconvolution* of emission line profiles. While tedious and difficult to automate, this procedure is capable of providing spatial information using spectra of lower S/N.

3 ULTRAVIOLET SPECTRAL IMAGING TECHNIQUE

To illustrate the power and the limitations of this technique, I summarize the analysis of a series of spectra of AR Lacertae (=HD 210334). AR Lac is an eclipsing RS CVn binary system (K0 IV + G2 IV). The two stars rotate synchronously with the orbital period (1.98 days), yielding an equatorial velocity of 72 km s^{-1} for the K star and 39 km s^{-1} for the G star. Because the system is bright (m$_v$=6.1) and because both stars are active, the system is an ideal candidate for ultraviolet spectral imaging.

We observed AR Lac over 80% of an orbital cycle in Sep. 1985 (Neff, Walter, and Rodonò 1986). Eighteen high-resolution (R$\sim$20,000) spectra of the Mg II k (2795 Å) line and 18 low-resolution (6 Å) spectra covering the range 1150 to 1950 Å were obtained.

The 18 observed Mg II k line profiles can, at first glance, be characterized by two emission components at radial velocities roughly corresponding to the predicted orbital velocities of the two stars plus an unresolved interstellar absorption line. In the initial round of the analysis, three gaussian components are fitted to the observed profiles. The width of the interstellar component is constrained to be the instrumental width.

In order for these three-component fits to provide an appropriate model of the line profiles, (1) the velocities of the two emission components must match the predicted stellar velocities, (2) the widths of these components must be consistent with the stellar rotational velocities, and (3) the equivalent width of the interstellar absorption component must remain roughly constant. None of these conditions are met by the three-component fits.

Emission arising from a *uniform* distribution on a stellar surface will produce a *symmetric* profile centered at the stellar velocity. Large-scale *non-uniformities* will produce emission components that vary in wavelength due to the Doppler shift produced by the star's rotation. The net profile will be asymmetric, and a single-component fit to this profile will not lie at the stellar velocity. Therefore by fitting the observed profiles with a symmetric component centered at the stellar velocity plus additional components to account for the residual emission, the non-uniform distribution can be mapped. Neff (1987a) and Walter *et al.* (1987) describe this procedure in detail. In practice, the minimum number of additional components required to match the observed profiles are determined iteratively, using an interactive multiple-component fitting routine. The primary constraints in this procedure are the positions of the uniform stellar components and the interstellar absorption equivalent width, which should remain constant. Secondary constraints are the rough constancy of the stellar emission line widths and smooth point-to-point variation of the stellar flux.

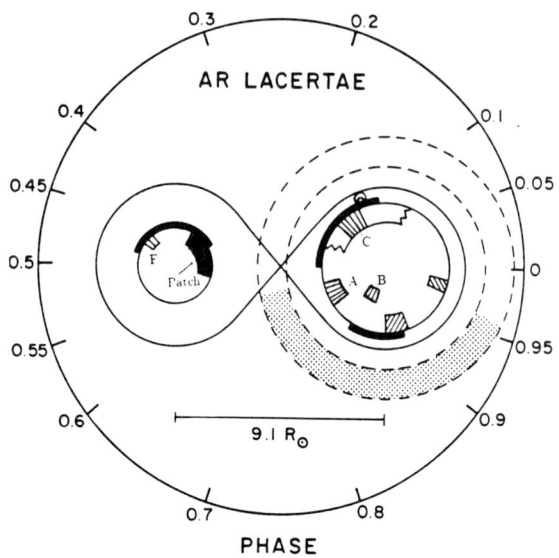

Figure 1: The latitudes, longitudes, and sizes of three plages on the K star (right), a flare on the G star (left), and the inactive region on the G star are depicted schematically to the scale of the binary system. The unlabelled plages are regions identified in a series of spectra obtained in Oct. 1983 (Walter *et al.* 1987). The compact and extended coronal structure was determined by the x-ray light curve obtained during eclipses in June 1980 by Walter, Gibson, and Basri (1983).

4   THE SPATIALLY RESOLVED CHROMOSPHERE OF AR LACERTAE

The resulting system model for AR Lac in Sep. 1985 (Figure 1) reveals five discrete chromospheric regions: (1) Plage A is located at longitude 205°, latitude 0° on the K star (longitude 0° corresponds to orbital phase 0.0, when the G star is totally eclipsed). It likely extends well above the stellar photosphere. The region covers 6% of the stellar surface. (2) Plage B is located at longitude 240°, latitude±50° on the K star. It covers 2% of the stellar surface. (3) Plage C is located at longitude 130°, latitude 0° on the K star. It likely extends well above the photophsere. It covers 9% of the stellar surface. (4) The G star is chromospherically *inactive* between longitudes 335° and 75°. (5) A flare was observed on the G star. The flaring region covered at most 2% of the stellar surface. In addition, the global (uniform) flux from the K star is 40% brighter from the trailing hemisphere than from the leading hemisphere. This information is presented as a series of images by Neff (1987a, 1987b).

5 CONCLUSIONS

Very detailed images of stellar photospheres can be derived by synthesizing a line profile from an assumed brightness distribution and then varying this distribution until the synthetic profile matches the observed profile. Unfortunately, a similar procedure can not be used to image stellar chromospheres. Nevertheless, a non-uniform chromospheric distribution manifests itself in the observed profiles. Because these non-uniformities are large scale and because the photospheric background is fainter in the ultraviolet, chromospheres can be imaged by a careful deconvolution of the line profiles. Higher quality data obtained with the Goddard High Resolution Spectrograph on the Hubble Space Telescope will allow more detailed images to be derived. It should also be possible to apply the same techniques to ground-based spectra of the Ca II K lines. Because of poor instrumental throughput and detector sensitivity, this is currently a difficult task. The Hα emission line is poorly suited for spectral imaging, primarily because it is substantially broadened in these stars by non-rotational mechanisms.

REFERENCES

Gondoin, P. 1986, *Astr. Ap.*, **160**, 73.
Hatzes, A.P. 1988, these proceedings.
Jankov, S. 1987, unpublished thesis, Universite Paris-Meudon.
Neff, J.E. 1986, in *Proc. SHIRSOG Science Working Group*, NSO.
Neff, J.E. 1987a, Ph.D. thesis, University of Colorado, Boulder.
Neff, J.E. 1987b, in *Proc. Fifth Cambridge Workshop on Cool Stars, Stellar Systems, and the Sun*, eds. J.L. Linsky and R.E. Stencel, Springer-Verlag, in press.
Neff, J.E., Walter, F.M., and Rodonò, M. 1986, in *New Insights in Astrophysics: Eight Years of Astronomy With IUE*, **ESA SP-263**, p. 153.
Vogt, S.S. 1988, these proceedings.
Vogt, S.S., and Penrod, G.D. 1983, *Pub. Astr. Soc. Pacific*, **95**, 565.
Vogt, S.S., Penrod, G.D., and Hatzes, A.P. 1987, *Ap. J.*, submitted.
Walter, F.M., Gibson, D.M., and Basri, G.S. 1983, *Ap. J.*, **267**, 665.
Walter, F.M., Neff, J.E., Gibson, D.M., Linsky, J.L., Rodonò, M., Gary, D.E., and Butler, C.J. 1987, *Astr. Ap.*, in press.

DISCUSSION

   JUDGE A general comment about chromospheric modelling. There has been much discussion about the inhomogeneous structure of stellar chromospheres for which there is much evidence in the sun and in some parts of the HR diagram, Jim has presented a graphic demonstration of the inhomogeneous chromosphere of the AR LAC K-star component. This star is an active star. To take the devil's advocate I suggest that, for 'normal' late-type stars one can still usefully apply one-component modelling techniques, <u>provided</u> one is careful not to over-interpret the data in particular, for the least active stars (single giants later than ≈ K0) there is some indication that acoustic, and not magnetic heating could be responsible for the observed chromospheric emission (I'm thinking of the 'basal fluxes' identified by the Utrecht group) which probably could not lead to the kind of discrete structures discussed here. The assumptions of one-component plane parallel models are clearly oversimplifications which nevertheless remain to be properly tested by observations in most cases, therefore these modelling techniques should still be pursued as useful first attempts at understanding spectra which, taking the case of cool, 'non-solar' stars, as an example, are only beginning to be understood.

   LINSKY The Sun is a very inactive star yet many components are needed to explain the range of spectroscopic features observed in integrated light. While it may be true that the cool single giants have simple, nonmagnetically heated atmospheres, this is not proven in any way. I agree that it makes sense at this time to model the atmospheres of these very quiet stars with one atmosphere component, however we should keep in mind that this way be a gross oversimplification due to the possible presence of radial and nonradial oscillations, acoustic waves, and large scale convective cells even if magnetic structures do not exist.

   FOING What is the accuracy in velocity shift and width determination and how does it translate in terms of accuracy in longitude, latitude and size of plage areas?

   NEFF I have derived the longitudinal extents of the plages by assuming that the deconvolved line width at the narrowest plage profile is due solely to rotational smearing. In fact, there is probably a substantial intrinsic width of the plage profile, so my size determination is an upper limit. The assumption of circular plages introduces further uncertainty. Therefore, the formal uncertainties in the analysis are much smaller than the real uncertainties. The longitudes are well determined, say ± 10°, by the phase of their central meridian passage. Their latitudes are determined by the amplitude of their radial velocity curves. Unfortunately, the height of the plage and differential rotation also affect the amplitude.

   MARCY What fraction of the stellar surfaces are covered by active regions in the AR Lac system ?

   NEFF If you assume that the K star is uniformly covered by plages smaller than but otherwise similar to the three that are resolved, then the total filling factor is about 25%. The global G star emission cannot be due to a distribution of similar plages, because the line ratios (particularly C IV/C II) are different.

LINSKY I would like to point out that the Doppler imaging technique provides information on the longitude, latitude, and area of active regions on stars. The last quantity is extremely important because it permits us to infer the surface brightness of emission lines in active regions and thus to infer chromospheric models of the active regions.

NEFF None.

FITZPATRICK How is the strength of the interstellar absorption component constrained in the fitting of the multiple line components ?

NEFF The strength of the interstellar absorption is not known <u>a priori</u>. However, the equivalent width should remain constant. This provides an important constraint on the fitting procedure. If the interstellar equivalent width in a given fit is much smaller than the mean determined by all other fits, then the emission level at that wavelength from the system must be underestimated, so the fit is rejected.

APPLICATIONS OF THE DOPPLER IMAGING TECHNIQUE TO THE ANALYSIS OF HIGH-RESOLUTION SPECTRA OF THE 3 OCTOBER 1981 FLARE ON V711 TAURI

Jeffrey L. Linsky[1] and James E. Neff
Joint Institute for Laboratory Astrophysics
University of Colorado
Boulder, Colorado 80309-0440 U.S.A.

ABSTRACT. An unconstrained four gaussian fit to the Mg II profile near the flare peak indicates that the flare occurred near the central meridian of the K1 IV star, perhaps above a spot. A more likely fit to the same data places the flare at $+90 \pm 30$ km s^{-1} relative to the K1 IV star, indicating significant downflowing plasma.

We obtained high resolution IUE spectra with both the SWP and LWR cameras of a bright flare on the RS CVn-type system V711 Tauri. Optical photometry at this epoch (Rodono et al. 1986) implies a two-spot model with central meridian passage of spot 1 occurring at phase 0.17. A flare was detected first in a low resolution, short wavelength spectrum (SWP 15161) at 0512 UT (ϕ=0.183) and then in a high resolution, long wavelength image (LWR 11668) at 0635 UT (ϕ=0.204). Subsequently four LWR high resolution spectra and two SWP spectra were obtained during the flare including a 2 hour high resolution, short wavelength spectrum (SWP 15103) centered at 1208 UT (ϕ=0.285). The peak radiative loss in the temperature range $4.0 < \log T_e < 5.2$ was 3×10^{31} ergs s^{-1}, and the estimated total emission was 6×10^{35} ergs.

The Mg II k lines were fit with four unconstrained gaussian profiles as described by Walter et al. (1987) and Neff (this volume). Examples of these fits are shown in Fig. 1 for two flare spectra and two non-flare spectra near the other quadrature (ϕ=0.680 and 0.677), where the radial velocities of the G and K stars are reversed. During the flare the flux in the G star remained constant while the flux and width of the K star profile both increased. At flare peak (ϕ=0.204) the centroid of emission was shifted $+15 \pm 5$ km s^{-1} relative to the K star, and the FWHM of the profile (after correction for instrumental broadening) was 85 km s^{-1}, compared to the nonflare value of 60 km s^{-1} and vsini = 38 km s^{-1}.

---
[1] Staff Member, Quantum Physics Division, National Bureau of Standards.

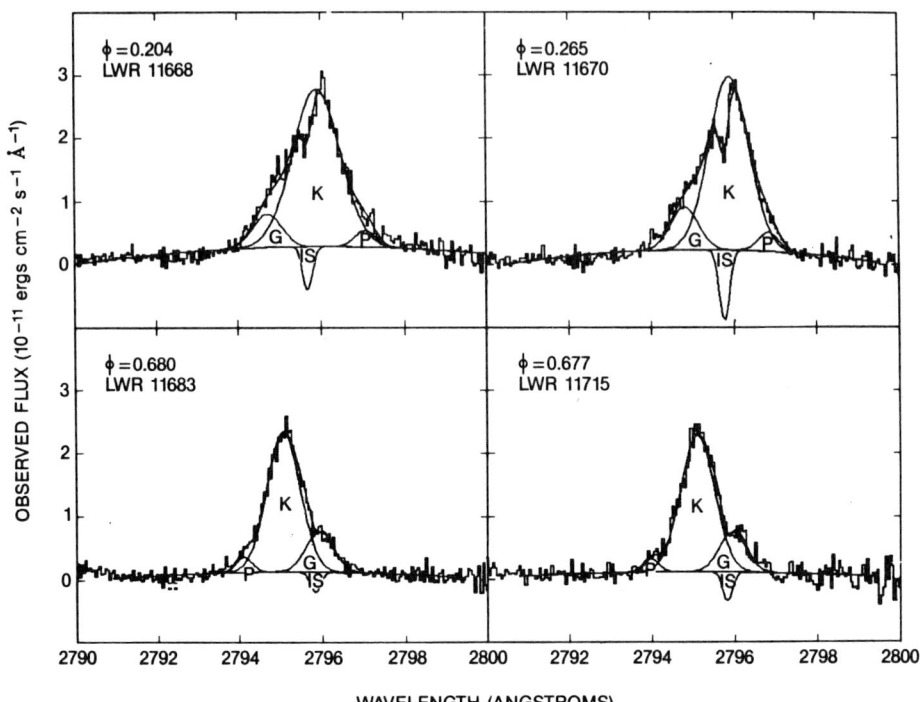

Fig. 1. Observed and four gaussian fits to the Mg II k line profiles for two flare spectra (ϕ=0.204 and 0.265) and two subsequent quiescent spectra near the opposite quadrature (ϕ=0.680 and 0.677). The gaussian profiles are identified with the K1 IV star (K), G5 V star (G), a plage (P), and Mg II interstellar absorption (IS).

In the preceding analysis the known radial velocities of the G5 V and K1 IV stars were not assumed but rather the gaussian profiles with centroid velocities close to the two stars were presumed to represent each star. The natural interpretation of this analysis is that
 - the flare occurred near the central meridian of the K star at the time of light minimum and perhaps above spot 1,
 - the flaring plasma showed little line-of-sight motion,
 - the plasma was highly turbulent with additional broadening corresponding to a gaussian FWHM of about 60 km s^{-1}.

Analysis of the rather noisy C IV 1548 Å profile obtained from the high resolution SWP spectrum is consistent with the flare occurring near the central meridian and perhaps over spot 1 of the K1 IV star, and thus with solar analogy in which flares occur over spots.

There is an alternative and we believe more plausible interpretation
of the same data. Outside of the flare the integrated Mg II fluxes
are nearly constant for each star. We have therefore refit the flare
peak profile (Fig. 2), constraining the Mg II fluxes and widths to be
the mean nonflare values for each star and the centroids of the
gaussians for the G and K stars to lie at the known radial velocities
(relative to the interstellar medium) at ϕ = 0.204. The residual
emission is then fit by a fourth gaussian with its centroid at +90±30
km s^{-1}. The flaring plasma must have a significant line-of-sight
systematic flow and be turbulent. This flow would be about +90 km
s^{-1} downward if the flare occurred over spot 1 or about +50 km s^{-1}
(away from the observer) if the flare occurred near the K star
receding limb. The Mg II profiles after flare peak did not have
sufficient flux in the flare to be fit with this technique.

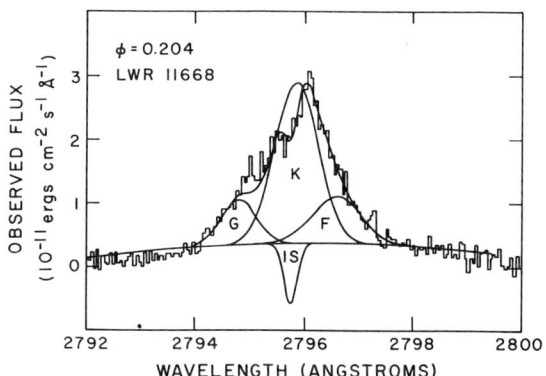

Fig. 2. The observed flare peak Mg II k line profile (ϕ=0.204) and a
four gaussian fit for which the strength and width of three gaussians
are constrained to be the mean quiescent values for the G star, K
star, and interstellar medium and the radial velocities are those predicted. The parameters of the fourth gaussian (F), selected to fit
the resulting residuals, can be ascribed to the flare. It is centered
at 90 ± 30 km s^{-1} relative to the K star.

We believe that the alternative interpretation (Fig. 2) is more
likely, since during the flare most of the K star should be quiescent
and thus contribute a mean quiescent profile. During a bright flare
on UX Ari, Simon, Linsky and Schiffer (1980) also observed red-shifted
Mg II emission but with larger line of sight velocities. A complete
analysis of this flare will be published elsewhere. We acknowledge
NASA grant NAG5-82 to the University of Colorado.

REFERENCES
Rodono, M. et al. 1986, Astron. Ap. 165, 135.
Simon, R., Linsky, J. L., and Schiffer, F. H. III 1980, Ap. J. 239,
911.
Walter, F. M. et al. 1987, Astron. Ap. (in press).

ABUNDANCE STRATIFICATIONS IN THE ATMOSPHERES OF Ap STARS:
THE CASE OF GALLIUM

G. Alecian and M.-C. Artru
Observatoire de Paris-Meudon
F-92195 Meudon Principal Cedex
France

ABSTRACT. The abundances are usually determined assuming homogeneous concentrations for all elements in the line forming region. If diffusion driven by radiation occurs, this condition is not verified and the element stratification affects the equivalent widths as well as the line profiles. We shall illustrate these problems by a study of the gallium case. A synthetic spectrum of the Ga II resonance line λ 1414.401 Å has been computed assuming that gallium is stratified according a recent theoretical study of the diffusion of this element.

1. INTRODUCTION

The Ap stars are known to have strong abundance anomalies: overabundances may be, in some cases up to 10^5 for metals and sometimes underabundances are detected (for instance for helium). These anomalies are explained by invoking the diffusion process: elements migrate through the outer stellar layers pushed up by the radiation field. This process is more or less efficient according to the element. Many papers have appeared on the subject and one may refer, as a first reading, to the basic paper of Michaud (1970) and also to the review by Bonsack & Wolff (1980) and the one by Alecian & Vauclair (1983).
 The diffusion process is fundamentally time-dependent. In about 10^4 - 10^5 years, it leads to the appearance of abundance stratifications in the Ap stars' atmospheres. In this case, an element may be strongly enhanced in some places and deficient in an other places. According to where the spectral lines of this precise element are formed, it may be detected over-, under-abundant or... normal. Of course, this description is very schematized, the process is much more complex since other effects like magnetic fields, macroscopic motions of the whole matter may interact with diffusion.
 Apart the abundance anomalies which are generally inferred from the equivalent widths, the stratification may change (for stratified elements) the spectral line profiles (see for instance, the Ca study by Borsenberger et al, 1981), and also curves of growth (Alecian, 1982). In the present paper we shall consider the effects of stratification, on the Ga II resonance line λ 1414.401 Å.

2. THE GALLIUM STRATIFICATION

The gallium abundances have been recently determined in a relatively large sample of Ap stars by Takada-Hidai et al (1986) on the basis of IUE observations of the UV resonance lines of Ga II and III. These data complete and improve previous determinations by Heacox (1979) and Guthrie (1984). They confirm that the largest gallium overabundances occur in Hg-Mn stars and that they are generally smaller in Ap-Si stars. A comparison of these stars, using the photometric index δ_{1400} of Jamar et al (1978) which measure the well known flux depression at 1400 Å (identified as to be related to silicon overabundance by Artru, 1986), shows that the overabundance of gallium may be anti-correlated with the overabundance of silicon.

2.1. Recent results on the gallium diffusion

A recent theoretical study of the gallium diffusion motivated by these new data, have been made by Alecian & Artru (1987) using up-to-date gallium atomic data. They concluded that, if mass loss effects are negligible, the gallium overabundances should appear weaker in the presence of horizontal magnetic field. Schematically, in Ap-Si stars one might expect that patches of gallium are located where magnetic field is vertical while patches of silicon are located at places where magnetic field is horizontal. On the other hand, they have shown that the upper limit of the observed gallium overabundances are well explained in the framework of the diffusion theory (see their Fig.6).

2.2. The effect of stratification on the Ga II resonance line λ 1414.401 Å

In order to test further the diffusion model concerning gallium, we tried to establish more precisely the effect a stratification of gallium on its spectral lines. We chose to study the λ 1414.401 Å resonance line of Ga II in an atmosphere like those of the most typical Hg-Mn star of the sample studied by Takada-Hidai et al (1986): HD 175640 ($T_{eff} \approx 12000$ K, log g $\approx$ 3.9, v.sin i < 5 km.s^{-1}). The equivalent width measured by these authors for the λ 1414.401 Å line is about 1.56 Å. Taking this equivalent width, the standard abundance determination (assuming homogeneous gallium) gives that gallium is overabundant by a factor of 3000 with respect to the solar value.

We computed a synthetic spectrum of the λ 1414.401 Å line, using the NLTE model of atmosphere of Borsenberger & Gros (1978) (T_{eff} = 12000 K, log g = 4). The line was computed in LTE with *inhomogeneous* abundance of gallium throughout the atmosphere.

Firstly, we determined what kind of stratification of gallium may be expected according to the radiation forces published by Alecian & Artru (1987). Actually the building of stratification is a time-dependent process which must be computed numerically (Alecian & Grappin, 1984). However, one may simply assume that equilibrium (zero diffusion velocity everywhere) is reached at the end of this process. This leads to the stratification of gallium shown in Fig.1 where a cloud-like accumulation occurs around τ_{5000} = 1. Using this stratification deduced a priori from the diffusion model, we synthesized the absorption of the λ 1414.401 Å line of Ga II.

2.2.1. The equivalent width

The computation of the synthetic spectrum gives an equivalent width of 1.58 Å which is very close to the observed one 1.56 Å. This is explained by the fact that the overabundance of Ga in the cloud-like accumulation shown in Fig.1, is not far from a factor of 3000 near the depth τ_{5000} = 1 where the line forms.

Other results obtained with various shape and positions of the cloud, have shown that the equivalent width of this line is very sensitive to the position of the higher edge of the cloud: to rise up this cloud would increase strongly the equivalent width since the maximum overabundance in it is stronger than $1.2\ 10^4$ times the solar abundance.

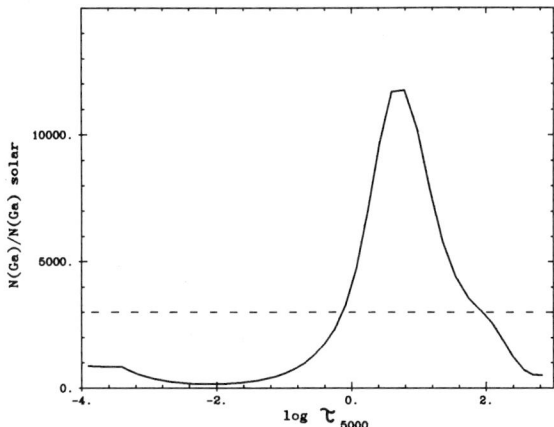

Figure 1. The cloud-like accumulation of gallium (continuous line) deduced from diffusion calculation for T_{eff} = 12000 K, log g = 4. The plot shows the overabundance of gallium (with respect to solar) vs the logarithm of the optical depth at 5000Å. The dashed line is the overabundance deduced from the standard homogeneous analysis.

2.2.2. The line profile

The abundance stratification also affects the line profile. In Fig.2. We show the computed profile of the Ga II resonance line λ 1414.401 Å assuming the same stratification than before.

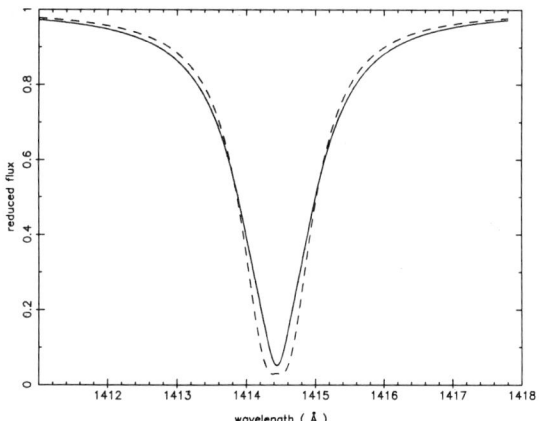

Figure 2. The line profile obtained through a synthetic spectra calculation. The continuous line corresponds to the stratified case (see Fig.1), the dashed line to the homogeneous case.

We compare it to the profile obtained without stratification (dashed line). In this case we have considered an homogeneous overabundance of x4200 with respect to solar, in order to have exactly the same equivalent width (1.58 Å) in both cases (it is of course slightly higher than the measured value). The profiles have been broaden at 0.15 Å which corresponds to a rotational velocity of 5 km.s^{-1}.

3. CONCLUSION

In the present work we have considered the case of a typical Hg-Mn star with good observational data and we have studied a gallium line for which the atomic data are reliable. This study has allowed to test a parameter-free model of diffusion: we emphasize that the cloud like accumulation of Fig.1 has been obtained through an approximate diffusion calculation but without any free parameter.

The equivalent width calculated by assuming the cloud-like accumulation of gallium is very close to the observed value.

Concerning the line profile shown in Fig.2, the difference between the stratified and homogeneous cases cannot be detected with IUE spectra, and needs higher resolution and better S/N. One cannot exclude other ions for which it might be more easy to put into evidence such an effect with present technics.

REFERENCES

Alecian, G.: 1982, Astron.Astrophys. **107**, 61
Alecian, G. and Artru, M.-C.: 1987, Astron. Astrophys., in press
Alecian, G., Grappin, R.: 1984, Astron.Astrophys. **140**, 159
Alecian, G., Vauclair, S.: 1983, Fundamentals of Cosmic Physics, **8**, 369, Gordon and Breach Science Publishers Ltd.
Artru, M.-C.: 1986, Astron. Astrophys., **168**, L5
Bonsack, W.K., Wolff, S.C.: 1980, Astron. J., **85**, 599
Borsenberger, J., Gros, M.: 1978, Astron.Astrophys.Suppl. **31**, 291
Borsenberger, J., Michaud, G., Praderie, F.: 1981, Astrophys. J., **243**, 533
Guthrie, B.N.G.: 1984, Mon. Not. R. Astr. Soc. **206**, 85
Heacox, W. D.: 1979, Astrophys. J. **41**, 675
Jamar, C., Macau-Hercot, D. and Praderie, F.: 1978, Astron. Astrophys. **63**, 155
Michaud,G.: 1970, Astrophys. J. **160**, 641
Takada-Hidai, M., Sadakane, K. and Jugaku, J.: 1986, Astrophys. J., **304**, 425

DISCUSSION

JUGAKU: Have you made a similar line profile computation for the Ga III (and possibly Ga I) resonance line?
ALECIAN: We have only considered two Ga II lines: the UV resonance line at 1414.401Å and also the visible line 4255.722-.92. Actually the best atomic data and the best observational data are for the UV resonance line.

MARCY: What values of gf, collisional broadening did you use? And what atmosphere did you use?
ALECIAN: The gf of the UV resonance line is 1.8 (it is known within 10% from recent determination). The collisional broadening is estimated from a semi-classical calculation and may be neglected for the UV line.

We have used the NLTE model atmosphere computed by Borsenberger and Gros (1978). Our synthetic spectrum does not take into account the NLTE effects on the gallium line formation, however we have verified that this approximation is justified.

STELLAR GRANULATION AND PHOTOSPHERIC LINE ASYMMETRIES

Dainis Dravins
Lund Observatory
Box 43
S-22100   Lund
Sweden

ABSTRACT. The fine structure of stellar photospheric convection (the stellar equivalent of solar granulation) can be analyzed with the aid of high-resolution spectroscopy. Photospheric absorption lines are slightly asymmetric and wavelength-shifted due to unequal photon contributions from bright and systematically Doppler-shifted granules and from darker intergranular areas. Numerical simulations of stellar surface convection in different stars have now been carried out, and such three-dimensional and time-dependent models predict the detailed stellar line profiles (including asymmetries and wavelength shifts), thus enabling a direct confrontation between observations and theory.

1. WHY STUDY STELLAR GRANULATION?

The study of stellar granulation, physical processes in stellar convection, and the ensuing inhomogeneities on stellar surfaces has considerable potential not only for the analysis of stellar atmospheres *per se*, but also in applications for other astrophysical problems. Examples of relevant topics are:

• <u>Structure of stellar surfaces</u>. How do stars look like? One very important goal for stellar astrophysics is to obtain high resolution images of stellar surfaces, to ultimately enable studies of stars as surface objects. Until such images are available, one should at least deduce the statistical properties of stellar surface features.

• <u>Physical processes in stellar convection</u>. Convection is a central but poorly understood parameter in the construction of stellar models and the determination of stellar ages, influencing both the energy transport through the atmosphere, and the replenishment of nuclear fuels in the core. What is the validity (if any) of the classical 'mixing-length' concept and other approximations in stellar atmosphere theory?

• <u>Spectral line broadening</u>. The understanding of small-scale motions in stellar photospheres will elucidate the mechanisms of spectral line broadening and show the relation (if any) to the classical 'micro-' and 'macro-turbulence' parameters.

- Accurate radial velocities. Since photospheric spectral lines are asymmetric and wavelength shifted due to unequal photon contributions from rising and sinking elements, spectroscopic determinations of stellar motions require a correction for granulation effects if accuracies much better than 1 km/s are required. Such effects constrain studies of e.g. the velocity dispersion in open galactic clusters.

- Radial velocity variations. Activity-cycle variations may modulate the structure of stellar granulation analogous to what is seen on the Sun during its 11-year activity cycle. Such variations may change somewhat the asymmetry of photospheric lines and thus mimic e.g. radial velocity variations expected from the star's motion in conjunction with a low-mass companion.

- Stellar photometric variability. The finite number of evolving granules on a stellar surface induces photometric variability in integrated starlight, and sets physical limits to the interpretation of very accurate stellar photometry. This may complicate critical photometric tasks such as the identification of planetary transits across stellar disks.

- Accurate abundance determinations. Models that assume homogeneous photospheres may yield inaccurate abundances. This will result from the circumstance that the population of a particular atomic or molecular species in a certain stage of excitation or ionization is a nonlinear function of temperature, and hence an atmosphere with temperature fluctuations will produce lines of different strength than an atmosphere without such fluctuations. Further, there may well be real abundance inhomogeneities across a normal star, where molecules dissociate in the hotter regions. Significant effects may be expected for lines of high excitation potential and for molecular ones.

- Magnetic flux density in stellar photospheres. Irrespective of how stellar magnetic fields originate, their flux density at the surface will be strongly dependent on the horizontal velocity fields and the associated horizontal pressure fluctuations in stellar granulation. These constrain the magnetic flux density to which magnetic fields can be compressed in the photosphere. Knowledge of magnetic flux densities is a prerequisite for further analysis of e.g. magnetohydrodynamic wave generation and propagation into higher atmospheric layers.

- Acoustic energy flux. The motions in stellar granulation generate pressure waves with probable effects both above and beneath the photosphere. In higher layers one may expect acoustic waves to dissipate, thus contributing to e.g. chromospheric heating. In deeper layers one should expect an interaction with large-scale oscillations, perhaps even global pressure modes.

- Comparisons to solar granulation. Last (but perhaps not least), studies of stellar granulation enable comparisons to be made with the much-studied solar counterpart. Credible theories of solar granulation should be able to predict not only observed solar properties, but also those of other (at least solar-type) stars.

2. HOW TO DETECT STELLAR GRANULATION?

Since, at the present time, it is not yet possible to obtain high spatial resolution images and spectra of stellar surfaces, we have to rely on more indirect methods to detect effects of stellar granulation. The visibility of solar granulation in integrated sunlight (the Sun seen as a star) gives a first indication of the magnitude of the effects likely to be encountered in other stars. A number of different methods should allow one to observe effects of stellar granulation:

- Photospheric line asymmetries. The different amplitudes and area coverages of upward and downward atmospheric motions, coupled with the correlation between photospheric temperature (brightness) and the sign of vertical motion (hot elements are rising) cause an intrinsic asymmetry of photospheric absorption lines. Such asymmetries are perhaps best described by the line bisector (median), which shows the apparent radial velocity at each depth in the line. For integrated sunlight, the amplitude of this bisector corresponds to some 200 m/s, only some percent of the line-width. Consequently, very high resolution observations of unblended lines are required to reveal these effects. The bisector shapes differ for lines of different parameters, reflecting different conditions and heights of formation. The asymmetries depend in particular on line strength, excitation potential, and ionization level.

- Photospheric line wavelength shifts. This has proven to be a useful tool for solar studies, although its stellar application is somewhat more complicated. The larger and brighter ascending granules give a slightly larger photon contribution than the darker and smaller intergranular lanes. This statistical bias of blueshifted photons causes a convective blueshift such that the wavelengths of solar lines (after corrections for solar motion and gravitational redshift) are slightly shorter than the corresponding laboratory values. The magnitude of this effect depends on the line parameters but is typically some 300 m/s or a few percent of the line-width. To study differential apparent radial velocities among lines of different parameters in one star requires high-quality spectra, a very accurate stellar wavelength calibration, and accurate laboratory wavelengths. For a determination of the absolute value of such convective lineshifts, the true stellar radial motion must also be known, as well as its gravitational redshift.

- Time variability of photometric irradiance. The finite number of granules on a stellar surface implies that random changes in the granulation structure will be accompanied by some finite change in the integral properties of the star, such as its bolometric irradiance.

In the solar case, with ≃ 20% intensity contrast per granule, and assuming all granules independent, the variability in integrated sunlight should be roughly this number divided by the square root of the number of solar granules (≃ 10^6), which comes out to a few times 10^{-4}. Such variations are observable from space and have probably already been measured as part of the variations of solar irradiance, although not yet conclusively identified as granular in origin. In giant stars, the increased pressure scale heights may cause there to be only a modest number of granules on the stellar surface, and the resulting effects from their time evolution could be greater. Such irradiance fluctuations might well be correlated with line-profile changes and could also include polarization effects.

- Interferometric imaging of stellar surfaces. The disks of a few giant stars can be resolved already with existing large telescopes or interferometers. Since convection cells on some giant stars might subtend a significant fraction of a stellar diameter, their spectral features should soon be detectable through e.g. speckle spectroscopy. Solar-type granules, however, have sizes only about one thousandth of the solar diameter, and their imaging requires optical baselines a thousand times longer than those required to resolve the stellar disk. To achieve this might require space-based interferometers and remains an interesting challenge for the future.

A more detailed discussion of different plausible schemes to detect stellar granulation is in Dravins (1987a). For a general review of in particular solar granulation signatures in the spectrum of integrated sunlight, see Dravins (1982). Solar granulation is reviewed in the monograph by Bray et al. (1984).

Among these different schemes to detect stellar granulation, the most promising seems to be the study of photospheric line asymmetries. In spectra of very high photometric quality and a spectral resolution $\lambda/\Delta\lambda$ at least about 100,000, the signature of (at least solar amplitude) granulation will appear. Such studies of line asymmetries indeed form the current basis for analysis of stellar granulation.

3. STELLAR PHOTOSPHERIC LINE ASYMMETRIES

A fair number of stars has now been observed for photospheric line asymmetries by different authors, revealing considerable asymmetry changes across the Hertzsprung-Russell diagram. The asymmetry of typical solar lines is characterized by a bisector in the shape of the letter 'C' (in spectral plots with wavelength increasing to the right), i.e. the middle part of the line is somewhat blueshifted relative to its top and bottom portions. Gray (1980) identified the sense of line asymmetry in the K giant Arcturus to be opposite that in the Sun, and similar lineshapes in other cool giants were seen by Ridgway and Friel (1981). In a survey of F, G, and K stars, Gray (1982) found line asymmetries to be generally present. More detailed lineshape observations for several stars are in Dravins (1987b). Particularly large asymmetries (with the asymmetry opposite to the Sun´s) are seen in F giants (Gray and Toner 1986; Dravins 1987b).

To reliably measure these subtle line asymmetries requires a good understanding of the instrumental profile in the spectrometer, a careful selection of as unblended lines as possible, and (in general) an averaging over several lines to take away residual noise. In Figures 1 - 2, examples of line asymmetry patterns are shown for two G2 V stars: the Sun (central main sequence) and α Cen A (upper main sequence).

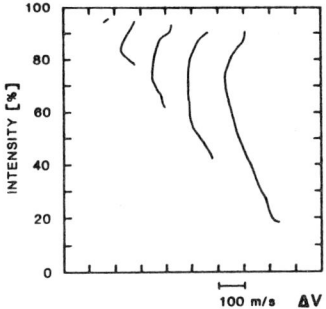
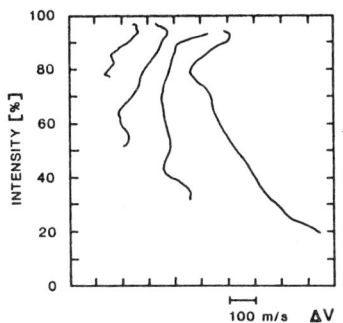

Figure 1. A signature of solar granulation in integrated sunlight. Each curve is the average of bisectors for Fe I absorption lines, grouped according to line-depth. Intensity is in units of the spectral continuum (100%) while wavelength increases to the right. Adopted from Dravins et al. (1981).

Figure 2. Bisectors for differently strong Fe I lines in α Cen A (G2 V), recorded with the coudé echelle spectrometer of ESO at La Silla using a double-pass scanner mode at spectral resolution $\lambda/\Delta\lambda \simeq 200,000$ (Dravins, 1987b). Although generally similar to the Sun, there is a clear difference in bisector slope for the strongest lines.

4. NUMERICAL SIMULATIONS OF STELLAR GRANULATION

Numerical supercomputer simulations of stellar surface convection in different stars have now been carried out (Nordlund and Dravins, 1988). These time-dependent models of the 3-dimensional, radiation-coupled compressible convection contain only three free physical parameters: effective temperature, stellar surface gravity, and chemical abundance. The models are stepped forward in time to reveal the properties of the convective elements (granules) in stellar photospheres: their sizes, the patterns in their gas flows, and their time evolution. These models are based upon the experience from the solar case (e.g. Nordlund 1985a), and have been completed for four different non-solar cases: T_{eff} = 6600 K, corresponding to Procyon (F5 IV-V); T_{eff} = 5800 K, one half solar surface gravity, corresponding to α Cen A; T_{eff} = 5800 K, one quarter solar surface gravity, corresponding to β Hyi (G2 IV), and T_{eff} = 5200 K, corresponding to α Cen B (K1 V).

By using these models as a set of spatially and temporally varying model atmospheres, synthetic spectral line profiles have been computed as averages over the simulation sequence (Dravins and Nordlund, 1988), analogous to previous solar work (Dravins et al., 1981, 1986). These profiles do *not* contain any arbitrary fitting parameters such as 'micro-' or 'macro-turbulence' and can thus be directly compared to observations to verify the degree of realism in the models. There is a very satisfactory agreement concerning the line asymmetry patterns in α Cen A and Procyon (the only stars for which extensive high-quality observations are available). For example, the large bisector curvature in α Cen A is predicted for the deepest lines only, as observed in Figure 2. Compared to the Sun, this effect can be traced back to the larger velocity amplitudes that develop on α Cen A due to its lower surface gravity. Such examples demonstrate the sensitivity of line asymmetries to minute changes in stellar atmospheric structure.

Other theoretical granulation models, including spectral line synthesis for both solar and non-solar conditions, are being pursued by a number of other authors, e.g. Steffen (1987). Besides more detailed simulations, there is a strong need for simpler parametrized models in order to understand basic properties in stars of spectral types that are not yet accessible to detailed modelling, to give guidance to observers as to what types of effects to look for, and also to give a connection to classical stellar model atmospheres (involving fitting parameters such as 'mixing-length', 'turbulence', etc.). Such work is being pursued by several authors.

5. OUTSTANDING PROBLEMS

New methods have now made stellar surface structure accessible to detailed observational and theoretical investigation. However, none of these methods is trivial. As can be seen in numerical experiments of gradually degrading the spectral resolution of solar spectra, already a spectral resolution $\lambda/\Delta\lambda$ = 100,000 (a resolution often considered 'high') is actually only marginally adequate to permit meaningful studies of bisector shapes in normal stars (Livingston and Huang, 1986; Dravins, 1987a). Rather, a resolution of 200,000 or 300,000 or even more is desirable to analyze line asymmetry patterns in different types of spectral lines. The numerical simulations of granulation predict a series of line asymmetry and wavelength shift phenomena in different stars (as function of excitation potential, wavelength region, etc.) that are beyond current observational capability, and that probably will require such resolutions for reliable detection. An ideal instrument should combine this very high resolution with a very low straylight level, a very high wavelength accuracy, and an extended spectral coverage. For brighter stars, such an instrument would by far not be limited by the photon flux collected by any reasonable telescope, but rather by inefficiencies in the spectrometer. Since neither current double-pass spectrometer scanners, nor Fourier transform spectrometers appear to give adequate performance, this remains an important instrumental and observational challenge for the future.

Current theoretical models suffer limitations in particular due to the physical approximations enforced by the finite computing power available. Future models should not be limited to a small volume near the stellar surface, but rather embrace a large fraction of the star. Computer codes that allow waves to propagate should allow an examination of e.g. the interaction between events in the granulation and excitation of large-scale pressure waves. Granulation in chromospherically active stars and in stellar active regions will further require inclusion of effects from magnetic fields (Nordlund, 1985b). This probably will require computations to be carried out to smaller spatial scales than in the non-magnetic case. Since already some of the present models need several tens of supercomputer hours for each run, the inclusion of all desired parameters seems difficult as a simple extrapolation of present methods. Either vastly more efficient algorithms are to be found, or computing power is to be enhanced by many orders of magnitude. Possibly, the most promising way to achieve such increased performance would be to use novel computer architecture, custom-designed for the stellar convection problem.

ACKNOWLEDGEMENT

This work is supported by the Swedish Natural Science Research Council.

REFERENCES

Bray, R.J., Loughhead, R.E., Durrant, C.J.:1984, *The Solar Granulation*, 2nd ed., Cambridge University Press, Cambridge
Dravins, D.:1982, *Ann.Rev.Astron.Astrophys.* **20**, 61
Dravins, D.:1987a, *Astron.Astrophys.* **172**, 200
Dravins, D.:1987b, *Astron.Astrophys.* **172**, 211
Dravins, D., Larsson, B., Nordlund, Å.:1986, *Astron.Astrophys.* **158**, 83
Dravins, D., Lindegren, L., Nordlund, Å.:1981, *Astron.Astrophys.* **96**, 345
Dravins, D., Nordlund, Å.:1988, *Astron.Astrophys.*, to be submitted
Gray, D.F.:1980, *Astrophys.J.* **235**, 508
Gray, D.F.:1982, *Astrophys.J.* **255**, 200
Gray, D.F., Toner, C.C.:1986, *Publ.Astron.Soc.Pacific* **98**, 499
Livingston, W., Huang, Y.R.:1986, in M.S.Giampapa, ed. *The SHIRSHOG Workshop*, National Solar Observatory, Tucson, p.1
Nordlund, Å.:1985a, *Solar Phys.* **100**, 209
Nordlund, Å.:1985b, in H.U.Schmidt, ed. *Theoretical Problems in High Resolution Solar Physics*, Max-Planck-Institut für Physik und Astrophysik, München, p.101
Nordlund, Å., Dravins, D.:1988, *Astron.Astrophys.*, to be submitted
Ridgway, S.T., Friel, E.D.:1981, in C.Chiosi, R.Stalio, eds. *Effects of Mass Loss in Stellar Evolution*, IAU coll.**59**, Reidel, Dordrecht, p.119
Steffen, M.:1987, in proc. workshop *The Role of Fine-Scale Magnetic Fields on the Structure of the Solar Atmosphere*, Cambridge University Press, Cambridge, in press

DISCUSSION

MARCY How are your models affected by the lack of magnetic fields, especially those between the granules ?

DRAVINS Such effects can be studied on the Sun : granulation in active regions (in so-called plage or facular areas) is penetrated by magnetic fields which are largely concentrated to inter-granular regions. In such regions, the photospheric line asymmetries are indeed different from those in quiet areas. For solar levels of activity, however, the area coverage of such active regions is small enough not to influence the average line profile very much : it is a second-order effect. Nevertheless, the periodic variation of active region area coverage over a stellar activity cycle is likely to induce a periodic variation in line profile parameters that might mimic the expected periodic radial velocity variation induced by an extrasolar planet. High-activity stars might be dominated by magnetic regions, and the study of convection in magnetic fields will probably be required to understand their photospheric line profiles. Some theoretical work has been done for solar conditions by Nordlund, but the stellar area is unexplored as yet.

VOGT From your detailed simulations, will you now be able to give those of us who do line profile synthesis a better parameterization for micro and macroturbulence ?

DRAVINS One of the aims of the detailed simulations is indeed to understand which parameters are important for simpler line profile modelling, and what relation there is between deduced 'turbulence' parameters and physical gas velocities in the stellar atmosphere. This work is in progress.

APPENZELLER You correctly stressed that your method avoids arbitrary and unphysical assumptions. However, does not the limited volume (of the star) which can be handled by the computer restrict the manifold of solutions which you can obtain, excluding e.g. all effects having large scales, such as larger scale motions or effects of global oscillations ?

DRAVINS Yes, the present simulations cover only a small volume near the stellar surface. A long-term goal is indeed to have models that embrace the entire star, and include processes such as wave generation and dissipation. With such models one should ultimately be capable of studying e.g. the excitation of global oscillations by events in the convective motions. The present simulations aim at understanding the origin of photospheric line profiles for solar-type stars. As observed on the Sun, the most significant line fluctuations occur on granular scales, and these are the only ones presently modelled.

RUTTEN I want to rephrase Appenzeller's question. Of course, while your simulations are free of fudge parameters as turbulence and mixing length, they do contain implicit limitations on which the results may sensitively depend. Apart from the neglect of waves and magnetic field that you mentioned, there are intricate assumptions in the radiative transfer that we perhaps shouldn't go into here, and there is the important choice of your Fourier grid. With regard to the solar simulation the comment has been raised that its largest scale is yet too

small, effectively forcing all of the mechanical energy flux to flow upward at granular scales, so that the computed granulation is too vigorous.
Now, how did you select the largest scale or smallest spatial frequency of the stellar grids ? And can you test whether it is sufficient ?

DRAVINS The present computations were made for a spatial grid of 32x32 Fourier coefficients on the stellar surface. Scaled to solar conditions, these cover linear scales between approximately 100-3000 km. This range was chosen to correspond to granular features as observed on the Sun. For a given stellar luminosity and surface gravity, there will be one characteristic scale of granulation, constrained by energy considerations (too small granules would not transport enough energy ; too large granules would break apart under their own pressure). This scale can be roughly estimated for modest departures from solar conditions but, due to the complexity of the problem, such estimates become very uncertain elsewhere in the Hertzsprung-Russell diagram. This is one of the main reasons why the present simulations have been carried out for solar-type stars only. There is no known reason why the present computer programs could not be applied also for other stars, if only the computing power was available. The present 32x32 resolution is a significant improvement over the early solar work with 16x16 resolution (Dravins, Lindegren and Nordlund, 1981), but significant increases in this resolution become very expensive in computing time due to the four-dimensional nature of the problem (3-dimensional space, and time).
The crucial test of whether the models are adequate comes from comparisons with observed line profiles, asymmetries, and wavelength shifts. Taken together, these form very sensitive tools for segregating different types of granular motion. Of particular interest would be the availability of very accurate wavelength measurements for stellar lines : a parameter that hitherto has been lacking.

OBSERVATIONAL EVIDENCE BY FOURIER TRANSFORM SPECTROSCOPY OF CONVECTIVE MOTION IN LATE-TYPE STELLAR ATMOSPHERES

Jean-Pierre Maillard
Institut d'astrophysique de Paris
98bis Blv. Arago
75014 Paris
France

Daniel Nadeau
Département de physique
Université de Montréal
CP 6128A, Montréal, QC
Canada H3C 3J7

ABSTRACT. We have obtained FTS spectra of 11 stars of spectral types from M0 to M4 III in the infrared K band. The shifts at line bottom of the lines of CO and Fe I show a dependence on excitation energy and line depth similar to that caused by convective motions in the photosphere of the Sun but with a larger amplitude. Time variability is small but important differences appear between individual stars of similar types. From these data it should be possible to obtain quantitative estimates of the convective velocities and horizontal temperature inhomogeneities.

1. INTRODUCTION

The determination of the parameters of convective motions (gas velocities, horizontal temperature inhomogeneities granulation scale) is of considerable interest for the study of red giants as these motions may be related to the isotopic enrichment of the atmosphere, to chromospheric excitation, and to the onset of large mass-loss. For stars far from the solar type, the hydrodynamical calculations are hampered by the lack of suitable initial conditions necessary to follow the evolution of the system towards equilibrium.

To find observational clues to motions in cool stellar atmospheres that could be linked to the parameters of convection, we measure the small differential shifts between the line positions of CO and metals (principally iron) as a function of excitation energy, line depth, and wavelength; we look for a dependence on spectral type and for differences among stars of the same type, and we monitor stars for time variations.

2. METHOD

Convective motions can be measured in the asymmetry and shift of spectral lines (Dravins 1982, Gray 1982). The severe blending of lines in the spectra of late-type stars implies that it is difficult to measure the asymmetry of the profiles. The shift at line bottom is less affected by blending than higher up in the line. Its measurement does not require as high a spectral resolution as the measurement of line asymmetries so that it is possible to obtain data on a larger sample of stars. By measuring the differential shift of many lines (⩾100), it is possible to obtain its dependence on excitation energy and line depth. Fourier Transform Spectrometry is well suited for this task because of its large spectral coverage and high internal accuracy.

The $\Delta v=2$ and $\Delta v=3$ bands of CO and numerous lines of Fe I can be observed in the infrared K (2.2 μm) and H (1.6 μm) bands, and cover a large range of excitation energy and line depth. While the absorption in the fundamental band of CO may come from a circumstellar envelope, the overtone bands are almost certainly of photospheric origin. The laboratory frequencies of the near-infrared lines of CO and Fe I are almost unique in having an absolute accuracy of 0.001 cm^{-1} with respect to a common frequency standard and being essentially free of pressure effects (Nadeau and Maillard 1988).

The data of Biémont et al. (1985) on the infrared spectrum of iron have made it possible to determine the shift of the Fe I lines in the Sun as a function of line depth and excitation energy with an accuracy of 30 m s^{-1} (Nadeau 1988). The solar lines of CO have velocities in agreement with this relation within the 0.001 cm^{-1} uncertainty.

3. DIFFERENTIAL LINE SHIFTS IN RED GIANTS

K-band spectra of 11 giants with spectral types from M0 to M4 III were obtained with the CFHT FTS spectrometer at a resolution of 5 km s^{-1}. The stars are selected on the basis of their brightness and their low variability, and double stars are avoided.

The stars HR 0045 and HR 3950 were observed twice, at 23 and 16 month interval respectively. For each of these observations we have plotted in Figure 1 the shifts between the average velocity of the lines of CO with excitation energy below 1 eV, the lines of CO with energy above 1 eV and the lines of Fe I with average energy near 5.5 eV. The reproducibility of the data obtained at different epochs is excellent. It gives an idea of the accuracy of the method and shows that there is little variation of the differential shifts on a time scale of a few years. On the other hand these two stars of similar spectral types have very different relative shifts. There is a large shift between the high and low energy lines of CO in HR 0045 but essentially none in HR 3950.

The data plotted in Figure 2 represent the same differential shifts as in Figure 1 but for all observed stars. A plot of the quantitative relation obtained from the Fe I lines in the Sun, evaluated at average energies and line depths corresponding to the giants data, is shown for comparison. Most red giants show a blueshift of the high excitation energy lines larger than in the Sun, indicating larger convective velocities or stronger temperature inhomogeneities. Some trends do appear as a function of spectral type but there are important individual variations. There is a correlation between the excitation energy and line depth of the observed lines so that it is not straightforward to separate their individual effects. The ratio of the drop in line depth to the increase in energy is high between the low and high energy lines of

CO, but much smaller between the high energy lines of CO and the lines of Fe I. One can deduce from this relation that the blueshifts appear more sensitive to excitation energy than line depth for earlier types while the opposite seems true for later types among the observed stars.

From these data we intend to get quantitative estimates of the convective velocities and horizontal temperature inhomogeneities in the photospheres of the red giants. We are also looking for the origin of the large differences between individual stars, which may indicate that additional parameters have an influence on the observed motions.

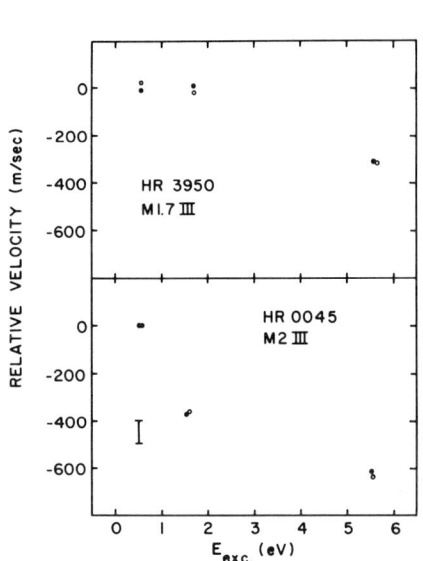

Figure 1. Differential shifts of the lines of CO and Fe I observed at different epochs. The error bar shows the uncertainty of the laboratory frequencies of Fe I with respect to CO.

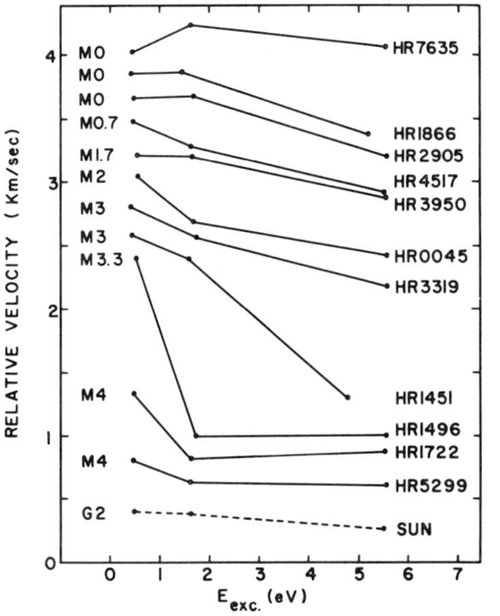

Figure 2. Differential shifts of the lines of CO and Fe I for the observed stars. The velocity scale refers to the amplitude of the relative shifts between lines of different energies.

REFERENCES

Biémont, E., Brault, J. W., Delbouille, L., and Roland, G. 1985, Astr. Ap. Suppl., 61, 107.
Dravins, D. 1982, Ann. Rev. Astr. Ap., 20, 61.
Gray, D. F. 1982, Ap. J., 255, 200.
Nadeau, D. 1988, Ap. J., 325, in press.
Nadeau, D., and Maillard, J.-P. 1988, Ap. J., submitted.

DISCUSSION

RUTTEN: Are there any Fe II lines in your data? Even if they are very weak, it may be valuable to include them. In the theoretical simulations, the ionization equilibrium of iron plays a very important role by influencing the near-ultraviolet and blue line haze, to which the computed granules are quite sensitive. Direct observational comparisons between Fe I and Fe II will be very interesting when comparisons with simulations will be made.

NADEAU: We have not looked for the lines of Fe II because their laboratory frequencies are not known to high accuracy. There may also be a chromospheric contribution to these lines that would make the interpretation of the shifts at line bottom ambiguous. But we can certainly measure them if the lines appear in our spectra, and it is possible that improved laboratory data will be available in the near future.

DOPPLER IMAGES OF SPOTTED LATE-TYPE STARS

Steven S. Vogt
Lick Observatory, Board of Studies in Astronomy and Astrophysics
University of California, Santa Cruz, CA 95064

ABSTRACT. Doppler imaging is a technique for deriving resolved images of rapidly rotating stars from a detailed analysis of very high signal-to-noise high resolution spectral line profiles. An improved version of this technique is presented, which now uses principles of maximum entropy image reconstruction to invert the line profile information. The effects that noise, finite resolution, and uncertainties in the assumed stellar physics have on the resultant images were explored through various test simulations. The technique is found to be efficient, accurate, and robust at deriving images of certain classes of stars from realistic quality data. Doppler images are presented of two spotted late-type stars, the RS CVn-type star HR 1099, and the FK Com-type star HD 199178. Both stars show surprisingly similar spot distributions. In each case, there is a single large cool spot straddling the pole, and a number of small cool spots at low latitudes. We expect that the small low latitude spots on each star will migrate poleward to join the polar spot, and suspect that the observed long-lived polar spots are the result of the poleward migration and merging of many active region complexes. If true, the poleward migration of starspots suggests that magnetic activity on very rapidly rotating stars is qualitatively different than that seen on our Sun. We suggest that the observed rotational trigger velocity for the appearance of large spots on late-type stars marks the transition from solar-type boundary layer dynamos to distributed dynamos, which occur only in more rapidly rotating stars. The sizes, locations, and migrations of the spots, however, may be more a result of the convective flow patterns than of any dynamo action, since the spots are quite long-lived.

1. INTRODUCTION

We have developed a technique for deriving resolved images of certain types of rapidly rotating stars. Collaborating with me in this task are my graduate students G. Donald Penrod and Artie Hatzes. We call the method 'Doppler imaging' since it exploits the correspondence between wavelength position across a rotationally Doppler-broadened spectral line and spatial position across the stellar disk to derive a 2-d image of the star. Cool spots and/or abundance peculiarities on a rotating star produce distortions in the star's spectral line profiles. If the line profiles are dominated by rotational broadening, a high degree of correlation exists between the position of a given distortion in a line and the longitude on the star of the feature which produced it. A high resolution spectrum of the line is thus, to first order, a 1-d image of the star in longitude, but completely blurred in latitude. As the star is observed from other aspects (other rotation phases) other 1-d images are obtained. Then, if the inclination, rotation period, and other basic physical parameters of the star are known, the set of 1-d images can be combined into a 2-d image of the star in much the same way as the technique of CAT scanning in medical imagery.

The term 'Doppler imaging' has apparently come to mean different things to different people, and is now being used by a number of different researchers to describe various procedures by which spatial information on a star is derived from Doppler signatures in spectral lines. However, there is an important distinction to be drawn. In all other cases of which I am aware, these methods involve just mathematically parameterizing the expected image with some simple function and then solving for the appropriate image parameters by model fitting (for example, fitting spherical harmonic models to non-radially oscillating stars to describe the surface velocity fields, or fitting spotted star line profiles with simple circular spot models). Such approaches, while often informative, are not actually 'imaging' methods since they make highly artificial *a priori* assumptions about the appearance of the expected image. The technique we refer to as Doppler imaging makes no such *a priori* assumptions about the image. The resultant image can be arbitrarily complex, and its information content is determined by the size and quality of the data set.

The problem of deriving maps of abundance peculiarities on Ap stars has been around for many years, and a variety of solutions to the problem have been developed (Deutsch 1958, 1970; Pyper 1969; Falk and Wehlau 1974; Megessier 1975). However, all of these approaches involved parameterizing the surface distribution as a truncated series expansion of spherical harmonics or the line profile data as a truncated series expansion of residual intensities. To keep the problem tractable, only a few terms were carried in the expansions. Thus, while the solutions were often mathematically elegant, they were highly constrained artificially and thus were not true 'imaging' methods. More recently, Khokhlova and co-workers (Khokhlova 1975; Khokhlova and Ryabchikova 1975; Goncharsky *et. al.* 1982; Khokhlova 1985; Khokhlova, Rice and Wehlau 1986) used an approach which could more suitably be called 'imaging'. Their method involved representing observed spectral lines by an integral equation of local equivalent width integrated over the stellar disk, and used a regularizing algorithm to solve for the local equivalent width map. However, their maps generally failed to recover any latitude information about surface features, and their solutions were always strongly biased to the sub-observer latitude on the star.

We have been studying a number of rapidly rotating spotted late-type stars which show time variable distortions in their line profiles. The distortions are attributable to dark spots being carried into and out of view by stellar rotation, as explained by Vogt and Penrod (1983a, 1983b). It was clear from the quite complex nature of these distortions that a much higher degree of spatial information was present in the data than any of the previous line profile inversion methods could cope with. Also, to extract useful astrophysics from images of the spotted stars, we needed a method which gave accurate and reliable latitude information, so that spot locations and migration in latitude could eventually be used to measure differential rotation and spot migration rates. Thus we set out to develop a true 'imaging' solution to the line profile inversion problem, one that made no *a priori* assumptions about the nature of the features to be imaged, and which did not constrain the image solution artificially. The technique we developed incorporates principles of maximum entropy image reconstruction and is proving to be quite robust, efficient, and accurate at deriving resolved images of certain types of rapidly rotating stars.

We are presently using Doppler imaging to study cool starspots on RS CVn and FK Com stars, and surface abundance distributions on Ap stars. The Ap star imaging is being done principally by Mr. Artie Hatzes as part of his Ph.D. dissertation research and will be presented elsewhere in this meeting. Here, I will try to give a brief review of the technique of Doppler imaging and what it has revealed about the nature of giant starspots on certain late-type stars. A more detailed account of the method is currently in press and will appear later this year (Vogt, Penrod, and Hatzes 1987 - here after referred to as VPH). Previous discussions leading up to the present version of the technique can be found in Vogt and Penrod (1983a, 1983b). I was also asked to present a review of Doppler imaging last week at the 27th Liege International Astrophysical Colloquium, so parts of the present discussion of the method will be largely a duplication of that review.

2. WHAT KIND OF STAR CAN BE DOPPLER IMAGED?

In order for a star to be suitable for Doppler imaging, it must be rotating fast enough that its line profiles are dominated by rotational broadening. The number of resolution elements across the disk is roughly 2 ($v \sin i$) / W where $v \sin i$ is the star's projected equatorial rotation velocity and W is the full width at half depth of the star's average specific intensity profile (i.e. that which would be obtained from a flux spectrum of the star in the absence of stellar rotation). For active late-type dwarfs and subgiants, W is typically about 10 km s^{-1}, so a minimum $v \sin i$ of 20 to 30 km s^{-1} is needed to get any useful amount of resolution. The larger the $v \sin i$, in principle, the better the resolution. However, since line equivalent width is conserved, the lines become shallower as the $v \sin i$ is increased, and the distortions, which are always some fraction of the line depth, become weaker. Above 80 to 100 km s^{-1}, the distortions become very difficult to detect and disentangle from line blends and weak terrestrial features. So projected velocities of 40 to 60 km s^{-1} are about ideal for late-type stars. For hotter stars with less line blending and/or stronger lines, larger $v \sin i$'s are usable.

In order to be able to assemble line profile data from many different rotational aspects of the star, the star's rotation period must be known. Fortunately, this is almost always well determined by broadband photometry. However, we have encountered cases where the published photometric periods were slightly incorrect, and the Doppler imaging technique could not converge on a solution. In such cases, the imaging technique itself was used to provide a much more accurate period as part of the solution.

The stellar inclination must also be at an intermediate value. Pole-on orientations have no line-of-sight components of surface velocity and thus no Doppler information. Equator-on orientations suffer a north/south hemispheric ambiguity and yield Doppler images with mirror-like symmetry about the equator. However, quite a wide range of inclinations are possible, and the method is fortunately quite insensitive to errors in the assumed inclination. In practice, we have found inclinations from 20° to 70° to be feasible.

The basic physical properties of the stellar atmosphere must also be known, but, fortunately, not to any high degree of precision. We have found that an LTE-level of approach to the line synthesis is quite adequate. We treat the limb angle dependence of line strengths, shapes, and continuum brightness explicitly in the method.

3. DATA REQUIREMENTS

Doppler imaging requires extremely high quality spectral data. The data consist of a set of high resolution, high signal-to-noise spectral line flux profiles obtained at different rotation phases. For the spotted late-type stars, we use either the Fe I line at 6431 Å, or the Ca I line at 6439 Å. Both are relatively unblended and reasonably clear of terrestrial absorption features. For Ap stars, we use either the Cr II line at 4824 Å or the Si II line at 6347 Å. In general, we only use a single atomic species of spectral line for each image, but the method can be set up to use many different lines simultaneously. While this complicates the process, it could provide additional constraints on the image. To take full advantage of the resolution available at the star, the instrumental profile must be no larger than the star's intrinsic line width. A resolving power of at least 40,000 is thus preferred for the spotted late-type stars, though some useful information may still be obtainable at lower resolutions.

The line profile distortions are typically only 1% or less of the continuum, so quite high signal-to-noise data are required to achieve reliable images. We aim for a S/N $\geq$ 400 per 0.13 Å resolution element, though our simulations (VPH) indicate that useful information is still obtainable down to S/N of about 150-200 per resolution element. To obtain complete images of the visible portion of the star, line profiles must be well distributed over rotational phase space. A minimum of 8 to 10 profiles distributed evenly over phase space is required for a good image. Images derived from uneven phase coverage may still be useful, but will lose resolution and become less unique.

The line profile data are obtained with the coudé spectrometer of the Lick Observatory Shane 3-m telescope. We use a Texas Instruments 800×800 3-phase CCD (Robinson and Osborne 1986), at the focus of the 80" coudé camera. With the 9.5" collimator and Grating VII in second order at 6435 Å, this set-up results in a dispersion of 38 mÅ per CCD pixel, and spectral coverage of 30 Å. We also use a Bowen-Walraven image slicer which not only increases throughput at the slit, but also 'cools' the image down in one dimension at the focal plane by spreading the light perpendicular to dispersion over many rows of the CCD. Thus, less dynamic range is required from the CCD and high signal-to-noise exposures are easier to obtain. The slicer accepts light from a 3" diameter entrance aperture, and slices it down to a projected slit of 3.4 pixels at the focal plane or 0.129 Å. The resultant resolution is 50,000 at 6435 Å.

The line profiles must be carefully reduced before attempting any imaging. Besides the usual flat fielding and zero point corrections which are necessary with most array detectors, the data must be thoroughly cleaned of telluric absorption features and, if the star is in a binary system, the spectra of the companion star(s) must also be carefully removed. Fortunately, with modern digital data and computers, this is relatively straightforward, but must be done with great care as the Doppler imaging signatures are quite small.

4. INVERTING THE SPECTRAL LINE FLUX PROFILES

Each line profile is basically a 1-d marginal brightness distribution or projected image of the stellar disk. The task of 'inverting' the line flux profiles to derive the brightness distribution across the stellar surface is thus basically a deprojection problem, and is best handled in matrix form on a computer. We have then an *image* space and a *data* space, with some matrix which transforms between them in one direction.

4.1 Set Up The Data Vector

Once each profile is fully reduced as described above, the set of all profiles are phased using the known rotation period and assembled end-to-end with just enough separation between to reach the continuum between profiles. Figure 1 shows a complete data set for the star HR 1099 from the Fall 1985 observing season, using the Ca I 6439 Å line. The crosses indicate the actual data, and the solid line represents our imaging software's final fit to these data (to be discussed later). Since the spots change significantly on a time scale of 4 to 5 months on these stars, all the line profiles were obtained in less than a 3 to 4 month interval. This set of data points defines our data vector D_k, and typically contains up to 1000 points.

4.2 Set Up The Image Vector

We create an image space by subdividing the star up into a grid of surface brightnesses, and expressing these surface brightness values as a simple 1-d image vector I_j. Figure 2 shows how the star is sectioned. We use a grid of zones whose edges lie along lines of longitude and latitude, and vary the number of zones at each latitude such that the true area of each zone is roughly constant over the star. Thus, zones at high latitude span a larger range in longitude than zones at low latitudes. The minimum number of zones required is constrained by the need to adequately resolve features on the star. The maximum number allowed is constrained by the need to keep the problem of manageable size for our VAX 11/780 computer. In practice, zone sizes near the equator of about 5° by 5° are sufficient to provide adequate sampling of the disk. So the number of zones used is typically 1200 to 2500. Here, the surface is divided up into 40 latitude bands. At this inclination of 40°, a total of 2310 zones are required to cover the visible portion of the star. Each zone is assigned a corresponding surface brightness. The set of all such surface brightnesses comprises our one-dimensional image vector.

4.3 Create The Response Function Matrix

We then set up a response function matrix **R** that transforms from image space to data space. The creation of this matrix is essentially the same procedure one goes through to synthesize a line flux profile from a star of known atmospheric and geometrical parameters. It involves working the problem *forward* using an LTE model of the stellar atmosphere and knowledge of the star's inclination, and

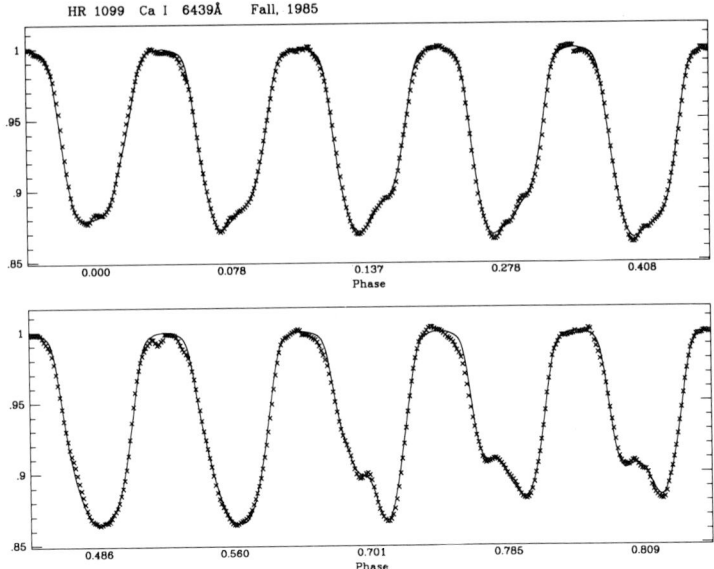

Figure 1. Data vector for HR 1099 in 1985.

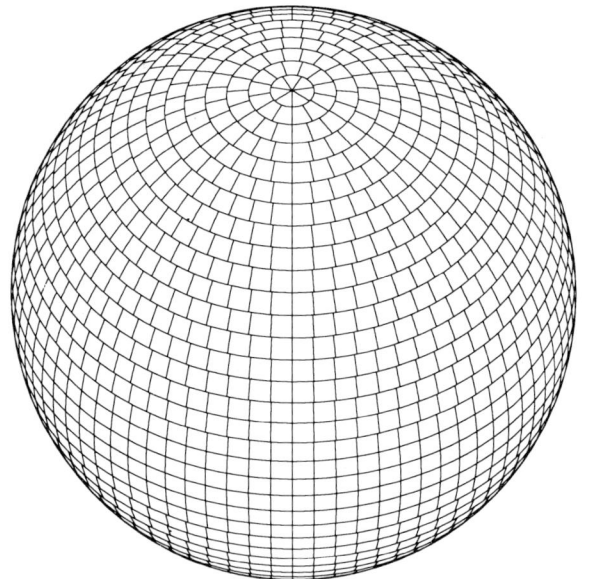

Figure 2. Zonal division of the stellar surface.

ically different in the 'spots' than in the photosphere, and in which both spots and photosphere make a significant contribution to the observed line, then some more elaborate prescription may need to be used. For our case, however, since the spot does not make a significant contribution to the observed flux, we set dI_λ/dI_c equal to I_λ/I_c (where I_c is the continuum surface brightness and the I_λ's the specific intensities across the spectral line), which much simplifies the computing.

The radial velocity of the zone due to rotation is then calculated. The appropriate values of dI_λ/dI_c across the line profile are then multiplied by the effective area of the zone, and shifted in wavelength according to the correct radial velocity. These values then become the response matrix elements for the data pixels obtained at that phase for that particular zone on the star. They tell how changing the surface brightness of that zone will affect the rotationally broadened line profile of the star at that particular phase. The entire transformation matrix, **R**, consists of all these values, for each zone on the star and for each data phase.

4.4 Inverting the Response Function Matrix

Once the problem is set up in the form $\mathbf{I\,R = D}$, one could ideally just solve directly for the image vector by inverting **R** and solving for **I**. Thus, $\mathbf{I = D\,R^{-1}}$. There are, however, a number of problems with this. **R** is not generally square and thus any given solution to the matrix equation is not unique. Furthermore, its rows are not independent. Effects such as projection along chords of constant radial velocity, finite width of the star's intrinsic line profiles and of the instrumental profile, incomplete phase coverage, and phase smearing during exposures all contribute to yield rows which are not independent. Also, even if **R** were well enough conditioned to be directly invertable, the data always contain some noise. Any noise features in the data would thus tend to produce spurious features in the image.

What is needed is a way to iteratively transform between image and data space, searching among all possible image vectors until one is found which fits all the observed data to within the known noise level of those data. The number of possible image vectors consistent with the data set can be large, and one must set some criterion for selecting the 'best' solution. One likely criterion is to choose the simplest or smoothest image, *i.e.* the one with the least amount of information, which is still consistent with all the data to within the known noise level of the data.

This type of problem is commonly referred to as digital image reconstruction, and is often encountered in radio astronomy, and medical imaging. A number of techniques have been developed for reconstructing an image from an incomplete and noisy data set. Perhaps the most widely used and powerful of these is the maximum entropy method (Skilling and Bryan 1984 and references therein), realized in software as MEMSYS, a computer program developed by Drs. J. Skilling and S. F. Gull. Of the various known digital image reconstruction techniques, maximum entropy has been shown to be the *optimal* technique of image reconstruction, and the *only* consistent regularization technique for images (Shore and Johnson 1980, 1983; Gull and Skilling 1983). In addition, the MEMSYS implementation of image reconstruction has proven to be extremely robust and efficient, and has seen wide application in astronomy, medical imaging, forensic imaging, and elsewhere. Subsequently, the MEMSYS software was kindly made available to us by Dr. Gull through Dr. R. Padman of the University of California at Berkeley Radio Astronomy Laboratory.

rotational velocity to set up a system of some 1000 equations of approximately 2000 terms each which relates all 2000 values of surface brightness on the star to all 1000 data points produced by the spectrometer. Thus,

$$I_1 R_{11} + I_2 R_{21} + \ldots\ldots\ldots + I_n R_{n1} = D_1$$
$$I_1 R_{12} + I_2 R_{22} + \ldots\ldots\ldots + I_n R_{n2} = D_2$$
$$\vdots \qquad \vdots \qquad \vdots$$
$$I_1 R_{1m} + I_2 R_{2m} + \ldots\ldots + I_n R_{nm} = D_m$$

or **I R=D**.

The matrix **R** represents the *response* of our measurement apparatus, and is analogous to the point spread function of an imaging system. In our case, the 'imaging' system is a high resolution spectrometer which measures stellar flux profiles and is situated at some inclination with respect to the star's rotation axis, working at a particular wavelength, and observing at a specific set of rotation phases. Each $R_{jk} = \frac{\partial D_k}{\partial I_j}$ and represents the response of data pixel k to changes in the surface brightness of image zone j. The **R** matrix is of course n by m in size, so for our typical cases, with $n \approx 2000$ and $m \approx 1000$, it requires (in byte form) about 2 Megabytes of memory. Each set of line profiles of a given star from a given epoch requires its own unique **R**, since, although the basic stellar parameters do not change, the exact phases, choice of lines, and instrumental resolution generally vary from one season to the next.

The continuum intensity and intrinsic shape of the spectral line profile from any localized spot on the star are both functions of limb angle. We first calculate the local specific intensity profiles for 30 limb angles using the ATLAS 5 subroutines from Kurucz (1979) and model atmospheres from a variety of sources. These profiles are then convolved with a typical radial-tangential macroturbulent velocity function and with the instrumental profile. At different phases, a typical zone on the stellar surface will have different (1) projected area, (2) limb angle, and (3) radial velocity (due to stellar rotation). Since we want to know how changing the surface brightness in a given zone will affect the observed rotationally-broadened profile at each phase, each of these quantities must be calculated for each zone at each phase. At each phase, the program first calculates the x-y position on the stellar disk corresponding to each latitude-longitude zone. At many phases (if the zone is not circumpolar) a given zone will be on the back side of the star and will not be visible. The effective areas of these zones are set equal to zero. For the zones that are visible, the limb angle is calculated. This then tells the code both which set of specific intensities to use and the projected effective area of that zone at that phase.

For simplicity, we assume that the shape and strength of the line profile at each limb angle is the same in the 'spots' as in the 'photosphere'. In our case, where we expect the spots to be virtually black anyway, this assumption is not critical, and much simplifies the computational process. It should be noted, however, that if the line equivalent width is different in the spot than it is in the photosphere, what we are really mapping is a combination of surface brightness and line strength. In cases in which the *shapes* of the spectral lines are dramat-

Maximum entropy image reconstruction involves finding the image with the greatest configuration entropy:

$$S = -\sum_j p_j log(p_j)$$

where p_j is a normalized *positive* dimensionless image quantity (surface brightness in our case) of the jth image pixel, and the sum is taken over all n image pixels. The maximum entropy image is the one with the least amount of spatial information, and is thus the smoothest or simplest image. In our case, the 'smoothest' image means the one with the least contrast between spots and stellar photosphere. An immaculate star, with no spots, will thus be the image with the maximum entropy but it will not fit the observed line profiles of a spotted star. MEMSYS uses a multi-dimensional search algorithm to find its maximum entropy image subject to the additional constraint that the image produces data which fit all the real data to within some level χ_0^2 where χ^2 is a measure of the misfit:

$$\chi^2 = \sum_k (g_k - d_k)^2 / \sigma_k^2.$$

Here, g_k is the data set produced by the image solution and d_k is the actual data, whose uncertainty at any point is σ_k.

MEMSYS accomplishes the constrained maximization by the method of Lagrange multipliers, using **R** and its transpose **R'** to iterate back and forth between image and data space, seeking to maximize the functional:

$$Q = S - \lambda\chi^2$$

where λ is the Lagrange multiplier. It accomplishes its search very efficiently through the use of conjugate gradient techniques and a 6-dimensional subspace of search directions, rather than doing simple line searches. For linear image reconstruction problems, surfaces of constant χ^2 are convex ellipsoids in N-dimensional image space. Since the entropy surfaces are strictly convex, the maximum entropy reconstruction is thus unique (Skilling and Bryan 1984).

In practice, the S/N in our actual line profile data is so high that the image solution is well defined by the χ^2 term only. That is, by the time the iteration has succeeded in fitting the data to the required level of accuracy, further improvements in maximizing the entropy have no appreciable effect on the image. Thus our Doppler images are well constrained by the data set alone and do not rely heavily on entropy considerations. To the level allowed by the data's inherent noise, they are thus unique.

Of course, the star does not care about maximum entropy and there may well be finer structure present in the true image, but it will never be uniquely determinable from the data set. The real beauty about the maximum entropy image and the reason it is particularly informative is that, because it does have minimum configurational information, one is assured that any structure present in the image is actually *required* by the data. Thus, the signal-to-noise, and quantity of data ultimately determines the level of unique detail allowed in the reconstructed image, as it must.

5. A TEST OF THE METHOD

We have tested the Doppler imaging software extensively by giving it synthesized spectral line flux data from test images of stars. The reader is referred to VPH for a more complete discussion of these tests. The following test case illustrates how well the method works in the best of all possible worlds using an artificial image of an RS CVn star. The $v \sin i$ of the artificial star is 40 km s^{-1} and its inclination is 40°, values typical of our real program objects. The spectral type of the artificial star was taken to be K1 IV (T_{eff} = 4750 K), and the spectral line used to generate the synthetic data set was Fe I λ6431 Å .

The test image chosen was a rather complicated image consisting of the letters V-O-G-T written in dark spots around our imaginary star. The test image of the 'Vogtstar' is shown in Figure 3 for eight equidistant rotation phases. The phantom star at lower right illustrates the location of the rotational pole and inclination. This image was then used to generate a data set of 16 line profiles of the Fe I λ6431 Å line at equidistant phases. The synthetic data were generated using an explicit disk integration scheme, using a totally different zone geometry than that used in the reconstruction software, and with the true spectral line shape and strength appropriate for the dark spots explicitly included in the calculations. In this case, however, we made two assumptions intended to maximize the information in the line profiles. First, we assumed that the spectral resolution used was infinite, and second, we assumed that the spots were extremely cool, with $T_e \simeq 2700$ K. Both effects tend to make the line profile variations rather more dramatic, and thus give the code more information to work with.

The 16 profiles were assembled end-to-end into a 512 point data vector, and the Doppler imaging software was then given the appropriate physical parameters for the star (inclination, rotation velocity, and specific intensity profiles as a function of limb angle), and asked to derive an image of the star using only these synthetic line profiles. The input line profiles are shown in Figure 4 (crosses), and the fit to those profiles after 30 iterations of the inversion software is shown by the solid line. We let the program proceed until the profiles were fit to a very high level of accuracy.

The image derived by the inversion process, also after 30 iterations, is shown in Figure 5 at the same eight phases as the input image, and gives a good view of the level of the success of the inversion procedure. Though there is some loss of resolution, as expected, the test image has been recovered very well, certainly well enough to recognize all the letters, and to see a distinct difference between the 'O' and the 'G' letters. The effective resolution is approximately 12°. The recovered image is free from any systematic limb brightening or limb darkening effects, and also shows no systematic latitudinal brightening or darkening effects. Clearly, in this best of all worlds, the inversion software works very well. The main features of the solution are very well constrained by the data, and do not rely heavily on entropy assumptions.

Figure 3. Test input image of the 'Vogtstar'.

6. DOPPLER IMAGES OF SPOTTED STARS

We are currently observing a number of RS CVn, FK Com, and Ap stars with Doppler imaging. For the RS CVn and FK Com stars, we are attempting to monitor the appearance and evolution of large starspots on a select sample of stars from year to year for information on the nature of magnetic dynamo processes within these stars. For the Ap stars, we are studying the nature of abundance patterns of various ions across the stellar surface. Magnetically confined diffusion of certain atomic species seems to be at work in the atmospheres of the peculiar A stars, and our Doppler imaging studies are aimed at providing observational confirmation of the diffusion theories as well as yielding information about the magnetic fields of stars which rotate too rapidly for conventional Zeeman analysis techniques.

Figure 6 shows a Doppler image of HR 1099 in 1981. This star is one of the brightest members of the RS CVn class of spotted late-type binaries and is very well-suited to the Doppler imaging method. The orbital period was used for the phasing of this and all subsequent HR 1099 images. The 1981 spot configuration consisted of a spot situated almost directly at the pole, with an attached protuberance descending to lower latitudes. Another small isolated spot lies below, at the equator. It is curious that the longitude of the protuberance on the polar spot coincides precisely with the extrapolated longitude of the migrating

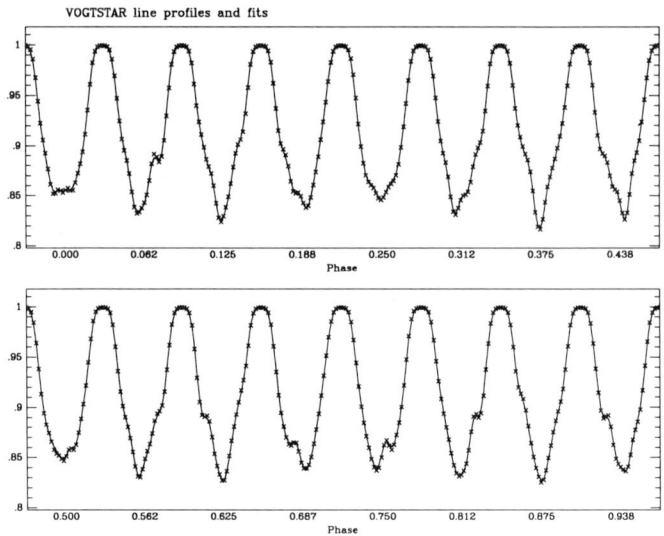

Figure 4. Synthetic line profiles for 'Vogtstar'.

large active region complex which emerged near the equator of this star in 1978. Though it is still too early to draw definitive conclusions from such coincidences until more is known about spot migration patterns, one is tempted to conclude that the equatorial complex of 1978 migrated poleward for 3 years, and, in 1981, was seen merging with the polar spot.

Figure 7 shows HR 1099 in 1984. The polar spot still has an attached protuberance though it's not yet clear if it's the same one as in 1981. There are also a number of low latitude spots, one pair of which bears a striking resemblance to a classical bi-polar sunspot group, with leading and trailing spots. Clearly, the spot distribution has changed dramatically between 1981 and 1984, though the polar spot is basically unchanged and appears to be a long-lived feature. Figure 8 shows HR 1099 in 1985. The polar spot is still basically unchanged, but now has a very large attached protuberance. Since we generally can manage to get only one image per year, it is difficult to know which features are the same and how they have moved from one year to the next. At the rate of one image per year, we may very well be undersampling the changing spot morphology.

We currently have imaging data on HR 1099 from 1981, 1982, 1984, 1985, and 1986. The 1983 season was lost to bad weather. When all these images are in final form, they will be combined and used to (hopefully) track the various spots from year to year. In this way, we hope to get a direct measure of latitudinal shear (differential rotation) and gain further understanding of cyclic dynamo processes at work within such stars. A very preliminary look at the present set of images on HR 1099 seems to suggest that spots initially form at low latitudes an

Figure 5. Doppler image recovery of 'Vogtstar'.

then migrate poleward to join a long-lived polar spot.

HR 1099 is about 120 light years away, and totally unresolvable with existing telescopes. To put things in perspective, the resolution achieved here with Doppler imaging (about 10^{-4} arc seconds) is equivalent to resolving Abraham Lincoln's ear on a penny viewed from 3000 miles. Achieving the same resolution with an optical telescope would require a diffraction-limited space-borne telescope of about 1 mile aperture. Providing the photon flux is adequate, the linear resolution of our Doppler images is, of course, independent of stellar distance.

Figure 9 shows a Doppler image of the star HD 199178 obtained in 1985. This star is a rapidly rotating late G giant and a member of the FK Com class. Stars of this class are rapidly rotating late G-K giants. They are apparently single stars with a high degree of surface activity, and may well be recently coalesced binaries. Again, one sees a very large polar spot situated almost directly at the pole, and a single smaller isolated spot at lower latitude. As with HR 1099, the polar spot may be quite long-lived, and is probably the result of the merging of many smaller spots which originally appeared at lower latitudes and migrated poleward to join the polar spot.

7. DISCUSSION

These first Doppler images of starspots revealed (to us at least) a striking resem-

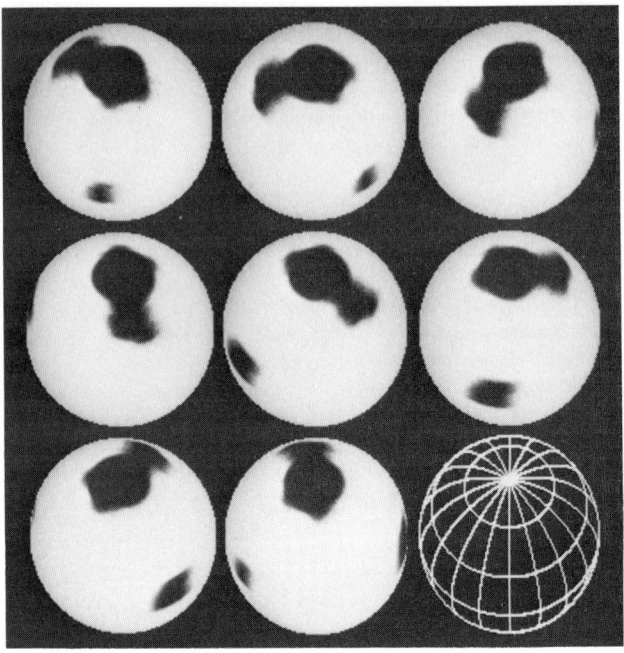

Figure 6. Doppler image of HR 1099 in 1981.

blance to x-ray images of solar coronal holes. In both cases, one sees the geometry of a long-lived polar spot, often with an attached protuberance descending to lower latitudes. Smaller spots (holes) first appear at lower latitudes, and in the solar case, migrate poleward to join the polar spot (hole). We believe (Vogt and Penrod 1983b) that starspots are perhaps more analogous to solar coronal holes than to sunspots. Solar coronal holes are prominently visible only in the x-ray, since their magnetic fields are only 10-20 G, and too weak to have any significant effect on radiation transport in the photosphere. It is only in the corona, where the densities are much lower that these weak fields can significantly affect the plasma motions, leading to lower temperatures and lower densities as a plasma wind flows outward above the hole. Thus, coronal holes are almost inconspicuous photospheric structures, but they dominate the structure of the corona, and appear prominently in x-ray coronal images. However, the magnetic flux inside starspots is probably several orders of magnitude larger than in solar coronal holes, yielding kilogauss fields over a convective cell. Such large fields are strong enough to dominate convective heat transport in the photosphere, and the region above the spot turns cool, leading to spots which are prominent in photospheric light. The blocked energy from such a starspot is readily absorbed by the convective envelope, and reradiated over the entire surface of the star over a many decade time scale. So, starspots are probably like sunspots, in that they are cool because magnetic fields dominate convective motions, but they are like coronal

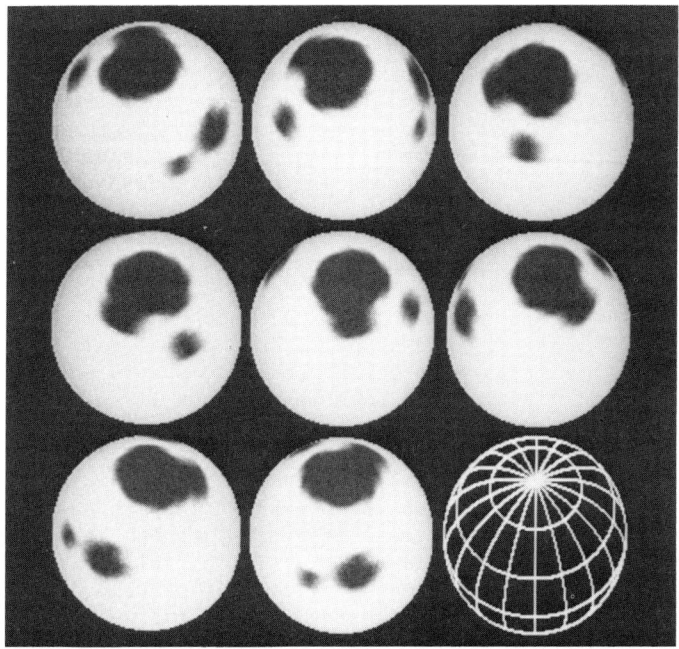

Figure 7. Doppler image of HR 1099 in 1984.

holes in terms of their sizes, and global migration patterns. In particular, they do not appear to show the familiar equatorward migration pattern of sunspots (the butterfly diagram), but rather a poleward migration pattern.

Starspots probably first appear at low latitudes as scaled-up analogs of solar complexes, with many active regions merging to form what we call a starspot. The filling factor of dark umbrae is very much larger than typical of solar complexes, however, because of the much larger emergent magnetic flux. The appearance of new magnetic flux is accompanied by much flaring, enhanced radio and X-ray emission, and chromospheric emission. With time, the activity declines as the fields become organized through magnetic reconnection, annihilation, and diffusion. The resulting unipolar open field regions then migrate to the poles, where they accumulate to form large long-lived polar spots. The size of the polar spot probably varies cyclically, as occurs with polar coronal holes on the Sun.

If the hypothesis of poleward migration of starspots is born out by further analysis of Doppler images, it may be an indication that magnetic activity on these rapidly rotating, deeply convective stars is qualitatively different than that traced by sunspots on our Sun. Models of distributed $\alpha - \omega$ dynamos operating in rapidly rotating convective shells predict poleward migration. Recent models of the solar dynamo, which attempt to explain the observed equatorward migration of sunspot formation zones, suggest that the solar dynamo operates primarily in the layer immediately beneath the convective zone, as a 'boundary

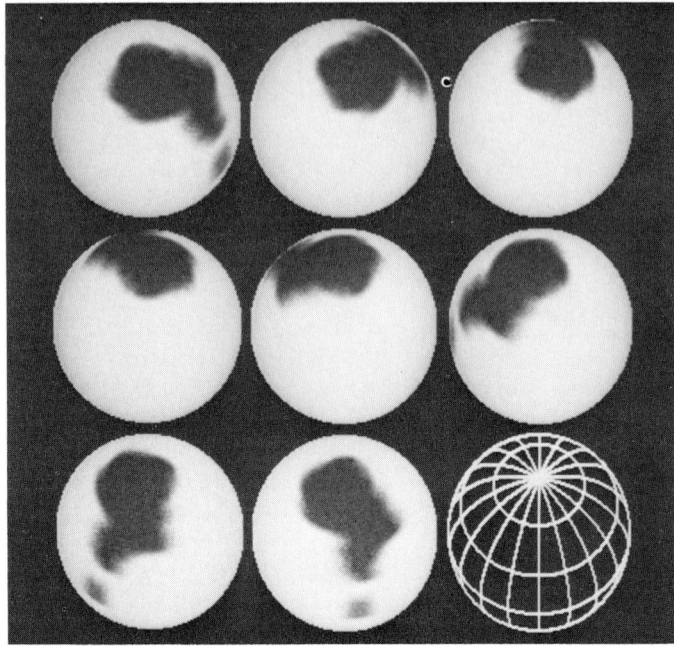

Figure 8. Doppler images of HR 1099 in 1985.

layer' dynamo. It is possible that the solar rotation is not rapid enough to operate a true distributed dynamo, while the more rapidly rotating stars that we observe do exhibit such a dynamo. It is also possible, however, that spot locations and migrations may be determined by the locations of large convective rolls in the convection zone, rather than by anything connected to the dynamo waves, as recent models of the solar activity cycle suggest. In stars with very deep convection zones, such as the ones we observe, it is possible that each hemisphere is dominated by a single convective roll, with magnetic flux rising at the equator and then being transported poleward, to appear as a large polar spot centered over the polar subduction zone.

Hints of periodicity in mean light level variations and in the amplitudes of light curves for a number of spotted stars suggest some evidence for starspot cycles. Periodic variations in the size of the polar spot would yield periodic mean light level variations quite naturally. Figure 10 is an attempt to combine various widely scattered reports of 'cycles' for RS CVn and BY Dra type spotted stars and to merge these with the systematic studies of periodicities of active single G and K dwarfs done by Vaughan et. al (1981). The figure and data are from Vogt (1983) and are perhaps now a bit out of date, but they do serve to illustrate an interesting point. Among the active late-type stars, the cycle period is essentially constant at around 10-12 years for all stars with rotation periods longer than about 6-7 days, including the Sun. The stars show little evidence of starspots,

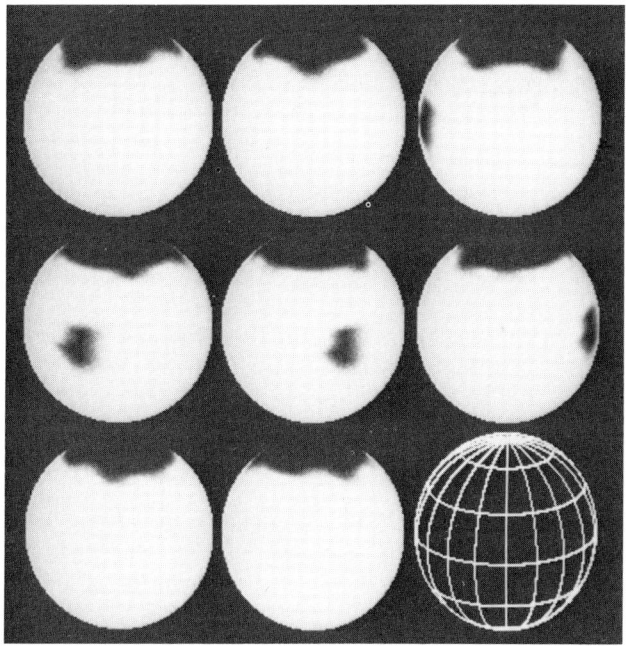

Figure 9. Doppler image of HD 199178 in 1985.

as would be the case for the Sun as viewed from a similar distance, and the cycle period is determined strictly from variations in the level of chromospheric emission. For more rapidly rotating stars, giant starspots appear, and the 'cycle period' increases linearly with decreasing rotation period. There appears to be a threshold rotation rate, corresponding to a 6-7 day rotation period, above which a star forms large starspots. The connection of these 'cycle periods' for the rapidly rotating stars to true dynamo periods is, however, very uncertain.

We suspect that the appearance of large spots on the rapidly rotating late-type stars marks the transition from a feeble solar-type boundary-layer dynamo to a more powerful distributed dynamo which operates throughout a deep rotating convective shell. The sizes, locations, and migrations of the starspots, as well as the 'cycle times', may well, however, be telling us more about the subsurface convective pattern in these deeply convective stars than anything about the true dynamo activity. Doppler imaging studies of cyclic variations in the size of the polar spot, of the frequency of spot formation, and of spot migration patterns should provide a useful tool for unraveling the nature of the very powerful magnetic activity on rapidly rotating stars, and hopefully will lead to a much better understanding of the activity of the most local dynamo, the Sun.

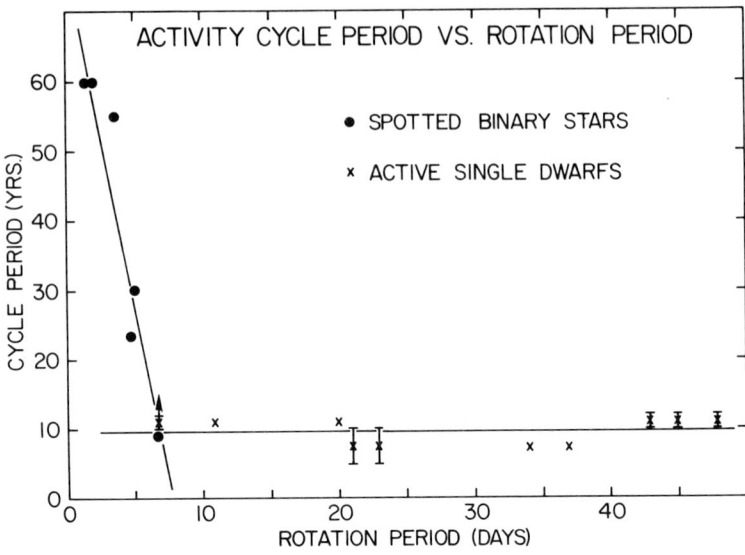

Figure 10. Cycle period vs. rotation period for a sample of active late-type stars.

8. WHAT'S NEXT FOR DOPPLER IMAGING?

The technique of Doppler imaging is still in its infancy and further development is underway at Lick. The true line synthesis problem is not a simple linear process with specific intensity line profiles which are independent of temperature or equivalent width. The response function matrix really needs three dimensions to handle the problem correctly. With larger computers than our VAX 11/780, the method could be extended to handle multiple lines simultaneously. An obvious goal would be to try to obtain simultaneous images of active stars using photospheric, chromospheric, and transition region lines from the UV to the IR. Also, with large enough computers, large pieces of the spectrum could be imaged at once, including perhaps hundreds of lines. Including so many lines at once might help ease the present high S/N per pixel requirements. There is the possibility that Doppler imaging could be done using linearly and/or circularly polarized light. Such 'Stokes parameter imagery' could provide information on magnetic fields in starspots, and on Ap stars.

Besides Ap stars, and spotted late-type stars, there are other potentially interesting classes of objects which may be amenable to the Doppler imaging technique. Imaging of accretion disks in cataclysmic variable systems has already been done using maximum entropy and broadband photometry. Several groups are at work using spectroscopy to Doppler image the disks. It should be possible to use the technique to derive accurate images of contact and near-contact binary systems. With eclipsing binaries, the occulting star offers a further probe which could be incorporated into the imaging technique. Finally, the increased light gathering power of the next generation of large telescopes will bring many of the T Tauri and other pre-main sequence spotted stars within reach of the Doppler imaging technique, allowing us to obtain direct images of the some of the spectacular activity which occurs on such stars.

I would like to thank Dr. Keith Horne for bringing MEMSYS to our attention, and Dr. Robert Kraft and the Lick Time Allocation Committee for generous amounts of observing time in support of this effort. Much of the instrumentation which makes this work possible was funded by NSF block grants AST-8320396 and AST- 8614510. The Doppler imaging research effort at Lick is funded entirely by NSF grant AST-8210202 whose support we gratefully acknowledge.

REFERENCES

Deutsch, A.J. 1958, in *IAU Symposium 6, Electromagnetic Phenomena in Cosmical Physics*, ed. B. Lehnert (Cambridge: Cambridge University Press), p. 209.
Deutsch, A.J. 1970, *Ap. J.*, **159**, 985.
Falk, A.E., and Wehlau, W.H. 1974, *Ap. J.*, **192**, 409.
Goncharsky, A.V., Stepanov, V.V., Khokhlova, V.L., and Yagola, A.G. 1982, *Astr. Zh.*, **59**, 1146 (English transl. in *Soviet Astr.*, **26**, 690.)
Gull, S.F., and Skilling, J. 1983, in *Indirect Imaging* ed. J.A. Roberts (Cambridge: Cambridge University Press), p. 267.
Khokhlova, V.L. 1975, *Astr. Zh.*, **52**, 950 (English transl. in *Soviet Astr.*, **19**, 576.)
Khokhlova, V.L. 1985, *Ap. Space Phys. Rev.*, **4**, 99.
Khokhlova, V.L., and Ryabchikova, T.A. 1975, *Astrophys. Space Sci.*, **34**, 403.
Khokhlova, V.L., Rice, J.B., and Wehlau, W.H. 1986, *Ap. J.*, **307**, 768.
Kurucz, R.L. 1979, *Ap. J. Suppl.*, **40**, 1.
Megessier, C. 1975, *Astr. Ap.*, **39**, 263.
Pyper, D.M. 1969, *Ap. J. Suppl.*, **18**, 347.
Robinson, L.B., and Osborne, J. 1986, *S.P.I.E.*, **627**, 492.
Shore, J.E., and Johnson, R.W. 1980, *IEEE Trans.*, **IT-26**, 26.
Shore, J.E., and Johnson, R.W. 1983, *IEEE Trans.*, **IT-29**, 942.
Skilling, J., and Bryan, R.J. 1984, *M.N.R.A.S.*, **211**, 111.
Vaughan, A.H., Baliunas, S.L., Middlekoop, F., Hartmann, L., Mihalas, D., Noyes, R.W., and Preston, G.W. 1981, *Ap. J.*, **250**, 276.
Vogt, S.S. 1983, in *IAU Colloquium 71, Activity in Red-Dwarf Stars*, ed. M. Rodono and P. Byrne (Dordrecht: Reidel), p. 137.
Vogt, S.S., and Penrod, G.D. 1983a, in *IAU Colloquium 71, Activity in Red-Dwarf Stars*, ed. M. Rodono and P. Byrne (Dordrecht: Reidel), p. 379.
Vogt, S.S., and Penrod, G.D. 1983b, *Pub. A.S.P.*, **95**, 565.
Vogt, S.S., Penrod, G.D., and Hatzes, A. 1987, *Ap.J.*, in press.

DISCUSSION

FOING: In your constraint on the solution, you maximize $Q = S - \lambda\chi^2$. How and what is the choice of the Lagrange multiplier between the constraint of maximum entropy and minimal X^2 distance to the data. How arbitrary is it for the stellar case?

VOGT: The entropy and χ^2 terms can be weighted individually. We set up the method to give very high weight to the χ^2 term, thus forcing the method to fit the data to very high accuracy (as is appropriate for high S/N data).

FOING: Do you need simultaneous photometry and how do you correct for the secondary if it is variable, or if you do not know the changes of continuum level of the primary?

VOGT: We do not use simultaneous photometry. If the lines of the secondary were variable, the deblending would become impossible. The principal effect of changing stellar brightness will be a variable equivalent width of the lines, a quantity we do not really exploit in the imaging of spotted late type stars.

LINSKY: How secure is the indentification of the large polar spot on HR 1099?

VOGT: The polar spot is very well determined, if our input physics is correct. The presence of large polar spots on these stars is further supported by a correlation between line profile shapes and inclination. HD 199178 (i > 60°) does not show very pronounced polar spot effects. HD 26337 (i ~ 45°) always has quite flat-bottomed profiles, and HR 1099 (i = 33°) also has very pronounced profile variations from the polar spot.

FURENLID: Do you feel that the polar spots are determined with the same confidence as the low-latitude spots, considering that there are no Doppler shifts for the polar spots?

VOGT: Yes, in fact they are generally better determined. They occupy the region of the star which is always in view, and thus has the largest amount of information in the line profiles. One does not need a Doppler shift to get information. The lack of Doppler shift contains the same information. Spots at very low latitudes have large shifts, but are not visible for as long, and thus (with finite phase coverage) are often less well determined.

SODERBLOM: To achieve high S/N with a CCD, must one spread the spectrum over a number of pixels? What is the best S/N per pixel achievable, in your experience?

VOGT: It certainly helps to spread the light out, but I have never done any tests to see how well one could do using only a few rows. One would quickly run into the dynamic range limit of the horizontal shift register. The highest S/N we routinely achieve with the Texas Instruments 800 × 800 CCD is somewhere around 400-500. Pixel-to-pixel variations calibrate out much better than variations on scales of 10's to 100's of pixels. We have not tried to push S/N further, but I would not be surprised that, with some care, much higher S/N's could be achieved with this CCD.

ABUNDANCE SIGNATURES OF INTERNAL STELLAR STRUCTURE IN F STARS

Ann Merchant Boesgaard

Palomar Observatory and Department of Astronomy
California Institute of Technology

ABSTRACT. We have determined Li abundances from high resolution, high signal-to-noise spectra near 6700 Å of the F dwarfs in several galactic clusters and in the field. The Li–temperature profiles show large Li depletions in the mid–F stars indicating that matter has been circulated (or diffused) downward in the stars. This brings new observational evidence about the details of stellar envelope structure. Li depletions occur in stars with T = 6400 – 6900 K, with depletions of over an order of magnitude for those with T = 6500 – 6800 K. The relationship to rotation, galactic Li, and the halo star Li content is discussed.

1. INTRODUCTION

The surface contents of both Li and Be provide sensitive probes to the internal structure of stars. For example, the Sun destroys Li by (p,α) type reactions at $\sim 2.5 \times 10^6$ K and has lost 99 percent of its Li, while it burns Be at $\sim 3.5 \times 10^6$ K and has lost none of its original Be apparently. The temperature at the base of the solar convection zone in solar models is only 1.9×10^6 K (D'Antona and Mazzitelli 1984), not enough for Li (or Be) destruction. For a hotter star, 1.25 $M_\odot$, the stellar model is only 6×10^5 K at the bottom of the surface convection zone. Thus, according to the models, convection alone does not circulate Li and Be to regions of the star where they can be destroyed. However, some main sequence stars of ≥ 1 $M_\odot$ do show depletion of Li or of both Li and Be, so other mixing mechanisms are required.

2. OBSERVATIONS

Spectra for a number of F main-sequence stars in open clusters and in the field have been obtained near the 6700 Å region at the 3.6 m Canada–France–Hawaii telescope on Mauna Kea and at the 5 m Hale telescope at Palomar. Those from the CFHT are Reticon spectra from the f/8.2 coudé camera and cover 135 Å with a spectral resolution of 0.11 Å and typical S/N values of 350 – 600. Those from Palomar are TI CCD spectra from the 72–inch coudé camera and cover 110 Å with 0.21 Å resolution and 300 – 500 in S/N. Because of the high signal–to–noise ratios, highly accurate Li abundances can be determined and, for stars with no detectable Li line, very low upper limits can be ascertained which correspond to very large (2

− 3 orders of magnitude) Li depletions.

The spectra have been flat–fielded, wavelength–adjusted, and continuum–flattened according to the usual procedures. (See Boesgaard and Tripicco 1986b and Boesgaard, Budge, and Burck 1988) for details.) Equivalent widths or upper limits have been measured for Li I λ6707 and for several Fe I lines. A weak Fe I line (2 − 4 mÅ) blends with the Li I line on the short wavelength side; to account properly for that in the calculations, the abundance of Fe/H must be found.

3. ABUNDANCE RESULTS

Stellar temperatures have been determined through use of several photometric indices and calibrations. Models of Kurucz (1979) have been used with a model atmosphere abundance routine to predict the Li I (and Li I + Fe I) equivalent widths for various abundances.

Figure 1 shows the Li–temperature profile for the F field stars (data from Boesgaard and Tripicco 1986b), the Hyades F dwarfs (primarily from Boesgaard and Tripicco 1986a), the Coma F dwarfs (from Boesgaard 1987a), and for the UMa Group F stars (from Boesgaard, Budge, and Burck 1988). The three open clusters

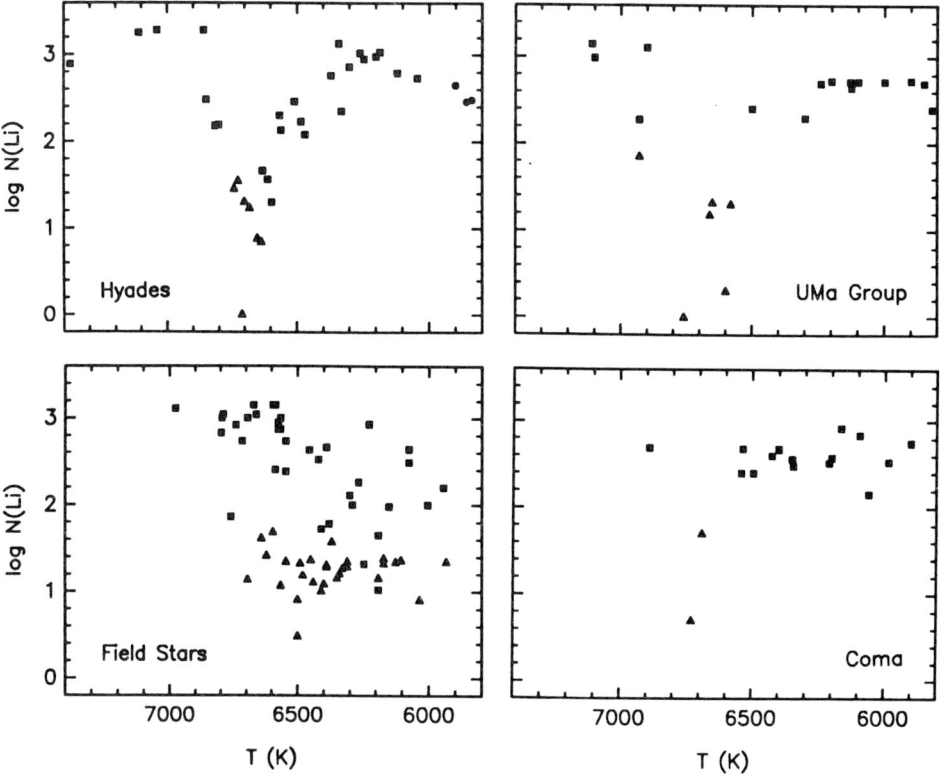

Figure 1. The Li–temperature profile for F field dwarfs, the Hyades F dwarfs, the Coma F dwarfs and the UMa F dwarfs. Triangles represent upper limits.

(ages 7 x 10⁸, 5 x 10⁸, and 3 x 10⁸ yr, respectively) all show a pronounced drop in Li content in the middle F star range, near temperatures 6400 – 6900 K, or in stars about 20 percent more massive than the sun. The field stars are not all affected by this phenomenon: about half of the stars in this temperature regime show the normal Pop I abundance, Li/H = 10^{-9} or log N(Li) = 3.0, and about half show large Li depletions.

An interpretation of the Hyades F star Li depletions in terms of diffusion has been put forward by Michaud (1986). His calculations show that chemical separation takes place below the convection zone because the upward radiative acceleration on Li is less than the downward acceleration of gravity for the middle-F stars. For the cooler F stars the diffusion time scale is too long to see the effect in the 7 x 10⁸ yr lifetime of the Hyades cluster. His calculations do not match the observations in two ways: they predict that Li will be supported in the early F stars resulting in overabundances of Li and they predict only an order of magnitude or so for Li depletions while more than two orders of magnitude are found. Both of these effects can be reconciled with the observations if the stars are losing mass at rates of about $10^{-14} - 10^{-15}$ $M_\odot$ yr^{-1}.

It is possible that complex circulation patterns are present in the stellar envelopes relating to rotation, differential rotation and magnetic fields. Figure 2 shows the drop in the measured $v \sin i$ values (from Kraft 1965) with temperature over the Li "chasm" in the Hyades. As one goes toward cooler temperatures in the F dwarfs the stellar rotation declines and the outer convection zone deepens. The interplay between these two phenomena may result in the sudden drop in the surface Li content followed by the slower rise. In the Li chasm it appears that the rapid rotators are more severely depleted in Li.

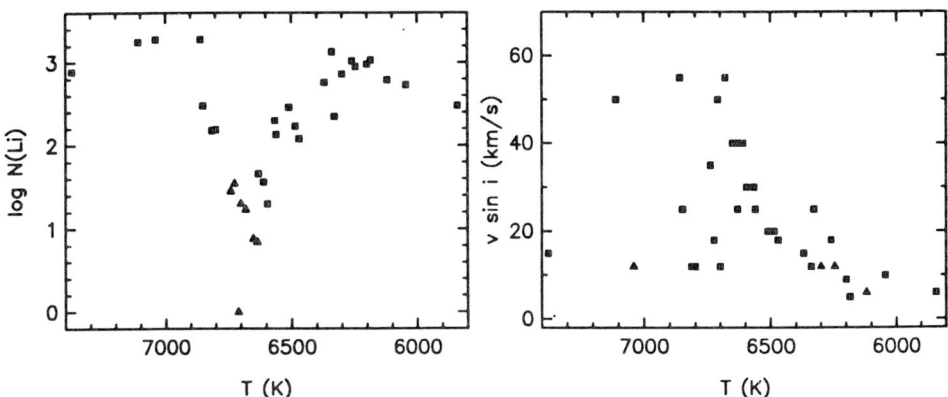

Figure 2. The Li–temperature and $v \sin i$–temperature profiles for the Hyades F dwarfs.

4. DISCUSSION AND CONCLUSIONS

All the galactic clusters appear to have the same amount of Li as shown by the F dwarfs that are hotter than the mid–F star Li gap, log N(Li) ~ 3.0. This is the same amount as seen in the chondrites (Nichiporuk and Moore 1974), the F field stars (Boesgaard and Tripicco 1986b), F and G visual binaries (Boesgaard and Tripicco 1987) and T Tauri stars (Mundt et al. 1983). These objects span the age range from $\sim 10^6$ to 5×10^9 yr. From this we can infer that there has been no measurable enrichment of Li in the galactic disk since the formation of the solar system.

The four galactic clusters in which Li in the F dwarfs has been recently studied – UMa at 3×10^8 yr (Boesgaard, Budge, and Burck 1988), Coma at 5×10^9 yr (Boesgaard 1987a), Hyades at 7×10^8 yr (Boesgaard and Tripicco 1986a and Boesgaard 1987b) and NGC 752 at 2×10^9 yr (Hobbs and Pilachowski 1986) – all show evidence of the deep Li depletions in the mid–F stars first shown by Boesgaard and Tripicco (1986a) in the Hyades. In addition, visual binaries with ages of $8 \times 10^8 - 3 \times 10^9$ yr also show this effect. The "Li chasm" occurs at $B - V = 0.40 - 0.47$, or $T \approx 6400 - 6900$ K. The chasm has a width of only 300 K at the log N(Li) = 2.0 level.

In the Hyades the bottom of the Li gap is at $T \sim 6650$ K where the depletions of Li are at least 3 orders of magnitude. As the the temperature decreases, the value of log N(Li) increases and $v \sin i$ decreases. For the stars in the Hyades Li chasm there appaers to be an inverse relation between log N(Li) and $v \sin i$ such that the more rapidly rotating stars are more depleted in Li.

On the cool side of the Li dip the three clusters that have ages near 5×10^8 show a flat distribution of Li *vs.* temperature with a log N(Li) value of ~ 2.7, slightly below the initial value found in the F stars on the hot side of the Li chasm. These stars have apparently undergone some Li depletion, the amount of which may be age related but is not very mass dependent. This flatness from $T \sim 5800 - 6300$ K is reminiscent of the flatness in the Li–temperature plot for the halo dwarfs. (See Figure 3.) The similarities in the flatness and in the spread of log N(Li) values at a given temperature indicate that it is possible that the halo stars have undergone Li depletion (diffusion?) during their long lifetimes from an initial level higher than that presently observed.

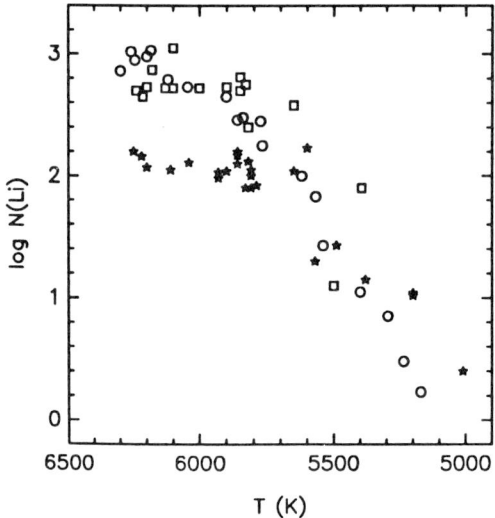

Figure 3. The Li–temperature profile for the late F and G dwarfs of the Hyades (open circles), the UMa group (open squares) and the halo dwarfs(filled stars).

The data for the halo stars in Figure 3 are only for those with velocities (relative to the local standard of rest) $\geq \pm 100$ km s^{-1} and with [Fe/H] ≤ -1.4; the temperature and Li values are from Spite, Maillard, and Spite (1984), Spite and Spite (1986), and Hobbs and Duncan (1987). The Hyades values are from Boesgaard (1987b) and those for UMa from Boesgaard et al. (1988).

The complex Li–temperature profile found for F and G dwarfs in galactic clusters indicates that the internal stellar structure and processes are more complicated than had been recognized. Calculations of the internal circulation of matter (certainly of Li atoms) caused by stellar rotation, especially differential rotation, are needed as are more sophisticated calculations of the effects of diffusion in both young stars and in halo stars.

I am happy to acknowledge support from NSF grants, AST-8216192 and RII-8521715, and a Guggenheim Fellowship and the concomitant hospitality of the California Institute of Technology.

REFERENCES

Boesgaard, A. M. 1987a, *Astrophys. J.*, **321**, in press.
Boesgaard, A. M. 1987b, *Publ. A. S. P.*, **99**, in press.
Boesgaard, A. M., and Trippico, M. J. 1986a, *Astrophys. J.*, **302**, L49.
Boesgaard, A. M., and Trippico, M. J. 1986b, *Astrophys. J.*, **303**, 724.
Boesgaard, A. M., and Trippico, M. J. 1987, *Astrophys. J.*, **313**, 389.
Boesgaard, A. M., Budge, K. G., and Burck, E. E. 1988, *Astrophys. J.*, **325**, in press.
D'Antona, F. and Mazzitelli, I. 1984, *Astr. Ap.*, **138**, 431.
Hobbs, L. M. and Pilachowski, C. A. 1986, *Astrophys. J.*, **309**, L17.
Hobbs, L. M. and Duncan, D. K. 1987, *Astrophys. J.*, **317**, 796.
Kraft, R. P. 1965, *Astrophys. J.*, **142**, 681.
Kurucz, R. L. 1979, *Astrophys. J. Suppl.*, **40**, 1.
Michaud, G. 1986, *Astrophys. J.*, **302**, 650.
Mundt, R., Walter, F. M., Feigelson, E. D., Finkenzeller, U., Herbig, G. H., and Odell, A. P. 1984, *Astrophys. J.*, **269**, 229.
Nichiporuk, W. and Moore, C. B. 1974, *Geoch. et Cosmoch. Acta*, **38**, 1691.
Spite, M. Maillard, J.-P., and Spite, F. 1984, *Astr. Ap.*, **141**, 56.
Spite, F. and Spite, M. 1986, *Astr. Ap.*, **163**, 140.

DISCUSSION

SNEDEN   Could you please expand on your points suggesting a) a constancy in Li over the age of the galactic disk, and b) the apparent depletion of Li in halo stars.

BOESGAARD   The "initial" Li abundance, which all the stars in a cluster are born with, must be determined from the F dwarfs on the hot side of the Li dip in the mid-F dwarfs. These values -log N(Li)=3.0- are the same from cluster to cluster and the same as the T Tau stars and the solar system values. On the cool side of the Li dip, T=5800-6400 K, there seems to be only very weak dependance of Li on temperature ; the stars in the three clusters, Hyades, Coma, and UMa (with ages ~ 5×10^8 yrs.) show depleted Li -log N(Li)=2.7. The older cluster, NGC 752 (with age ~ 2×10^9 yrs.) shows -log N(Li)=2.4 in this temperature regime. The halo stars at log N(Li)=2.1 and age ~ 12×10^9 yrs. may be showing a similar depletion with age in this temperature region.

F. SPITE   First my congratulations for the beautiful work of Ann Boesgaard.
About the comparison of the flat part of clusters and of pop.II dwarfs : if it is expected to find a plateau with small scatter for a cluster, in which all stars have the same age and metallicity, it is more difficult to understand for pop.II dwarfs which may have the same age but widely different metallicities. And up to now, the theories of lithium depletion predict depletion rates dependent on metallicities.

BOESGAARD   However, there does seem to be a real spread in the Li abundances at a given temperature in the halo stars. This indicates that some depletion may have occured. All the halo stars plotted have [Fe/H]<-1.4.

DUNCAN   Comment on F. Spite's question.
I suggest waiting until this afternoon, when I will discuss if star to star variations are real, and the effect of [Fe/H] differences on Li destruction rate.

NISSEN   There has also ben observation of Li in F stars in the old, open cluster M67 (Pilachowski et al., and Spite et al.). Do these results fit your Li - age relation for the plateau around $T_{eff} \approx$ 6000-6400 K ?

BOESGAARD   The hottest main-sequence stars (late F dwarfs) left in M67 barely reach the plateau region. However, these authors have observed six stars in this temperature range with log N(Li) values all near 2.5.

LIGHT ELEMENTS AS PROBES OF STELLAR INTERIORS

A. Baglin and P.-J. Morel
Observatoire de Nice
B.P. n° 139
06003 NICE Cedex
France

For stars cooler than 6400°K, the lithium is observed to be depleted with respect to the universal value for Population I stars. The depletion depends on mass and age and occurs, both during the main-sequence (Hyades) and the pre-main-sequence (Pleiades).

The physical model of this depletion is the following: light elements are nuclearly destroyed at low temperature ($2 \sim 5 \times 10^6$ °K) i.e. close to the star surface. The surface convective zone, eventually somehow extended by overshooting, plays the role of a fully mixed reservoir (FMR) of light elements. For $T_{eff} \lesssim 5000$°K this FMR embodies the nuclear destruction region (NDR) and the light elements are destroyed. For $T_{eff} \gtrsim 5000$°K the depletion is due to some transport process acting between the bottom of the FMR and the NDR (fig. 1). The observed abundances of light elements hence, are indicators of the nature of the transport process at work between the photosphere and the NDR.

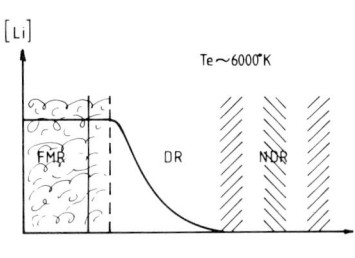

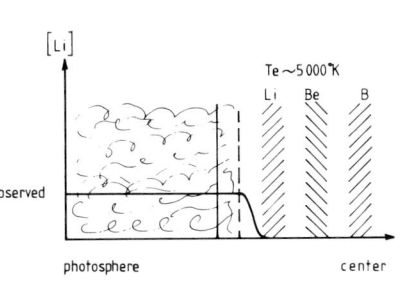

Using envelop models computed with the available physical data, Baglin et al. (1985) referred hereafter as BMS - showed that the depletion of lithium in the Hyades as observed by Cayrel et al. (1983), was consistent with a turbulent difussion coefficient derived from the instability

Figure 1. Scheme of the physical process.

of the shear flow due to meridional circulation. However these results have to be revisited:
- New physical data are available, they allow to compute more reliable envelop models.
- The solar seismology has started to map the structure of the solar envelop, hence more sensitive observational tests are available.

Molecular opacity sources increase the opacity in the outerlayers and reduce the extend of the convective zone. This effect is larger for lower masses.

Heliosismology has shown that the bend of the temperature gradient of the solar convective zone occurs at $\sim 0.7\ R_0$. (Christensen-Dalsgaard et al., 1985). But solar envelopes computed with opacities including molecular sources, even with large mixing length have too shallow convective zones (Lebreton and Maeder, 1986).

The proposed modelization depends on two main free parameters:
- $\alpha = \ell/H_p$ the ratio of the mixing length to the pressure scale height.
- ϕ the "efficiency" of the diffusion process.
Different scenarios of the turbulent diffusion generation have been tested:
- Molecular analog (Schatzman, 1977) :

$$D = \phi \nu$$

- Shear of the meridional circulation (Zahn, 1984):

$$D = \phi \frac{r}{g(\nabla_{ad} - \nabla_{rad})} \frac{\Omega^2 R^2 L}{M^2}$$

- Simplified version (BMS, 1985):

$$D = \frac{\phi}{(\nabla_{ad} - \nabla_{rad})}$$

- Saturation by dissipative process:

$$D = \phi = cte$$

1- The Hyades sequence.
The lithium depletion for the Hyades sequences has been predicted as a function of α and ϕ (fig. 2). Errors bars, both in mass and abundance, define zone of solutions in the (α, ϕ) place (fig. 3).
  The scenario: "Saturation by dissipative process", with an overshooting of 0.23 H_p, appears to be the only one for which the intersection of these zones is not empty. It corresponds to a value of the constant diffusion coefficient of $D \sim 2500$ and to a mixing length coefficient of $\alpha \sim 2.8$. With these values, the computed abundances are consistent with the observations (fig. 1).

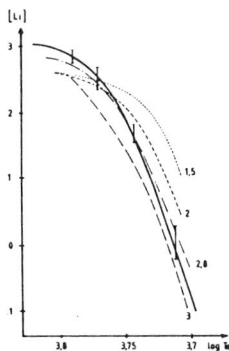
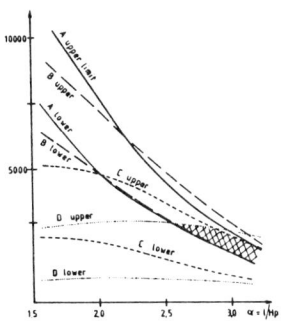

Figure 2. $|Li|$ versus $\log_{10} T_{eff}$ for different values of α compared to the observations (vertical bars) The best fit is obtained for the bold curve which corresponds to $\alpha = 2.8$, an overshooting region of .23 H_p and D = 2500.

Figure 3. The (α, Φ) plane for the scenario "Saturation by dissipative process". Models ABCD upper and lower limits correspond to the four observations shown in fig. 2 with their error bars.

2- **The Sun.**
The observed lithium abundance is: $|Li| = 0.96$. The computation with $\alpha = 2.8$ and D = 2500 = cte gives $|Li| = -1.5$. If the mixing length ratio of 2.8 obtained for the Hyades is an universal value, a smaller D = 800 is needed in the calculations to get the Sun lithium abundance of $|Li| = 0.96$. The observed beryllium abundance is $|Be| = 1.3$ (no depletion) and the computed value with $\alpha = 2.8$ and D = 800 is $|Be| = 0.87$ Since the beryllium is destroyed deeper than the lithium, the diffusion coefficient has to decrease with depth: the scenario "D = cte" is not entirely consistent.

3- $\underline{\alpha \text{ Cen A and B.}}$
The relevant physical parameters for this visual binary are:

α Cen A: $T_{eff} = 5780°K$, $M/M_\odot = 1.1$, $L/L_\odot = 1.6$, $|Li| = 1.28$

α Cen B: $T_{eff} = 5320°K$, $M/M_\odot = 0.9$, $L/L_\odot = .46$, $|Li| = 0.00$

The age is $\sim 4.25 \times 10^9$ years (Demarque et al., 1986). With $\alpha = 2.8$ and D = 2300 = cte, one obtains results consistent with the observations: $|Li|_A = 1.28$ and $|Li|_B = -7.7$.

CONCLUSIONS.

Light elements are powerful tools to probe stellar interior theory:
. Turbulent diffusion seems, up to now, the only mechanism which can reproduce the observed temperature dependence of lithium abundance in the Hyades.
. Convective zones seem to be deeper than actually predicted.
. The value of the diffusion coefficient obtained for α Cen is close

to the Hyades value.
Improvements of the stellar interior theory need accurate abundance observations of light elements in homogeneous sets of stars like clusters or pairs of well known binaries.

ACKNOWLEDGEMENTS.

Yveline Lebreton, Roger Cayrel, Giusa Cayrel, Evry Schatzman, Nicolas Grevesse, Claude Burkhart.

REFERENCES.

Baglin, A., Morel, P.-J., Schatzman, E.: 1985, Astron. Astrophys. **149**, 309.
Cayrel, R., Cayrel de Strobel, G., Campbell, B., Däppen, W.: 1984, Astrophys. J. **283**, 205.
Christensen-Dalsgaard, J., Duvall, Jr, T.L., Gough, D.O., Harvey, J.W., Rhodes Jr, E.J.: 1985, Nature **315**, 378.
Demarque, P., Guenter, D.B., van Altena, W.F.: 1986, Astrophys. J. **300**, 773.
Lebreton, Y., Maeder, A.: 1986, Astron. Astrophys. **161**, 119.
Schatzman, E.: 1977, Astron. Astrophys. **56**, 211.
Zahn, J.-P.: 1984, in "Instabilités hydrodynamiques et applications astrophysiques", p. 389, eds. A. Baglin et M. Auvergne, S.F.S.A. Paris.

DISCUSSION

BECKMAN A remark: In to morrow's session, Rebolo and I will present a study of Li depletion in the lower temperature part of the Hyades main sequence, i.e. for temperatures below 6300K. We have separated the effect of rotation from the effect of temperature, and show that at lower rotational velocity there appears to be less lithium. These observations offer us an additional means to probe the nature of the physical mechanism which drive the depletion, i.e. we can see how it depends on rotation for stars of equal temperature.

CHROMOSPHERIC VELOCITY FIELDS DIAGNOSTICS FROM CaII AND MgII EMISSION PROFILES*

Vladilo,G.[1], Crivellari,L.[1], Castelli,F.[1],
Beckman,J.E.[2], and Foing,B.H.[3]

1. O.A.T., Trieste, Via Tiepolo 11, Italy
2. I.A.C., La Laguna, Tenerife, Spain
3. L.P.S.P., 91370 Verrieres-Le-Buisson, France

ABSTRACT. We discuss the present limits to the velocity field diagnostics in stellar chromospheres achievable with ESO CAT+CES and IUE high resolution spectra.

1. INTRODUCTION

Standard line formation theory - including hydrostatic equilibrium (HE) - explains the emission profiles of CaII H and K, and MgII h and k in late type stars as an effect of the outward temperature rise in the chromosphere on these collision dominated lines. Realistic stellar atmosphere models must take into account the breakdown of HE, revealed in the form of velocity fields. High resolution and very high S/N spectra are required to obtain quantitatively useful information on these velocity fields, which may have different physical mechanisms of origin, different length scales, and different senses of motion. On the basis of our experience with ESO CAT+CES and IUE high resolution spectra (Table I), here we describe the methods employed and the accuracy achievable in velocity field diagnostics from the CaII H and MgII h and k emissions. The general presence of interstellar contamination in MgII h and k, even for nearby stars (Vladilo et al. 1986), is taken into account.

2. WAVELENGTH CALIBRATION

After the standard wavelength calibration of the spectra to the laboratory rest frame by means of calibration lamps, we reduce the wavelength scale to the the stellar photosphere in order to measure chromospheric motions relative to that frame. The accuracy achievable using the stellar radial velocity for producing a photospheric wavelength scale is not enough for our purposes, so we recalibrate the

----
* Based on ESO observations collected at La Silla (Chile) and IUE observations collected at Villafranca del Castillo (Spain)

initial (laboratory) wavelength scale using a set of n photospheric unblended absorptions covering the spectral range around the chromospheric line under study. We use Gaussian fits to the selected lines to measure their central wavelengths, λ_i^{FIT}, in the laboratory scale. Typical errors on λ^{FIT}, for CAT+CES spectra in the CaII region, are of about 0.5 mÅ (40 m/s).

The average $\langle \delta\lambda \rangle = 1/n \; \Sigma_{i=1,n} (\lambda_i^{FIT} - \lambda_i^{LAB})$, where λ^{LAB} is the laboratory wavelength of the selected photospheric line, is assumed to be a measure of the mean photospheric velocity, and the wavelength scale is accordingly corrected.

The dispersion $\sigma = \{\Sigma_{i=1,n} (\delta\lambda_i - \langle\delta\lambda\rangle)^2/(n-1)\}^{1/2}$, with $\delta\lambda_i = (\lambda_i^{FIT} - \lambda_i^{LAB})$, gives an estimate of the wavelength calibration error and of the accuracy of the assumption of a single photospheric velocity. From solar measurements integrated over the disk, the wavelength shifts and asymmetries of weak metallic lines are of several 10^2 m/s (Dravins et al., 1981; Gray, 1988), and this gives a limit to the precision available using photospheric lines as a reference. Typical values of σ for CAT+CES spectra in the CaII region are 2 to 3 mÅ (150 to 250 m/s), of the same order as the velocity dispersion in the solar photosphere. For IUE spectra in the MgII region, σ is of the order of 30-40 mÅ (3-4 km/s).

3. TWO-COMPONENTS MODEL

After the correction to the photospheric rest frame we measure the central wavelengths of the emission and the self-reversal - $\delta\lambda_{EM}$ and $\delta\lambda_3$ respectively - by a least square fit with two Gaussian components, one in emission, the other in absorption. If we postulate that the emission peak is formed in a single layer just above the photosphere, and the self-reversal in a specific layer higher in the chromosphere (two-components model), then $\delta\lambda_{EM}$ and $\delta\lambda_3$ give an estimate of the bulk radial velocities of these layers. Use of Gaussian profiles is generally justified by the resulting goodness of the fit.

We measured $\delta\lambda_{EM}$ and $\delta\lambda_3$ for a sample of 18 CaII H spectra of 4 late-type dwarfs and found a good correlation between $(\delta\lambda_3-\delta\lambda_{EM})$ and the blue-to-red intensity ratio, I_{2V}/I_{2R} (Crivellari et al. 1987, Fig.4). The accuracy achievable using the two-components model with Gaussian fits is shown by the fact that the range in $(\delta\lambda_3-\delta\lambda_{EM})$ spanned by that correlation is less than 2 km/s, i.e. one half the CAT+CES FWHM$_{INSTR}$.

TABLE I

	CaII in dwarfs	MgII in dwarfs and giants
References	Crivellari et al.(1987)	Vladilo et al. (1987)
Instruments	1.4m CAT ESO+Reticon	0.45m IUE+SEC Vidicon
FWHM$_{INSTR}$	3.75 km/s	20 km/s
S/N (continuum level)	few x 10^2	~ 20
S/N (bottom photospheric abs.)	~ 40	few units

We have measured $\delta\lambda_{EM}$ in the MgII h and k lines for 7 dwarfs and 15 giants with emission wings uncontaminated by interstellar absorption (Vladilo et al., 1987). We do not find any significant displacement of the emission with respect to the photosphere in dwarfs. This can be explained by the fact that the error on $\delta\lambda_{EM}$ induced by the wavelength calibration (3-4 km/s) is greater than the range of chromospheric velocities inferred from our CaII H measurements in dwarfs. A peculiar phenomenon is present in the giants, where the MgII k emission does show a significant red-shift (almost 5 km/s on the average), whereas the h emission does not.

We have also measured (same ref.) $\delta\lambda_3$ in a group of 8 dwarfs and 7 giants with self-reversals uncontaminated by interstellar absorption. A mean red-shift of about 7 km/s of the self-reversals with respect to the photosphere is present in both dwarfs and giants. Gaussian fits to the self-reversals in giants cannot reproduce the observed profile, which is asymmetric.

4. ASYMMETRY MEASUREMENTS

To measure the asymmetry of the emission component we adopt the method of the median bisector (Kulander and Jefferies, 1966). The emission profile is divided into n equally spaced intensity levels. For each level the quantity $m_{EM} = (\lambda_V + \lambda_R)/2 - \lambda_0$ is evaluated, where λ_V and λ_R correspond to the intersection between the intensity level and the blue and red emission wing respectively, and λ_0 is the photospheric rest wavelength. The optimal choice of n depends on the number of resolution elements included in each side of the emission profile, R. Typically, for dwarfs, $R \simeq 4$ for CaII H (CAT+CES spectra), while $R \simeq 1$ for MgII h and k (IUE spectra) so that asymmetry measurements are meaningless with IUE data. We estimate a conservative error of ±4mA (±300 m/s) for each single m_{EM} measurement in the CaII H CAT+CES spectra. In fact, we found evidences of curved bisectors in CaII H spectra of ξ BooA and 70 OphA with amplitudes of the shifts smaller than 1 km/s (Crivellari et al. 1987), confirming the accuracy and the utility of the asymmetry measurements.

REFERENCES

- Crivellari,L., Beckman,J.E., Foing,B.H., Vladilo,G.: 1987, Astron. Astrophys. **174**, 127
- Dravins,D., Lindegren,L., Nordlund,A.: 1981, Astron. Astrophys. **96**, 345
- Gray,D.: 1987, these Proceedings
- Kulander,J.L.,Jefferies,J.T.: 1966, Astrophys.J. **146**, 196
- Vladilo,G.,Crivellari,L.,Molaro,P.,Beckman,J.E.,Genova,R.: 1986, New Insights in Astrophysics ESA SP-263, 233
- Vladilo,G., Molaro,P., Crivellari,L., Foing,B.H., Beckman,J.E., Genova,R.: 1987, Astron. Astrophys. in press

SPECTROSCOPIC VARIABILITY IN LATE-TYPE DWARFS USING HIGH S:N SPECTRA

B.H. Foing[1], J.E. Beckman[2], G. Vladilo[3]
1. LPSP/IAS Verrieres, France.
2. Instituto de Astrofisica de Canarias, Tenerife, Spain.
3. Osservatorio Astronomico di Trieste, Trieste, Italy.

ABSTRACT. Monitoring of active late-type dwarfs, spectroscopically with high resolution ($\lambda/\Delta\lambda \sim 10^5$) and high S:N (>300) round their activity cycles, principally in the CaII H resonance line, offers techniques to explore (a) plage filling factors (b) 3-dimensional chromospheric velocity fields (c) maps of surface activity via "Doppler Imaging". In this paper we deal with the use of spectral signatures to derive plage cover.

1. INTRODUCTION

Chromospheric emission flux variability of CaII H and K in solar-like stars was studied by O.C. Wilson (1978) in his classic work on magnetic activity cycles (periods of years). Short term modulation (periods of days) of this flux was observed by Vaughan et al. (1981) and used by them to compute rotation periods, and hence to investigate the relation period-activity-age. The use of high S:N line profiles instead of fluxes allows us in principle to obtain three new types of information:

(1) The proportion of active (i.e. plage and network) to quiet chromospheric surface cover, given a knowledge of the intrinsic profile emitted by each component.

(2) The presence of velocity fields in three dimensions. The presence of an asymmetry, or the position of a bump in the line profile, and the speeds with which those features change should allow us to distinguish vertical motions from the projected rotational motions of plages.

(3) Combined density and velocity can be combined ("Doppler imaging") to yield maps of chromospheric surfaces, specifying active and quiet regions.

2. COMPARISON OF CaH FOR α Cen A AND α Cen B

In Fig. 1 we show an example of the kind of spectroscopic change that can occur due to the change in plage cover during a rotational period. The spectra are of α Cen B, taken with the CAT+CES (ESO) at a resolution $\lambda/\Delta\lambda = 10^5$, and with signal to noise at the H1 and K1 minima of 100 and in the continuum of 300. They span a period of 11 days in June 1985, and are in two groups: 8/9 June and 17/19 June, which show little change within each pair. The photospheric spectrum is constant, but the chromospheric emission profile shows a very clear change. The contrast with α Cen A is striking, as shown in Fig. 2, where difference spectra for the two objects are shown. In α Cen A there is no measurable difference to within an impressive 0.3% RMS error across the spectrum, while in α Cen B there is a "difference profile" which corresponds to a 21% increment in the H index over minimum phase, and to a FWHM of 33 km s-1.

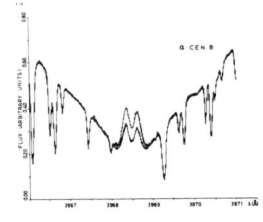

Fig 1: Changes in the Ca H emission from the chromosphere of α Cen B. Four spectra are superposed. There is no photospheric variation but clear chromospheric changes between 8/9 June and 17/19 June 1985 (2 pairs).

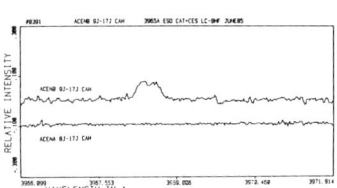

Fig. 2: Differences between spectra of α Cen B and α Cen A from spectra of 9 and 19 June 1985. The clean low-noise subtration for α Cen A gives confidence in the variability data for αCen B.

3. USE OF PROFILE CHANGES TO SEPARATE ACTIVE FROM QUIET CHROMOSPHERE

Assuming a chromospheric disc to be composed of three components: a quiet virtually magnetic field-free background, a homogeneously distributed network structure, and magnetically concentrated plages at specific positions we can develop the formalism to distinguish these.

If the resulting line profile Φ_λ is composed of terms due to the quiet chromosphere with profile Φ_λ^Q, due to the network with profile Φ_λ^N, and filling factor f^N and due to plages, with profile Φ_λ^A and filling factor f^P the profile will be espressed as

$$\Phi_\lambda(t) = \Phi_\lambda^Q (1 - f^N - f^P(t)) + \Phi_\lambda^A (f^N + f^P(t))$$
$$= \Phi_\lambda^Q (1 - f^N) + \Phi^A f^N + f^P(t) (\Phi^A - \Phi^Q) \qquad (1)$$

Of course, one cannot use equation (1) directly to find separately all of the unknown parameters. However, careful observations such as those exemplified in Fig. 1 for α Cen B show that one can make very useful approximations. Modulation of the observed profile round a couple of stellar rotation periods shows that $\Phi_\lambda(t)$ takes a base-line form, and modulations occur above this; we can assume that this is due to the time variations of the 3rd term on the L.H.S. of (1) as plages appear and disappear via rotation. Terms 1 and 2 remain constant. For truly solar-like stars (G2V) we can then assume a known network filling factor f^N and known quiescent and active profiles ϕ^A and ϕ^Q to obtain plage filling factors $f^P(t)$. For adjacent spectral types, we have to assume similar values for f^N ϕ^A and ϕ^Q and look for self-consistent solutions. These can be cross-checked by using lines other than Ca H, e.g. Ca K or the Ca IR triplet.

One must remember that maximum variability in the chromospheric emission core is not associated with maximum plage cover and indeed that total plage cover would yield an invariant "high" profile. This contrast effect is shown in the highly active star 70 Oph A, whose Ca H profile is shown at high S:N in Fig. 3. In this, as in Fig. 1, a logarithmic registration technique, embodied in a programme NEWSUM (Crivellari et al., 1983) allows variable features to be detected against a constant background. 70 Oph A which shows small variations, must be virtually plage-covered, and in principle allows us to obtain "pure" plage profiles for its spectral type (K0V). In practice one must be careful to eliminate local interstellar medium effects at H3, which has not been done for 70 Oph A.

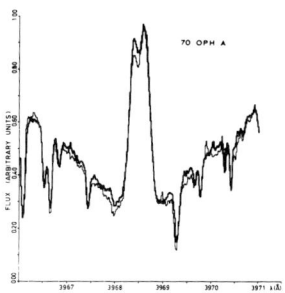

Fig 3: Spectra of 70 Oph A in Ca H, separated by 10 nights. This shows strong plage cover and low modulation as plage filling factor is high.

The totally quiescent behaviour of α Cen A points to the final piece in the jigsaw. Its Ca H profile, as well as being constant in time, shows similar characteristics to a solar "quiet disc" or integrated solar minimum profile, and we can use it as a reasonable estimate of quiet profiles for stars of type G2V. Our full methodology for filling factor measures is thus: seek out stars with time independent Ca H fluxes and "low" profiles. High S:N spectra yield Φ_Q. Seek out stars with small variations in fluxes and "high" profiles. High S:N spectra yield Φ^A. For stars with flux-varying spectra, monitoring around a period can then give $f^P(t)$. Care is needed with asymmetries due to LISM and to the presence of restricted individual plages.

References

Wilson, O.C.: 1978, Astrophys. J. **226**, 379.
Vaughan, A., Baliuna, S.L., Middelkoop, F., Hartmann, L.W., Mihalas, D., Noyes, R.W., Preston, G.W.: 1981, Astrophys. J. **250**, 276.
Crivellari et al.: 1983, Astron. Astrophys. Suppl. Ser. **52**, 135.

SHORT PERIOD OSCILLATIONS IN ACTURUS, ALDEBARAN, AND POLLUX

Peter H. Smith and Robert S. McMillan
Lunar and Planetary Laboratory
University of Arizona
Tucson, AZ 85721 USA

ABSTRACT. A total of 48 nights of time series data have been obtained for the K giants: Arcturus, Pollux, and Aldebaran. A careful analysis of both single and multi-night sets using the earth's motion as a velocity calibrator has yielded stellar velocity time series accurate to $\pm$ 3 m/s per observation. Periodogram analyses of these sets have revealed the existence of oscillations with periods near 2.5 hrs and amplitudes of $\pm$ 5 m/s for both Pollux and Aldebaran, but not for Arcturus. Preliminary analysis of a 5-night set for Pollux using the CLEAN algorithm suggests at least three modes separated by about 35 microHertz.

1. OBSERVATIONS

Since the discovery of 5-minute oscillations on the Sun, there has been a great deal of interest in the search for small amplitude radial velocity variations in nearby stars. Because the accurate determination of stellar radial velocity requires high spectral resolution, high signal-to-noise ratios, and short exposure times, the few measurements that have been reported in the literature have been made on bright stars. Noyes et al. (1984) found a 10-minute period in epsilon Eri with modal spacings of 172 microHertz, Fossat (1987, this Symposium) reports modal spacings in the power spectra of alpha Cen A and Procyon, and M. Smith (1983) published oscillation periods for Aldebaran and Arcturus. All these claims are unsubstantiated and so near the detection limits that it is fair to say that this work represents a brave beginning into what promises to be a new and exciting way of studying stars: asteroseismology.
   We are using an instrument developed at the University of Arizona which is permanently installed at the 0.9-meter Steward Observatory telescope on Kitt Peak (McMillan et al., 1986). Serkowski et al. (1979) originated the concept of using a stabilized, carefully calibrated Fabry-Perot in transmission to provide longterm wavelength calibration during the search for planetary systems. This ensures excellent short term stability. The etalon can be accurately tilted to provide for the

tracking of spectral features as they drift in wavelength in response to the earth's motions and to impose a known velocity shift onto the signal for use in analysis of the exposures.

By imposing a known 'velocity' signal onto several exposures and finding the fractional intensity change for each transmission order, an approximate slope (defined as the percentage intensity change in an order per km/s) can be determined for each order. The orders having small slope values, mostly continuum points, are used for normalizing the different exposures.

After normalization, the orders which fall on steeply sloped features in the spectrum ('active' orders) are used to derive velocity estimates. For each active order the relative intensity change between each observation and a reference observation chosen near the middle of the series is plotted against time. The known earth motion curve is fit to this series with only two unknowns, the slope value and a bias; these values provide accurate velocities for each active order and each observation. An average of all the active orders in a observation gives the best estimate of the velocity relative to the reference observation. Advantages of this technique are that the ensemble used in the average also gives accurate error bars for each observation and that the curve fit gives an opportunity to reject outliers caused mainly by decay events within the CCD chip. The final step in the reduction is the transform of the series by a slow Fourier technique to a normalized periodogram.

2. RESULTS AND DISCUSSION

Periodograms from single nights and sets of contiguous nights have been used to search for oscillations. It is also helpful to average the periodograms from separate nights to reduce the noise. Before concluding that an oscillation was present in the power spectrum, it was required that it be observed on several individual nights. The normalized periodogram provides a false alarm capability (Horne, and Baliunas, 1986) so that peaks which fail to exceed a minimum threshold can be safely rejected.

Arcturus has a period of 1.842 ± 0.005 days and an amplitude of 160 m/s peak-to-peak (Smith et al., 1987). This periodic behavior is not the type of p-mode oscillation observed for the sun. When time series within a night are examined, the peaks generally fall short of the detection threshold and, when they do exceed the threshold, there does not appear to be any substantiation on subsequent nights. We conclude that for time scales between a few minutes and several hours that there are no oscillations of amplitude greater that 2-3 m/s. This result contradicts the 97 ± 28 min period claimed by M. Smith (1983).

Although the number of data sets on Aldebaran is much smaller than for Arcturus, on several nights significant peaks appear near 2.5 hours' period. These peaks are only marginally significant in themselves, but there is strength in the multiple observations. The amplitudes are on the order of ± 5 m/s. We have obtained one set of 4 contiguous nights which has been transformed as one continuous set and also reveals the peak at 2.5+0.5 hours in agreement with M. Smith's 110+16 min period.

Unlike the previous two stars, Pollux shows a very significant peak in the periodogram which has been observed on all 14 nights currently reduced. On 1987 Jan 13, a time series greater than 10 hours in length clearly shows almost 4 cycles of oscillation. The amplitude of the oscillation can be accurately measured from the velocity plots at $\pm$ 6 m/s; however, there seems to be a different amplitude on different nights. The same sort of phenomenon occurs with the position of the peak in the periodogram: there are significant displacements between nights centered about a period near 2.5 hours. In an effort to understand these variations, a set of 5 contiguous nights has been analyzed as a single time series. The characteristic shape of the peak near 2.5 hours is a broad energy distribution about 3 times the width of the spectral window. Using the CLEAN algorithm (Roberts et al., 1987) to remove the confusing sidelobes caused by the day gaps in the time series sample, there appears to be 3 equally separated modes spaced by 35 microHertz. If these modes are real and caused by low order p modes, then the true order separation is twice this value or about 70 microHertz.

The predictions for K-giant oscillation are in a primitive state since they are extrapolations from solar models and may not include all the relevant physical processes. However, Frandsen (private communication) predicts time scales greater than 50 min for giant stars and amplitudes greater than 1 m/s per mode. He is drawing on unpublished results of the study done by Christensen-Dalsgaard and Frandsen (1983) for solar-type stars; the modal spacings they report decrease with increasing luminosity from the solar value of 136 microHertz to values less than 80 microHertz for the giant stars. The time scale is also supported by the older study of Ando (1976), who ran models specifically for K giants and stressed that the periods must be longer than the acoustic cut-off period of about 1.5 hours. It would be very helpful if more detailed models were computed for these interesting and observable stars.

We wish to acknowledge the help and support of William Merline, Marcus Perry, and Jack Frecker; without them the project would not be possible.

3. REFERENCES

Ando, H. 1976, Publ. Astron. Soc. Japan, 28, 517.
Christensen-Dalsgaard, J., and Frandsen, S. 1983, Solar Phys., 82, 469.
Horne, J. H., and Baliunas, S. L. 1986, Ap. J., 757.
McMillan, R. S. et al. 1986, Proc. S. P. I. E., 627, 2.
Noyes, R. W. et al. 1984, Ap. J., 285, L23.
Roberts, D. H. et al. 1987, A. J., 93, 968.
Serkowski, K. et al. 1979, Ap. J., 228, 630.
Smith, M. A. 1983, Ap. J., 265, 325.
Smith, P. H., McMillan, R. S., and Merline, W. J. 1987, Ap. J., 317, L79.

DISCUSSION

LINSKY Your data clearly point out the need for a network of instruments at different longitudes to study oscillations of stars without time gaps.

SMITH I agree entirely, in order to resolve the predicted mode splittings this will be essential.

MEASUREMENTS OF MAGNETIC FIELDS ON COOL STARS

Steven H. Saar
Joint Institute for Laboratory Astrophysics
University of Colorado and NBS, Boulder, CO 80309, USA

ABSTRACT. I discuss the some of techniques used (and problems involved) in measuring stellar magnetic fields on cool stars, and detail how these measurements are broadening our understanding of stellar magnetic activity.

1 INTRODUCTION AND A BRIEF HISTORY

Magnetic fields lie at the heart of the so-called "solar-stellar connection", playing a crucial role in the structure, energy balance, and evolution of the atmospheres of cool stars. The detailed physics of the these interactions, however, remains elusive, in part due to the lack of information about stellar magnetic parameters. The need for direct measurements of magnetic field strengths and the fraction of the stellar surface that cover is clear.

Unfortunately, magnetic fields on cool stars are quite difficult to measure. The detection of magnetic fields on cool stars is hampered by the locally bipolar topology of the fields themselves, which effectively cancels the circular polarization signal from the unresolved stellar disks (e.g., Borra, Edwards, and Mayor 1984). Linear polarization, which does not cancel in integrated starlight, has been recorded in broadband measurements for a few stars (Huovelin et al. 1985), but is difficult to interpret (Landi Degl'Innocenti 1982). Consequently, efforts to detect the magnetic fields of solar-like stars through polarization in spectral lines have been largely unsuccessful.

A breakthrough came when Robinson (1980) devised a method of measuring stellar magnetic fields in *unpolarized* light by studying the subtle Zeeman broadening of magnetically sensitive line profiles relative to insensitive reference lines. Such an analysis can provide an estimate of the fraction of the stellar surface that is covered by magnetic fields in addition to the field strength itself. Qualitatively, a line profile is modeled as $F = fF_{mag} + (1-f)F_{quiet}$, where F_{mag} and F_{quiet} are the line flux profiles in the magnetic (with a field strength equal to B) and quiet (B = 0) regions, and f is the magnetic area filling factor. For simplicity, the models used so far assume that the the thermodynamic properties of the magnetic and quiet regions are identical. The resulting magnetic parameters f and B therefore refer to elements of the stellar surface analogous to "bright" magnetic regions on the Sun such as network and plage.

The effects of a magnetic field on an unpolarized line profile are subtle, however, and difficult to measure accurately. Differences between magnetically sensitive (high Lande g) and insensitive (low g) lines are generally only a few percent of the continuum, requiring high signal-to-noise (S/N) spectra. Typically S/N $\geq$ 50 is necessary, although the exact figure depends on the values of f and B as well as other observational and stellar parameters (Marcy 1982; Saar 1987). The magnitude of the Zeeman broadening itself is quite small, since the splitting of the magnetic components is only $\Delta\lambda_B = 4.2 \times 10^{-3}(g/2.5)(\lambda/600$ nm$)^2$(B/1000 G) nm . Thus, high spectral resolution is also needed. The minimum spectral resolution ($\lambda/\Delta\lambda$) required is approximately $2\Delta\lambda_B$, which corresponds to at least 75,000 at 600 nm and 40,000 at 2 μm. Ideally, S/N = 100 - 200 and a resolution of 100,000 should be obtained. Stellar rotation imposes further limits magnetic measurements, since rotational line broadening can overwhelm the magnetic broadening signal for v sin i > 10 km s^{-1}.

Unrecognized blends can also significantly effect the accuracy of the derived magnetic parameters (e.g., Gondoin, Giampapa, and Bookbinder 1985, Linsky 1985). Blends can cause broadening that mimics the Zeeman effect, leading to inaccurate or even spurious magnetic field detections. The ubiquitous molecular opacity sources in K and M dwarf atmospheres, for example, render magnetic field measurements at optical wavelengths extremely difficult for these stars.

In spite of these difficulties, numerous measurements of magnetic fields on cool stars have been made over the past several years. Following the initial detection of ξ Boo A by Robinson, Worden, and Harvey (1980), subsequent measurements of the star by Marcy (1981) found no evidence for Zeeman broadening, the first indication of magnetic variability on an active dwarf. At about the same time, attempts to study correlations between simultaneous measurements of magnetic, chromospheric, and coronal fluxes were made (Basri, Walter, and Marcy 1981). Giampapa, Golub, and Worden (1983) discovered magnetic fields on an active giant (λ And) in the first use of infrared spectra for magnetic field determinations. A major accomplishment was Marcy's (1983, 1984) publication of the results of the first extensive survey of magnetic field parameters for 29 late-type dwarfs Gray (1984) serendipitously discovered Zeeman broadening in several more dwarfs during the course of studies of stellar rotation. Marcy and Bruning (1984) searched for magnetic broadening in 8 late-type giants and subgiants, but found none.

Some of the results of these early studies were rather suprising, however. Some stars with only moderate levels of activity showed filling factors of nearly 90 % (e.g., ϵ Eri) Other, quite inactive stars showed similar amounts of magnetic flux (e.g., 61 Cygni A, τ Ceti). Enormous swings in the surface magnetic field and filling factor appeared to take place on timescales of a day. The total flux ($\propto$ fB), however, remained roughly constant in time. Indeed, Gray (1985), in an examination of all magnetic measurements on cool stars published to that date, noted that the product fB was a constant independant of spectral type and rotational velocity. Thus, the early magnetic measurements seemed to indicate that all stars produced the same amount of magnetic flux, contrary to observations of stellar "activity" indicators and to the expectations of dynamo theories.

2 NEW METHODS, RESULTS, AND FUTURE PROSPECTS

It now appears that the rather curious results of the initial magnetic field surveys were the result of simplifying assumptions used in the Zeeman broadening analyses. In particular, the early methods assumed that all lines were simple Zeeman triplets on the linear part of the curve-of-growth, and could therefore be constructed by essentially adding together three appropriately shifted low g line profiles. As one might anticipate, this approximation is inappropriate for the moderately strong lines employed in the Zeeman broadening studies, and its use leads to systematic errors in the derived magnetic parameters (Saar 1987; Hartmann 1987). Weak line blends, which will affect the cores and wings of stronger lines in different ways, also introduce systematic errors in the derived f and B values.

To help remedy this situation, I have developed some new methods for deriving magnetic parameters from spectra of cool stars (Saar, Linsky, and Beckers 1986, Saar 1987) which include magnetic radiative transfer effects (Unno 1956), the full Zeeman patterns, and some compensation for line blends. The new technique models differences between line profiles, either comparing magnetically sensitive and insensitive lines from the same spectrum, or by comparing the same high g line in two stars of the same spectral type, one of which is known to be magnetically inactive. The latter, differential approach is used to eliminate the effects of blends to first order. The number of free parameters are minimized by determining the non-magnetic broadening parameters independently (from low g lines) and applying these results to the high g line models.

The new magnetic analysis methods have now been applied to a considerable body of data, and some preliminary trends can be discerned. 1) The product fB is not constant (Saar and Linsky 1986). Rather, fB and f increase with stellar angular velocity, consistent with simple ideas of the dynamo mechanism and the observed increase of chromospheric and coronal emission with rotation (Saar and Linsky 1986; Linsky and Saar 1987). There is some evidence for a saturation in f at high Ω (Saar, Linsky, and Giampapa 1987). 2) B increases with decreasing T_{eff} and increasing gravity and gas pressure down the main sequence. A possible cause of this is pressure equilibrium between B and the quiet photosphere (B $\propto P_{gas}^{0.5}$; Saar and Linsky 1986). 3) f $\propto t^{-0.6}$ while B is constant in time, in agreement with the observed dependence of Ω on t and suggesting that f is the dominant magnetic parameter governing stellar activity (Linsky and Saar 1987). 4) The mean strength of the surface field (= fB, the unsigned magnetic flux density) correlates with outer atmospheric emission such that the X-ray flux, $F_x \propto$ (fB)$^{0.9}$ and the residual Ca II flux (Schrijver 1983), $\Delta F_{CaII} \propto$ (fB)$^{0.6}$, consistent with relations derived for the Sun, and with flux-flux relations derived for stars (Saar and Schrijver 1987). Rotational modulation of chromospheric and transition-region line fluxes with magnetic flux for the active dwarf ξ Boo A support this picture, and when combined with measurements of broadband linear polarization, permit a rough determination of the spatial distribution of active areas on the star (Saar et al. 1987).

These results must be regarded as somewhat preliminary, however, since not all the line profiles have been modeled differentially to remove blends. Also, the data have been fit using convolutions for the velocity broadeneing, which is only an approximate method (Bruning 1984). Tests show that no single intensity profile can reproduce the shape of the disk-integrated flux profile (Fig. 1), implying the convolution approach could lead to systematic errors in the derived f and B values. We have therefore added full–disk integrations to the Zeeman line modeling codes to properly account for the rotational and turbulent line broadening, and are in the process of reanalyzing the data.

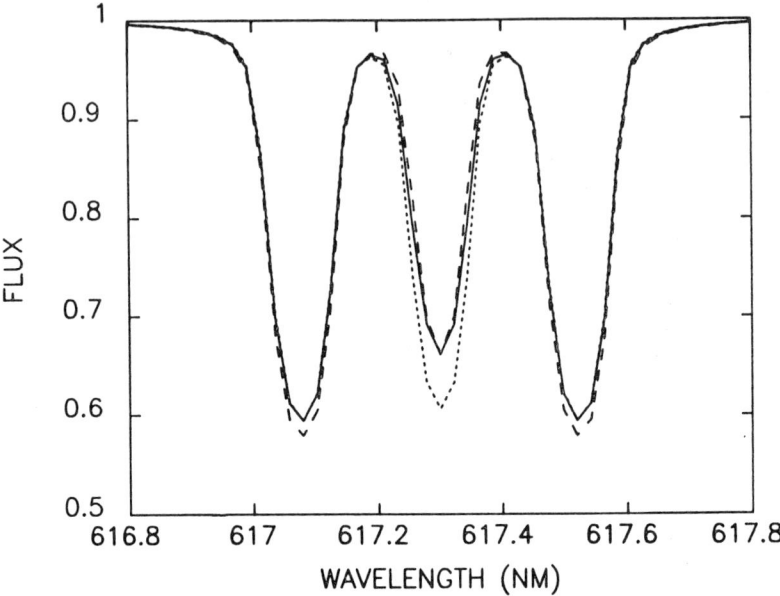

Figure 1. Comparison of computed disk-integrated magnetic flux profile (for a line-to-continuum opacity ratio of 10, g = 2.5, B = 5000 G, f = 1.0, v sin i = 0, and a limb darkening coefficient of 0.6; solid line) with intensity profiles computed at magnetic field to line–of–sight angles of $\theta = 31°$ (dashed line) and $\theta = 48°$ (dotted line). Note that the central π and the shifted σ components of the flux profile cannot be *simultaneously* matched by a single intensity profile.

Several further improvements to stellar Zeeman analysis are also on the horizon (and will be discussed in the following talks). Basri and Marcy (1986) and Marcy and Basri (1988) have developed codes which use the Unno formulation with a more realistic model atmosphere. Mathys (1987) and Mathys and Solanki (1988) are using a multiline correlation approach (after Stenflo and Lindegren 1977) which may yield information on the thermodynamic differences between the quiet and magnetic regions on the stars. Thus, the future promises to bring ever more accurate measurements of stellar magnetic parameters and with it, better understanding of the stellar "activity" phenomenon.

This research is supported by NASA grant NGL-006-03-057 to the University of Colorado.

REFERENCES

Basri, G. S., and Marcy, G. W. 1986, *Bull. A. A. S.*, **18**, 984.
Basri, G. S., Walter, F. M., and Marcy, G. 1981, *Bull. A. A. S.*, **13**, 828.
Borra, E. F., Edwards, G., and Mayor, M. 1984, *Ap. J.*, **284**, 211.
Bruning, D. H. 1984, *Ap. J.*, **281**, 830.
Giampapa, M. S., Golub, L., and Worden, S. P. 1983, *Ap. J.* (Letters), **268**, L121.
Gondoin, Ph., Giampapa, M. S., and Bookbinder, J. A. 1985, *Ap. J.*, **297**, 710.
Gray, D. F. 1984, *Ap. J.*, **277**, 640.
Gray, D. F. 1985, *Pub. A. S. P.*, **97**, 719.
Hartmann, L. 1987, presented at *The Fifth Cambridge Workshop on Cool Stars, Stellar Systems, and the Sun*, Boulder, CO, July 8 - 11, 1987.
Huovelin, J., Linnaluoto, S., Piirola, V., Tuominen, I., and Virtanen, H. 1985, *Astr. Ap.*, **152**, 357.
Landi Degl' Innocenti, E. 1982, *Astr. Ap.*, **110**, 25.
Linsky, J. L. 1985, *Solar Phys.*, **100**, 333.
Linsky, J. L., and Saar, S. H. 1987, , presented at *The Fifth Cambridge Workshop on Cool Stars, Stellar Systems, and the Sun*, Boulder, CO, July 8 - 11, 1987.
Marcy, G. W. 1981, *Ap. J.*, **245**, 624.
Marcy, G. W. 1982, *Pub. A. S. P.*, **94**, 989.
Marcy, G. W. 1983, in *Solar and Stellar Magnetic Fields: Origins and Coronal Effects*, ed. J. O. Stenflo (Dordrecht: Reidel), p. 3.
Marcy, G. W. 1984, *Ap. J.*, **276**, 286.
Marcy, G. W., and Basri, G. S. 1988, these proceedings.
Marcy, G. W., and Bruning, D. H. 1984, *Ap. J.*, **281**, 286.
Mathys, G. 1987, presented at *Observational Astrophysics with High Precision Data*, Liegé, Belgium, June 23-26, 1987.
Mathys, G., and Solanki, S. K. 1988, these proceedings.
Robinson, R. D. 1980, *Ap. J.*, **239**, 961.
Robinson, R. D., Worden, S. P., and Harvey, J. W. 1980, *Ap. J. (Letters)*, **236**, L155.
Saar, S. H. 1987, *Ap. J.*, in press.
Saar, S. H., Huovelin, J., Giampapa, M. S., Linsky, J. L., and Jordan, C., 1987, presented at *Activity in Cool Star Envelopes*, Tromso, Norway, July 1-8, 1987.
Saar, S. H., and Linsky, J. L. 1986, *Advances in Space Physics*, **6**, No. 8, 235.
Saar, S. H., Linsky, J. L., and Beckers, J. M. 1986, *Ap. J.*, **302**, 777.
Saar, S. H., Linsky, J. L., and Giampapa, M. S. 1987, presented at *Observational Astrophysics with High Precision Data*, Liegé, Belgium, June 23-26, 1987.
Saar, S. H., and Schrijver, C. J. 1987, , presented at *The Fifth Cambridge Workshop on Cool Stars, Stellar Systems, and the Sun*, Boulder, CO, July 8 - 11, 1987.
Schrijver, C. J. 1983, *Astr. Ap.*, **127**, 289.
Stenflo, J. O., and Lindegren, L. 1977, *Astr. Ap.*, **59**, 367.
Unno, W. 1956, *Pub. Ast. Soc. Jap.*, **8**, 108.

DISCUSSION

    MATHYS I would like to make the comment that radiative transfer effects are important not only in the case of saturated lines but also whenever the lines are not optically thin. As soon as you depart from the weak line limit, simple atomic parameters such as the effective Landé factor may no longer be sufficient to characterize the magnetic broadening.

    SAAR Yes, I agree completely. We include both radiative transfer effects and the full Zeeman patterns for all the lines we model.

MAGNETIC FIELD MEASUREMENTS ON LATE-TYPE STARS: A NEW TECHNIQUE

Geoffrey W. Marcy
San Francisco State University

Gibor Basri
Univ. of California, Berkeley

ABSTRACT. A new technique for detecting and measuring magnetic fields on cool stars is presented, incorporating both an improved observational approach and a rigorous LTE theoretical treatment. We have identified two lines in the near infrared, one of which is very sensitive to Zeeman broadening and the other relatively insensitive, that are easily accessible to CCD and Reticon detectors. The lines are modelled by solving numerically the equations of transfer for the Stokes parameters in a full model atmosphere, permitting computation of all relevant depth-dependent physics such as polarized line opacities and broadening sources. We have obtained high resolution, high S/N spectroscopic observations of several G and K dwarfs and have synthesized the two lines using two free parameters, namely, the average field strength, B, and the fraction, f, of the surface covered by fields. For the chromospherically active stars ε Eridani and Xi Boo A, we find B=1000, f=35% and B=1200, f=40%, respectively. A careful study has been made of various sources of broadening to determine their ability to mimic Zeeman broadening, and none is capable of doing so.

1. INTRODUCTION

It has become increasingly clear during the past decade that a variety of characteristics of cool stars, including spots, chromospheres, flares, and coronae, are qualitatively similar to corresponding solar phenomena that are spatially associated with magnetic fields. (For a review see Hartmann and Noyes, 1987.) Thus, several attempts have been made to detect the Zeeman effect in stellar spectra, notably, Robinson et al.(1980), Marcy (1984), Gray (1985), and Saar (1988). Generally, these workers have tried to detect the tiny extra broadening in line profiles due to the Zeeman effect. The effort therefore requires predicting, with extreme accuracy, the magnitude of all other sources of broadening in a line profile and then properly extracting the residual Zeeman signal.
    However, these past attempts have all involved various assumptions about the line formation. The most sophisticated effort (Saar, 1988)

includes the transfer of the Stokes parameters through an artificial atmosphere that has a linear source function with optical depth and a constant ratio of line to continuous opacity. This work, and the previous efforts, have all yielded field-strength estimates on late-type dwarfs of between 1000G and 2000G, similar to those on the solar surface. But, the fraction of the stellar surface covered by fields is often found to be remarkably high, typically 30% to 50% for the more chromospherically active stars.

Here we present a Zeeman analysis which treats the line transfer of the Zeeman effect in a realistic model atmosphere, and we include all depth-dependent physics, with special attention given to the mechanisms that broaden the profiles. Observationally, we adopt two lines in the near infrared: $\lambda 8468.41$ (FeI) having a Landé-g value of 2.5 and $\lambda 7748.28$ (FeI) having a Landé-g value of 1.1 . The former line is considerably more sensitive to the Zeeman effect than optical lines since Zeeman splitting increases as the square of the wavelength.

2. THE THEORETICAL CALCULATION

The differential equations of transfer for the Stokes parameters through an atmosphere containing Zeeman-split transitions were developed by Unno (1956) and were shown to be analytically solvable under the assumption of a linear source function (taken to be the Planck function) and a constant ratio, η, of line to continuous opacity. Clearly, this analytical approach will not yield profiles that are directly comparable to real stellar flux profiles within the required accuracy of about 1%. Saar (1988) has addressed this difficulty by estimating effective fixed values for η and the line-broadening parameters through the use of observed Zeeman-insensitive lines. This approach (along with several other important improvements) has apparently confirmed the existence of magnetic fields covering substantial fractions of the surfaces of active stars. However, since no two lines are formed in the same way, the estimated effective stellar and line parameters represent good approximations at best, and, as we have found, moderate errors will result from even the best choices for those parameters.

A significant improvement is possible by solving numerically the Unno equations of transfer in a realistic atmosphere, in which all depth-dependent quantities are included. The three coupled transfer equations for I, Q, and V can be integrated upward through the atmosphere, starting with the boundary conditions at the base, namely, I = Planck function, and Q = V = 0. A second-order Runge-Kutta scheme is employed and is found to be stable and accurate as long as the step sizes in optical depth are no more than about one third of one optical depth. To meet this requirement, all physical quantities are interpolated onto a dynamic depth grid, resulting in between 100 and 350 depth points, depending on the angle of the line-of-sight to the surface normal. A Gaussian quadrature method is used to construct a flux profile from intensity profiles at three points on the stellar surface. The splitting and relative intensities of the Zeeman components of the lines of interest were computed assuming LS coupling (Condon and Shortely, 1970).

The entire calculation is carried out in LTE with all atomic physics computed at each depth point. We used scaled solar models based on the HSRA such that the temperature structure is scaled as the effective temperature and all particle densities are recomputed to satisfy the equation of hydrostatic equilibrium and the Saha equation. Both thermal and collisional broadening are computed at each depth and we have carefully considered different values of the collisional broadening coefficient, C_6, in an effort to examine its ability to mimic Zeeman broadening. The final flux profiles were broadened to simulate macroturbulence (an exponential form was used), rotation, and the instrumental effects. As a final check, profiles were synthesized assuming a linear source function and constant ratio of line-to-continuous opacity. The resulting profiles agreed well with the analytic solutions of Unno.

It is worth noting that for field strengths of about 1000G, the theoretical intensity profiles often exhibit central reversals and odd bends in the shape of the line profile. These are also seen in the exact analytic solutions for certain values of η and are due to the presence or absence of overlap between the displaced Zeeman components, each having different intensities and polarizations. These abnormalities disappear upon broadening.

Finally, we have included one line blend in the theoretical calculation, namely that due to Ti I, that is 0.05A longward of line center of $\lambda 8468.41$. We have taken the gf value from Kurucz and Peytremann (1975) which yields a feature of equivalent width 16 mA, in the absence of the principal line.

3. OBSERVATIONS

Spectroscopic observations were made of $\lambda 8468.41$ and $\lambda 7748.28$ at the coudé focus of the 3-m telescope at Lick Observatory using a CCD detector. The resolution was about 130,000 and S/N ratio per pixel was typically 200. A small stellar sample was initially chosen consisting of the Sun (daytime sky), Xi Boo A, ε Eridani, and HD 166620, the last being a chromospherically inactive K2 dwarf. Two weak blends, due to MgI and terrestrial water, in the red wing of 8468.41 were carefully removed using neighboring similar lines.

To apply the theoretical calculations, a number of atomic and stellar parameters must be determined. The gf values for both lines were taken from Kurucz and Peytremann (1975) and we chose to augment the value of the collisional damping constant, C_6, by a factor of seven as this yielded excellent fits to solar profiles. We explored extensively the role played by C_6 in the detection of residual Zeeman broadening in the observed profiles. In short, as collisional broadening acts primarily on the wings, one can easily determine upper limits to C_6 and its broadening effect cannot mimic that due to the

Zeeman effect. The abundance of Fe was determined for each star by matching the equivalent width of the Zeeman-insensitive line, $\lambda 7748$, and the shape of that line was also used to determine the macroturbulence. Values for rotational broadening were derived from the literature (Soderblom 1981, Noyes et al. 1984) and all macro-broadening was accomplished by standard convolution (Gray 1976, but see Bruning 1984).

The observed solar profiles are fit well, within one percent, by assuming that there is no magnetic field on the surface. For the chromospherically inactive K2 dwarf, HD166620, the Zeeman-insensitive line, $\lambda 7748$, is also fit well, as seen in Fig.1. Note that no scaling has been done to any of the profiles. The observed Zeeman-sensitive line for HD166620 appears slightly broader than that calculated, assuming B=0. This appears to be weak evidence for Zeeman broadening in that inactive star. A similar synthesis of the two lines of ε Eridani is also shown in Fig.1. Here, the observed Zeeman-sensitive line appears significantly broadened relative to the theoretical line computed with B=0, though the Zeeman-insensitive line is well fit. We consider this strong evidence that Zeeman broadening is playing a significant role in the line formation on ε Eri. Similarly, the Zeeman-sensitive line from Xi Boo A is considerably broader than that computed with B=0 (Fig.1).

For the three stars mentioned above, theoretical profiles were computed for a variety of assumed magnetic field strengths. These profiles were added with various weights to those constructed with no magnetic field to simulate the flux profiles from a stellar surface having a fraction, f, covered by fields. The final theoretical profiles providing a best fit are shown in the right column of Fig.1. There is a clear indication that the red wing of $\lambda 8468$ is still contaminated by some blend, especially at K2. However, the blue wing is fit well in all cases. The resulting magnetic field parameters are, Xi Boo A: B=1200G, f=40% ; ε Eri: B=1000G, f=35%; HD166620: B=1500G, f=15%.

An estimate of the uncertainties can be gained qualitatively by varying the input magnetic parameters until the fit of the theoretical profile becomes unacceptable. Such tests indicate that the uncertainties in the magnetic field strength are about 250G and those in the covering factor are about 1/5 of the measured value. However, the true uncertainties are greater owing to our poor knowledge of the atmosphere in the magnetic regions. Indeed the large covering fractions derived here imply that all determinations of surface parameters, such as gravity and effective temperature, represent rough averages over the inhomogeneous surface.

4. DISCUSSION

The important question is whether or not the Zeeman broadening implied here for the three late-type stars is real. We have carried out an exhaustive search for alternative mechanisms that could provide differential broadening of the Zeeman-sensitive line over the insensitive line. In particular, we considered the effects of varying

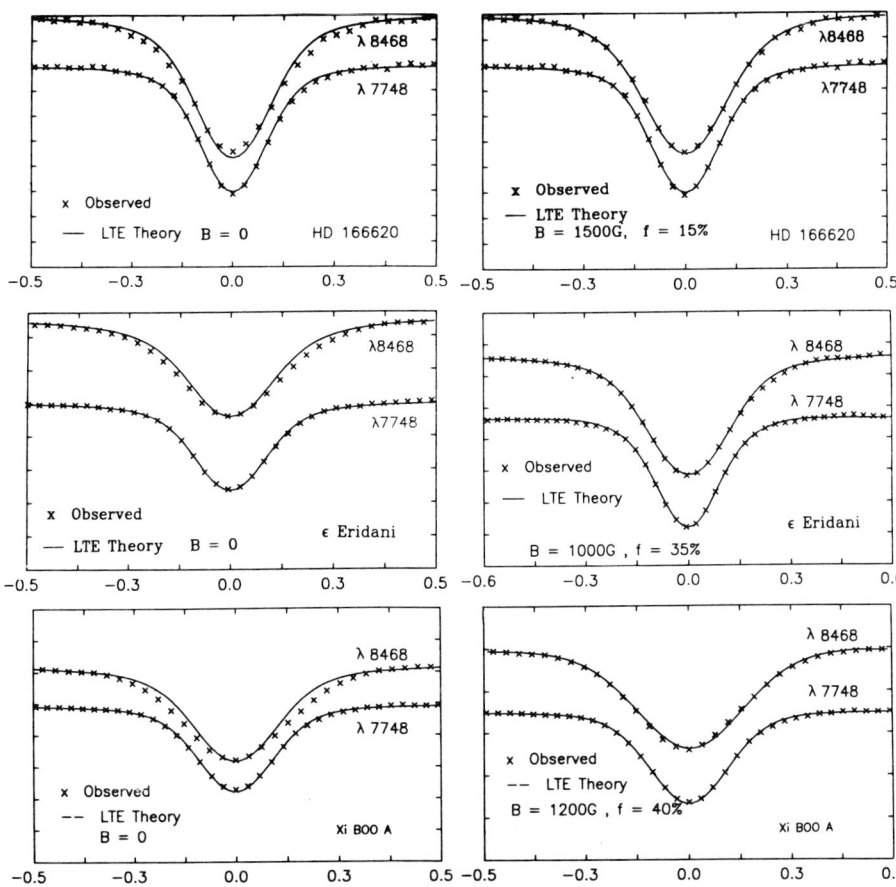

Figure 1. Comparison of the observed profiles to those computed theoretically for λ8468 (Zeeman-sensitive) and λ7748 (Zeeman-insensitive). The three panels on the left show theoretical profiles constructed with no magnetic field. Note that the observed profile λ8468 is apparently broadened due to the Zeeman effect in all three cases. The right panels show the fit obtained when a magnetic field is included in the synthetic profiles. B is the average field strength assumed, and f is the fraction of the stellar surface covered by the fields.

a number of input parameters to the line synthesis, including gf, C6, the atmospheric structure, Fe abundance, and macro-broadening. While some variation in these parameters can be shown to be consistent with the observed profiles, they can, in no way, account for the observed excess broadening of the Zeeman-sensitive line. We conclude that the observed broadening is indeed due to the Zeeman effect.

Finally, one wonders whether the large covering fractions derived for active stars are real. We have assumed that the observed profile is accurately represented by the weighted sum of profiles from magnetic and non-magnetic regions. We cannot account, however, for the relative continuum brightness of the two types of regions, nor have we compensated for the different line strengths from the two regions. Thus the true covering factors may be considerably different from those derived here. One may ask whether the magnetic regions on active stars are crudely similar to solar spot umbrae, penumbrae, or faculae. To resolve this, future stellar magnetic work should include simultaneous photometric and spectroscopic observations to search for rotationally modulated diagnostics of analogous solar magnetic regions.

We would like to acknowledge helpful discussions with L. Hartmann, D.Bruning, and S.Saar. This research was supported in part by NSF grant AST 86-03979.

REFERENCES

Bruning, D.H. 1984, Ap.J.,281,830.
Condon,E.U. and Shortley,G.H. 1970, The Theory of Atomic Spectra Cambridge Univ. Press, London.
Gray, D.F., 1976, The Observation and Analysis of Stellar Photopheres, Wiley and Sons, NY, p394.
Hartmann,L.W. and Noyes,R.W. 1987, 'Rotation and Magnetic Activity in Main Sequence Stars', to appear in Annual Reviews of Astronomy and Astrophysics, Vol.25.
Kurucz,R.L. and Peytremann,E. 1975, Smithsonian Astroph. Obs. Special Report 362.
Marcy,G.W. 1984, Ap.J.,276,286.
Noyes,R.W., Hartmann,L.W., Baliunas,S.L., Duncan,D.K., Vaughan,A.H. 1984, Ap.J.,279,763.
Robinson,R.D., Worden,S.P., and Harvey,J.W. 1980, Ap.J. Letters, 236,L155.
Saar, S.H. 1988 preprint.
Soderblom,D.R. 1981, Ap.J.,263,239.

DISCUSSION

LANDSTREET You have gone to some trouble to obtain accurate local line profiles. Another important aspect of the problem is the assumed magnetic geometry, which one would expect to have substantial effect on the resulting profiles. Could you describe your assumed magnetic geometry, and tell us how much your results depend on that geometry ?

MARCY (As I understood it) The magnetic geometry assumed is an average of radial magnetic fields at three different values of μ. Other magnetic geometries should be explored.

LINSKY It is important to recall that the analyses by Marcy and Saar both assume that the model atmospheres and broadening parameters for magnetic and nonmagnetic regions are the same. This is unlikely to be true if the Sun may be considered a useful guide. For example, if the continuum brightness is larger in magnetic, than nonmagnetic regions then the filling factor will be smaller than derived by assuming no difference. Also if the broadening parameters are different in the magnetic and nonmagnetic regions, then the derived magnetic parameters may be quite uncertain.

MARCY This is an important point. Not only are the atmospheres different in and out of magnetic regions, but the velocity fields may well be different. A two-components stellar atmosphere, perhaps based on solar faculae, should be considered for future work.

SNEDEN How do you set your continuum and how does it affect the accuracies of your results ?

MARCY The continuum is set by using the highest points in the 80 angstroms of spectrum that we obtain. Tests show that misplacement of the continuum can indeed affect the derived fields. The percent error in the field is approximately equal to the error in continuum placement divided by the amount of enhancement of the line wing due to the Zeeman effect.

HOLWEGER You mentioned the strange line profiles you have got in your model atmosphere calculations, and you have given an explanation in terms of an interplay of different Zeeman components. I suggest that these emission cores are due to the use of a model with a chromosphere, and assuming LTE. An emission core is easily generated with the HSRA in this manner.

MARCY We tested your concern by a trial in which the source function decreased linearly with optical depth. The oddly shaped profiles persist. They can be understod entirely by considering the LTE line transfer of both linearly -and circularly- polarized radiation, in the presence of Zeeman absorption components.

BASRI A remark following Holwegger's question: as an old fan of chromospheric activity, it immediately occured to me that your point might be the case. We re-ran the computation with the chromosphere removed, with the result that all features of the profile remained, including the "central reversal". Examination of the contribution functions showed the line is really fully formed below the temperature minimum.

ABUNDANCE AND MAGNETIC FIELD GEOMETRIES OF HELIUM-STRONG AND HELIUM-WEAK STARS

David A. Bohlender and J. D. Landstreet[*]
Department of Astronomy
University of Western Ontario
London, Ontario   N6A 3K7

1. INTRODUCTION

The helium-weak and helium-strong stars are main sequence stars with anomalously weak and strong helium lines for their spectral types respectively. Many members of the two classes have strong, globally ordered magnetic fields (Thompson and Landstreet 1985; Bohlender <u>et al.</u> 1987) and are currently thought to represent high temperature extensions of the Ap stars. In collaboration with C. T. Bolton (U. of Toronto), we have obtained high S/N phase resolved spectra of several stars using the coudé reticon detector at CFHT. One of the principle goals of this work is to determine abundance and surface magnetic field geometries of several helium peculiar stars with large, well-determined effective fields. We employ a line synthesis program (Landstreet 1987) that incorporates the effects of surface magnetic fields and non-uniform abundances on the observed line profiles of a star. Since these stars are rapid rotators the surface magnetic field strength must be inferred from differential magnetic intensification of lines with different magnetic sensitivities. Of the few lines with suitable strengths in these hot stars we have decided that the Si III multiplet 2 lines are best suited for this aspect of our investigation. We have also modelled the unblended He I line $\lambda 4437$, ignoring magnetic effects for the time being. Individual results are discussed below.

2. HD 64740

This star's effective field ranges from -900 at $\phi=0.0$ to 500 G at $\phi=0.5$. We find helium to be overabundant in two spots: one near the negative pole with radius 70° and $\varepsilon_{He}=0.25$ and one at the opposite pole with radius 60° and $\varepsilon_{He}=0.20$. The silicon profiles are reproduced by a band of silicon between colatitudes 40° and 70° from the positive

---

[*]Visiting Astronomers, Canada-France-Hawaii Telescope operated by the National Research Council of Canada, the Centre National de la Recherche Scientifique of France and the University of Hawaii.

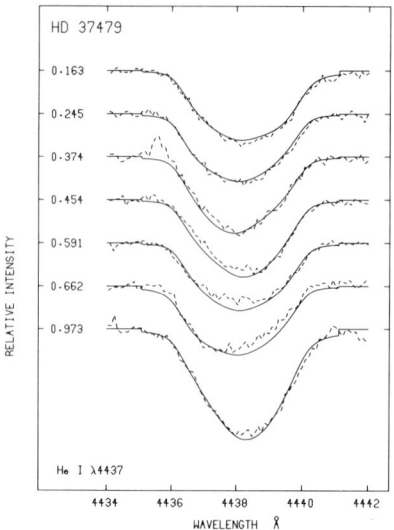
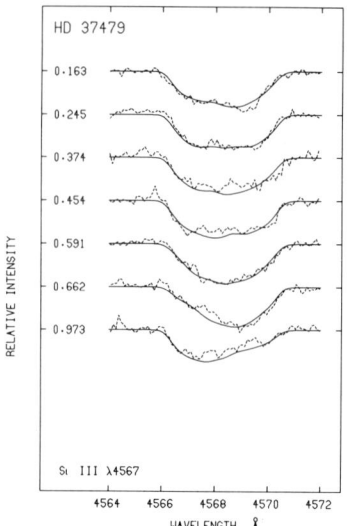

Figure 1. Observed (dashed line) and model (solid line) He and Si line profiles of HD 37479. We have employed a model atmosphere with T_{eff}=25000 K, i=58°, β=64° and a vsini of 150 km s^{-1}. The phases to the left of each spectrum are determined from the ephemeris 2442778.819 + 1.19081E. In each figure individual spectra are separated by 5% of the continuum intensity.

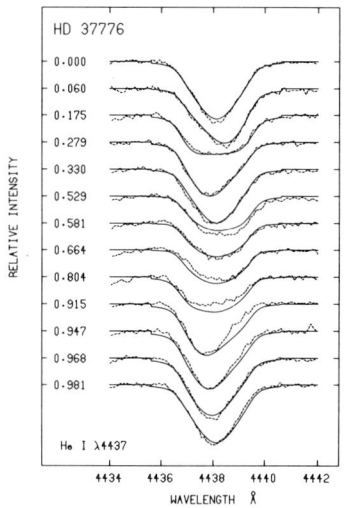
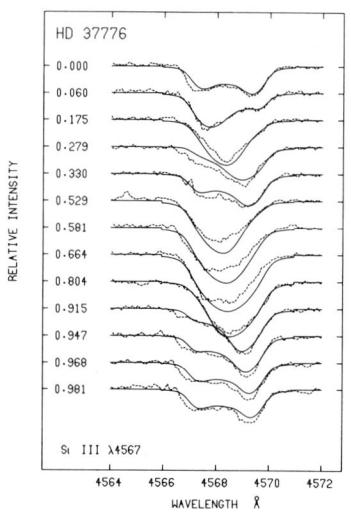

Figure 2. Observed and model He and Si line profiles of HD 37776. Here, T_{eff}=22500, i=β=90°, and vsini=100 km s^{-1}. Phases are given by the ephemeris 2445724.669 + 1.53869E.

pole with $\log(N_{Si})=-4.0$. The remainder of the surface has $\log(N_{Si})=-4.7$. Unfortunately, the magnetic field is not strong enough to enable us to uniquely define the surface field, and therefore the inclinations of the rotation and magnetic axes, i and β.

3. HD 37479 (σ Ori E)

The prototypical helium strong star has an effective magnetic field that varies from +2800 to -1500 G. Fits for the helium and silicon line profiles are illustrated in figure 1. Two patches of helium give a reasonable fit to the observed profiles: one with radius $40°$ and $\varepsilon_{He}=0.70$, $70°$ from the positive pole, and the other on the opposite side of the star with radius $30°$ and $\varepsilon_{He}=0.40$. Helium is slightly overabundant elsewhere. Silicon has solar abundance in a band between $30°$ and $90°$ from the negative magnetic pole and in a $50°$ radius spot $45°$ from the positive pole, but is underabundant elsewhere by a factor of ten.

4. HD 37776

This may well be the most interesting of the entire class of helium peculiar objects. Its unique magnetic field variation has been discussed by Thompson and Landstreet (1985), and can be reproduced with a superposition of colinear dipole, quadrupole, and octupole components. Model profiles for He I λ4437 and Si III are given in figure 2. These models do not include a magnetic field, but lend some support for the above field model in that the He and Si abundance geometries are approximately axisymmetric with respect to the location of the proposed magnetic pole (φ=0.175). The helium line variations are reproduced with a band of helium with width varying between $40°$ and $60°$ and with $\varepsilon_{He}=0.30$. The band crosses the sub-solar point at phases 0.0 and 0.330. Helium is slightly underabundant elsewhere, with $\varepsilon_{He}=0.07$. The preliminary silicon geometry consists of two patches of silicon centered at phases 0.175 and 0.675. In these spots $\log(N_{Si})=-2.7$. The abundance is normal on the rest of the surface.

5. Other Objects

High quality CFHT spectra have also been obtained for several other stars, including the helium weak star HD 175362, the helium variable HD 125823, and the peculiar Si star HD 32633. Analysis of these data is in a very early stage.

REFERENCES

Bohlender, D. A., Brown, D. N., Landstreet, J. D., and Thompson, I. B. 1987, to appear in Ap. J., 323, December 1.
Landstreet, J. D. 1987, paper submitted to Ap. J.
Thompson, I. B., and Landstreet, J. D. 1985, Ap. J. (Letters), 289, L9.

DISCUSSION

PRADERIE What model atmosphere and source function do you use for each of your lines at each phase ?

BOHLENDER We use Kurucz's (1979) LTE blanketed model atmospheres. We assume the lines are formed in LTE. At each phase the star is divided into 60 equal area elements, and for each area the local abundance and line-to-continuum opacity ratio is determined for each Zeeman component of the line. The local field strength determine the extent of the line splitting for each area element. A disk integration is then performed and a series of line profiles for the four Stokes parameters and for various v sin i is the final result.

SURFACE MAGNETIC FIELD MEASUREMENTS IN HOT CHEMICALLY PECULIAR STARS

Pierre DIDELON
Strasbourg Observatory
11, Rue de l'université
67000 STRASBOURG
FRANCE

ABSTRACT. The first results of magnetic field measurements are presented here for HD 187474, a slowly rotating Ap star. From resolved Zeeman pattern the strength of the field and its mean inclination were obtained. From differential magnetic broadening a second value of the field strength has been deduced, which is compatible with the previous one. The "Robinson" method has been tested and a good agreement is found between observed and calculated Zeeman broadening of FeII lines. This method can therefore certainly be used to measure the surface field in slow rotating chemically peculiar stars.

1. INTRODUCTION

The magnetic field present in some Chemically Peculiar (CP) stars, seems to play an important role in the phenomena observed in these stars (variations, diffusion,...). The knowledge of the surface field (Hs) is therefore of great interest.
    Hs is deduced from line splitting or broadening measured in unpolarized light on classical spectra. Measurements are difficult and the Zeeman pattern is often not resolved in stellar spectra. To go futher and measure Hs in more stars the differential magnetic broadening must be studied. The application of different methods to slow rotating magnetic star first, will show if they are consistent and reliable.
    I present here the results concerning the CP star HD 187474. It has a rotational period of 6.7 years and so is a very good candidate to perform the first tests.

2. OBSERVATIONS

High resolution spectra (R=100000) at high S/N ($\approx$300) were obtained with the CES and a Reticon at the ESO Coudé Auxiliary Telescope. The spectra cover a wavelength range of about 50Å. Five spectra of HD 187474 were obtained in October 1986, centered on the following wavelengths:
$\lambda\lambda$ 4505Å, 5020Å, 5295Å, 6240Å and 7400Å.

3. Hs MEASUREMENTS FROM RESOLVED ZEEMAN PATTERN

I have selected 16 "nice" Resolved Zeeman Pattern (RZP) to obtain a well suited Hs value from their splitting. 11 have been identified and the corresponding mean is Hs=5.1±0.6 KGauss (Didelon,1987).

The good quality of the data allows not only to measure the displacement of the components, but also to measure their individual intensities. These intensities are a function of the angle θ between the magnetic field axis and the line of sight (Gray,1984). The measurements of the equivalent width give: $I\pi/I\sigma=0.98\pm0.2$, which corresponds to $\theta=55°\pm5°$. This one of the rare cases where the direct determination of the magnetic field orientation was possible.

4. DIFFERENTIAL MAGNETIC BROADENING.

I selected in the spectra at $\lambda c=4505\text{Å}$ the strongest identified lines, which have the same intensities. Then the line widths must be realted to magnetic field broadening. I plotted the Full Width at Half Maximum (FWHM) of the lines versus their Landé factor z (Fig.1).

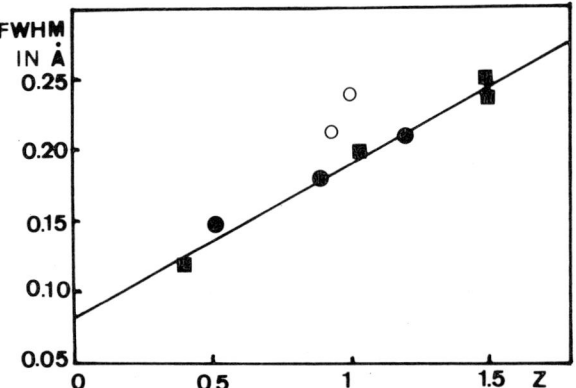

Figure 1. Magnetic broadening of lines

The FWHM seems to increase linearly with z. This is expected if Zeeman splitting is simply added to a mean profile observed at z=0. Empty circles represent lines which are certainly blended and so they have been excluded from the regression analysis. Fill squares give the position of the FeII(37) lines used in Robinson method. Note that they all lie near the mean relation. The straight line of fig.1 has been obtained by least square fit. Its slope gives a Hs value of 5.5 KGauss; which is compatible with that deduced from RZP.

5. TEST OF THE ROBINSON METHOD

Differential Magnetic Broadening (DMB) can be studied more precisely with the so-called "Robinson method", which has been applied only to cool stars(Sun et al,1987; Gray,1984). This method uses the division of the Fourier Transform (FT) of two lines with different magnetic sensitivity, which gives the Zeeman Broadening Function (ZBF). I have studied the DMB effects on FeII(37) lines. I used the line at λ4491Å (z=0.4) as unsensitive magnetic line of reference. The wavelengths and z values of the three other lines are respectively; λ4489Å,z=1.5; λ4515Å, z=1.0; λ4520Å,z=1.5. The division of the FT of one of these lines by the FT of the "unsensitive" reference line gives the observationnal ZBF. The values of the field strength and its inclination given by RZP are used to calculate the expected ZBF.

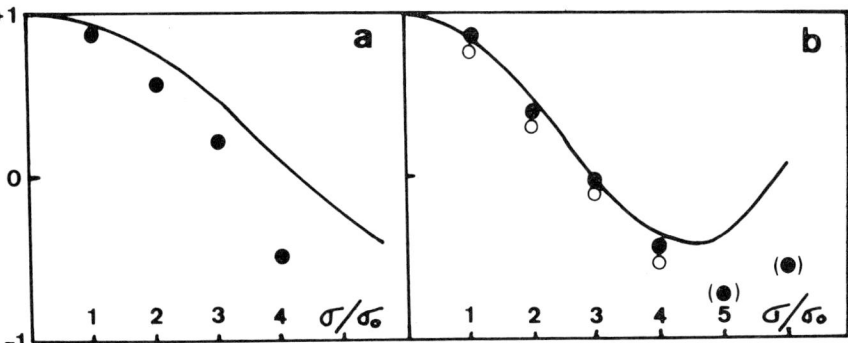

Fig.2: Zeeman broadening functions associated with different FeII Lines.

The figures 2a and 2b show the comparison of the observed and calculated ZBF. The full lines correspond to the calculated ZBF, the points give the observed ZBF. The ZBF obtained with the sensitive line λ4515Å is plotted in figure 2a. The calculated ZBF didnot fit the observed one, which is much stepper. This effect is due to additional broadening certainly due to a small undetected blend. This blend did not affect the FWHM of the line (see fig.1). In figure 2b i plotted together the ZBF obtained with the 2 other sensitives lines, which have the same z values (1.5). The observed ZBF of λ4520, respectively λ4489, is represented by full dots, respectivelly empty dots. The agreement between the calculated and the observed ZBF of these lines is satisfactory. Moreover the two observed ZBF have approximately the same values, which confirms the reliability of the data. The discrepancy at high frequency is due to noise contamination.

Finally, for the lines with z=1.5, a good agreement exists between observed and calculated ZBF. The Robinson method can therefore certainly be used to determine Hs, at least in slowly rotating CP stars. However several ZBF are necessary to get a field value with a good confidence level, to avoid undetected additionnal sources of broadening.

REFERENCES

Didelon, P.:1987, *The Messenger* n°49
Gray,D.F.:1984, *Astrophys.J.277,640*
Sun,W.H., Giampapa,M.S. and Worden,S.P.: 1987,*Astrophys.J.312,930*

DISCUSSION

MATHYS Have you performed, or do you intend to perform tests of the Robinson method at different phases of variation ? I do indeed think that the geometric aspect of the field may critically influence the success of the Robinson technique for the determination of the surface field of Ap stars.

DIDELON
- Yes I plan to pursue tests of the Robinson method, and follow some stars through their variation period.
- The geometrical aspect of the field is only a problem when mixed with the geometrical aspect of abundances patches. This effect is not so crucial in slow rotators which I observed for the moment but it will certainly limit the method in rapid rotators.

SPECTROPOLARIMETRY OF MAGNETIC STARS: HD 125248 *

G. Mathys
Geneva Observatory, ch. des Maillettes 51, CH-1290 Sauverny, Switzerland

J.O. Stenflo
Institute of Astronomy, ETH–Zentrum, CH–8092 Zurich, Switzerland

ABSTRACT. We present preliminary results about the magnetic field of the Ap star HD 125248, from spectra recorded in RCP and LCP light with the Zeeman analyzer of the CASPEC at ESO.

In this paper we present a preliminary report about the diagnostic contents of spectra of the A0p star CS Vir (= HD 125248) simultaneously recorded in right (RCP) and left (LCP) circularly polarized light with the Zeeman analyzer of the CASPEC at the ESO 3.6m telescope. The instrumentation and its performance have been described by Mathys and Stenflo (1986). The journal of observations is given in Table 1, where the phases are computed according to the ephemeris (Babcock, 1960) HJD (positive extremum of the longitudinal field) = 2 430 143.07 + 9.2954E. We have analyzed the set of Fe II lines listed in Table 2. The identification of the transitions is from Johansson (1978). In columns 3 and 4, the values of the Landé factors of the lower (g_l) and upper (g_u) levels are the experimental ones (Reader and Sugar, 1975) whenever possible; computed values (see Mathys and Stenflo, 1986) are quoted between parentheses. We employ the new parameterization of the line profiles introduced by Mathys (1987); the reader is referred to this paper for a detailed description of the notations. In this preliminary work, we make use of averages, either over the various phases (which we will denote by $\{\ldots\}_{av}$) or over the various lines (for which we use the notation $[\ldots]_{av}$) in order to reduce the scatter due to measurement errors, etc. in the relations that we try to evidence.

The average over the phases of the absolute value of the wavelength shift between the centres of gravity of the RCP and LCP lines, $\{|\langle\lambda_R\rangle - \langle\lambda_L\rangle|\}_{av}$ is expected for weak lines to be proportional to $\bar{g}\lambda_0^2\{|\langle H_z\rangle|\}_{av}$, where $\bar{g}$ is the effective Landé factor of the transition, λ_0 is its wavelength, and $\langle H_z\rangle$ is the longitudinal field, i.e. the line-intensity weighted average over the visible stellar hemisphere of the component of the magnetic field vector along the line of sight. Figure 1 shows a plot of $\{|\langle\lambda_R\rangle - \langle\lambda_L\rangle|\}_{av}$ vs. $\bar{g}\lambda_0^2$. The lines represented by squares nicely match the predicted linear dependence of $\{|\langle\lambda_R\rangle - \langle\lambda_L\rangle|\}_{av}$ on $\bar{g}\lambda_0^2$. The dashed line is a linear regression (forced through the origin) defined by these points. Its slope corresponds to $\{|\langle H_z\rangle|\}_{av} = (1449 \pm 39)$ G. The cross (×) in Fig. 1 represents the line λ 6446, which could not be measured in one of the polarizations on the spectrum taken on HJD 2 446 896.594 because of the superimposition of a radiation event ("cosmic ray"). The three lines represented by plus signs (+), $\lambda\lambda$ 5955, 6149 and 6416, definitely lie below the straight line defined by the other lines. These three lines happen to be the three strongest lines of the sample, and it is thus tempting to conclude that their behaviour in Fig. 1 reflects the fact that the proportionality breaks down for strong lines. However it should also be pointed out that: (i) these lines are also peculiar in their Zeeman pattern and (ii) they may, as well as λ 6175, suffer

---

* Based on observations collected at the European Southern Observatory, La Silla, Chile

Table 1. Spectra of HD 125248 recorded with the Zeeman analyzer of the CASPEC

Date of mid-exposure HJD 2 440 000.+	Duration of exposure (s)	Phase	$\langle H_z \rangle$ (G)
6219.581	300	0.513	-1816 ± 207
6547.621	360	0.803	980 ± 202
6548.601	420	0.909	1647 ± 128
6549.614	390	0.018	1952 ± 136
6894.559	330	0.127	1933 ± 179
6894.901	360	0.164	2080 ± 117
6895.557	420	0.234	1273 ± 112
6895.896	600	0.271	1004 ± 148
6896.594	900	0.346	-417 ± 258
6896.901	480	0.379	-1007 ± 193
6897.680	480	0.463	-1382 ± 166
6897.891	420	0.485	-1830 ± 177

Table 2. Sample of Fe II lines used for the study of the magnetic field

λ_0 (Å)	Transition	g_l	g_u
5952.525	$3d^7 \, d\,^2D_{5/2} - (a\,^3P)\,4p\,z\,^2P^\circ_{3/2}$	(1.200)	1.329
5955.700	$(^5D)\,4d\,e\,^4F_{5/2} - (^5D_3)\,4f\,^2[3]^\circ_{7/2}$	(1.029)	(1.357)
5961.706	$(^5D)\,4d\,e\,^4F_{9/2} - (^5D_4)\,4f\,^2[6]^\circ_{11/2}$	1.29	(1.244)
5991.368	$(^3G)\,4s\,a\,^4G_{11/2} - (^5D)\,4p\,z\,^6F^\circ_{9/2}$	1.237	1.43
6060.991	$(a\,^3F)\,4p\,x\,^4D^\circ_{7/2} - (^5D)\,5s\,e\,^4D_{7/2}$	1.385	(1.429)
6084.099	$(^3G)\,4s\,a\,^4G_{9/2} - (^5D)\,4p\,z\,^6F^\circ_{7/2}$	1.15	1.399
6149.246	$(^3D)\,4s\,b\,^4D_{1/2} - (^5D)\,4p\,z\,^4P^\circ_{1/2}$	(0.000)	2.70
6175.138	$(b\,^3F)\,4s\,c\,^4F_{7/2} - (^3G)\,3p\,x\,^4F^\circ_{7/2}$	(1.238)	1.21
6383.721	$(^5D)\,4p\,z\,^4D^\circ_{5/2} - 4s^2\,c\,^4D_{5/2}$	1.35	(1.371)
6416.921	$(^3D)\,4s\,b\,^4D_{5/2} - (^5D)\,4p\,z\,^4P^\circ_{5/2}$	1.327	1.592
6442.951	$(^5D)\,4p\,z\,^4F^\circ_{7/2} - 4s^2\,c\,^4D_{7/2}$	1.29	(1.429)
6446.402	$(b\,^3F)\,4s\,c\,^4F_{7/2} - (^3G)\,4p\,x\,^4G^\circ_{9/2}$	(1.238)	(1.172)

from unrecognized blends. Therefore it seems safer to determine the longitudinal field using only the "weak" lines. Thus the values of $\langle H_z \rangle$ given in table 1 were derived using the lines of table 2 except for $\lambda\lambda$ 5955, 6149, and 6416. (At phase 0.346, λ 6446 was not employed either.)

For weak lines and a strong randomly oriented field, the second order moment of the intensity profile about the line centre λ_0, $R_I^{(2)}(\lambda_0)$, is expected to be in a first approximation proportional to $\left(\frac{2}{3} C_2^{(-1)} + \frac{1}{3} C_2^{(0)}\right) \lambda_0^4 \langle H^2 \rangle$, where $C_2^{(-1)}$ and $C_2^{(0)}$ are atomic parameters describing the Zeeman pattern of the transition (Mathys and Stenflo, 1987a, b), and $\langle H^2 \rangle$ is the line-intensity weighted quadratic average over the stellar disk of the modulus of the magnetic field. The field is probably not randomly oriented at a given phase, but when averaging over the phases, this should be a good approximation. A plot of $\{R_I^{(2)}(\lambda_0)\}_{av}$ vs. $\left(\frac{2}{3} C_2^{(-1)} + \frac{1}{3} C_2^{(0)}\right) \lambda_0^4$ is shown in Fig. 2. Except for $\lambda\lambda$ 5955 and 6175, a quite nice correlation appears. Though this may be purely coincidental (in particular in view of the arbitrariness involved in the choice of a random field geometry), we performed a linear regression of $R_I^{(2)}(\lambda_0)$ as a function of $\left(\frac{2}{3} C_2^{(-1)} + \frac{1}{3} C_2^{(0)}\right) \lambda_0^4$, for all the lines but $\lambda\lambda$ 5955 and 6175. The result is the dashed line of Fig. 2. The slope corresponds to $\{\langle H^2 \rangle\}_{av} = 4.98\ 10^7\ \text{G}^2$, or $\sqrt{\{\langle H^2 \rangle\}_{av}} \approx 7\,\text{kG}$ (which implies $\sqrt{\{\langle H_z^2 \rangle\}_{av}} \approx 4\,\text{kG}$ for an isotropic distribution of field vectors), which are quite plausible values.

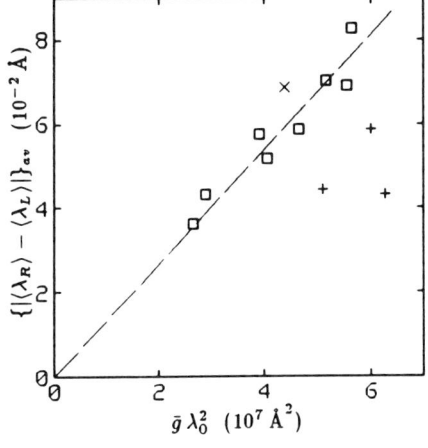 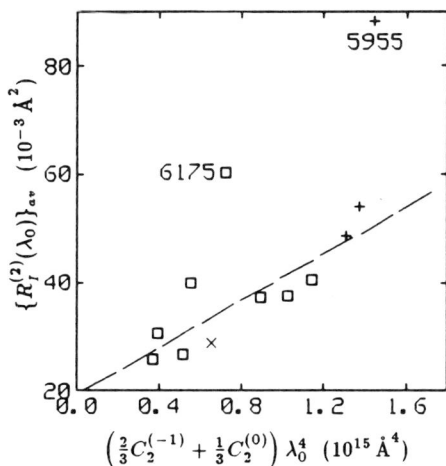

Fig. 1. $\{|\langle\lambda_R\rangle - \langle\lambda_L\rangle|\}_{av}$ vs. $\bar{g}\lambda_0^2$

Fig. 2. $\{R_I^{(2)}(\lambda_0)\}_{av}$ vs. $\left(\frac{2}{3}C_2^{(-1)} + \frac{1}{3}C_2^{(0)}\right)\lambda_0^4$

The difference between the second order moments of a line in the RCP and LCP spectra about the corresponding centre of gravity ($\langle\lambda_R\rangle$ and $\langle\lambda_L\rangle$), $\Delta R^{(2)} \equiv R_R^{(2)}(\langle\lambda_R\rangle) - R_L^{(2)}(\langle\lambda_L\rangle)$, provides a quantitative measure of the crossover effect (the oppositely polarized line components have different widths, Babcock, 1956). For weak lines, $\Delta R^{(2)}$ is expected to be proportional to $\bar{g}\, v_e \sin i\, \lambda_0^3 \langle xH_z\rangle$, where $v_e \sin i$ is as usual the projected equatorial velocity and $\langle xH_z\rangle$ is the line-intensity weighted first order moment of the line-of-sight component of the field about the plane defined by the line of sight and the rotation axis of the star (x is in units of the stellar radius). Excluding the lines $\lambda\lambda$ 5955, 6149 and 6416, and λ 6446, for the reasons given above, we have performed a linear regression, forced through the origin, of $\{|\Delta R^{(2)}|\}_{av}$ vs. $\bar{g}\lambda_0^3$. If we adopt a value of $10\,\mathrm{km\,s^{-1}}$ for $v_e \sin i$, the slope of the regression line corresponds to an average (over the phases) absolute value of the first order moment (about the plane defined by the line of sight and the stellar rotation axis) of the line-of-sight component of the field $\{|\langle xH_z\rangle|\}_{av} = (640 \pm 50)$ G, not unreasonable.

Finally we have computed the value of $\Delta R^{(2)}$ averaged over the line sample, $[\Delta R^{(2)}]_{av}$, at the various rotation phases. Only those lines that were included in the above regression are used, but the shape of the variation curve of $[\Delta R^{(2)}]_{av}$ is not significantly different when the whole sample is considered. A smooth variation appears, which is compatible with the previously reported observation that the crossover effect is maximum near phases 0.4 and 0.7.

REFERENCES

Babcock, H.W.: 1956, *Astrophys. J.* **124**, 489
Babcock, H.W.: 1960, in *Stellar Atmospheres*, ed. J. Greenstein, University of Chicago Press, Chicago, p. 282
Johansson, S.: 1978, *Phys. Scr.* **18**, 217
Mathys, G.: 1987, *Astron. Astrophys.* (in press)
Mathys, G., Stenflo, J.O.: 1986, *Astron. Astrophys.* **168**, 184
Mathys, G., Stenflo, J.O.: 1987a, *Astron. Astrophys.* **171**, 368
Mathys, G., Stenflo, J.O.: 1987b, *Astron. Astrophys. Suppl.* **67**, 557
Reader, J., Sugar, J.: 1975, *J. Phys. Chem. Ref. Data* **4**, 353

DISCUSSION

SAAR Blends will cause systematically different effects on line cores and wings, so that even for many lines and a random distribution of blends, there will be systematic effects. Do you have any preliminary idea of how blends may affect your measurements ?

MATHYS We have not looked at this point in detail yet. I believe that the effect is less critical than in a two-line method, such as Robinson's, but you may actually know more than I do about this problem, since you have studied it more completely.

BASRI Do you determine the fitting parameters separately for each spectral type, or just use the solar values ?

Remark following : As you have already mentioned, you can profitably use the profile information you have to do a more detailed physical analysis later, and perhaps assign physical meaning to your empirical parameters.

MATHYS We used the same form for the regression equation for all stars as well as for the Sun, but perform the calculation of the regression coefficients for each star individually. We have also tested different regression equations, but either they hardly affect our conclusions about the magnetic field, or they yield a much larger scatter about the fit and can thus be discarded.

RUTTEN In your regression analysis, you also obtain solutions for non-magnetic line formation parameters, e.g. excitation energy, wavelength, Doppler width. I think it is necessary to interpret these dependences in detail to be certain they don't backfire on the magnetic parameters. Do you understand the behaviour of these other parameters in detail ?

MATHYS Since we derive _empirically_ the dependences that we introduce in our regression equation, we do not interpret them in detail. However, the dependences that we introduce seem physically reasonable and the conclusions that we draw about the magnetic field are not very sensitive to the exact form of the regression equation that we use.

THE MAGNETIC FIELD GEOMETRY AND ABUNDANCE DISTRIBUTIONS OF IRON-PEAK ELEMENTS IN THE AP STAR 53 CAMELOPARDALIS

J.D. Landstreet[+]
Department of Astronomy
University of Western Ontario
London, Ontario   N6A 3K7
Canada

ABSTRACT. Spectra and magnetic measurements of the strongly magnetic Ap spectrum variable 53 Cam have been modelled using a line synthesis programme that calculates LTE line profiles, including magnetic and polarization effects and non-uniform distribution of elements over the star. The magnetic field is found to be reasonably well modelled with colinear dipole, quadrupole and octupole components of strengths (at the strong negative magnetic pole) of -16.3, -7.3, and +4.9 kG respectively. Chromium and iron are almost uniformly distributed with abundances of about 10^2 and 10 times solar. Ti and Ca, in contrast, are quite non-uniformly distributed; Ti is more than 10 times overabundant near the negative pole and probably underabundant elsewhere, while Ca is overabundant around the positive pole and slightly underabundant elsewhere.

1. INTRODUCTION

Magnetic Ap stars often exhibit variations in observed magnetic field and in the strengths of spectral lines. The magnetic variations are usually interpreted as being produced by rotation of the star, which has a strong, roughly dipolar magnetic field inclined at some angle to the rotation axis; the observed variations are simply due to changing aspect. Similarly, the spectrum variations are believed to be due to non-uniform abundance distributions over the stellar surface, with the observed changes again due to changing aspect as the star rotates.

The surface abundance variations are widely suspected to be produced in some way by diffusion of trace elements under the competing influences of gravity and radiation pressure, with some additional effect due to the magnetic field that leads to variations over the

---

[+] Visiting Astronomer, Canada-France-Hawaii Telescope operated by the National Research Council of Canada, the Centre National de la Recherche Scientifique of France and the University of Hawaii.

surface. To test theories of how diffusion may be modified by a strong magnetic field, it is important to have good maps of both the magnetic field structure and of the distribution of chemical abundances with respect to this field structure for a few stars. At present there are no stars for which good maps of both abundances and of the field structure are available.

The star 53 Cam is a well-known Ap star with a large magnetic field (+4 kG $\gtrsim B_e \gtrsim$ -5 kG, 17 kG $\gtrsim B_s \gtrsim$ 10 kG) and strong variations of lines of Ca, Eu, and Ti (Babcock 1958; Preston 1969; Huchra 1972; Faraggiana 1973; Borra and Landstreet 1977). Both the inclination i of the rotation axis to the line of sight and the obliquity β of the magnetic axis to the rotation axis are large, so the observer sees nearly the whole star in one rotation period. 53 Cam is thus well suited to the kind of mapping that is required, and so a series of high resolution (0.1Å), high signal-to-noise (~300) spectra of the star, spaced at about $\Delta\phi$ = 0.12 in phase through the 8.03 day rotation period, were obtained using the coudé Reticon system at the Canada-France-Hawaii telescope. These data, together with Balmer line Zeeman analyzer measurements of the longitudinal field previously obtained, have been modelled to obtain information about the magnetic field geometry and abundance distribution of Ca, Cr, Fe, Sr, and Ti over the surface of the star, using a new line synthesis programme. This programme calculates accurate LTE intensity (and polarization) profiles of spectral lines, taking into account the magnetic splitting and radiative transfer effects for the full Stokes vector, as well as allowing a non-uniform abundance distribution over the stellar surface. This programme is used to model in the "forward" direction; an initial magnetic field geometry and simple abundance distribution is assumed, profiles of various spectral lines are calculated and compared with observations, and the assumed models are modified until reasonable agreement is obtained. In practice, this method seems to converge to satisfactory models.

2. RESULTS FOR 53 CAM

For 53 Cam, this modelling has been carried out using visual inspection of results, and guesses to improve the models. Such a procedure can only cope with a small number of parameters, and so the magnetic field has been limited to colinear dipole, quadrupole, and octupole components of adjustable polar strength and obliquity. The inclination and obliquity (which are interchangeable) have values of about 64° and 82°; the polar strengths of the field components are -16300, -7500, and +4900G at the strong negative pole. The dipole component provides the large observed effective field; the quadrupole introduces an asymmetry such that the field strength around the negative pole is considerably larger than around the positive pole; and the octupole increases the magnitude of the field strength around the magnetic equator relative to that at the poles.

Similarly, a simple few-parameter model of the abundance distribution has been adopted. After considerable experimentation, it was found that all line profile variations could be approximately reproduced assuming three zones, each of constant abundance: a cap extending 60° from each magnetic pole, and a belt, also 60° wide, around the magnetic equator. For each element, the abundance of each of the three zones is adjusted to obtain best agreement with observation.

This simple abundance distribution geometry gives a good model of Cr and Fe lines. Both of these elements are found to be slightly more abundant (about +0.4 dex) in the polar caps than in the equatorial belt; the polar abundance of Cr is about 2.0 dex above solar, while Fe is about 1.0 dex overabundant.

Sr is represented in my spectra by only one, somewhat blended, line, but seems to have an overabundance of about +3.4 dex around the negative pole, and an overabundance of about +2.8 dex elsewhere.

Ca and Ti are still more non-uniform in abundance. Ti seems to be overabundant by about 1.4 dex in the cap around the negative pole, but underabundant by an uncertain factor elsewhere. Ca is apparently rather overabundant around the positive magnetic pole and somewhat underabundant elsewhere. (Models of Ti and Ca are less well-determined than those for Cr, Fe, and Sr because at some phases the observed lines are weak enough to be dominated by rare earth blends or to vanish.)

A paper describing these results in more detail has been submitted to the Astrophysical Journal.

ACKNOWLEDGEMENTS: This work has benefitted from discussions with Drs. Baschek, Cowley, Fuhr, Michaud, Stenflo, and Wehrse. Financial support from the National Sciences and Engineering Research Council of Canada and from the Deutsche Forschungsgemeinschaft is gratefully acknowledged.

REFERENCES

Babcock, H.W. 1958, *Ap. J. Suppl.* **3**. 141.
Borra, E.F. and Landstreet, J.D. 1977, *Ap. J.*, **212**, 141.
Faraggiana, R. 1973, *Astr. Ap.*, **22**, 265.
Huchra, J. 1972, *Ap. J.*, **174**, 435.
Preston, G.W. 1969, *Ap. J.*, **157**, 247.

DISCUSSION

MEGESSIER What are the constraints on the determination of the geometry of the elements overabundances according to your procedure ? How is it correlated to the field geometry ?

LANDSTREET The abundance distribution is assumed to be symmetric around the <u>magnetic</u> axis. The star is simply divided into a small number of zones of constant abundance (caps or rings); the abundance in each zone is adjusted to give the best agreement with observed spectral line profiles. This simple model gives a reasonably good description of Cr and Fe, which are roughly uniform in abundance with slightly (x2) higher abundances at both magnetic poles than around the equator. Ti is not as well modelled, but appears to be strongly concentrated around the negative magnetic pole and depleted elsewhere on the star.

MAGNETIC FIELD OF LATE-TYPE STARS: A NEW APPROACH *

G. Mathys
Geneva Observatory, ch. des Maillettes 51, CH-1290 Sauverny, Switzerland

S.K. Solanki
Institute of Astronomy, ETH-Zentrum, CH-8092 Zurich, Switzerland

ABSTRACT. Magnetic field diagnosis in three late-type stars is performed using the multiline technique originally developed by Stenflo and Lindegren (1977) to study spatially unresolved magnetic features on the sun.

We present in this paper a preliminary report of the application of a new method to the diagnosis of magnetic fields in late-type stars: the multiline technique first developed by Stenflo and Lindegren (1977) for the study of the spatially unresolved structure of solar magnetic fields. In this approach, the profiles of a large number of simultaneously recorded spectral lines are characterized by some simple parameters whose mutual correlations and dependences on atomic parameters such as the excitation potential of the lower level of the transition or the effective Landé factor are empirically determined through a regression analysis.

The observations have been performed with the ESO Coudé Echelle Spectrometer fed by the Coudé Auxiliary Telescope. The long camera was used with the RETICON to achieve a spectral resolution of 100 000. Each star was consecutively observed in several wavelength ranges over the same night, so that a sufficient number of lines are available to apply the statistical analysis sketched above. For each star, we study the largest possible sample of apparently unblended Fe I lines. Details about the observations are given in Table 1. Columns 1 and 2 give the star's name and HR number, and Col. 3 the V magnitude. The duration of observation listed in Col. 5 refers to the interval of time elapsed between the beginning of the exposure of the first wavelength range and the end of the exposure of the last wavelength range, and the date of the observation in Col. 4 corresponds to the middle of that time interval. In Col. 6 we give an estimate of the achieved S/N, for the wavelength domains with the lowest and the highest quality recording.

Table 1. Observations of late-type stars with the CES at the CAT

Name	HR	V	Sp.	Date HJD 2 440 000.+	Duration (hours)	S/N
τ Cet	509	3.50	G8V	6718.70	8.4	500–700
ϵ Eri	1084	3.73	K2V	6722.76	4.7	250–400
40 Eri A	1325	4.42	K1V	6723.75	5.8	300–400

The parameters characterizing the line profiles are (Stenflo and Lindegren, 1977): (i) the line depth d, (ii) the chord length $v_D(z)$ at level zd ($0 < z < 1$) above the line bottom (expressed in

---

* Based on observations collected at the European Southern Observatory, La Silla, Chile

velocity units in terms of the formal Doppler width of a Gaussian profile that has the same width as the considered line at the given chord level), and (iii) the area S of the line below the half level chord (a line strength parameter, less sensitive to blends than the equivalent width).

The following regression equation proves to be suitable for the line width at any level in the line:

$$v_D(z) = x_0 + x_1 v_m^2 \lambda^2/v_0 + x_2 \langle v_m^2 \rangle \lambda^2/v_0 + x_3 S + x_4 S^2 + x_5 \chi_e v_0, \tag{1}$$

where $x_0, \ldots, x_5$ are the regression coefficients, λ is the line wavelength, χ_e is the excitation potential of the lower level of the transition, $v_0 = y_0 + y_1 S^2$ is an approximation of v_D (see Stenflo and Lindegren, 1977, for more details), and

$$v_m^2 = (\bar{g}^2 + X_\sigma) \frac{1 + \cos^2 \gamma}{2} + X_\pi \frac{\sin^2 \gamma}{2} \tag{2}$$

for a magnetic field at an angle γ to the line of sight (Mathys and Stenflo, 1987a). $\bar{g}$ is the effective Landé factor of the transition, and X_π and X_σ are the second order moments about their centre of gravity of the π and the σ_+ components of the Zeeman pattern, respectively, expressed in units of the Zeeman splitting of a normal triplet having a Landé factor equal to 1 (Landi Degl'Innocenti, 1982, 1985; Mathys and Stenflo, 1987a, b). The regression coefficient x_1 is related to the magnetic field strength B through:

$$x_1 = k^2 c^2 \alpha \delta_c \delta_l B^2, \tag{3}$$

where $k = 4.67\ 10^{-13}\ \text{Å}^{-1} \text{G}^{-1}$, c is the velocity of light, α is the filling factor, δ_c is the continuum contrast of the magnetic regions relative to the nonmagnetic regions, and δ_l is the average ratio of line strength in magnetic regions to that in nonmagnetic regions.

The regression equation for the line depth d is similar to Eq. (1). However tests have shown that the line width regression is a better diagnostic for magnetic field properties. One reason for this is that the noise is greater in the line cores, which are formed at different levels in the atmosphere, than in the line flanks, which are all formed at nearly the same height.

From Eq. (3), one sees that the relation between x_1 and B involves three other a priori unknown quantities, α, δ_c and δ_l. Let us now discuss to which extent these quantities and the field strength can be disentangled using the Stenflo-Lindegren technique. For a filling factor less than 1, the part of the line nearer the continuum is more strongly affected by the magnetic field than the line core, thus e.g. the ratio $x_1(z = 0.7)/x_1(z = 0.5)$ is a measure of the product $\alpha \delta_c \delta_l$ as has been illustrated in the case of the sun by Brandt and Solanki (1987). On the other hand, the multiline approach also provides the possibility of deciding whether the magnetic fields are concentrated in hot or in cool regions of the stellar surface, i.e. in stellar plages or spots. Indeed, in the former case, the low excitation lines are weakened with respect to the high excitation lines in the magnetic regions compared with the nonmagnetic regions. The opposite will be the case if the magnetic fields are concentrated in cool elements. Thus, sorting the lines into a high excitation and a low excitation group, δ_l will be different for these two groups. Since the product $\alpha \delta_c B^2$ is approximately the same for both groups, one has:

$$\frac{x_1^{\text{high}}}{x_1^{\text{low}}} = \frac{\delta_l^{\text{high}}}{\delta_l^{\text{low}}}. \tag{4}$$

Hence performing the regression for both groups of lines separately, the difference in the obtained coefficient x_1 is ideally a measure of the difference in temperature between the magnetic elements and their nonmagnetic surroundings (provided that the S/N and statistics are good enough). On the other hand, it has recently been shown in the case of the sun (Grossmann-Doerth et al., 1987; Schüssler and Solanki, 1987) that in the absence of resolved magnetic features, no method exists (either polarimetric or not) that permits the determination of α directly: only $\alpha \delta_c$ is defined.

The results that we obtain are summarized in table 2. Column 3 gives the number of Fe I lines used in the analysis. The quantities listed in Cols. 4 to 7 were derived assuming a mean

Table 2. Derived magnetic field values

Name	HR	No.of lines	$\sqrt{\alpha\delta_c\delta_l}\,B$ (G)	$\dfrac{\sqrt{\alpha\delta_c\delta_l}\,B}{\sigma(\sqrt{\alpha\delta_c\delta_l}\,B)}$	$\alpha\delta_c\delta_l$	B (G)
τ Cet	509	65	220	1.2		
ϵ Eri	1084	45	800	4.6	0.10–0.20	1790–2530
40 Eri A	1325	44	460	2.6	$\gtrsim 0.15$	$\leqslant 1190$

angle between the line of sight and the magnetic field $\gamma = 34°$ in Eq. (2). Note that in principle the magnetic term of the regression equation (1) could be split into two and γ could be diagnosed independently, taking advantage of the presence of various Zeeman patterns in the line sample. In HR 509, there is only marginal evidence of detection of a magnetic field. For the other two stars, HR 1084 and HR 1325, for which the presence of a magnetic field is more certain, we also set some constraints on the value of the product $\alpha\delta_c\delta_l$ and on the related value of B. Finally, in HR 1084, sorting the lines into two categories according to the excitation potential ($\chi_e^{\text{low}} \leq 3$ eV, $\chi_e^{\text{high}} \geq 4$ eV), we obtain $\delta_l^{\text{high}}/\delta_l^{\text{low}} \approx 2.0$, indicating that the observed magnetic field is concentrated in regions that are hotter than the nonmagnetic part of the atmosphere. These various results are compatible with those obtained previously by other techniques (Marcy, 1984; Gray, 1984; Saar, 1987). This is discussed in more detail by Solanki and Mathys (1987).

To summarize, we have presented here a preliminary account of the application of the Stenflo–Lindegren technique to the diagnosis of the magnetic field of late-type stars. The advantages of this method over the previously used ones are its relative unsensitivity to line blends and its ability to set constraints on the temperature difference between magnetic and nonmagnetic regions of the atmosphere. For the future, the following developments of the present work are envisaged:
(i) To compare the results that we obtain with those of other methods, applied to the same data.
(ii) To investigate possible systematic effects of the line blends on the results.
(iii) To study the influence of the time spread of the observations (note in this respect that since the various wavelength ranges were not recorded in order of increasing or decreasing wavelength or effective Landé factor, rapid changes in $\sqrt{\alpha\delta_c\delta_l}\,B$ essentially increase the scatter about the regression).
(iv) To try to set constraints on the magnetic geometry.
(v) To observe other stars, in particular in the southern hemisphere where no systematic survey has up to now be performed.

References

Brandt, P.N., Solanki, S.K.: 1987, *Astron. Astrophys.* (to be submitted)
Gray, D.F.: 1984, *Astrophys. J.* **277**, 640
Grossmann-Doerth, U., Pahlke, K.-D., Schüssler, M.: 1987, *Astron. Astrophys.* **176**, 139
Landi Degl'Innocenti, E.: 1982, *Solar Phys.* **77**, 285
Landi Degl'Innocenti, E.: 1985, *Solar Phys.* **99**, 1
Marcy, G.W.: 1984, *Astrophys. J.* **276**, 286
Mathys, G., Stenflo, J.O.: 1987a, *Astron. Astrophys.* **171**, 368
Mathys, G., Stenflo, J.O.: 1987b, *Astron. Astrophys. Suppl.* **67**, 557
Saar, S.H.: 1987, Ph. D. Thesis, JILA, Boulder
Schüssler, M., Solanki, S.K.: 1987, *Astron. Astrophys.* (in press)
Solanki, S.K., Mathys, G.: 1987, in *Proceedings of the Midnight Sun Conference on Activity in Cool Star Envelopes*, Tromsø, Norway, in press

DISCUSSION

NISSEN How faint magnitudes can you reach with CASPEC using this technique ?

MATHYS You can in principle go as faint with the Zeeman analyzer as in the standard mode of the CASPEC, expect for a factor of about two (as a result of the splitting of the beam into two polarizations). That is, for a S/N of 100 or higher, it should be possible to reach magnitude 13. Our main limitation comes rather from the mild resolving power of the CASPEC.

MAGNETIC FIELD AND SILICON DIFFUSION IN Bp-Si STARS

C.Mégessier[1], T.Lanz[2], J.D. Landstreet[3]
[1]Observatoire de Paris-Meudon, place J.Janssen
 92195 Meudon France
[2]Institut d'Astronomie de l'Université de Lausanne
 CH-1290 Chavannes-des-Bois, Switzerland
[3]Department of Astronomy University of Western Ontario
 London, Ontario, Canada N6A 3K7

ABSTRACT. SiII lines of magnetic Bp-Si stars in open clusters have been observed with the CAT (ESO) in order to get a mapping of the Silicon abundance distribution over the stellar surface, in the frame of the oblique rotator model. We point out the influence of the Zeeman splitting and of the abundance inhomogeneities on the line profiles.

1. THE ASTROPHYSICAL CONTEXT

The main sequence Bp-Ap Si stars are remarkable by their strong magnetic field (some kilo-Gauss) and their silicon overabundance which is not constant over the stellar surface. Theoretical works on the diffusion in the stellar atmosphere, in the presence of a magnetic field, predicts that the amount of the silicon overabundance is tied to the magnetic field geometry (Michaud et al.,1981). Moreover the silicon overabundance is expected to move from the magnetic equator to the magnetic pole in times shorter or comparable to the main sequence stage of these stars. Accordingly the geometry of the silicon overabundance would depend on the stellar age (Mégessier,1984).

Observed silicon lines, effective magnetic field and photometric variations occur with the same period, which is the rotational period of the star. These variations arise from the inclination of the magnetic axis to the rotational one, and from the inhomogeneous silicon distribution over the stellar surface.

To get a mapping of the silicon over the stellar surface it is necessary to determine the abundance at several phases distributed over the stellar rotational period. However the Zeeman splitting due to the strong magnetic field affects the line profiles. This effect has to be taken into account to get a correct value of the abundance. Accordingly spectra with a high signal-to-noise ratio are required to enable us to get the shape of the lines.

2. THE OBSERVATIONS

We selected stars in open clusters so that their age is known. Three Bp-Si stars, the magnetic field of which was measured by Landstreet, has been observed in March 1987 with the CES at the CAT (ESO). We chose the SiII multiplet λ6347-6371 A which is in the region of high sensitivity of the receptor. Simultaneous photometric observations have been performed in the Geneva photometric system to determine the stellar rotational period and the phase of the spectroscopic and magnetic data.

Here we present the preliminary results for HD 92664 which has a moderate magnetic field (-1200≤He≤-200 Gauss) and which $v_e \sin i$ is 66 km/s and for HD 147010 which magnetic field is very strong (-5800≤He≤-3200 Gauss) but which $v_e \sin i$ is small (less than 20 km/s or even less than 10 km/s).

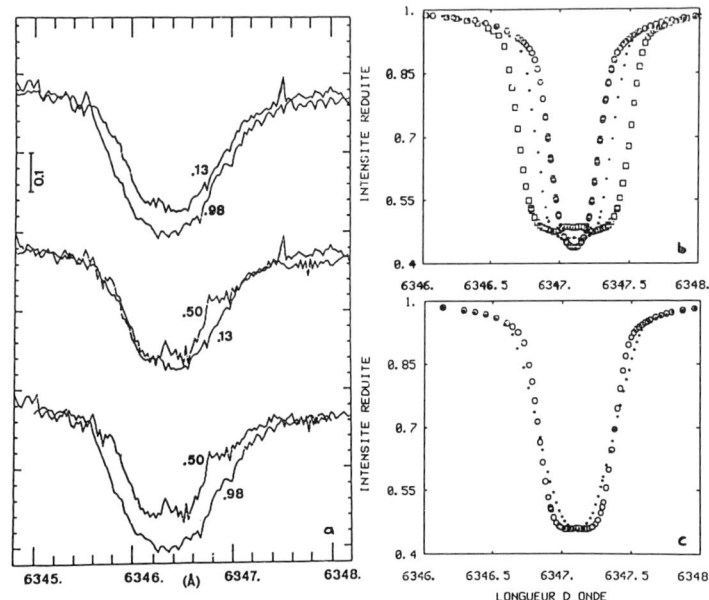

Figure 1 a) Variations of the SiII line λ6347 (mult. 2) for HD 147010. The spectra are normalized to unity. The phase is indicated near each spectrum. The SiII line has a remarkable box shape. b) LTE spectrum synthesis of the λ6437 SiII line for $v_e \sin i$=5 km/s and for three values of the mean local magnetic field 3kGauss(o), 6kGauss(.) and 10kGauss(□).

c) When convoluted by a rotational profile the line profile has no more a box shape. Open circles: Be=6 kGauss and $v_e \sin i$=5 km/s, dots :Be=6 kGauss and $v_e \sin i$=10 km/s

3. MAGNETIC FIELD INTENSITY AND ROTATIONAL VELOCITY EFFECT ON THE LINE PROFILES.

SiII lines have been calculated with a Kurucz model (T_{eff}=11700°K, log g=4, metal=10x☉). The Zeeman pattern of the transition was taken into account. Three values of the mean local magnetic field : 3, 6 and 10 kilo-

Gauss, have been considered. Then the spectra have been convoluted by a rotational profile. In case of a small rotational velocity (5 km/s) (Fig 1b) the line becomes square shaped and with a 10 kilo-Gauss field it is splitted into two components. The equivalent width of the line increases with the field intensity. The SiII line is saturated but a large field allows a desaturation via the Zeeman splitting. With a somewhat larger rotational velocity (10 km/s) (Fig 1c) the line is no more square shaped but its width is larger than in the absence of a field.

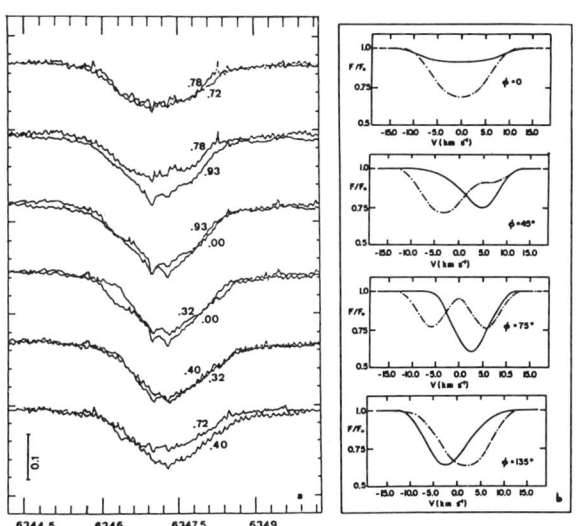

Figure 2 a) Variations of the SiII line λ6347 for HD 92664. The line shapes are dominated by a rotational profile, and not by the Zeeman splitting. b) Predicted variations and asymetries of the lines (from Michaud et al.,1981) for a concentration of elements at the magnetic poles (point dash line) or in a ring (full line)

4. THE OBSERVED LINE PROFILES AND THEIR VARIATIONS.

The profile of the SiII lines in the spectra of HD 147010 has the box shape which arises from the Zeeman splitting (Fig 1a). It also implies a slow rotational velocity (≤10 km/s). The main changes in the line profiles at different phases affect the line width rather than its symmetry. These spectral variations originate then from different sights over the magnetic field geometry during the rotational period.

In the case of HD 92664 (Fig 2a) the line shapes are dominated by a rotational profile, and not by the Zeeman splitting. Asymmetries are clearly present at several phases. So the variations are primilary due to the inhomogeneous distribution of the Silicon. Fig 2b shows the predicted variations and asymmetries (from Michaud et al.,1981) for a concentration of elements at the magnetic poles (point dash line) or in a ring (full line) where the field is horizontal (around the magnetic equator for a pure dipolar field). A comparison between these theoretical predictions and the observed variations suggests that Silicon is concentrated near the magnetic equator of HD 92664.

5. CONCLUSION

Line profiles allow to determine the Silicon overabundance geometry on the stellar surface better than the only line intensities. A spectral resolution as high as possible is required to separate the geometric effects from the magnetic field effects on the lines. We could point out these two effects in our spectra, although their signal-to-noise ratio are rather limited.This limitation arise from the necessary temporal resolution compared to the rotational period.

Both high spectral resolution and high signal-to-noise ratio are necessary to map these stars and then better understood the link between the magnetic field and the local overabundances.

REFERENCES.

Mégessier,C.:1984,Astron.Astrophys. 138,267
Michaud,G.,Mégessier,C.,Charland,Y.:1981,Astron.Astrophys.103,244

DISCUSSION

ANDERSEN   I should like to see theoretical diffusion calculations trying to reproduce the element distribution on a few specific well-observed stars (preferably of different types). In the field of close binary evolution, models could produce Algol or Sirius-type systems in a general way very early, but as soon as it was attempted to model specific well-observed systems, the difficulties became exposed and the theoretical limitations became apparent (non-conservative mass transfer,etc..). I suspect that the situation with respect to the diffusion models is somewhat analogous.

MEGESSIER   Before to be able to reproduce theoretically the element distribution on specific stars, we need more observational constraints. In the present state of the available observations, we may say that the diffusion computations do reproduce the observed features, since these latter give a crude representation only of the star. We need to know the magnetic field geometry and its local intensity and direction, the local element overabundances on the stellar surface with a better accuracy and a better confidence. In that respect several problems are still to be solved. This only will allow to constraint more the parameters to be introduced in the diffusion computations for specific stars.

PHYSICAL INPUT FOR THE DETERMINATION OF STELLAR ABUNDANCES

B. Gustafsson
Stockholm Observatory*
S-133 00   SALTSJÖBADEN
Sweden

ABSTRACT. The fundamental parameters and the physical modelling and data used in the analysis of high-resolution high S/N spectra of stars are discussed. Particular emphasis is led on recent developments in these respects and the importance of further improvements is stressed.

1.   INTRODUCTION

For weak stellar spectral line the equivalent width is approximately proportional to the chemical abundance of the element in question and the oscillator strength. The factor of proportionality is, however, dependent on stellar parameters, atmospheric structure and other data and processes that directly affect the line formation.
   The equivalent widths can today, in high S/N spectroscopy at sufficient resolution, be measured with an accuracy of 10% or better. The f values can in many cases be determined with an even higher accuracy (cf. Huber's review in the present volume). Thus, one could hope that abundances could be estimeted with a corresponding accuracy. This is, in general, not possible; typical external errors in abundances are in good cases about 30%, and often greater than that. The reason for this is our lack of precise knowledge about the fundamental parameters, atmospheric structures and line formation. A clear-cut separation between the errors in chemical abundances caused by uncertainties in these different respects is rather artificial. In particular, the approximation of the atmospheric structures by a grid of standard model atmospheres also affects the choice of fundamental parameters (if made spectroscopically or photometrically

---

* Present address: Uppsala University,
   Astronomical Observatory, Box 515, S-751 20, UPPSALA, Sweden

by comparisons with model spectra or colours) and the data and approximations involved in the line calculation may affect both models and choices of fundamental parameters. Moreover, the choice of f values may also be affected by these circumstances; namely, if the f values are obtained in a differential analysis with respect to some other star, such as the Sun. Nevertheless, we shall here follow a schematic approach and in separate sections discuss the abundance effects of errors in fundamental parameters, in model atmospheres and in synthetic spectra, respectively. The vital problem of the uncertainties in f values is not treated here; the reader is referred to Huber's paper in the present volume for a discussion of those.

2. FUNDAMENTAL PARAMETERS

The determination of fundamental parameters for stars was discussed at IAU Symposium 111 (Hayes et al., 1985); for an overview of the use of models for deriving such parameters, see Gustafsson and Jørgensen (1985). Here, I shall only list some recent developments and comment on the abundance errors resulting from errors in the parameters.

A number of recent studies have contributed to a significantly increased accuracy in effective-temperature determinations. Among these studies are the lunar-occultation measurements of red giants by Ridgway and collaborators (see Schmidtke et al. 1986, and references given there), the systematic application of the infrared-flux method of Blackwell and Shallis (1977) to different types of stars (early-type stars : Underhill et al. 1979, Underhill 1982, Remie and Lamers 1982; solar-type dwarfs: Saxner and Hammarbäck 1985, Magain 1987; G-K giants: Bell and Gustafsson 1987; M and C giants: Tsuji 1981 a and b), the use of line spectra and ionization equilibria, supported by non-LTE calculations for O-stars by Kudritzki and collaborators (e.g. Kudritzki et al. 1983) and the use of Balmer-line profiles for solar-type stars by Cayrel et al. (1985). With optimal methods the temperatures are now obtainable with an accuracy of about 2-3% for early-type stars and even better for solar-type stars. For the O-type stars and for the coolest stars, of spectral type M and N, the errors are, however, about two times greater.

Which are the consequences of the errors in effective temperatures on the abundance analyses? Very schematically, we find from a glance at recent analyses that the temperature errors lead to errors in the interval 0.00 - 0.10 dex in absolute abundances. Exceptions from this are the CNO elements in early-type stars and in the M stars, where the errors may amount to two times as much or even more than that (cf., e.g., Smith and Lambert, 1985). Note also the more indirect coupling between temperature errors and errors in log g, which may in turn cause severe errors in abundances; examples thereof are provided by the O and B-type stars, and also by cooler stars if molecular lines with their great temperature sensitivity are being used for gravity determination (cf. Bell et al. 1985).

The surface gravities may be determined from Balmer lines for early spectral types, from the Balmer discontinuity (e.g. measured by the Strömgren δc_1 index) for late A and F-type stars, from pressure-broadened lines for solar-type dwarfs and sub-giants (cf. Edvardsson 1987) and from estimated absolute bolometric magnitudes (and guessed masses) for cool giants. (The non-LTE overionization effects could lead to severe systematic errors in determinations based on ionization equilibria, see below.) Typical errors in current gravity estimates amount to 0.1 - 0.2 dex for early-type stars, 0.3 dex for G-K-type giants and 0.5 dex for cooler giants. This corresponds to typical errors in abundances of 0.00 - 0.10 dex; exceptions are the CNO elements for red giants where the errors are about 0.2 dex, and the metals for late K, M and N giants where the errors range in the interval 0.1 - 0.2 dex.

Also errors or uncertainties in the chemical abundances themselves cause errors in the abundance analysis, notably in studies where the analysis is confined to particular elements or is not brought to complete self-consistency. This effect occurs through blanketing in the models, through blends or "veils" in the spectra and, for late-type stars, through the electron pressure which is determined by the abundance of metal ions and which in turn determines the continuum opacity. A typical uncertainty in the over-all metal abundance of 0.2 dex leads, however, to errors less than about 0.05 dex in the abundances for most elements. Exceptions are the CNO abundances, in particular in cool stars; an error in [Me/H] of 0.2 dex thus leads to typical errors in the CNO abundances of K and N stars of about 0.1 dex.

The effects of uncertainties in the microturbulence parameter for contemporary high-quality analyses of many elements in solar-type stars are minor since weak lines, the equivalent widths of which are little affected by microturbulence, may be measured. For certain elements, however, such as rare earths, only a few and often saturated lines are available, and the resulting abundances may be in error by 0.2 dex or so. For hot stars, with fewer suitable lines, the microturbulence uncertainties may be more problematic. Baschek et al. (1982) ascribe the dominating uncertainties in their analysis of an early subdwarf to the uncertainties in the microturbulence parameter. For cool stars these uncertainties may also have severe effects, since blends and veils of weak molecular lines make the use of saturated lines necessary in the analyses. E.g., from Smith and Lambert (1985) we find that uncertainties in the microturbulence parameter of 1 km/s lead to errors in their determinations of carbon and nitrogen abundances for M giants of about 0.2 dex.

In conclusion we find that the total abundance uncertainties, due to errors in the fundamental parameters, amount to typically 0.1 dex for many metals in most types of stars. Exceptions are the O and B stars, as well as the M and N stars, where a realistic error is about twice as great. For the CNO elements in stars of spectral type late G, K, M or N the errors are about two times greater than those of metal abundances. In particular, the error in the nitrogen abundances amounts to typically 0.4 dex for the M and N stars, which mainly reflects the temperature sensitivity of the CN line strengths.

3. MODEL ATMOSPHERES

As regards the modelling of stellar atmospheres it is worth noting that, even within the frame-work of classical plane-parallel LTE models with mixing-length convection it is not certain that specification of the fundamental parameters discussed above leads to a unique model atmosphere. A bifurcation in the upper solar photosphere, currently ascribed to the surface cooling of CO (cf. Ayres and Testerman 1981, Ayres 1986, Kneer 1983, Muchmore and Ulmschneider 1985, Muchmore 1986) may well be reflected in double solutions of the classical problem (cf. in particular Nordlund, 1985). A similar effect, due to SiO, may exist for cooler stars (cf. Muchmore et al. 1987). Recently, we have shown that polyatomic opacities in the upper layers of carbon-star models may, under certain conditions, lead to drastically different solutions to the model-atmosphere problem (Eriksson Gustafsson and Jørgensen, unpublished research).

Evidently, we have two kinds of uniqueness problems to worry about for classical models. As was mentioned above, models with the same sets of fundamental parameters may be quite different and give quite different spectra. On the other hand, models with different fundamental parameters may have almost identical spectra. The classical example of the latter case is the very difficult problem of simultaneously determining helium abundance and gravity in late-type spectra.

The groups still calculating grids of classical models work on refinements of the blanketing treatment and synthetic spectra by including still weaker atomic and molecular lines (cf. Kurucz 1985) as well as, for the cooler stars, polyatomic absorption (Eriksson et al. 1984, Jørgensen et al. 1985, 1987). The polyatomic line absorption, being non-correlated in frequency with the diatomic absorption at greater depths, leads to errors when the Opacity-Distribution Function Method is used, while the Opacity Sampling Method requires very many frequency points to a considerable cost in computing time (Ekberg et al. 1986). Thus, a new efficient and more reliable blanketing algorithm is needed for the coolest stars.

A vigorous activity now takes place in work on models where at least one or more of the classical basic assumptions have been relaxed. Some examples will be given here. For the early-type stars Anderson (1985), using his multi-frequency/multigray algorithm, and Werner (1986), using Scharmer's operator perturbation technique, extend the non-LTE models of Mihalas and Auer (1972) and Kudritzki and collaborators (Kudritzki 1979) to include blanketing from carbon and, in on-going work, from other heavy elements. The effects of departures from LTE on solar-type models are now possible to study (an early attempt was that of Saxner 1985) but still rather uncertain due to uncertain cross sections for inelastic collisions between hydrogen and metal atoms and uncertain uv-fluxes. The wind-blanketing for early-type stars is considered successfully (cf. Abbott and Hummer 1986, Bohannan 1987) and found to lead to significant T_{eff} revisions for O-type supergiants. Anelastic convection is solar-type stars, with consideration of inhomogeneities in 3D radiative transfer and overionization effects for iron, is simulated numerically by Nordlund (1984). These models reproduce observed line widths and asymmetries in a gratifying way (cf. Dravins 1988, and references quoted therein). Nordlund estimates a correction to the solar iron abundance, based on planeparallel models, of about a factor of two but stresses the great uncertainty in this estimate. Spherically symmmetric models for red giants and super-giants have been studied extensively by SchmidBurg, Scholz, Wehrse and collaborators (see Scholz 1985, and references cited therein). These models show very interesting coupling between extension and molecular formation which not only complicates the task of the spectroscopist but also enables the determination of stellar gravities and radii (and thus masses) independently. Recently, Bessell et al. (1987) have also attempted calculations of blanketed spherically symmetric Mira models with shocks.

What are the errors made when grids of classical model atmospheres are used in abundance analysis? One may attempt to answer this question by comparing analyses made with classical models to those made with models with partly improved physics. One may also compare analyses with grid models to those made with temperature structures tailored to exactly reproduce observed strong-line profiles or continuous fluxes for the star in question. (Such semi-empirical models were, e.g., discussed by Ruland et al.

1980 or Magain 1985). An example of such a study of effects of model errors is that by Gustafsson (1983), who discussed the uncertainties in current analyses of Pop. II stars. Studies where effects of model uncertainties on abundances have been investigated seem to lead to typical model errors in abundances of 0.1 - 0.2 dex. For the N stars Lambert et al. (1986) find characteristically 0.3 dex. One may fear that these estimates, being more or less ad hoc but always incomplete, are underestimates.

4. CALCULATION OF SPECTRA

Even if departures from LTE would not be important for the atmospheric structures of certain types of stars, they significantly affect the spectra of many elements. This has been recognized since long for the early spectral types - more recently evidence for non-LTE effects in late-type stars has been accumulated. Thus, Ruland et al. (1980) found significant inconsistencies for Pollux (K 0 III) in abundance of Fe, Ti and Cr, when derived from lines with different excitation energies. This effect has been confirmed by several others for red giants (e.g., Kovacs 1983, 1985) and similar effects were found by Steffen (1985) for Procyon (F5 IV-V) and by Magain (1988) for several elements in Pop. II dwarfs. In fact, it cannot be excluded that trends in relative abundances with changing [Fe/H] for metal-poor stars could be ascribed to these effects (cf. Gustafsson 1987, Magain 1988). Brown et al. (1983) traced an inconsistency in the Zr/Ti ratio of red giants, when derived from neutral atoms and ions, respectively. The effects have been interpreted as primarily due to over-ionization in combination with different depths of formation for lines of different excitation, and some have been at least qualitatively reproduced in statistical equlibrium calculations (Steenbock 1985); although these are highly uncertain as a result of uncertainties in collision cross sections, in particular from inelastic collisions with hydrogen atoms (see also the study of Li by Steenbock and Holweger 1984).

For the early-type stars a great number of theoretical studies of specific elements and ions have been made (among the more recent ones are the detailed studies of Fe, Mg and Ba in Vega, A0 V, by Gigas 1986, 1987). Much work is, however, needed before all important aspects of the physics of line-formation has been properly included and all relevant elements and transitions have been treated in such studies. In particular, the electron collision cross sections and the photo-ionization cross sections need to be known with a higher accuracy.

A number of more trivial aspects of the calculation of spectra should be remembered. The identification of lines, of blending lines and of veiling lines across the lines of interest and/or overlying the "continuum regions" is vital, not the least for cooler stars. Here, there are many severe uncertainties. The proper consideration of hyperfine structure and isotopic shifts may cause revisions of abundances by almost one order of magnitude in certain cases (as compared with if the line-splitting is neglected). The line broadening is not properly understood for many lines of interest for high-precision spectroscopy. Dissociation energies for the molecules may be uncertain by embarrassing amounts; a well-known example is $D_o^o(CN)$ which is uncertain by more than 0.1 eV, leading to errors in the nitrogen abundances for the coolest red giants by 50% or more (cf. Lambert et al. 1986).

5. CONCLUSIONS

Our schematic discussion leads to the following conclusions:
(1) Errors in the estimated fundamental parameters introduce abundance errors ≥ 0.1 dex. The errors are greatest for stars of spectral type O and for M and N stars, and greater for supergiants than for giants and dwarfs.
(2) The errors in model atmospheres, when studied, are found to lead to errors in abundance determinations of about the same order of magnitude as those caused by fundamental parameter errors.[1]
(3) The departures from LTE, not treated properly as yet, may again lead to systematic errors of the same order of magnitude as those discussed above.
(4) For the analysis of certain types of stars (not the least cool stars) and certain chemical elements (not the least those represented by few spectral lines) uncertainties due to blends and veils, hfs, dissociation energies etc. may again cause uncertainties of the same order of magnitude.
   From this one may conclude that
(I) it is important to improve the determinations of T_{eff} and $\log g$ for most stars if improved abundance estimates are wanted;
(III) it is important to improve the model atmospheres and the calculation of spectral lines further. This also requires the determination of a vast number of cross sections and other atomic and molecular quantities of importance.

---

[1] cf. the so-called Robin effect: Errors difficult to estimate are generally not supposed to exceed those more easily estimated.

One should note that high S/N spectroscopy at sufficient resolution offers vital support to the development of methods for accurate T_{eff} and log g determinations, as well as to the improvement of model atmospheres and spectrum calculations. This requires, however, that accurate comparisons between observed and calculated spectra are made, not only for the quick determination of abundances and other parameters but also for tracing inconsistencies and studying these further (e.g., as a function of stellar parameters). Evidently, the development of the efficient high S/N spectrometers puts such extra obligations on the shoulders of the observer if he/she wants to contribute to a development where the accuracy in equivalent widths and f values can be fully exploited in the abundance determinations.

Fortunately, since most error sources discussed above are systematic, a very high accuracy may be reached in spite of these errors in <u>differential studies</u> where ratios between abundances for different but similar elements, or abundance differences for different but similar stars, are being measured. Although this fact is almost a truism and often quoted and used, the <u>extent</u> of these cancellation effects has not been studied very much and the differential studies are not very often designed such that the cancellatio is maximized. With the presently rapidly growing qualitative understanding of model errors, departures from LTE etc., and the widening of the accessible frequency region of stella spectra, such "self-compensating" high precision methods for stellar quantitative spectral analysis could be further developed.

REFERENCES

Abbott, D.C., and Hummer, D.G. 1985, Astrophys. J. 294, 286.
Anderson, L.S. 1985, Astrophys. J. 298, 848.
Auer, L.H., and Mihalas, D. 1972, Astrophys. J. Suppl.Series 24, 153.
Ayres, T.R., and Testerman, L. 1981, Astrophys. J. 245, 1124.
Ayres, T.R., Testerman, L., and Brault, J.W. 1986, Astrophys. J. 304, 542.
Baschek, B., Kudritzki, R.P., Scholz, M., Simon, K.P. 1982, Astron. Astrophys. 108, 387.
Bell, R.A., and Gustafsson, B. 1987, in preparation.
Bell, R.A., Edvardsson, B., and Gustafsson, B. 1985, Monthly Notices Royal Astron. Soc. 212, 497.
Bessell, M.S., Brett, J.M., Scholz, M., and Wood, P.R. 1987, Astron. Astrophys., in press.
Blackwell, D.E., and Shallis, M.J. 1977, Monthly Notices Royal Astron. Soc. 180, 177.
Bohannan, B., this volume.
Brown, J.A., Tomkin, J., and Lambert, D.L. 1983, Astrophys. J. Letters 265, L93.
Cayrel, R., Cayrel de Strobel, G., and Campbell, B. 1985, Astron. Astrophys. 146, 249.
Dravins, D. 1988, this volume.
Edvardsson, B. 1987, Astron. Astrophys., in press (see also this volume).
Ekberg, U., Eriksson, K., and Gustafsson, B. 1986, Astron. Astrophys. 167, 304.
Eriksson, K., Gustafsson, B., Jørgensen, U.G., and Nordlund, A. 1984, Astron. Astrophys. 132, 37.
Gigas, D. 1986, Astron. Astrophys. 165, 170.
Gigas, D. 1987, Astron. Astrophys., in press (see also this volume).
Gustafsson, B. 1983, Publ. Astron. Soc. Pacific 95, 101.
Gustafsson, B. 1987, in 'ESO Workshop on Stellar Evolution and Dynamics in the Outer Halo of the Galaxy', ed. M. Azzopardi, in press.
Gustafsson, B., and Jørgensen, U.G. 1985, in Hayes et al. (1985), p. 303.
Hayes, D.S., Pasinetti, L.E., Davis Philip, A.G. 1985, 'Calibration of Fundamental Stellar Quantities', IAU Symp. 111, D. Reidel Publ. Co.
Jørgensen, U.G., Almlöf, J., Gustafsson, B., Larsson, M., and Siegbahn, P. 1985, J. Chem. Phys. 83, 3034.
Jørgensen, U.G., Almlöf, J., Larsson, M., and Siegbahn, P. 1987, in prep.

Kneer, F. 1983, Astron. Astrophys. 128, 311.
Kovacs, N. 1983, Astron. Astrophys. 124, 63.
Kovacs, N. 1985, Astron. Astrophys. 150, 232.
Kudritzki, R.P. 1979, Proc. of 22nd Liège Conf., p. 295.
Kudritzki, R.P., Simon, K.P., Hamann, W.-R. 1983, Astron. Astrophys. 118, 245.
Kurucz, R.N. 1985, in Highlights of Astronomy 7, 827.
Lambert, D.L., Gustafsson, B., Eriksson, K., and Hinkle, K.H. 1986, Astrophys. J. Suppl.Series 62, 373.
Magain, P. 1985, Astron. Astrophys. 146, 95.
Magain, P. 1987, Astron. Astrophys. 181, 323.
Magain, P. 1988, this volume.
Muchmore, D. 1986, Astron. Astrophys. 155, 172.
Muchmore, D., Ulmschneider, P. 1985, Astron. Astrophys. 142, 393.
Muchmore, D.O., Nuth, J.A. III, Stencel, R.E. 1987, Astrophys. J. Letters 315, L141.
Nordlund, Å. 1984, in 'Small Scale Dynamic Processes in Quiet Stellar Atmospheres', ed. S.L. Keil, Sacramento Peak, Sunspot, New Mexico, p. 30.
Nordlund, Å. 1985, in 'Theoretical Problems in High Resolution Solar Physics', ed. H.U. Schmidt, Max-Planck-Institut für Astrophysik, München, p 1.
Remie, H. and Lamers, H.J.G.L.M. 1982, Astron. Astrophys 105, 85.
Ruland, F., Holweger, H. Griffin, R.&R., Biehl, D. 1980, Astron. Astrophys. 92, 70.
Saxner, M. 1985, Thesis, Uppsala University.
Saxner, M., and Hammarbäck, G. 1985, Astron. Astrophys. 151, 372.
Scholz, M. 1985, Astron. Astrophys. 145, 241.
Schmidtke, P.C., Africano, J.L., Joyce, R.R., and Ridgway, S.T. 1986, Astron. J. 91, 961.
Smith, V.V., Lambert, D.L. 1985, Astrophys. J. 294, 326.
Steenbock, W. 1985, in 'Cool Stars With Excess of Heavy Elements', eds. M. Jaschek and P.C. Keenan, D Reidel Publ. Co., p. 231.
Steenbock, W., and Holweger,H. 1984, Astron. Astrophys. 130, 319.
Steffen, M. 1985, Astron. Astrophys. Suppl.Series 59, 403.
Tsuji, T. 1981a, Astron. Astrophys. 99, 48.
Tsuji, T. 1981b, J. Astrophys. Astron.2, 95.
Underhill, A.B. 1982, Astrophys. J. 263, 741.
Underhill, A.B., Divan, L., Prévot-Burnichon, M.-L., Doazan, V. 1979, Monthly Notices Royal Astron. Soc. 189, 601.
Werner, K. 1986, Thesis, University of Kiel, and Astron. Astrophys. 161, 177.

DISCUSSION

    LINSKY Your point concerning thermal bistability or multi-stability is very important and we need to determine empirically whether or not this perverse phenominon occurs in different classes of stars. One method for doing so is to check whether the abundances derived from molecular spectra agree with abundances derived from neutral and ionized species.

    GUSTAFSSON I agree.

    BECKMAN Could you please re-state your estimates of the numerical values of the uncertainties into abundance analyses of solar type stars by inhomogeneities and by non-LTE effects ?

    GUSTAFSSON Nordlund's calculations suggest that the iron abundance of the Sun may be overestimated by 0.2 dex or even somewhat more, if plane-parallel models are used instead of his 3D models. This estimate must be regarded as crude and preliminary.

    M. SPITE You said that the error on the abundances, due to the models, in Pop II stars is about 0.1-0.2.
Is the error the same for the dwarf stars and for the giant stars ?

    GUSTAFSSON It is hard to judge, indeed. One would expect that analyses of Pop II dwarfs were more affected by uncertainties in convection than the giants are while the converse may be fine as regards departures from LTE. None of these statements is free from reservations, however. Thus, the brighter UV fluxes of the dwarfs (assumed to be hotter) may lead to more severe departures from LTE there, and the lower densities in the giants may enable the convection-generated inhomogeneities to survive to higher layers in spite of the fact that the convective instability sets in at greater depths.

DATA ANALYSIS

R. CAYREL

Observatoire de Paris, France

ABSTRACT

The relationship between the r.m.s. photometric accuracy on the data points of a stellar spectrum and the final accuracy on abundances and physical parameters derived from these data is discussed.

1. INTRODUCTION

Modern detectors allow to achieve nominal signal/noise ratio of several hundreds in stellar spectroscopy. However, the final accuracy on the physical parameters in which we are interested does depend not only upon the photometric accuracy of the spectra, but of what is happening during several steps of reduction, involving most of the time substantial losses in accuracy. This make, in practice, the accuracy on abundances, for example, only 15% good, even if one starts with spectra obtained with a signal to noise ratio of 200.

It is not our intention to treat the whole subject in a few pages, so we shall focus on just a few critical points. The first one has to do with the accuracy reachable on the equivalent width of weak lines with a given signal/noise ratio in the continuum. The second point is the accuracy obtainable from the analysis of Balmer lines on temperature differences between two stars. Because the equivalent width of weak neutral lines are very unsensitive to gravity, for F, G and K stars, the accuracy on equivalent widths, plus the accuracy on temperature determination largely control the accuracy obtainable on abundance ratios between two stars belongings to these spectral classes.

2. FROM THE FLUX DATA POINTS TO THE LINE PARAMETERS

Assuming that instrumental effects have been properly corrected for, by dark exposures subtraction and flat-fielding (not forgetting that these operations degrade the row signal/noise ratio by an appreciable amount) the data consist of a set of ordinates s_i running from one to the number of pixels of the detector, n= 1872 for most reticons in use). We assume that we know the statistical mean error on each s_i and, for simplicity, we consider first the academic case where we can found a spectrum subinterval in which the spectrum can be described by a continuum of constant level and a single absorption line of gaussian profile :

$$(1) \quad S(x) = S_c - S_c \, u . \exp\left\{ -\frac{1}{2}\left(\frac{x-v}{w}\right)^2 \right\}$$

We also assume, which is even more academic, that Sc is known, from unperturbed regions of the spectrum.

Then the characteristics (v wave-length of the line center, w standard width, u central depth of the line) of the line must be derived from the usually over-determined system (solved by least square)

$$(2) \quad r_i = 1.0 - \frac{S_i}{S_c} = u . \exp\left\{ -\frac{1}{2}\left(\frac{x_i-v}{w}\right)^2 \right\}$$

$$i = i_{min}, i_{max}$$

Where i_{min} is the data point of smallest wave-length and i_{max} the data point of the largest wave-length used in the least-square procedure. Of course ($i_{max} - i_{min}$) must be at least equal to 2 (three points) and preferably larger. This immediately poses the problem of the pixel size :

$$(3) \quad \delta x = x_{i+1} - x_i$$

with respect to w. The common sense requests that the data point must be influenced by the presence of the line at a level larger than the noise. So if ϵ is the r.m.s. relative photometric accuracy on the flux :

(3) $\epsilon = \dfrac{1}{S_c} <\delta S^2>^{1/2} = <\delta r^2>^{1/2}$

one should have :

$$u \exp\left\{-\dfrac{1}{2}\left(\dfrac{1.5\,\delta x}{w}\right)^2\right\} \gg \epsilon$$

The factor 1.5 holds because the most central data point is, at the most, at a distance of half a pixel to the real center of the line.
This gives for $u = 0.1$ and $\epsilon = 0.01$

$$\delta x < 1.43w$$

or :

$$\delta x < 0.6 \times FWHM$$

as a function of the full width at half maximum, FWHM
However this condition, which is necessary to find the center and the width of the line is <u>not necessary</u> if one is only interested in its equivalent width, as we shall see later.
Landman, Roussel-Dupré and Tanigawa (1982) have solved the problem of finding the accuracy on u, v and w for a given photometric accuracy ϵ. The result is :

(3) $\quad <\delta u^2>^{1/2} = \left(\dfrac{3}{2}\right)^{1/2} \left(\dfrac{1}{\pi}\right)^{1/4} \left(\dfrac{\delta x}{w}\right)^{1/2} \epsilon$

(4) $\quad <\delta v^2>^{1/2} = \dfrac{\sqrt{2}}{\pi^{1/4}} (w.\delta x)^{1/2} \dfrac{\epsilon}{u}$

(5) $\quad <\delta w>^{1/2} = <\delta v^2>^{1/2} = \dfrac{\sqrt{2}}{\pi^{1/4}} (w.\delta x)^{1/2} \dfrac{\epsilon}{u}$

provided $\delta x \ll FWHM$. (In practice is it enough that $\delta x \leqslant w$).
An interesting consequence stems from these formulae : if the observation is photon noise limited, $\epsilon \sim \delta x^{-1/2}$ and the errors are all independent of the pixel size.
Landmann and al. (1982) have not given the error on the equivalent width. As δu and δw are correlated it is not possible to compute the error on the product uw using the above formulae. The exact computation gives :

(6) $\quad <\delta W^2>^{1/2} = \dfrac{\sqrt{3\pi}}{\pi^{1/4}} \; (w\delta x)^{1/2} \epsilon$

or :

(6') $\quad <\delta W^2>^{1/2} \simeq 2.3 \; (w.\delta x)^{1/2} \epsilon \simeq 1.5 \; (FWHM.\delta x)^{1/2} \epsilon$

A similar formula can be obtained in a much simpler way. The equivalent width is by definition :

$$W = \int_{-\infty}^{+\infty} r(x)\,dx = \sum_{i=i_1}^{i_2} r_i \delta x = \delta x \sum_{i=i_1}^{i_2} r_i$$

the sum Σ being extended to all pixels potentially affected by the line. For a gaussian profile it is fairly safe to sum over a width equal to 6 times the standard deviation, and the error on w comes as :

(7) $<\delta W^2>^{1/2} = \delta x . \epsilon \sqrt{\dfrac{6w}{\delta x}} = 2.45 \; (w\delta x)^{1/2} \epsilon$

$\simeq 1.6 \; (w\delta x)^{1/2} \epsilon$

So whatever is the method for getting the equivalent width, the root mean square error on it is the product a numerical factor close to 1.5, by the geometrical mean between the pixel size and the FWHM, and by the relative r.m.s. the photometric accuracy in the continuum.

If one applies this formule to conditions frequently met at the CAT reticon spectrograph at ESO one gets :

$FWHM = 0.1 \text{Å} \qquad \delta x = 0.035 \text{Å}$

and with :

$\epsilon = 1/200$

$<\delta W^2>^{1/2} = 0.4 \; m\text{Å}$

3. FURTHER PHYSICAL LIMITATIONS ON THE ACCURACY OF SMALL EQUIVALENT WIDTHS.

Repeated measurements of equivalent widths on several spectra of the same star show that formulae (6) or (7) are too optimistic. The scatter on such measurements is more frequently of the order of 1 to 2 mÅ rather than 0.4 mÅ. The principal source of additional error is the fact that the continuum is not known exactly, and worse, not even definable with an accuracy of 0.5 %. One is easily convinced of that by taking a look at a segment of the "Solar flux Atlas from 296 to 1300 nm by Kurucz et al. 1984 (see fig.1). In the numerical example at the end of section 2, if we assume an error of 0.5 % in the location at the continuum this induces an error of 6w x .005 in the equivalent width (no more square root cancellation !) or :1.3 mÅ on W, three times larger than the statistical photometric error.

In order to minimize this error we found that is better to requestion in the least square fit the exact position of the continuum i.e. to have S_c as unknown as will as a, w and v. We also found that v is usually not to be taken as unknown for weak lines, the wave-lengths differences being known with high accuracy from tables and not "negotiable".

In pratical cases it is found that the line under scrutiny is always more of least "blend" with other lines. It is easy to generalize the least square fit to n lines :

$$S(x) = S_c \left\{ 1 - \sum_{i=1}^{n} u_i . \exp\left(- \frac{1}{2}\left(\frac{x - v_i}{w}\right)^2 \right) \right\}$$

$$W_i = \sqrt{2\pi} . u_i w$$

Then the value of w become much more critical : if two consecutive lines are too close in wave-length, the least-square procedure collapse, or, if well handled allow only to determine the sum of the equivalent widths of the two nearby lines but not how much is ascribable to each of them.

So it may be necessary to increase the resolution mainly for separating the line from its nearest neighbours.

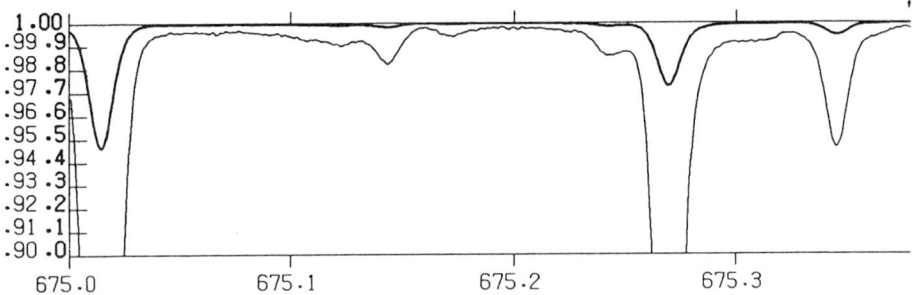

Fig.1- Structure of the "continuum" in the solar spectrum, from Kurucz et al. 1984. The wave-length scale is in nanometers. The thin line has an ordinate scale enlarged by a factor of 10 (left scale). One clearly sees that there is no well defined continuum at the 1% level.

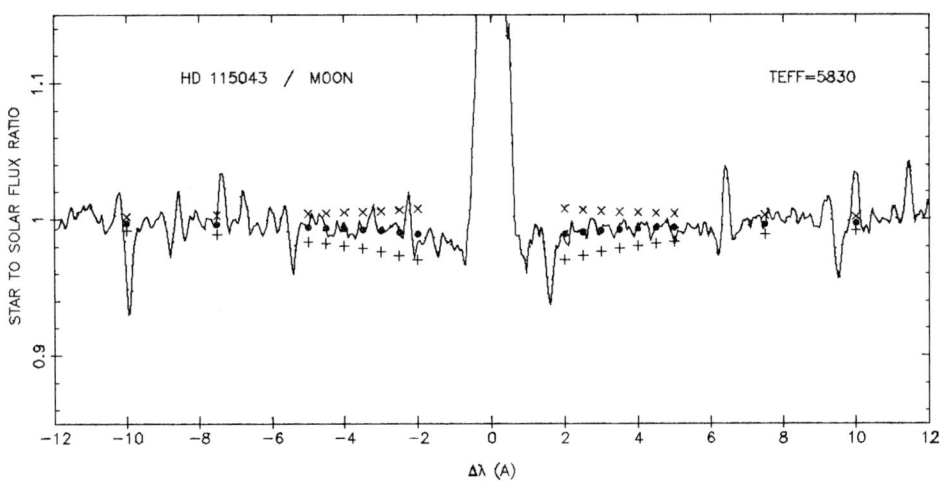

Fig.2-Ratio of stellar (HD 115043) to moon spectra in the Hα region. Comparison with theoretical ratios for effective temperatures 5830 K and 5830± 100 K. The precision of the determination is clearly better than 50K. The central (-2,+2 A) portion is strongly affected by chromospheric activity in HD 115043 (UMa stream) and of no use for temperature determination.

4. TRADE-OFF BETWEEN NOISE LEVEL AND SATURATION EFFECTS.

Taking now 1.5 mÅ as a reasonable r.m.s. error on equivalent widths (independent of W) what equivalent widths are most useful for abundance work ?

The relative error decreases when W increases ,but soon saturation effects become dominant and the error due to the uncertainty on the "microturbulence" takes over. A 15 mÅ equivalent width suffers a 10 % uncertainty and at this point an error by 0.1 dex on the Doppler width contributes by only 3 %. The balance between the two sources of error occurs at about 20 mÅ and we consider that this is a kind of optimal compromise between the need of keeping the saturation effects small and having good relative accuracy on the Ws.The relative accuracy is then only of the order of 10 % even if we started from a nominal S/N ratio 200 in the continuum. This example shows how difficult it is to take full advantage of very high S/N ratio of 500 or more. My conclusion is indeed very close to the one given by D. Gray : before making meaningful use of high photometric S/N noise one hits difficulties as undefinable physical continuum, slight residual instrumental effects due to a different collimation of stellar and flat-field or calibration beams in the spectrograph, etc ... The effort should then go as well towards mastering these effects (as exemplified by the HF cell for accurate wave length measurements) as to still improve the purely photometric accuracy, any-how unimprovable above the limit set by the granular nature of light.

5. EFFECTIVE TEMPERATURE DETERMINATION FROM BALMER LINE WINGS.

For abundance work it is as important, to derive an accurate effective temperature as to have accurate equivalent widths of weak lines. High-signal to noise spectra have allowed to use the wings of the Balmer lines, and more specifically of Hα which is located in a region where the spectrum is clean and the efficiency of reticons high. Fig.2 show the comparison between the ratio of a stellar spectrum to moon light in the Hα region, compared to the predicted ratio computed from Gustafsson's models.

The theoretical ratios are computed with Teff=5770 K for the solar (full disk) spectrum.

One sees from the figure that a temperature difference is easily detectable, even if as small as 50 K. This is just about what is needed if the error or the Bolzmann population factor for a neutral lines lying 5 *ev* below the ionization level is to be kept below 10 % at 6000K. Recent work on the broadening theory of Balmer lines (Stehlé et al. 1983) shows that the Vidal, Cooper and Smith (1973) prescription is valid at a few angströms from line center for Hα. Because the population of H⁻ and of neutral lines are both proportional to the electronic pressure, the equivalent width ratio of a weak neutral line in a star and in the sun does depend only on the abundance ratio and on the temperature difference. The abundance ratio may in principle be determined from on line with 0.1 *dex* accuracy, and if more lines of the same element are available, some $\sqrt{n}$ gain may be achieved, this is particularly true for iron.

6. CONCLUSION

- Simple formulae are given to obtain the rms errors on line position, line width, line depth, and line equivalent width for a given r.m.s. noise at continuum level.

One conclusion is that 1) the pixel size is not at all a critical parameter. 2) increasing the spectral resolution decreases the r.m.s. error on equivalent widths (as the square root of the line width) as long as the line width follows the spectral resolution. Once the intrinsic width of the line is reached no more gain is obtainable by further increasing the spectral resolution.

- At signal/noise ratio above 200 the final accuracy obtainable on equivalent widths of weak lines is more set by the difficulty of defining a continuum than by the photometric error on the signal. It is recommendable to use a very local determination of the continuum by including the continuum position in the least-square fitting.

- Differences in effective temperatures as small as 50K can easily be detected from the strength of the wings of $H\alpha$ in late F to mid-G dwarf star with high S/N ratio spectra.

- Abundance ratios between stars of similar spectral types can be determined with an accuracy of 0.1 dex, from a single line, from spectra with a S/N ratio of 200.

REFERENCES

- KURUCZ R.L., FURENLID I., BRAULT J., TESTERMAN L., 1984, Solar Flux Atlas from 296 to 1300 nm. National Solar Observatory, Sunspot New-Mexico USA

- LANDMAN D.A., ROUSSEL-DUPRE R., TANIGAWA G., 1982, Astrophys. J. 261, 732.

- STEHLE C., MAZURE A., NOLLEZ G., FEAUTRIER N., 1983, Astron. Astrophys. 127, 263.

- VIDAL CR., COOPER J., SMITH, 1973, Astrophys. J. Suppl. 25, 37.

DISCUSSION

   SODERBLOM I would like to ask Ingemar Furenlid to remind us how the continuum level was established for the solar flux atlas, since Dr. Cayrel pointed out the lock of true continuum windows.

   FURENLID The continuum of the Solar Flux Atlas (Kurucz et al., 1984) is determined by a global fit to high points over wavelength ranges of a couple of thousand angstroms. It should thus be emphasized that it is strictly speaking a pseudo-continuum.

   WALKER How do you correct for scattered light ? What kind of error does it introduce ?

   R. CAYREL The correction is simple : if there is 5% straylight the equivalent widths are decreased by this amount.

   G. CAYREL We have not applied the correction because our analyses were differential : the corrections cancel out.

IMPROVED DATA REDUCTION TECHNIQUES FOR THE ESO CES PLUS RETICON SPECTRA.(*)

B.H. Foing
Laboratoire de Physique Stellaire et Planetaire
F-91370 Verrieres-le Buisson
France

L. Crivellari
Astronomical Observatory
I-34131 Trieste
Italy

ABSTRACT.

We developed routines for intensity, equivalent width, and radial velocity measurements on CES plus reticon spectra obtained with the ESO 1.4m telescope. In order to achieve the optimal recovery of the signal, the noise has to be minimized by removing any parasitical effect. Special care has been devoted to the correction for remanence effects in the reticon dark counts. Typical results are presented and discussed.

INTRODUCTION.

Since October 1983 we have been carrying out a comprehensive study on the chromospheres of solar-like dwarfs, including (i) multi-component modelling; (ii) the investigation of chromospheric variability in active stars; (iii) the diagnostics of velocity fields and the detection of plages. (See Beckman et al., 1984; Crivellari et al. 1987.)
The most valuable observations for this purpose include H alpha, the H and K lines of Ca II, the infrared triplet of Ca II, and the h and k lines of Mg II.
All the visible observations relevant to our programme have been obtained using the 1.4 m Coude' Auxiliary Telescope (CAT) of the European Southern Observatory at La Silla (Chile), equipped with the Coude' Echelle Spectrograph (CES) and the reticon. High resolution (R up to 10**5) and high signal-to-noise (S/N) spectra are obtained,
Two different kinds of measurements are required by our

(*) Based on observations collected at the European
    Southern Observatory,La Silla, Chile.

program: (a) accurate spectrophotometry, both absolute to determining chromospheric emission fluxes, and relative for line variability detection; (b) precise wavelength calibration for velocity field diagnostics.

Among the spectral features under study, the H and K resonance lines of Ca II deserve special consideration for the wealth of physical information one can derive from them. On the other hand some specific problems arise from their intrinsic shape: their complex structure requires high spectral resolution; the intensity of the lines at the bottom of the broad photospheric absorption is very faint.

Both facts make long exposures necessary (some times as long as three hours) which necessarily produce contamination of the true signal by parassitic effects. Especially reticon dark counts are a severe source of errors.

AN ESTIMATE OF THE NOISE IN CES SPECTRA.

The recorded spectrum can be expressed as

(1) $$F'(x) = \frac{\beta}{a} s(x) f(x) + b(x) + D(x) ,$$

where $f(x)$ is the true stellar spectrum, $s(x)$ the instrumental sensitivity function and β/a the conversion factor from the incoming photons to the digitized recorded signal. $b(x)$ is the total read-out noise, due to the combined effect of the instrumental electronic noise and the periodic noise introduced by the use of four independent lines for recharging the diode array. $D(x)$ is the signal due to the dark counts of the reticon. The image lag phenomenon (remanence effect) can easily be identified as the prime cause of dark counts. (See Maurice and Maugis, 1981.) The independent variable x is a monotonic function of wavelength.

In order to extract $f(x)$ from the recorded signal $F'(x)$ one should know the quantities $s(x)$, $b(x)$ and $D(x)$. The instrumental response of the reticon can be derived experimentally by recording an uniform "flat-field" exposure of intensity C. The recorded flat field is in the form

(2) $$S'(x) = \frac{\beta}{a} s(x) C + b(x) .$$

An empirical estimate of $b(x)$ is also easily obtained. A mean read-out signal $\overline{b(x)}$, found by averaging a limited number (typically five) of independent read-out spectra, can be assumed as a good estimator for $b(x)$. To determine a good estimator for $D(x)$ is a much more difficult problem. A thorough discussion of the correction for remnant effects in the reticon dark counts is contained in the paper by Crivellari et al. (1987). On the basis of the results of our analysis, when correcting the recorded signal for the dark counts contaminations, we use as an estimator the quantity $\overline{D} = m \cdot t_{exp} + q$, where

m and q are experimentally determined.
The normalized signal i(x) can be expressed as

(3) $$i(x) = \frac{\beta/a\, s(x) f(x) + \eta(x) + \varepsilon(x)}{\beta/a\, s(x) C + \eta(x)}$$

where $\eta(x)$ and $\varepsilon(x)$ are defined as

(4a) $\eta(x) \equiv b(x) - \overline{b(x)}$ (4b) $\varepsilon(x) \equiv D(x) - \overline{D}$

If $\overline{b(x)}$ and $\overline{D}$ were the correct estimators for $b(x)$ and $D(x)$ respectively, the expectation values of $\eta(x)$ and $\varepsilon(x)$ would both be identically zero. Then we can conclude that $i(x)$ is the true stellar spectrum, but for a multiplicative constant. From eq. (3) it is easy to derive the error $i(x)$ in terms of each contributing variable. We have

(5) $$\Delta i(x) = \left\{ \frac{\Delta^2 F(x)}{\sigma^2(x)} + \frac{[S(x)-F(x)]^2 \Delta^2 \eta(x)}{\sigma^4(x)} - \frac{\Delta^2 \varepsilon(x)}{\sigma^2(x)} - \frac{F^2(x)\Delta^2 S(x)}{\sigma^4(x)} \right\}^{1/2}$$

where $\sigma(x)$ is defined as $S(x)+\eta(x)$. $F(x)$ and $S(x)$ are $\beta/a\, s(x)$ times $f(x)$, and C respectively.

In Table 1 we present a set of typical S/N values for a sample of Ca II H profiles, obseved in October 1985.

We asume that the errors $\Delta F(x)$ and $\Delta S(x)$ derive from the statistical fluctuations of the corresponding signal (Poisson statistics). The error on $\varepsilon(x)$ is directly related to the rms fluctuation of the read-out spectrum. From our study of the remanence effect we can derive $\Delta \eta(x)$. When evaluating the errors ΔF and ΔS, $s(x)$ and C must be known. Actually in the present analysis we had to determine them indirectly from the recorded $S(x)$, in lack of exact values from literature. In particular the estimate of C, necessary to detemine $s(x)$, is quite crude. This could have led to an overestimate of $\Delta i(x)$. Consequently the relative error and S/N values must be taken as conservative.

MEASUREMENTS ON THE CES SPECTROGRAMS.

a) Intensity measurements.

A calibration law to convert the CES arbitrary intensity scale into absolute flux units is not available. Indirect methods must be used when absolutely calibrated profiles are required, e.g. for comparison with theoretically synthetized spectral feature. (See Castelli et al., 1988, these proceedings.)

However the immediate request of our investigation is a reliable <u>relative</u> spectrophotometry. Either we intercompare spectra of stars with the same overall physical properties, but different level of chromospheric activity to define quantitative activity indexes, or we analyze spectra of the same star, taken at different phases, to look for variability brought about by stellar rotation.

To do that we choose the spectrum of one star as the reference spectrum and bring all the other into registration with it by means of a least-squares fit (program NEWSUM). The basic idea of the method is that after registration we have a set of spectra whose fluxes are expressed on a common (even if arbitrary) scale. A direct comparison is then possible, without the need of a normalization (not always reliable over the limited wavelength range of a single echelle order) to a refernce continuum.

NEWSUM also offers the possibility to co-add several spectra, irrespective of their signal level, to produce averaged spectra with improved S/N ratio.

When the object of the investigation is the search for variability, a statistical analysis of the spectrograms is automatically performed (program JERKA), to spot out locii of possible variations.

The detailed description of our suite of computer codes (namely NEWSUM and JERKA) has been already published. (Ramella et al., 1980; Beckman et al., 1980; Franco et al., 1984.)

Two kinds of measurements are usually performed on the spectrograms: i) The intensity of individual features in the profiles; ii) the integrated flux in selected pass bands to mimic narrow band spectrophotometry.

A good example of the application of these techniques is contained in the paper by Crivellari et al. (1987). Here we have the place only to recall that the scattering of the integrated fluxes around their average is typically of 0.5%. That proves the very high quality of CES plus reticon spectrograms.

b) Wavelength measurements.

Any profile of the spectral features under study is the result of integrating contributions from different layers of the stellar atmosphere. We present in a separate paper (Vladilo et al., 1988, these proceedings) our technique to determining the accurate position of selected features within the complex profile of the H (K) Ca II lines. The method is based on the reconstruction of the observed line profile with two gaussian bands.

Combining the errors which affect the measurement of each single step of the calibration procedure, starting from the standard IHAP routines, we claim a conservative error of +4 mÅ for each single measure.

CONCLUSIONS.

We intend to present this paper as a cue for a discussion about the real needs (in terms of spectral resolution and S/N) for detailed diagnostics from line profiles.

The choice of the Ca II H (or K) line as a test of our data reduction procedure has been deliberated. Among all the available examples, we selected a line which (i) is intrinsically faint, with a portion of its profile almost saturated, (ii) demands very high resolution to show up its complex structure, (iii) falls in a spectral region where the reticon sensitivity is low. Compared with these severe drawbacks, the performances of the CES are really outstanding. Let the reader go to Table 1, where the S/N values relevant to 9 Cet are reported. For a G2 star of m 6.4 it is possible to obtain a high resolution spectrum (R=8.10**4) with a relative error at the minimum H1 less than 5% within a still reasonable exposure time (240 min).

Our analysis proved the high stability of spectrophotometry with the CES plus reticon. After registration via NEWSUM, the scatter of individual spectra around the mean is a few percent of the signal level.

The results on variability and velocity field diagnostics, presented in our previous paper (Crivellari et al., 1987) clearly show that also in case of <u>low</u> S/N ratio a wealth of useful information can be obtained, when a careful data processing is performed. We claim that, by using proper measurement techniques, it is possible: (a) to detect variations in the normalized fluxes down to some 3% in the most critical cases; (b) to measure velocity gradients within an accuracy of 300 m*s-1.

Table 1.

Star	Sp.type	m	t (mn)	D	feat.	F'(x)	S'(x)	S/N	Star	Sp.type	m	t (mn)	D	feat.	F'(x)	S'(x)	S/N
β Myi	G2 IV	2.8	90	13	λ3950	830	665	190	ϵ Eri	K2 V	3.7	60	12	λ3950	380	265	85
					H1	150	830	30						H1	130	305	30
					λ3980	1010	890	230						H2	290	220	65
τ Cet	G8 V	3.5	120	14	λ3950	710	340	160						λ3980	460	220	100
					H1	170	410	40									
					λ3980	890	430	200	9 Cet	G2 V	6.4	240	19	λ3950	250	200	50
α CenB	K1 V	1.3	20	10	λ3950	450	260	100						H1	110	230	20
					H1	100	310	20						H2	150	230	30
					H2	160	310	35						λ3980	270	230	60
					λ3980	480	325	110	Sun	G2 V				λ3950	2335	2290	510
δ^2 Eri	K1 V	4.4	100	13	λ3950	420	190	95						H1	280	2220	65
					H1	125	210	25						λ3980	1910	890	430
					H2	150	210	30									
					λ3980	500	220	110									

REFERENCES.

- Beckman, J.E., Crivellari, L., Foing, B.H.: 1984, The Messenger, **38**, 24.
- Beckman, J.E., Crivellari, L., Morossi, C., Ramella, M., Vladilo, G.: 1980, IUE Data Reduction, Vienna, November 1980, W. Weiss et al. eds., p.207.
- Castelli, F., Gouttebroze, P., Beckman, J.E., Crivellari, L., Foing, B.H.: 1988, these proceeedings.
- Crivellari, L., Beckman, J.E., Foing, B.H., Vladilo, G.: 1987, Astron. Astrophys., **174**, 127.
- Crivellari, L., Beckman, J.E., Foing, B.H., Vladilo, G.: 1988, these procedings.
- Franco, M.L., Crivellari, L., Molaro, P., Vladilo, G., Ramella, M., Morossi, C., Allocchio, C., Beckman, J.E.: 1984, Astron. Astrophys. Suppl. Ser., **58**, 693.
- Maurice, E., Maugis, M.: 1981, ESO internal memorandum.
- Ramella, M., Morossi, C., Vladilo, G., Crivellari, L.: 1980, IUE Data Reduction, Vienna, November 1980, W. Weiss et al. eds., p.203.
- Vladilo, G., Crivellari, L., Castelli, F., Beckman, J.E., , Foing, B.H.: 1988, these proceedings.

DISCUSSION.

G. Cayrel. What are you doing with cosimc rays spikes if you have long exposure?

L. Crivellari. We check the spectrograms for possible cosmic events. Then we remove the bad pixel(s), if we have been lucky enough to spot out spikes.

J.E. Beckman. I believe that you did refer indirectly to the point that the long dark counts are not only dependent on the present exposure, but also on the history of the reticon chip, that is on the previous exposure.

L. Crivellari. Of course. The main source of the reticon dark counts is ascribed to the remanence effect.

PRECISE ATOMIC DATA

Martin C.E. Huber*
ESA/ESTEC, Space Science Department
Postbus 299, 2200 AG Noordwijk
The Netherlands

ABSTRACT

We discuss the three questions: Why do we need accurate atomic data for stellar spectroscopy? What accuracy is needed? And: How can the accuracy of oscillator strengths be assessed? In conclusion we comment on the state of the art, stress the importance of uncertainty estimates and also discern between precision and accuracy.

1. INTRODUCTION: WHY DO WE NEED ACCURATE DATA FOR STELLAR SPECTROSCOPY?

Abundance determinations require accurate atomic data for two interdependent reasons: (i) to derive a reliable abundance value from observed equivalent widths with the aid of a realistic atmosphere model and (ii) to check on the validity of this model - in particular, with regard to the actual (inhomogeneous) structure of the atmosphere and deviations from local thermodynamic equilibrium (LTE). Advances in both, instrumentation and interpretation have made it possible now to extend these latter tests from solar to stellar spectroscopy.

In instrumentation, the replacement of photographic film and plates by (imaging) photoelectric detectors (see, e.g., Timothy 1983) has resulted in an increase in signal-to-noise ratio - and thus also attainable resolving power - as well as in radiometric accuracy. Therefore, measurements of equivalent widths and line profiles are now limited in accuracy mostly by the uncertainty in the continuum position and extended instrumental line wings (Griffin 1969): currently, equivalent widths in high-resolution stellar spectra can probably be measured with an accuracy of three to four per cent. And in the solar case, an isolated, unblended line can probably be measured with an accuracy of two per cent. Here, double-pass grating spectrometers (Delbouille, Roland and Neven, 1973) or Fourier-transform spectrometers (FTS) (Kurucz et al., 1984) are now used and this results in instrumental profiles

* On leave of absence from Swiss Federal Institute of Technology Zurich (ETHZ), Institute of Astronomy.

with negligible wing intensities.

As far as interpretation is concerned, the construction of theoretical LTE model atmospheres for specific parameters, T_e, log g and metallicity is now almost routine. The MARCS program of Gustafsson et al. (1975), for example, is implemented on the UK STARLINK network. Nevertheless, Rutten and Kostik (1988) have shown, that LTE models lead to solar oscillator strengths for Fe I, that can be in error by 0.1 dex (where 0.1 dex is a logarithmic interval corresponding to a factor of $10^{0.1}$). Solanki and Steenbock (1987) also demonstrated that the use of homogeneous LTE models result in apparent abundance values that differ by a similar factor from those which consider non-LTE (NLTE) effects, and that even larger discrepancies must be expected, if parts of the inhomogeneous atmosphere (as, e.g., the chromospheric network) are considered separately. (These discrepancies are most pronounced for low-excitation lines of Fe I (E_l 3.5 eV), they are somewhat less important for higher excitation potentials and practically absent for Fe II.) NLTE atmospheres are now becoming more common and can be expected to be much developed in future.

2. WHAT ACCURACY IS NEEDED?

Oscillator strengths are not the only atomic data needed in interpreting solar and stellar spectra. As soon as a line is no more optically thin, damping and - in many cases - hyperfine splitting become of importance. Accuracy requirements for each of these quantities are treated in the following subsections.

2.1 Atomic oscillator strengths

To interpret properly the double-pass and FTS solar spectra, for which equivalent widths can be measured to an accuracy of two per cent, an accuracy in gf of one per cent or better is desirable. This accuracy is needed to test both, the extent of the applicability of LTE model-atmospheres and the improvements offered by NLTE models.

2.2 Collision damping

The effect of uncertainty in damping is usually underestimated and often completely overlooked (Blackwell, Calamai and Willis, 1972). As an example, a solar line with an equivalent width of 8 pm, a lower excitation potential of 4 eV and a wavelength of 600 nm, whose damping constant is in error by a factor of three will lead to an error of nearly 30 per cent in abundance. Although the error decreases quite rapidly with decreasing excitation and equivalent width, it is clear that, if equivalent widths of accuracy four per cent are to be interpreted using oscillator strengths of accuracy one per cent, great attention should be paid to the accuracy of the damping constants.

The rôle of damping in the solar spectrum is treated by Simmons and Blackwell (1982) and by Blackwell, Booth and Petford (1984a). The Fe-I

damping constants used are those measured by O'Neill and Smith (1980a
and 1980b). Further work referring to damping in the solar spectrum is
reported by Blackwell et al. (1987). Calculations of sufficiently
accurate damping constants for complex atoms will probably remain a very
difficult undertaking. Much more experimental work is therefore needed!

2.3 Hyperfine structure

Hyperfine structure is another important, much neglected topic.
Hyperfine splitting can be caused by both, the isotope shift (in the
case of a mixture of isotopes) and by the interaction of a (non-zero)
nuclear spin, I, with the total angular momentum, J, of the electrons.
This often results in a number of close-lying line components.
Hyperfine splitting makes its influence felt in laboratory measurements
as well as in solar and stellar spectra. Examples for the case of Mn I
are given by Booth et al. (1983a and 1984a) as well as by Booth et al.
(1984b and 1983b), respectively. In brief, profiles "diluted" by
hyperfine splitting change the behaviour of the curve of growth.

Much more work is needed here as well, but, fortunately, measurements
are straightforward. They are probably taken most easily and reliably
by use of a hollow cathode (or other narrow-line source) and a FTS.

3. HOW CAN THE ACCURACY OF OSCILLATOR STRENGTHS BE ASSESSED?

Assessing the accuracy of oscillator strengths is difficult. The
following two methods have been used most frequently: (i) intercomparing
values measured by different groups and (if ever possible) by different
techniques, and (ii) comparing solar abundances derived by use of lines
from the same multiplet.

An intercomparison of oscillator strengths for low-excitation lines of
Fe-I is shown in Figure 1. The dynamic range of the comparison -
extending over six decades - is remarkable. However, the methods used
to obtain the data shown in this figure (the absorption method by
Blackwell et al. (1979) and a combination of anomalous-dispersion and

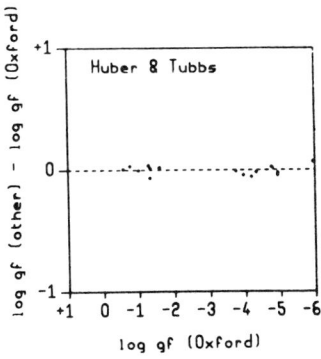

Figure 1.

Comparison of oscillator strengths
for low-excitation lines of Fe I by
Huber and Tubbs (1972) with those
by Blackwell et al. (1979).

absorption measurements by Huber and Tubbs (1972)) both depend on the population of the lower level of a given transition. A more thorough assessment can be made, if results from absorption and emission measurements are compared: the former depends on the population of the lower, and the latter on that of the upper levels of the lines in question. Tozzi, Brunner and Huber (1985) have compared their branching-fraction data for Cr I with the absorption data of Blackwell, Menon and Petford (1984b). To the best of our knowledge, this was the first comparison between laboratory transition probabilities where the claimed precision of the Oxford values could be confirmed - at least to the three-per-cent level. A formal procedure for such comparisons - now designated as the Ladenburg method (Huber and Sandeman 1986) - has been developed by Cardon, Smith and Whaling (1980).

If one compares derived solar abundances, the influence of NLTE effects can be minimised by exclusive use of lines belonging to one or several similar multiplets. The accuracy of such tests is somewhat degraded by the necessary solar equivalent-width measurements. Nevertheless, good results were obtained by Blackwell et al. (1987): sixteen Ti-I lines of excitation potential 1 eV, for example, resulted in an abundance value having a standard deviation of 3.4 per cent (for the most suitable microturbulent velocity of 0.85 km s^{-1}).

4. CONCLUSIONS AND OUTLOOK

We are now entering a realm, where transition probabilities with a 10 per-cent precision are no longer sufficient. And, indeed, a precision of a few per cent for measured oscillator strengths of complex spectra has now been confirmed. Once this precision is reached, one must also consider (and measure!) damping coefficients and the hyperfine splitting.

Above all, however, astronomers and laboratory physicists must estimate uncertainties carefully; for, strictly speaking, a value without an uncertainty is meaningless. While it is necessary to have at least one accurate oscillator strength to obtain a reliable abundance value, precise (i.e. self-consistent, but possibly off-the-mark) data will do for tests of atmospheric models.

It should also be noted, that abundance determinations converge to a constant and (one would hope) correct value as the excitation potential increases (cf., e.g., Blackwell et al., 1987). For accurate abundance work it is, therefore, important to measure transition probabilities for high excitation lines.

ACKNOWLEDGEMENT

Prof. D.E. Blackwell's very kind help in preparing this paper is greatly appreciated.

REFERENCES

Blackwell, D.E., Ibbetson, P.A., Petford, A.D. and Shallis, M.J., 1979, Mon. Not. R. astr. Soc., 186, 633.
Blackwell, D.E., Booth, A.J., Menon, S.L.R. and Petford, A.D., 1987, Astron. Astrophys. (in press).
Blackwell, D.E., Calamai, G. and Willis, R.B., 1972, Mon. Not. R. astr. Soc., 160, 121.
Blackwell, D.E., Booth, A.J., Menon, S.L.R., Petford, A.D. and Smith, G., 1983, Mon. Not. R. astr. Soc., 204, 141.
Blackwell, D.E., Booth, A.J. and Petford, A.D., 1984a, Astron. Astrophys., 132, 236.
Blackwell, D.E., Menon, S.L.R. and Petford, A.D., 1984b, Mon. Not. R. astr. Soc., 207, 533.
Booth, A.J., Shallis, M.J. and Wells, M., 1983a, Mon. Not. R. astr. Soc., 205, 191.
Booth, A.J. and Blackwell, D.E., 1983b, Mon. Not. R. astr. Soc., 204, 777.
Booth, A.J., Blackwell, D.E., Petford, A.D. and Shallis, M.J., 1984a, Mon. Not. R. astr. Soc., 208, 147.
Booth, A.J., Blackwell, D.E. and Shallis, M.J., 1984b, Mon. Not. R. astr. Soc., 209, 77.
Cardon, B.L., Smith, P.L. and Whaling, W., 1980, Phys. Rev. A, 20, 2411.
Delbouille, L., Roland, G. and Neven, L., 1973, Photometric Atlas of the Solar Spectrum from 3000 Å to 10 000 Å. Institut d'Astrophysique de l'Université de Liège, Belgium.
Griffin, R.F., 1969, Mon. Not. R. astr. Soc., 143, 319.
Gustafsson, B., Bell, R.A., Erikssen, K. and Nordlund, A., 1975, Astron. Astrophys., 42, 407.
Huber, M.C.E. and Tubbs, E.F. 1972., Astrophys. J., 177, 847.
Huber, M.C.E. and Sandeman, R.J., 1986, Rep. Prog. Phys. 49, 397.
Kurucz, R.L. and Peytremann, E., 1975, A table of semi-empirical gf-values, parts 1, 2, 3. Smithsonian Astrophys. Obs. Spec. Rep. 362.
Kurucz, R.L., Furenlid, I., Brault, J.W., Testerman, L., 1984, Solar Flux Atlas from 296 to 1300 nm. National Solar Observatory Atlas No. 1.
O'Neill, J.A. and Smith, G., 1980a, Astron. Astrophys., 81, 100.
O'Neill, J.A. and Smith, G., 1980b, Astron. Astrophys., 81, 108.
Rutten, R.J. and Kostik, R.I., 1988, Physics of Formation of Fe II Lines Outside LTE, Proc. IAU Coll. No. 94, (Dordrecht: Reidel).
Simmons, G.J. and Blackwell, D.E., 1982, Astron. Astrophys., 112, 209.
Solanki, S.K. and Steenbock, W., 1987, Astron. Astrophys. (submitted).
Timothy, J.G., 1983, Publ. Astron. Soc. Pacific, 95, 810.
Tozzi, G.P., Brunner, A. and Huber, M.C.E., 1985, Mon. Not. R. astr. Soc., 217, 423.
Unsöld, A. 1955, Physik der Sternatmosphären, 2nd edn. (Berlin: Springer).

DISCUSSION

GIRIDHAR Is there a correlation between the enhancement factor to C_6 and excitation potential of the lines ?

HUBER Yes, the enhancement factor tend to become higher for high excitation potential lines. (see, for example, Simmons and Blackwell: 1982, Astron.Astrophys.<u>112</u>, 209)

MAGAIN I wish to stress that the astrophysicists do not only need accurate oscillator strengths for high excitation neutral lines, but also for ion lines.

HUBER I agree ...

GRIFFIN You mentioned that the accuracy of stellar equivalent widths is in the best case about 3 or 4 percent. I should like to make the following comment : two groups of astronomers, both claiming access to high S/N, high resolution observations of high accuracy, have published equivalent widths for 25 weak lines of Fe I in the same star. Their measurements differ by 20 percent.

CAYREL Is there any laboratory measurements planned for getting the broadening of neutral metallic lines by neutral hydrogen ?

HUBER Such measurements are very difficult, nevertheless some have already been made in the fifties (Kush H.J.:1958,Z.Astrophys. <u>45</u>, 1). Modern measurements are made in He and then corrected for the case of H by use of a semi-empirical theory (see, for example, O'Neill and Smith :1980, Astron.Astrophys. <u>81</u>, 108).

OSCILLATOR STRENGTHS FROM THE HIGH S/N SOLAR SPECTRUM

Robert J. Rutten
Utrecht Observatory
The Netherlands

Most of modern solar physics is not concerned with high S/N spectrometry, but rather with high spatial resolution in narrow-band imaging as well as spectrometry, and with high Fourier resolution in helioseismology.

Nevertheless, atlases of the solar spectrum are still being made, of ever higher spectral purity including S/N ratio. Fig. 1 compares the latest disk-center intensity atlas (Neckel 1987, based on Kitt Peak Fourier Transform Spectrometer data) with the well-known Jungfraujoch Atlas (Delbouille et al. 1973). Such high-quality data are useful for sun-as-a-star calibrations of stellar radiative transfer methods, of stellar activity and granulation diagnostics, and of atomic parameters used in abundance determination. It is my task here to discuss an example of the latter: solar gf-values.

The classic example of deriving oscillator strengths by fitting solar lines is Holweger's (1967) thesis, more recently followed by Gurtovenko and Kostik (1981, 1982). Currently, these Kiev workers are preparing a solar gf compilation containing 2000 lines from 50 elements.

How precise are these solar gf-values? Let us briefly review the three major error sources: continuum errors, NLTE effects, and inhomogeneities. First the classical problem of continuum placement. Fig. 1 illustrates the rather arbitrary choices made in straightening and concatenating the short grating-spectrometer records of the Jungfraujoch Atlas. The third and fourth strips show the upper 5% of each atlas only, to display the continuum windows. The bottom strip shows their difference and represents a "Delbouille response function": it rather closely follows the 5-10 Å wide dips and humps present in the line connecting the highest HN peaks, implying that these were largely normalized away in the Jungfraujoch record adjustments. The broad-band HN continuum is not necessarily the "true" continuum either, but it has less adjustment noise in this frequency range. Such noise affects the determination of the "local" continuum from which line profiles should be measured if one does not go to the extent of full spectrum synthesis. In that case, the neglect of quasi-continuous opacities (unresolved blends, weak-line haze, overlapping

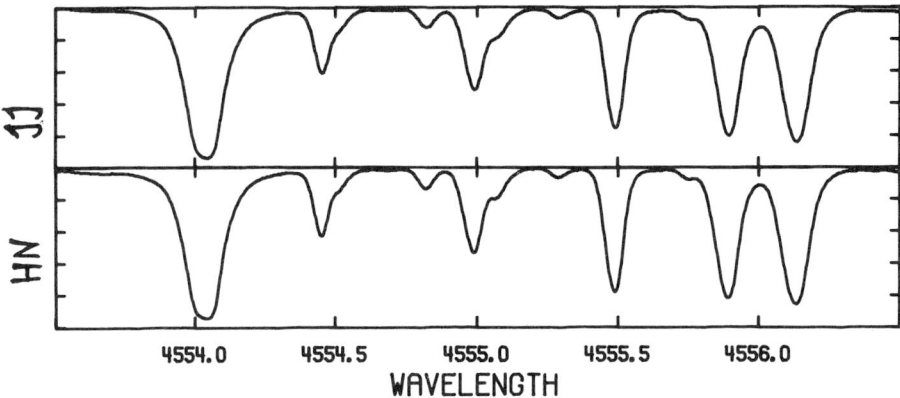

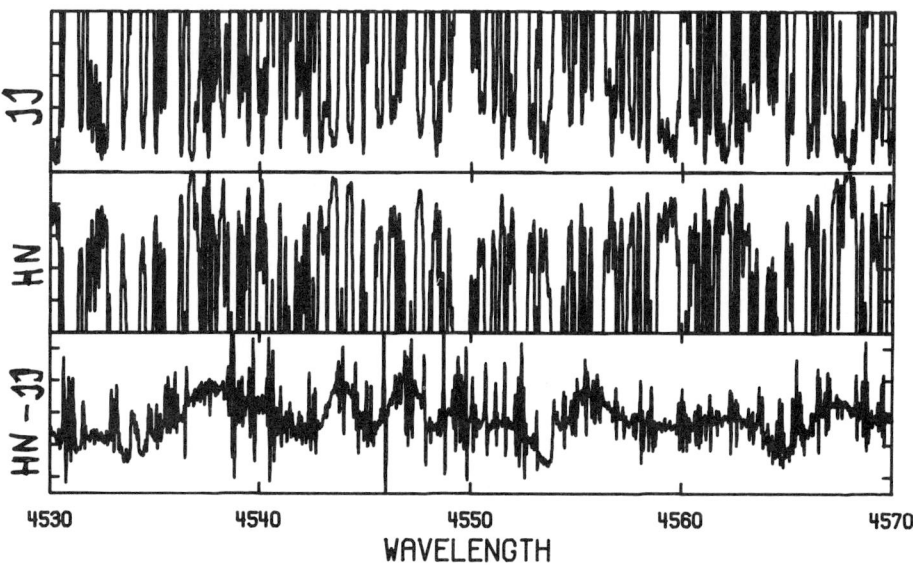

Fig. 1. Comparison of the Jungfraujoch Atlas (JJ) and the Hamburg-NSO Atlas (HN) of the solar disk-center spectrum. The higher resolution of HN is evident in the asymmetric core of Ba II 4554 and the blend of Cr II 4555. The third and fourth tracing show only the upper 5%, for JJ upside down. The asymmetry between them and their difference plotted at the bottom illustrate JJ normalization errors.

line wings) results in too small fitted gf-values because the computed line formation is located too deep, where the source function is steeper than at the actual height of formation. However, this deficit is partially cancelled by the steep inward decrease of the typical line-to-continuum opacity ratio. Solar modeling (Rutten and Van der Zalm 1984, Rutten and Kostik 1988) shows that for weak lines the error is typically halved: a 10% opacity deficit when fitting observed lines results in only a 5% (0.02 dex) gf underestimate. The error is much larger, however, if one measures from some "true" continuum or if one fits the depth of saturated lines.

The second issue is the effect of departures from LTE. These are usually neglected, but implicitly corrected for by assuming an LTE-masking model atmosphere based on LTE line fits (Rutten and Kostik 1982). The basic parameter is the steepness of the photospheric temperature gradient which sets the amount of NLTE overionization in the ultraviolet, and so affects the line optical depth scale even if the line source function obeys LTE (see Rutten 1988). Similarly, the often neglected overionization of hydrogen affects the continuum optical depth scale in the upper photosphere.

The temperature gradient is less steep in the newest Harvard modeling (Avrett 1985) than it used to be, because the inclusion of yet more Kurucz line-haze opacity has shifted the height of formation of the ultraviolet further outward, stretching the photospheric height scale. The empirical NLTE modeling has thus more or less retrieved the classical LTE Holweger-Müller (1974) model, and the predicted photospheric departures from LTE have become small. This is demonstrated in Fig. 2 for all clean solar iron lines: the differences between NLTE and LTE gf fits are indeed negligible (except for large-gf lines which have NLTE excitation, see Rutten 1988).

The third issue, inhomogeneities, provides the largest uncertainty. The solar granulation represents the dominant spoiler of the plane-parallel simplification for photospheric lines. Extensive supercomputer simulations by Nordlund (1984) lead to excellent fits of observed solar Fe I lines without invoking classical free parameters such as microturbulence, macroturbulence and damping enhancement factors. This is the first fudge-free reproduction of Fraunhofer lines! However, the lines fit only if the iron abundance (the only free parameter) is 0.4 dex below the standard value. This is a very important discrepancy. It arises because there is much neutral iron present in the cool intergranular lanes, and because spatial averaging differs from unresolved modeling because the ionization equilibrium is highly nonlinear in temperature.

The discrepancy would be yet larger (0.6 dex) if LTE would hold for the iron ionization equilibrium. It would also be larger if NLTE source functions had been adopted for the ultraviolet line haze, which contributes strongly to granular overshoot through radiative heating (Nordlund 1985).

In conclusion, solar gf-values for weak lines have better than 0.1 dex precision if one may trust plane-parallel modeling, even when assuming LTE. However, Nordlund's simulation of the solar granulation dramatically upsets this comfortable result.

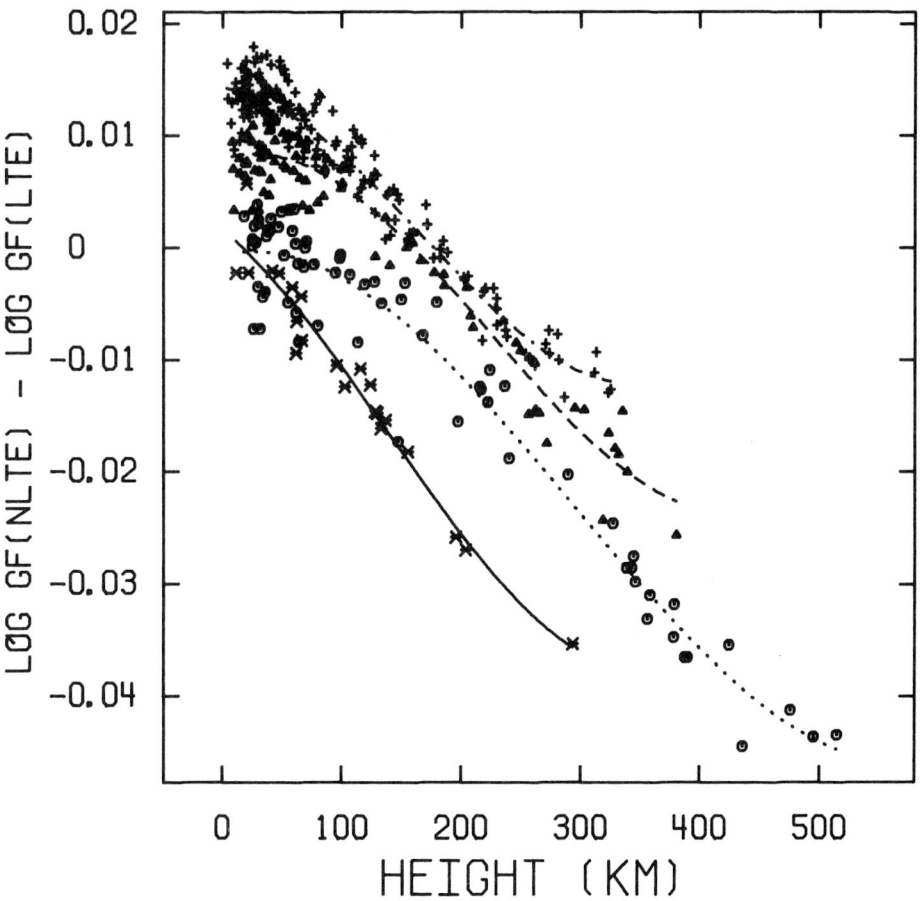

Fig. 2. Comparison of LTE and NLTE fits of the equivalent widths of all 254 Fe I and 22 Fe II lines in the solar clean-line list by Rutten and Van der Zalm (1984), against height of formation. NLTE fit: MACKKL model (Maltby et al. 1986), MACKKL turbulence, NLTE departure coefficients from Avrett (private communication). LTE fit: HOLMUL model (Holweger and Müller 1974), MACKKL turbulence. Fe II lines: asterisks, fitted with the solid curve. Fe I lines: split into low-excitation lines (circles, dotted fit), middle-excitation lines (triangles, dashed fit) and high-excitation lines (plusses, dot-dashed fit) as in Rutten and Kostik (1984). The separation of the various classes results from differences in the shape of the contribution function. The outward decrease results from model differences. The two descriptions are so close, however, that the differences shown here are negligible.

REFERENCES

Avrett, E.H.: 1985, in B.W. Lites (ed.), "Chromospheric Diagnostics and Modeling", National Solar Observatory Conference, Sacramento Peak, Sunspot, p. 67
Delbouille, L., Roland, G., Neven, L.: 1973, "Photometric Atlas of the Solar Spectrum from λ 3000 to λ 10000", Institut d'Astrophysique, Liège
Gurtovenko, E.A., Kostik, R.I.: 1981, Astron. Astrophys. Suppl. 46, 239
Gurtovenko, E.A., Kostik, R.I.: 1982, Astron. Astrophys. Suppl. 47, 193
Holweger, H., Müller, E.A.: 1974, Sol. Phys. 39, 19
Holweger, H.: 1967, Zeitschrift für Astrophysik 65, 365
Maltby, P., Avrett, E.H., Carlsson, M., Kjeldseth-Moe, O., Kurucz, R.L., Loeser, R.: 1986, Astrophys. J. 306, 284
Neckel, H.: 1987, private communication (preliminary tape version from Hamburg Observatory)
Nordlund, Å.: 1984, in S.L. Keil (ed.), "Small-scale dynamical processes in quiet stellar atmospheres", National Solar Observatory Conference Sacramento Peak, Sunspot, New Mexico, p. 181
Nordlund, Å: 1985, in H.U. Schmidt (ed.), "Theoretical Problems in High-resolution Solar Physics", Proceedings of the MPA/LPARL Workshop 1985, Max Planck Institut für Astrophysik, München
Rutten, R.J.: 1988, in R. Viotti (ed.), "Physics of Formation of Fe II lines Outside LTE", IAU Coll. 94, Reidel, Dordrecht
Rutten, R.J., Kostik, R.I.: 1982, Astron. Astrophys. 115, 104
Rutten, R.J., Kostik, R.I.: 1988, in R. Viotti (ed.), "Physics of Formation of Fe II lines outside LTE", IAU Coll. 94, Reidel, Dordrecht
Rutten, R.J., Van der Zalm, E.B.J.: 1984, Astron. Astrophys. Suppl. 55, 143

DISCUSSION

GUSTAFSSON: The value of a correction to the solar iron abundance of -0.3 dex $+0.1$ dex, the second number being the correction for extra overionization, was confirmed in a recent private communication with Åke Nordlund. I think he would agree, however, that further calculations with existing solar simulations and corresponding plane-parallel models are necessary before this correction is well-determined.

RUTTEN: The 0.4 dex I quoted results from subtracting his published value (Nordlund 1984, $A_{Fe} = 7.18$) from my plane-parallel one (Rutten and Van der Zalm 1984, $A_{Fe} = 7.63$). NLTE overionization was included in both. However, I agree that more computations are necessary, especially on the effects of departures from LTE in the Kurucz line-haze opacities.

THE ACCURACY OF HIGH S/N SPECTROSCOPIC MEASUREMENTS

Douglas K. Duncan
Space Telescope Science Institute
3700 San Martin Drive
Baltimore, Maryland 21218 USA

ABSTRACT. A careful intercomparison is made of several sets of high S/N spectrsocopic observations made by different astronomers, using different telescopes and detectors, observing the same objects. Spectra of moderately weak (10-50 mÅ) Li lines, observed at S/N of one to several hundred using the CFHT Reticon, Lick Observatory Reticon, McDonald Observatory Digicon, and the Intermediate Dispersion Spectrograph plus CCD of the Isaac Newton Telescope agree within the claimed measurement accuracies of about 3-6 mÅ almost without exception. CFHT Reticon observations of Hyades dwarfs are compared with unpublished observations taken with a CCD at Lick Observatory; again, the agreement is very good. However, abundance determinations based on these measurements agree less well. The discrepancy arises from different temperatures used by different investigators for the same stars. The accuracy of various methods of temperature determination is examined, including B-V, R-I, and V-K colors, continuum scans, Hα line profile fitting, and temperatures derived from fine analyses. With careful work and good data, temperatures still limit the accuracy of some of the programs discussed.

1. INTRODUCTION

The development of more sensitive and more linear detectors over the last several years has permitted spectra of higher S/N to be obtained. New questions are being addressed, the answers to which depend on great accuracy in spectroscopic and abundance measurements. Measurements of slight abundance differences between stars or small changes in line profiles with time in a given star [e.g. Doppler Imaging] are examples. Especially when results obtained at different observatories are compared, one wants to know whether the quoted S/N values are indicative of all the errors of measurement. More generally, are abundance determinations based on high S/N data achieving the accuracy claimed?

A recent investigation, done with L. Hobbs (Hobbs and Duncan, 1987), required a very careful intercomparison of abundances obtained at different observatories, with different equipment. The investigation was of the abundance of Li in halo stars. Standard models of the big bang (cf. the review of Boesgaard and Steigman, 1985) predict the formation of He, D, and ^{7}Li, in amounts which are a strong function of the baryon to photon ratio. Measurements of the primordial abundance of any of these isotopes thus have important cosmological significance.

Li is a trace element. In a wide variety of pop. I objects it is found with a logarithmic abundance of $n_{Li} \approx 3.0$ (where $n_H = 12.00$). However, solar type stars gradually destroy Li, and the current abundance in the sun is only $n_{Li} \approx 1.0$, 100X less than that with which it formed. It might be expected that halo stars, much older than the sun, would have destroyed essentially all their Li. Thus it was an important and surprising discovery when Spite and Spite (1982), and Spite, Maillard, and Spite (1984) showed that *all* the extremely metal-poor stars of temperature 5600K to 6300K they observed showed Li abundances of $n_{Li} = 2.05 \pm 0.2$. Apparently the lack of metals in these stars leads to thinner convection zones which do not mix Li to regions which destroy it as fast as in the sun.

The Li abundance observed in the halo stars is 10X less than is seen in pop. I objects, and this naturally raises the following question. Did the big bang produce an abundance $n_{Li} \approx 2$, and galactic sources (presently unknown) raise the abundance to the currently observed value $n_{Li} \approx 3$, or did the big bang produce $n_{Li} \approx 3$, and the halo stars deplete their Li about one order of magnitude, but not the two orders of magnitude seen in the sun. If the halo stars all truly have the same abundance, the former hypothesis is prefered, but if small star-to star differences exist it is an indication that at least some destruction has occurred, even in the halo stars.

2. A COMPARISON OF DIFFERENT LI ABUNDANCE DETERMINATIONS IN HALO STARS

2.1. Equivalent width (EW) comparisons

Hobbs and Duncan, first working independently and later together, obtained data on many additional halo stars and compared their results to those of Spite et al. The Duncan-Hobbs spectra were obtained mainly with a Lick Observatory coude camera and bare Reticon detector, and a McDonald Observatory coude camera and Digicon detector. The former has a resolution of 0.3 Å and the latter 0.26Å. Signal to noise was typically 80, and the claimed accuracy 3-7 mÅ, which represents 10-20% in these moderately weak lines. Five stars were observed both at Lick and McDonald. Their measured EWs showed very good agreement, with a mean difference of 0 ± 4 mÅ. Twelve stars already observed by Spite et al. were repeated in order to permit a detailed intercomparison of results.

The Spite et al. spectra were obtained at the coude focus of the Canada France Hawaii Telescope (CFHT). The detector was a bare Reticon, the resolution 0.26Å, and typical S/N 100. Figure 1 compares the two sets of results. Agreement is seen to be very good, with rms. difference for the 12 stars 0 ± 4 mÅ.

2.2. Abundance Determination

Figure 2 shows the Li abundances determined by Spite et al. vs. the Duncan-Hobbs values. Although the mean difference is close to zero, the scatter about the diagonal line of perfect agreement is large, std. dev. $\sigma = 0.15$. An EW error of 20% in these lines corresponds to an abundance error of 0.08 dex; this cannot be the main source of scatter.

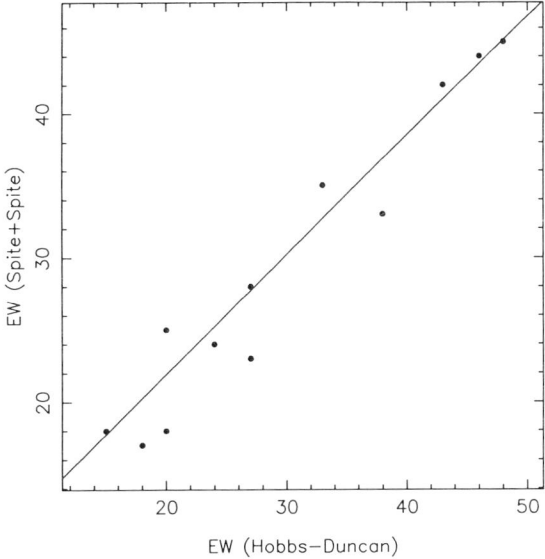

Figure 1. An equivalent width comparison shows very good agreement.

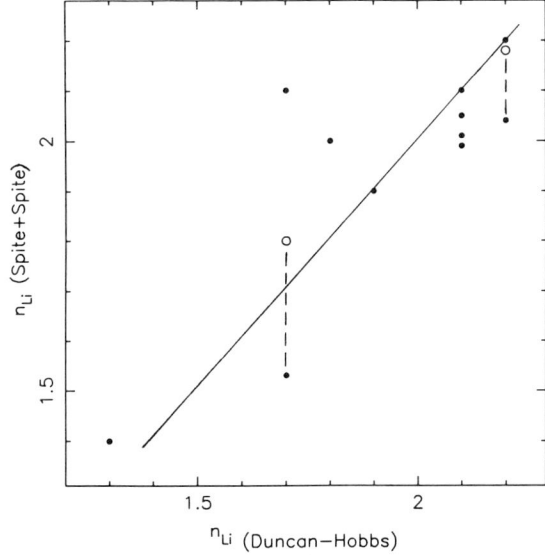

Figure 2. Comparison of Li abundance determinations. Two open circles show the effect of correcting temperatures as explained in the text.

2.3. Temperature Comparison

Figure 3 makes a comparison of effective temperatures used for the same stars in the two separate investigations. The scatter is relatively large, with an rms. difference of 120K. This corresponds to an abundance change of about 0.12 dex, and explains most of the scatter in Figure 2.

Temperatures in the Spite et al. investigations were determined from fine analyses and R-I colors. In an effort to get the most uniform possible temperatures, Hobbs and Duncan considered R-I and V-K colors, continuum scans made by Peterson and Carney (1979), and Stromgren colors. In two cases, reasons were found to doubt the temperatures taken from the literature by the Spites. If those two stars are re-analyzed with new temperatures, their abundances change as shown by the open circles in Figure 2. Hobbs and Duncan found that temperatures determined from fine analyses were the least consistent of all sources examined.

Comparing different sources of temperatures for all their halo stars, Hobbs and Duncan found an rms. difference of 55 K between T(continuum scan) and T(V-K), and an rms. difference of 80 K between T(scan) and T(R-I). *Random* errors, which are more important than the zero-point calibration of the temperature scale when one is interested in differential abundance comparisons between stars, should be reduced by averaging independent temperature determinations. Hobbs and Duncan conclude that their *relative* temperature errors are 60 K. This compares with the accuracy estimated by Peterson and Carney themselves, of 80 K, from combining R-I, V-K, and scan temperatures.

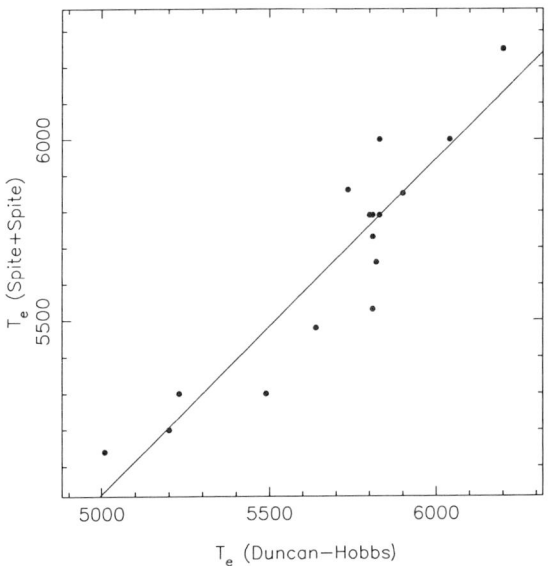

Figure 3. Comparison of temperatures used for the same stars in two separate investigations.

2.4. Another independent comparison

Quite recently Rebolo, Molaro, and Beckman (RMB, 1987) have finished a very comprehensive study of Li abundances in halo dwarfs. Their conclusions agree with those of Hobbs and Duncan and those presented here, and, for six stars which are in common, allow another external check of measurement accuracy. The RMB observations were made at the Cassegrain focus of the Isaac Newton telescope at the Observatorio del Roque de los Muchachos (La Palma). The intermediate dispersion spectrograph plus CCD yielded a resolution of about 0.44Å. Typically S/N of about 150 was achieved, leading to an estimated measurement accuracy of 10% in EW. Comparison of the RMB results with those of Hobbs and Duncan shows a mean EW difference of 0 ± 4 mÅ, in very good accord with the accuracy claimed by each set of authors. Since the RMB determinations of temperature are based on some of the same sources used by Hobbs and Duncan, they do not provide independent information.

3. ARE THE STAR-TO-STAR DIFFERENCES IN HALO LI ABUNDANCE REAL?

With the best-determined temperatures and equivalent widths in hand, we now confront the question whether the star-to-star abundance differences real. We focus attention on 6 very metal-poor stars of almost exactly the same temperature, listed in Table I.

TABLE I

HD or BD	R-I	V-K	T_{scan}	EW_{DH}	EW_{SS}	EW_{RMB}	n_{Li}
19445	.345	1.39	5820	38	33	35	2.1
94028	.33	1.39	-	33	35		2.1
201891	.33	1.42	5810	27	23	27	1.9
219617	.33	1.41	5800	43	42	40	2.2
+17 4708	.34	1.40	5810	25		25	1.9
+26 3578	.33			24	24		1.8

Contributions to the expected error in Li abundance determination include 0.08 dex, if the EW errors are 20% (they appear to be somewhat better here), and 0.08 dex from an effective temperature uncertainty of 80 K, for a combined error of 0.11 dex. The observed rms. scatter about the mean is 0.15 dex. This leads to a formal reduced chi-square value of 2.1, with a 5% probability of scatter as large as observed being due to chance. The point farthest from the mean is HD 219617, and if this star was not included in the analysis, the scatter would be 0.134, with a probability 20% of arising from chance. HD 219617 is a binary with nearly identical components, but there is no particular reason to think that this would effect its Li abundance. In the end, we must conclude as did Hobbs and Duncan: there is marginal, but not compelling evidence of real differences.

RMB make an important arguement concerning Li destruction. They point out that halo stars somewhat cooler than those of Table I, which do show Li destruction, actually have destroyed their Li much more slowly than pop. I stars of the same temperature. As one moves to higher temperatures in pop. I stars, the rate of Li destruction decreases rapidly (cf. the Hyades observations of Cayrel et. al 1984). If the same pattern obtains in pop. II, the stars in Table I could not have altered their Li abundances from that with which they formed.

4. POPULATION I OBJECTS

Equivalent width determinations are simplified in metal-poor objects by the smoothness of the continuum. Aadditional comparison was therefore made between two independent sets of abundance determinations in the same group of pop. I objects. These were Li abundance determinations in Hyades dwarfs. The first observations were those of Cayrel, Cayrel, and Cambell (C^3, 1985), who obtained spectra of S/N $\approx$ 250 and resolution 0.2Å with the CFHT Reticon. The others were unpublished observations made by D. Soderblom with a coude camera and CCD detector at Lick Observatory, at somewhat lower S/N. No systematic differences in EW were detected. Cayrel et al. derived temperatures from fitting Hα profiles, and they claim a random error of only 30 K in this method. Comparing their Hα temperatures with ones derived from B-V, R-I, and V-K colors (slightly adjusting scale zero points to match) results in scatter of 70 K, 70 K, and 40 K respectively. Again accurate temperatures are critical to the abundance determination.

5. CONCLUSION

Equivalent widths determined at high S/N with a number of different telescopes and spectrographs show every indication of being as accurate as the authors claim. All of the examined spectra were in the red part of the spectrum in solar type stars, or in metal-poor stars, where the continuum is relatively uncluttered. Abundances determined from the spectra often agree less well, due to different temperatures used in the analyses. These results emphasize the need for stellar parameters of accuracy commensurate with the high accuracy of the spectra themselves, to be used for input in abundance analyses. In some cases it may be advantageous to derive these directly from the spectra, e.g. temperatures from balmer-line profiles.

In the interesting case of the Li abundance in very metal-poor dwarfs, the most careful investigations produce no conclusive evidence that the Li observed is anything other than the uniform primordial abundance produced by the big bang.

I would like to than R. Rebolo and D. Soderblom for communicating work in advance of publication.

REFERENCES

Boesgaard, A.M., and Steigman, G. 1985, *Ann. Rev. Astr. Ap.*, **23**, 319.
Cayrel, R., Cayrel de Stroebel, G., Campbell, B., and Dappen, W. 1984, *Ap. J.*, **283**, 205.
Cayrel, R., Cayrel de Stroebel, G., and Campbell, B., 1985, *Astr. Ap.*, **146**, 249.
    Hobbs, L.M. and Duncan, D.K. 1987, *Ap. J.*, **317**, 796.
Peterson, R.C., and Carney, B.W., 1979, *Ap. J.*, **231**, 762.
Rebolo, R., Moloro, P., and Beckman, J.E. 1987, preprint.
Soderblom, D., 1987, private communication.
Spite, F., and Spite, M. 1982, *Astr. Ap.*, **115**, 357.
Spite, M., Maillard, J.P., and Spite, F. 1984, *Astr. Ap.*, **141**, 56.

DISCUSSION

PETERSON All the stars hot enough show the same lithium abundance, although their iron abundances range from $-3.0 <$ [Fe/H] < -1.4 does this imply, in your opinion, that the lithium abundance is cosmological ?

DUNCAN I think it suggests it, yes.

BOESGAARD Yes, lithium is decoupled from the Fe abundance in halo dwarfs. However, Li diffusion (as opposed to depletion by nuclear burning) could occur in a way that is not dependent on the metallicity during the $\sim$ 15 billion year lifetimes of these stars. I think the halo star Li content should be interpreted as a lower limit to the Big Bang Li production.

DUNCAN One cannot preclude this, and a conservative position is that the Li abundance is a lower limit. However, there is no reason to suppose this, either. I know of no theoretical calculations which predict Li depletion in Pop.II stars, independent of temperature. If I become convinced that star to star differencies are real, then I will believe in depletion, that is the best evidence.

MOLARO you said that, may be, there is a scatter in the Li abundances on the "plateau". Assuming this as real, does it show some trend with temperature ?

DUNCAN The scatter shows no trends with Te or [Fe/H], as long as [Fe/H] < -1.4.

SODERBLOM Ann Boesgaard suggested that diffusion, for example, could influence the apparent Li abundance of the Pop.II stars. Can anyone suggest tests to tell if a process like that has occured ? (no answer...)

LITHIUM ABUNDANCES OF SOLAR-TYPE STARS: A CRITICAL APPLICATION OF HIGH SIGNAL-TO-NOISE, HIGH RESOLUTION SPECTROSCOPY

David R. Soderblom
Space Telescope Science Institute
3700 San Martin Drive
Baltimore, Maryland 21218 USA

ABSTRACT. The study of lithium depletion in solar-type stars should help elucidate essential properties of convection in those objects. Recent high-quality observations have revealed extensive flaws in our understanding of this phenomenon. The state of our knowledge of Li depletion is reviewed, with a brief possible explanation of it all.

1. NEW LIGHT ON OLD PROBLEMS

The advent of highly efficient detectors capable of producing high quality stellar spectra has enabled more detailed study of phenomena that were limited by photographic data. One of these problems is the study of lithium depletion in solar-type stars.

Observationally, one is faced with the difficulty of having only one feature to measure. This is blended with a nearby ($\Delta\lambda = 0.3$Å), weak Fe I line, ($W_\lambda(\odot) = 8.4$ mÅ), and the Li line at 6708Å is frequently very weak itself (about 2 mÅ for the Sun). In stars cooler than the Sun, many weak CN lines add to the confusion. The benefit gained from high resolution and signal-to-noise is demonstrated by the many papers on Li abundances presented at this conference.

Having obtained a good spectrum, one wishes to analyze it to derive the star's Li abundance, log N(Li). This analysis is *very* temperature sensitive: an uncertainty in T_{eff} of 100 K leads to an uncertainty in the Li abundance of ±40%. This is because most of the star's Li is ionized, and cannot be observed directly.

Finally, one wishes to interpret the Li abundance to say something about the star. We are still uncertain how good Li is as an age indicator (Soderblom 1984), but even if a tight relationship is assumed, the time scale for Li depletion is exquisitely mass sensitive. An example is α Cen A, which is nearly indistinguishable from the Sun spectroscopically, is only slightly more massive, yet has a substantially greater Li abundance (Soderblom and Dravins 1984).

Despite all this, the Li problem is pursued because these observations should reveal essential properties of convection. The presence of a convective envelope is the distinguishing characteristic of a late-type star. Better knowledge of convection is necessary to better understand other related phenomena such as magnetic dynamos and their manifestations.

The photographic data were just uncertain enough to be consistent with a nice, simple picture: stars (at least those of the young disk) are formed with the same initial Li abundance. This is gradually depleted over the star's main sequence lifetime, at a rate determined mainly by a star's mass, as convection carries surface Li to a depth where the temperatures are high enough to destroy it ($T \gtrsim 2.4$ MK).

The details in this picture were poorly understood, but it made sense qualitatively. New, high-quality data are improving this picture, adding some embarassments, but also clarifying other areas.

2. WHAT DO WE REALLY KNOW?

(1) Over at least the past ~ 5 Gyr, stars have formed with the same Li abundance, $\log N(\text{Li}) \approx 3.0$. This is substantiated by the Li abundances of carbonaceous chondrites, T Tauris, and the most Li-abundant stars in young clusters (see, e.g., Boesgaard, Budge, and Burck 1988). If the interstellar medium is being further enriched in Li, it is at a slow rate. The precise range in Li abundance seen in different parts of our Galaxy is unknown.

(2) The Sun has $\log N(\text{Li}) \approx 1.0$ (Müller, Peytremann, and de la Reza 1975), about 1% of its original value. Note that these first two points provide unambiguous evidence that the Sun has depleted most of its initial Li (as opposed to having formed with low Li).

(3) For homogeneous groups of stars (same age and composition), Li is strongly mass-related. There is a precipitous dip among the mid-F dwarfs now so well delineated by Boesgaard and her co-workers (e.g., Boesgaard and Tripicco 1986a, b), and a more gradual decline from about G0V to lower masses (Cayrel et al. 1984).

(4) Even within such a homogeneous group there is substantial star-to-star scatter that is real, and which does not go away with advancing age. For example, see the observations of NGC 752 or M 67 of Hobbs and Pilachowski (1986a, b).

(5) Metal-poor stars (old disk and Population II) lose their Li either very slowly or not at all. Very metal-deficient stars ([Fe/H] $\lesssim -1$) have strikingly high abundances of Li (Hobbs and Duncan 1987), yet even mildly metal-deficient stars that are unarguably old (for example, β Hyi [Rebolo et al. 1986]) have embarassingly large amounts of Li.

(6) There is no evidence for any star having $^6\text{Li}/^7\text{Li}$ significantly greater than the solar value, despite what was thought from the photographic data (Andersen, Gustafsson, and Lambert 1984; Soderblom 1985; Rebolo et al. 1986).

3. WHAT DO WE THINK WE KNOW?

(1) Li depletion is related to convection, at least for stars at or below the sun's mass. This is strongly indicated by the more rapid depletion seen at lower masses, but other scenarios are possible. For example, one could postulate that the observed decline of Li with age reflects main sequence mass loss, and that that mass loss is greater in lower mass stars. (Mass loss would carry away the most Li-rich material at the surface, leading to dilution of the remainder.

(2) All stars in a cluster start with the same Li abundance. It seems highly unlikely that an appreciable spread would exist, especially since none is seen in [Fe/H], but it's not impossible, particularly if the sites of Li formation are within star-forming regions.

(3) The Li depletion rate is fixed by mass, composition, and age. If stochastic factors lead to appreciable star-to-star differences over time scales of billions of years, that would seem to violate the Vogt-Russell theorem. Although differences in a star's rotation rate could influence Li depletion (because magnetic fields can inhibit convection), for solar-type stars there appears to be a rapid convergence to a fixed value of rotation for stars of the same age and mass. Also, it doesn't seem right that modest differences in star's initial rotation rates should have observable manifestations nearly 10 Gyr later. It is also frustrating to see a solar temperature star in M 67 (a cluster of about the same age and composition as the Sun) that has much more Li than the Sun (Hobbs and Pilachowski 1986b).

(4) Chromospheric activity doesn't much affect the apparent Li abundance. Some of the star-to-star differences in Li could be explained by the influences of chromospheric activity on line formation. Both Giampapa (1986) and Boesgaard (this conference) have followed several very active stars over several rotation periods, and see no variation in the Li line at a level well below 1%. This suggests that chromospheric activity does not strongly affect the apparent Li abundance of a star, although it is still possible that *overall* chromospheric activity levels play a role.

(5) The Li abundances of halo stars ($\log N(\text{Li}) \approx 2.0$) represents the Li created in the Big Bang. There could well have been some depletion among these stars—perhaps 3.0 is the true Big Bang value.

4. WHAT DO WE NOT KNOW?

(1) The mechanism of Li depletion. However, improved observations have spurred efforts in this area (more below).

(2) The true dependence of depletion rate on mass, composition, and age— only better observations can help here.

(3) If stochastic factors play a significant role in Li depletion. Let's hope not.

(4) If autogenesis is important in determining the observed Li abundances. It's probably not, and tends to produce too much ^{6}Li anyway.

(5) If stars in clusters are really the same age.

5. WHAT DO WE NEED TO KNOW?

(1) Li in more clusters of different ages (*e.g.*, α Per and the Ursa Major Group). Such observations are in press.

(2) Li in clusters of the same or similar age, especially if they have different compositions. For example, Praesepe and Coma should be compared in detail to the Hyades.

(3) A detailed Li survey of F, G, and K dwarfs at high resolution and signal-to-noise. Is Li as rare as we think it is in stars less massive than the Sun? The distributions of Li abundances as a function of spectral type would also provide useful information.

(4) Uncertainties in Li abundances. High quality data demand high quality analyses, including good quantitative estimates of uncertainties. This is partic-

ularly important in analyzing differential effects, and for testing models against observations. Spectrum synthesis is arguably the best technique for extracting information from high quality spectra, but has not been used enough.

(5) If stars in visual binaries have consistent Li abundances.

(6) The trends in Li abundance among metal-poor stars. Although many such stars have abundant Li, not *all* of them do. Two notable exceptions are μ Cas and τ Cet. This gets back to the question of just what stellar parameters determine Li depletion.

6. MAKING SENSE OF IT ALL

Some recent calculations by Stringfellow, Faulkner, and Bodenheimer (1987) suggest a way of accounting for much of what's observed. Their work will appear separately; for these purposes it is sufficient to note that their results suggest that Li depletion in a star or cluster is painfully sensitive to metallicity. They were able to reproduce the Li vs. mass (or color) curve of Cayrel et al. (1984), but found that changes in [Fe/H] of only 10 to 20% produced very different curves.

If one were sufficiently confident of these models, they would enable the determination of cluster abundances to very high precision. More realistically, since depletion time scales are so dependent on metallicity in their models, it is probable that the internal spread in Li seen in clusters is due to different times of formation of individual stars. If this spread were due to star-to-star differences in [Fe/H], the spread would grow with age, which is not seen. An exponential process (which Li depletion is if a constant fraction is carried to a sufficiently high temperature in a given time) preserves any initial spread in Li, and that agrees with observation.

Their model also accounts for the substantial Li seen in mildly metal-deficient stars (Duncan 1981; Soderblom 1983). However, they would predict significant differences in Li between Hyades and the Coma cluster (same age but different metallicity), which isn't seen (Soderblom et al. 1987).

Better quality observations of stars in clusters should help delineate the trends well enough to make reasonable sense of Li depletion. One lesson from this, though, is reinforced by other high signal-to-noise studies: A full and accurate analysis of the observations requires a comparable level of knowledge of the relevant fundamental stellar parameters (mass, compositions, age, etc.). In fact, these parameters are generally known very poorly.

I remain convinced that we can understand the Li abundances of solar-type stars in a reasonably straightforward fashion. There is some equation that expresses how the instantaneous Li depletion rate of a star depends on its overall properties, from which one could calculate its Li abundance history. We are getting clues to the relative weightings of various coefficients, but are a long way from pinning down values. We are not even certain what all the relevant parameters are.

REFERENCES

Andersen, J., Gustafsson, B., and Lambert, D. L. 1984, *Astr. Ap.*, **136**, 65.
Boesgaard, A. M., Budge, K. G., and Burck, E. E. 1988, *Ap. J.*, **325**, in press.
Boesgaard, A. M., and Tripicco, M. J. 1986a, *Ap. J. (Letters)*, **302**, L49.
———. 1986b, *Ap. J.*, **303**, 724.

Cayrel, R., Cayrel de Strobel, G., Campbell, B., and Däppen, W. 1984, *Ap. J.*, **283**, 205.
Duncan, D. K. 1981, *Ap. J.*, **248**, 651.
Giampapa, M. S. 1986, private communication.
Hobbs, L. M., and Duncan, D. K. 1987, *Ap. J.*, **317**, 796.
Hobbs, L. M., and Pilachowski, C. 1986a, *Ap. J. (Letters)*, **309**, L17.
———. 1986b, *Ap. J (Letters)*, **311**, L37.
Müller, E. A., Peytremann, E., and de la Reza, R. 1975, *Solar Phys.*, **41**, 53.
Rebolo, R., Crivellari, L., Castelli, F., Foing, B., and Beckman, J. E., 1986, *Astr. Ap.*, **166**, 195.
Soderblom, D. R. 1983, *Ap. J. Suppl.*, **53**, 1.
———. 1984, in *Cool Stars, Stellar Systems and the Sun*, eds. S. L. Baliunas and L. W. Hartmann (Heidelberg: Springer), p. 205.
———. 1985 *Pub. A.S.P.*, **97**, 54.
Soderblom, D. R., and Dravins, D. 1984, *Astr. Ap.*, **140**, 427.
Soderblom, D. R., Oey, M. S., Stone, R. P. S., and Johnson, D. R. H. 1987, *Ap. J.*, in preparation.
Stringfellow, G., Faulkner, J., and Bodenheimer, P. 1987, private communication.

DISCUSSION

        R. CAYREL   In connection with the reported absence of modulation of the Lithium resonance line in active stars, I want to make the following remark :
The lithium line is enhanced in spots but weakened in plages. A simple computation shows that if the area ratio of plages/spots is as in the active Sun the expected modulation is almost nil.

        SODERBLOM   I agree that the sense of the modulation is not obvious, which is why simultaneous data for <u>both</u> Li and Ca II H and K or Hα must be obtained.

        REBOLO   Could you give us information about the type of the stars in which rotational modulation of the lithium line has been searched and the precision obtained in the work ?

        SODERBLOM   See the remark by Ann Boesgaard. She reports that the equivalent width of the lithium feature in χ^1 Orionis changed by less than $\approx$ 1/2% over its rotation period.

        BOESGAARD   I have made observations to look for rotational modulation of Li in 6 stars at CFHT for 4 nights. The stars included χ^1 Ori, κ Cet and others. (I can give you the list.). None showed any variation in the Li equivalent width to better than 0.5 <u>percent</u>. In χ^1 Ori the LiI line is 103mÅ±0.5mÅ over the four nights.

        SODERBLOM   Did you have any measurement of the chromospheric variations then ?

        BOESGAARD   Yes. I asked for simultaneous observations of CaII at Mt. Wilson ; they had poor weather and some equipment problems. I made observations at Hα in between the Li observations as a check on the chromospheric activity changes. I did find some modulation in the core of Hα for χ^1 Ori.
[See separate page for my other comment which followed this].

        SODERBLOM   This is very encouraging -- at least these stars are <u>consistently</u> fooling us, if they're fooling us at all. I still wonder if the <u>overall</u> enhancement of chromospheric activity in young stars can influence the apparent Li abundance, but this lack of modulation to such a fine level makes that highly unlikely.
I also thank G.Marcy for clarifying a rumor that some Pleiades had exhibited considerable Li variability over long time spans. They do not, which is also encouraging.

        BOESGAARD   You pointed the difficulties with the features which blend with the weak solar LiI line, the CN and the FeI line. These are not a problem for the early F stars where the FeI line at 6707.441 Å is $\leq$ 1 mÅ and CN is virtually non-existent.

        SODERBLOM   You are correct, of course-- the problem is simpler until one reaches the stars where rotation blends the lines.

Accurate spectroscopic surface gravities for 8 sub-giants

Bengt Edvardsson
Astronomiska Observatoriet
Box 515
S-751 20 Uppsala
Sweden

ABSTRACT. Logarithmic surface gravities between 2.9 and 3.8 (cgs) have been determined for eight G8IV - K2III stars from the analysis of pressure broadened wings of strong metal lines. Comparisons with trigonometrically determined surface gravities give support to the spectroscopic results. A thorough and rather conservative error analysis suggests that the errors are smaller than about 30% and confirms that the method is very insensitive to uncertainties in the fundamental stellar atmospheric parameters. Surface gravities of α Cen A and B and of Arcturus have also been obtained. Chemical abundances have been derived for 12 elements from Na to Ni, and for five heavier elements from Y to Nd have tentative abundances been derived from a single line of each element. Effective temperatures in the photometric system of Frisk (1983) are found to be consistent with spectroscopic temperatures from the excitation equilibrium of Fe I. Surface gravities determined from the ionization equilibria of Fe and Si are found to be systematically lower than the strong line gravities, which may be an effect of errors in the model atmospheres, or departures from LTE in the ionization equilibria.

A recommendation

I only wish to urge those who determine surface gravities from ionization equilibria, which are very sensitive to model structure parameters and to non-LTE effects, to check - when possible - their log g values with gravities determined from the wings of pressure broadened metal lines.

In order to save space the reader is referred to Edvardsson (1987).

Reference

Edvardsson, B.: 1987, *Astron. Astrophys.*, in press.

DISCUSSION

GUSTAFSSON In connection with Arcturus I just want to point out that the method, as designed by Edvardsson, was not ideal for obtaining the gravity of a low-gravity object like Arcturus. Thus, the errors expected in the gravity determination for the subgiants should be considerably smaller.

EDVARDSSON Yes, indeed, the most suitable lines between 5000–10000 Å which I have selected are getting too weak and less sensitive to the gas-pressure when we go from sub-giants to giants. For giants and low-metallicity stars one could go to stronger lines at shorter wavelengths (c.f. the paper by Pierre Magain in this volume), but then one will have to be very careful with continuum problems due to increased line-blending at shorter wavelengths and problems with the model opacities at these wavelengths, most recently discussed by R. Buser and R.L. Kurucz in this volume.

NLTE ANALYSIS AND CHEMICAL COMPOSITION OF HOT LOW-MASS STARS

U. Heber, K. Hunger and K. Werner
Institut für Theoretische Physik und Sternwarte
Olshausenstr. 40
D-2300 Kiel 1
Federal Republic of Germany

ABSTRACT. Spectroscopic analyses of hot pre-white dwarfs, i.e. subluminous O and B stars, are presented. In the B-type stars, the resulting abundance patterns are indicative of atmospheric diffusion (gravitational settling). Amongst the O-type subdwarfs, a new group of comparatively luminous stars is identified. Their position in the HR-diagram suggests that, unlike the "classical" sdOs, they are in a post-AGB stage of evolution. Spectroscopic evidence is presented showing that the born-again post-AGB star scenario of Iben et al. (1983) can explain their origin.

1. INTRODUCTION

The hot low-mass stars in question are the various types of pre-white dwarfs. The central stars of planetary nebulae (CSPN) have long since been recognised as immediate progenitors of the white dwarfs. In addition, two other groups of stars, the sdB and sdO stars, must be considered as pre-white dwarfs. Recent surveys have shown that the sdOs and sdBs are not just rare "freaks", but, in fact, dominate the population of faint blue stars (Green et al., 1986).

In their pioneering work on faint blue stars, Greenstein and Sargent (1974) carried out spectroscopic analyses of various O- and B-type subdwarfs. The evolutionary status of both groups, however, remained quite unclear. They appeared closely related to the horizontal-branch stars and an evolutionary link between the two groups seemed likely. However, growing evidence has recently been reported (Heber et al., 1984; Groth et al., 1985; Heber, 1986) suggesting that no such direct evolutionary link exists and that these groups of stars form two distinct evolutionary channels towards the white-dwarf stage. From these recent spectroscopic analyses, the following picture of pre-white dwarf evolution has emerged.

The three distinct evolutionary channels (CSPN, sdO, sdB) all start at the horizontal branch (HB). According to available evolutionary calculations, the evolution of a post-HB star depends strongly on the core-mass to total-mass ratio q. If q is small (red part of the

HB), a post-HB star will ascend the AGB, eject a nebula at its tip and evolve as a CSPN towards the white-dwarf domain. If q is very large (q > 0.95), we reach the extremely blue end of the HB, the so-called extended horizontal branch (EHB). An EHB star is left with only one energy source, the helium-burning core, since the hydrogen-rich envelope is inert. The internal structure of an EHB star bears great resemblance to a helium main-sequence star (of half a solar mass) with a thin hydrogen-rich shell on top of it. An EHB star will evolve (similar to a helium main-sequence star) directly to the blue and into a white dwarf. The subluminous B stars can be identified with these EHB models (Heber et al., 1984).

For slightly smaller values of q (q < 0.95), a post-HB star evolves towards the AGB but does not ascend it. Instead, the evolutionary track turns towards the blue and towards the white-dwarf domain. Sweigart et al. (1974) were the first to identify the sdO stars with such evolutionary tracks.

In this review we shall discuss mainly new spectroscopic results for the subluminous O and B stars but will come back to the CSPNs at the end of it.

2. ANALYSIS AND CHEMICAL COMPOSITION OF SDB STARS

Some thirty sdB stars were analysed by model atmosphere techniques (see Heber, 1987, for a more detailed review). Their effective temperatures range from 22000 K to 40000 K and their surface gravities from $\log g = 5$ to 6. All the stars were found to be helium deficient.

Metal abundances have been derived for 11 stars only. The analyses of these were aimed mainly at the nucleogenetically important elements, carbon and nitrogen, and at silicon.

It turned out that helium deficiency is accompanied by deficiencies of carbon and silicon in all except two stars, whereas N is almost normal in all objects studied so far. There seems to be a trend for the Si abundance to decrease with increasing T_{eff}: Si is extremely deficient (3 dex or more) in all stars hotter than 30000 K while it is only mildly deficient (if at all) at lower temperatures. The greatest carbon deficiency also occurs in the hottest sdBs. Greenstein and Sargent (1974) were the first to suggest that the helium deficiency in the sdBs can be naturally explained by gravitational settling (diffusion). If this is true, the metal abundances must also be affected by diffusion. Indeed, gravitational settling appears to be the only conceivable mechanism which could produce the large Si and C deficiencies observed.

Admittedly, most of what is known today about the abundances of metals in the atmospheres of sdBs is based on high-resolution UV spectra obtained with IUE. These spectra are quite noisy. High-S/N, high-resolution optical spectra of a dozen sdBs were obtained recently using the ESO-Cassegrain echelle spectrograph (CASPEC). Equivalent widths as low as 10 mÅ can be measured which will allow precise abundance patterns to be determined. Moreover, the high spectral resolution which can be achieved with CASPEC will also allow the isotopic ratio $^3\mathrm{He}/^4\mathrm{He}$ to be measured.

3. NLTE ANALYSIS OF VERY HOT SDO STARS

The ESO-CASPEC was also used recently to observe about 20 relatively bright sdO stars ($10 < V < 13$) at high spectral resolution (0.25 Å) and high S/N (30 to 100). Most of the objects were selected from the (low galactic latitude) survey by Drilling (1983).

IUE observations have indicated that these objects are hotter ($T_{eff} \gtrsim 60000$ K) than any of the sdO stars analysed previously (e.g. Hunger et al., 1980). The high-resolution optical spectra confirm the very high T_{eff} in many cases, since He I is often absent, the limiting equivalent width being ~ 30 mÅ.

A subsample of 9 stars seemed to be especially interesting because their spectra displayed emission lines of highly ionised metals and/or relatively narrow Balmer lines (see Heber and Hunger, 1987). The latter is regarded as evidence for a comparatively low surface gravity. We analysed the spectra of these nine objects using improved NLTE model atmospheres computed with the "accelerated-lambda-iteration" method described by Heber et al. (these proceedings). As an example of the quality achieved in the analysis, we display in Fig. 1 the line profile fit for the hottest programme star LSS 1362 ($V = 12.5$).

The results are summarized in Fig. 2, where the programme stars are designated by error bars. Other sdOs, analyzed previously from low resolution spectra are also plotted (without error bars). Three programme stars have normal helium abundances, whereas the others are enriched in helium. Four of the latter even do not show any hydrogen.

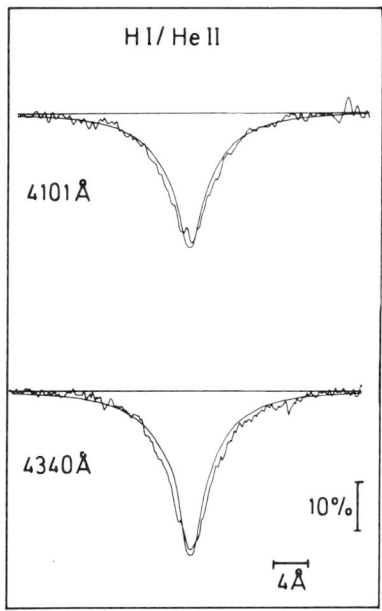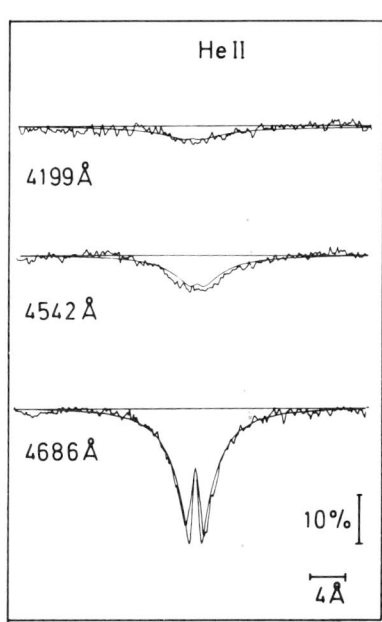

Figure 1: H- and He II- line profile fits for LSS 1362. The model parameters are: $T_{eff} = 100000$ K, $\log g = 5.3$, $n_{He}/n_H = 0.1$.

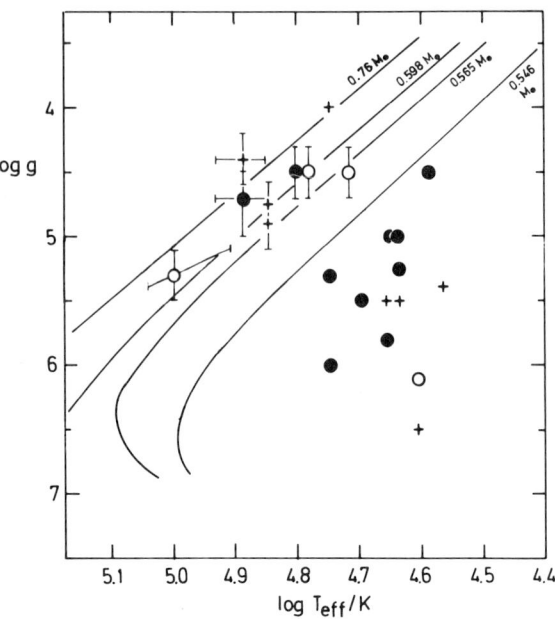

Figure 2: Position of the programme stars in the (g, T_{eff})-plane and comparison with evolutionary tracks (see text). Stars with normal helium abundance are shown as open circles, intermediate helium-rich ones as filled circles and extremely helium-rich stars as crosses.

Three of the extremely helium-rich sdOs (LSE 153, LSE 259, LSE 263) have also been analyzed for the abundances of carbon, nitrogen and magnesium (Husfeld, 1987). N is enriched in all three objects (0.9 to 1.8 dex.) as is C in LSE 153 and LSE 259 (by about 1 dex.). LSE 263, however, is C-deficient by 1 dex. (A magnesium overabundance might still be uncertain since it is based on the detection of only one line, Mg II, λ 4481Å.)

Also shown in Fig. 2 are evolutionary tracks descending from the asymptotic giant branch for masses between 0.546 $M_\odot$ and 0.76 $M_\odot$ (Schönberner, 1979, 1983; Wood and Faulkner, 1986). As can be seen from Fig.2 our stars do not lie in the region where the classical sdO stars are to be found. Instead, they can be identified with post-AGB tracks of about 0.6 $M_\odot$ and with a mean luminosity of about $10^{3.8}$ $L_\odot$. Effective temperatures and gravities of the programme stars (as well as their masses and luminosities) are typical for central stars of planetary nebulae (CSPN). Hence, spectroscopically, they may also be termed as CSPN except they lack the nebulae. Or perhaps the nebulae have not been noticed ? A careful inspection of the ESO sky survey plates did in fact reveal a very extended faint nebulosity around LSS 1362 (see Heber et al., 1987). All others do not show nebulosities on the sky survey plates. Why is this the case ?
There are three conceivable reasons why no nebulae can be detected:
i)   The stars simply left the AGB without ejecting a nebula. This is not a satisfactory answer (at least for helium rich stars) since it cannot explain the enrichment of helium, carbon and nitrogen.
ii) After ejection of a nebula, our stars evolved much more slowly than the other CSPNs and, therefore, the nebulae had already been

dispersed before the stars became hot enough to ionize them. As pointed out by Heber and Hunger (1987) this scenario is unlikely to be correct for the programme stars.

iii) The third and most interesting explanation is the concept of "born again" post-AGB stars. After ejection of a nebula, such a star crosses the HR diagram in the usual way, i.e. as a true CSPN, and finally reaches the hot end of the cooling sequence of white dwarfs. According to Iben et al. (1983), a last thermal pulse may occur in this phase. Such a pulse brings the star back to red giant temperatures and dimensions. During the pulse, most of the hydrogen left to the star at the onset of the pulse is mixed into the helium-burning convective shell and thus is completely burned. The star is now almost devoid of hydrogen and proceeds to burn helium in a shell. The evolutionary track in the (g, T_{eff}) diagram is approximately the same as that for hydrogen-burning post-AGB stars (CSPNs). However, no new nebula is expelled and the old one has long since disappeared. This scenario can also solve the riddle of the helium, carbon and nitrogen enrichment. A final hydrogen burning episode is predicted during the peak of the final thermal pulse and a mixing episode during the giant phase (see Iben et al., 1983). Both mechanisms can mix processed material (He and N from the CNO-cycle; C from the 3α-process) to the surface.

Hence we conclude that two different (evolutionary) subclasses of sdOs exist. A sdO might either evolve along a "Sweigart" track (see 1.) or as a "born-again post-AGB" star. As a consequence, sdO stars cover a wide range in absolute magnitude ($-1 \lesssim M_V \lesssim 7$).

4. REFERENCES

Drilling, J.S.: 1983, Astrophys. J. (Letters), **270**, L13
Green, R.F., Schmidt, M., Liebert, J.: 1986, Ap. J. Suppl. **61**, 305
Greenstein, J.L., Sargent, A.I.: 1974, Astrophys. J. Suppl. **28**, 157
Groth, H.G., Kudritzki, R.P., Heber, U.: 1985, Astron. Astrophys. **152**, 107
Heber, U.: 1986, Astron. Astrophys. **155**, 33
Heber, U.: 1987, IAU coll. no. 95, Davis press, in press
Heber, U., Hunger, K.: 1987, ESO messenger No. **47**, 36
Heber, U., Werner, K., Drilling, J.S.: 1987, Astron. Astrophys., submitted
Heber, U., Hunger, K., Jonas, G., Kudritzki, R.P.: 1984a, Astron. Astrophys. **130**, 119
Hunger, K., Gruschinske, J., Kudritzki, R.P., Simon, K.P.: 1981, Astron. Astrophys. **95**, 244
Husfeld, D.: 1987, IAU coll. No. 95, Davis press, in press
Iben, I. Jr.: Kaler, J.B., Truran, J.W., Renzini, A.: 1983, Astrophys. J. **264**, 605
Schönberner, D.: 1979, Astron. Astrophys. **79**, 108
Schönberner, D.: 1983, Astrophys. J. **272**, 708
Sweigart, A.V., Mengel, J.G., Demarque, P.: 1974, Astron. Astrophys. **30**, 13
Wood, P.R., Faulkner, D.J.: 1986, Astrophys. J. **307**, 659

Table I: Atmospheric parameters of the programme stars plotted in Fig.2

star	T_{eff}/K	log g	$n_{He}/(n_H+n_{He})$
ROB 162[a]	51000	4.5	0.1
BD+37°442[b]	55000	4.0	1.0
LS IV-12°1	60000	4.5	0.1
BD-3°2179	62000[c]	4.5[c]	0.2[c]
LSE 153[d]	70000	4.75	>0.9
LSE 263[d]	70000	4.9	>0.9
LSE 259[d]	75000	4.4	>0.95
KS 292	75000	4.7	0.25
LSS 1362	100000	5.3	0.1

[a] Heber and Kudritzki (1986, Astron. Astrophys. **169**, 244)
[b] based on photographic spectra, Giddings, 1980 Ph.D. thesis UCL
[c] preliminary values, analysis is not completed
[d] see Husfeld (1987)

DISCUSSION

JORISSEN Is it possible to have information about abundances for elements heavier than iron in these stars from UV spectrum ? It could provide some support to the "last thermal pulse" proposed scenario, if heavy elements appeared to be overabundant !

HEBER The high resolution IUE spectra display a crowding of lines in the 1200 Å to 1800 Å region. Many of these lines were identified as Fe V, Fe VI or Fe VII. However, more than half of the lines remained unidentified. It might well be that some of them arise from elements heavier than iron. Laboratory spectral analysis is quite incomplete for such highly ionized species.

THE CHEMICAL COMPOSITION OF VEGA

Detlef Gigas
Institut für Theoretische Physik
und Sternwarte der Universität Kiel
Olshausenstrasse 40
D-2300 Kiel 1
Federal Republic of Germany

ABSTRACT. A non-LTE abundance determination of magnesium, iron, and barium was carried out for the atmosphere of Vega. The results indicate marked underabundances of these elements relative to the solar values. First preliminary results of numerical simulations of convective and oscillatory phenomena in the atmosphere of Vega are briefly described.

1. INTRODUCTION

In spite of the importance of Vega (α Lyr, A0V) as a primary standard for flux measurements and as a reference for abundance studies, comparatively little is known about the chemical composition of this star. Contrary to earlier studies, more recent LTE abundance determinations suggest marked deficiencies of many elements relative to the Sun.

Our investigation, based on high-dispersion, high S/N photographic spectrograms recorded by R. and R. Griffin at Mount Wilson and covering a wavelength range from 3050 Å to 6850 Å, aims at a reliable assessment of element abundances for three elements: magnesium (representing the light elements), iron ('iron peak' elements), and barium (heavy elements). Departures from LTE are taken into account explicitly for these elements.

In what follows we briefly describe some of the results. Details are given elsewhere (Gigas, 1986 and 1987).

2. MODEL ATMOSPHERE AND COMPUTATIONAL PROCEDURE

The abundance determination is carried out with a line-blanketed ATLAS6 LTE model atmosphere with T_{eff} = 9500 K, log g = 3.90 (Lane and Lester, 1984), and an overall metallicity of [M/H] = -0.5 dex. A depth-dependent microturbulence with an average value of $\xi \sim$ 2.0 km/s was employed for the line analysis.

The non-LTE calculations are carried out with the code developed by W. Steenbock, which is based on the complete linearization scheme of Auer and Heasley (1976; see also Steenbock and Holweger, 1984).

3. MODEL ATOMS AND ATOMIC DATA

The model atoms were constructed with the intention of including all relevant energy levels and strong radiative transitions. In the case of sparsely populated ionization stages, special attention was paid to the inclusion of a relatively large number of high-lying energy levels, which may be important for the collisional coupling to the next ionization stage.

Our model atoms comprise 71+28+1 (Mg I/Mg II/Mg III), 79+20+1 (Fe I/Fe II/Fe III), and 42+1 (Ba II/Ba III) energy levels with 41+30 (Mg I/Mg II), 52+23 (Fe I/Fe II), and 36 (Ba II) line transitions treated explicitly. All atomic data were selected after a critical survey of recent experimental and theoretical results.

4. DEPARTURES FROM LTE AND NON-LTE ABUNDANCE CORRECTIONS

4.1. Magnesium

Except for the two lowest Mg I levels, which are slightly underpopulated for $\log \tau_{5000} < -0.5$, all Mg I level occupation numbers are close to their LTE values for optical depths larger than $\log \tau_{5000} \sim -1.5$. While the Mg II ground state and the Mg II 3p level are hardly affected by departures from LTE, higher levels show an underpopulation relative to LTE. For some levels (e.g. the 4f level) non-LTE effects become noticeable already at optical depths of $\log \tau_{5000} \sim -1.0$, while others (e.g. the 3d level) start deviating from LTE at $\log \tau_{5000} \sim -2.5$. The Mg III continuum is overpopulated relative to its LTE occupation number in the outer atmospheric layers.

Non-LTE abundance corrections are small for weak lines of Mg I and Mg II; corrections of -0.10 dex are present for stronger transitions.

4.2. Iron

Fe I levels are in LTE for $\log \tau_{5000} > 0.0$; towards smaller optical depths there is an increasing underpopulation due to the effects of photoionization. We note that in the range $-1.5 < \log \tau_{5000} < 0.0$ (a region important for the formation of many spectral lines) all Fe I levels are subject to similar departures from LTE due to their mutual coupling by electron collisions. Fe II levels are generally in LTE up to $\log \tau_{5000} \sim -2.5$. Some high-lying levels of Fe II and the Fe III continuum exhibit moderate departures from LTE in the outermost layers.

Non-LTE abundance corrections of the order of +0.32 dex are derived for Fe I lines. Except for some strong transitions in the near UV, they are in general negligibly small for lines of Fe II.

4.3. Barium

All Ba II levels are more or less underpopulated at optical depths smaller than $\log \tau_{5000} \sim -0.2$. High-lying levels exhibit only small departures while the underpopulation is most pronounced for the ground state. The Ba III continuum population number differs only slightly from its LTE value.

Our investigation yields non-LTE abundance corrections for the Ba II resonance lines comparable to those derived for lines of Fe I.

5. ELEMENT ABUNDANCES RELATIVE TO THE SUN

Comparing our results with the solar abundance values compiled by Grevesse (1984), we arrive at the following element abundances relative to the Sun:

Magnesium	[Mg/H]	=	(-0.58 ± 0.15) dex
Iron	[Fe/H]	=	(-0.55 ± 0.22) dex
Barium	[Ba/H]	=	(-0.21 ± 0.23) dex

Abundance values derived from both ionization stages of magnesium and iron agree within their mutual error limits. Our results thus indicate a definitely non-solar composition of the atmosphere of Vega.

6. HYDRODYNAMIC STABILITY OF THE ATMOSPHERE

Atmospheres of A0-type stars are generally assumed to be static. For Vega, mixing-length theory (ATLAS6 model) predicts maximum convective velocities of only $\sim (0.25 - 0.50)$ km/s. Nevertheless, a microturbulence (convective velocities? pulsational phenomena?) of ~ 2.0 km/s is derived from line analyses, even if non-LTE effects are accounted for.

To seek a possible explanation for this discrepancy, we are currently performing numerical simulations of convective phenomena in A-type stellar atmospheres. Our code is based on the method of bicharacteristics as described by Stefanik et al. (1984), which solves the time-dependent nonlinear equations of motion in two dimensions on the assumption of cylindrical symmetry.

First preliminary results indicate the presence of atmospheric oscillations with flow velocities considerably larger than those predicted by mixing-length theory and comparable to the observed microturbulence. Further work on this subject is in progress.

REFERENCES

Auer, L.H., Heasley, J.N.: 1976, *Astrophys. J.* **205**, 165
Gigas, D.: 1986, *Astron. Astrophys.* **165**, 170
Gigas, D.: 1987, *Astron. Astrophys.* (in press)
Grevesse, N.: 1984, *Physica Scripta* **T8**, 49
Lane, M.C., Lester, J.B.: 1984, *Astrophys. J.* **281**, 723
Steenbock, W., Holweger, H.: 1984, *Astron. Astrophys.* **130**, 319
Stefanik, R.P., Ulmschneider, P., Hammer, R., Durrant, C.J.: 1984, *Astron. Astrophys.* **134**, 77

DISCUSSION

   RUTTEN   Let me first comment that it is ridiculous to have to cram so much outstanding work into a ten-minute presentation ! I have many questions, but the one I want to ask now is how sensitive your NLTE underpopulations and abundance deficiencies are to the gradient of the model atmosphere. It might be much flatter in reality than you have assumed due to mechanical heating by the motions you observe, and also by yet stronger line blanketing in the ultraviolet.

   GIGAS   It is possible that alterations of the thermal structure may influence the derived departures. Since the amount of mechanical heating is still unknown, the influence of shock wave dissipation could not be checked. The impact of mechanical heating on the derived departures will also depend on the atmospheric layers in which mechanical energy is dissipated. To get an assessment of the importance of alterations of the atmosphere's thermal structure, I computed departure coefficients for atmospheric models with slightly different values of effective temperature and overall metallicity. The changes of the non-LTE abundances derived depend on the element investigated ; they are larger for Fe I, Mg I, and Ba II than for Mg II and Fe II.

   GRAY   Do you find any correlation between temperature and velocity in your oscillation calculations ? As you may recall from D. Dravins' talk, Sirius shows inverse line asymmetries. Admittedly Vega is not Sirius, but they are close enough so one would expect similar asymmetries in Vega. (A correlation of temperature with velocity is needed to produce asymmetries).

   GIGAS   The models calculated so far are characterized by an almost horizontally homogeneous temperature distribution while vertical temperature differences may occur. Whether there is a correlation between temperature and velocity fields has not been investigated by now. In any case, observations of line shape asymmetries should provide an important test for the accuracy of gas flows simulations in stellar atmospheres.

   R. CAYREL   I am surprised by your conclusion that elements are underabundant as a result of your NLTE analysis, if you have the departure coefficients b's smaller than one. You would then expect that the LTE abundances must be corrected upwards and not downwards.

   GIGAS   We are computing abundance values from observed equivalent widths. To account for an observed equivalent width in the presence of level underpopulations due to non-LTE effects, a higher element abundance will be derived if non-LTE effects are taken into account. In an LTE analysis, an even larger underabundance would result for Fe I and Ba II than in the case of a non-LTE abundance determination.

   ANDERSEN   Your oscillation computations seem to predict a radial velocity variation with a period of perhaps 10-20 min and an amplitude of 1-2 kms^{-1}. This should be readily detectable with present techniques (if the oscillations are in phase over the stellar surface).

   GIGAS   Since I could only present preliminary results of first computations at this Symposium, periods and amplitudes of these oscillations will have to be investigated in more detail.

GUSTAFSSON The fact that the Fe II lines also give this low iron abundance seems to indicate that the thermal-structure uncertainties and the uncertainties in the non-LTE calculations should not be too important for the main abundance results. So we have to face that this young star probably only has 1/3 of the solar Fe and Mg abundances. You must have some speculations concerning the reason for this.

GIGAS According to a recent paper by Michaud and Charland (Astrophys. J. 311, 326 (1986)), metal underabundances present in the atmospheres of λ Boo stars (a group of metal deficient A-type stars) may be explained by diffusion processes in the presence of (small) mass losses. Perhaps similar processes have been going on in Vega as well.

CHALABAEV Recently, J. Borsenberger computed a non-LTE model atmosphere for T_{eff} = 9500°K and log g = 4 (see Chalabaev et al. in this volume). He found that the thermal structure of the atmosphere is quite different from an LTE case. The different temperature profile in the atmosphere can change the derivate abundances.

GIGAS At present, fully line-blanketed non-LTE model atmospheres are not available for A-type stars. Thus, one has to choose between model atmospheres computed including departures from LTE for H and certain light elements (Hubeny, Astron. Astrophys. 98, 96 (1981)), but neglect of line-blanketing effects, and LTE model atmospheres computed with a detailed treatment of line blanketing (Kurucz, Astrophys. J. Supp. 40, 1 (1979)). The impact on the derived abundances will also depend on the optical depths in which LTE and non-LTE temperature profiles differ.

FURENLID Can you give some numbers for the NLTE departures in the ionization equilibra of the studied elements ?

GIGAS In principle, departure coefficients for ionization equilibria (e.g. Fe I/Fe II) can easily be derived from the departures of all levels involved in the computation. In the case of Fe I, almost all levels show uniform departures for log $\tau \geqslant$ -1.5. The ionization equilibrium in this region should thus be shifted by about the same amount.

JUGAKU
1) Which gf-values of Fe II lines did you use ?
2) Sadakane et al. (1985) found $v_{(micro-turbulence)}$ = 2.0±0.5 km/sec from Fe II lines in the region 2100-3000 Å. Combined with your analysis I am inclined to think that v is not dependent of optical depth. Would you comment on this point ?

GIGAS
1) Fe II transition probabilities were compiled after a critical survey of recent experimental and theoretical results. In particular, I consulted the recent critical NBS compilation of atomic gf-values (Martin et al., in preparation).
2) It is encouraging to see that a depth-independent microturbulence of v = 2.0 km/sec gives a reasonable fit to the line data as well. However, I feel that this fit can be improved by using a depth-dependent v like the one employed in my investigation, which, moreover, was derived taking into acount departures from LTE for Fe levels.

A NEW LOOK AT THE Am STARS

C. Burkhart
Observatoire de Lyon
69230 St Genis-Laval
France

M.F. Coupry and C. van't Veer
Institut d'Astrophysique de Paris
98 bis Bd Arago
75014 Paris
France

ABSTRACT. High resolution, high signal-to-noise observations are required for a better understanding of the Am stars and related topics. Two examples are shown : the relationship between lithium content and temperature in the Am stars of the Hyades cluster and the observational problem induced by multiple spectroscopic binarity frequent among the Am stars.

1. INTRODUCTION

Up to now, there are very few studies of Am stars performed with the new spectroscopy, that is high resolution, high signal-to-noise. Yet, the new spectroscopy allows:
- the study of the abundances of new elements such as He,Li,C,N,O, which appears very interesting especially as the best theory to explain metallic-line stars properties, the diffusion theory, leads to the best accurate computations in the case of the light elements,
- better abundance estimates thanks to better spectra making weak lines available, which is important because the atmospheric abundance anomalies of the Am stars are mild and a higher accuracy is necessary to hope for the construction of any diffusion model of a given star, to show up from star to star any correlation between elemental abundances, or any pattern of an atomic sequence (Sr,Y,Zr; the Rare Earths;...), and
- the discovery and the study of the multiple spectroscopic binarity whose occurence is very high among the Am stars.

Here, we exhibit some results on the lithium abundance in the Am stars of the Hyades cluster and how the binarity is an observational problem for these stars.

All the spectra were obtained in the region $\lambda\lambda 6675-6725$ with the coudé spectrograph of the Canadian-Franco-Hawaïan 3.6m telescope, the 1800 lines/mm holographic grating, and a cooled Reticon array of 1872 diodes. The dispersion is 1.97 Å/mm or 0.0295 Å/pixel. The signal-to-noise is generally greater than 300.

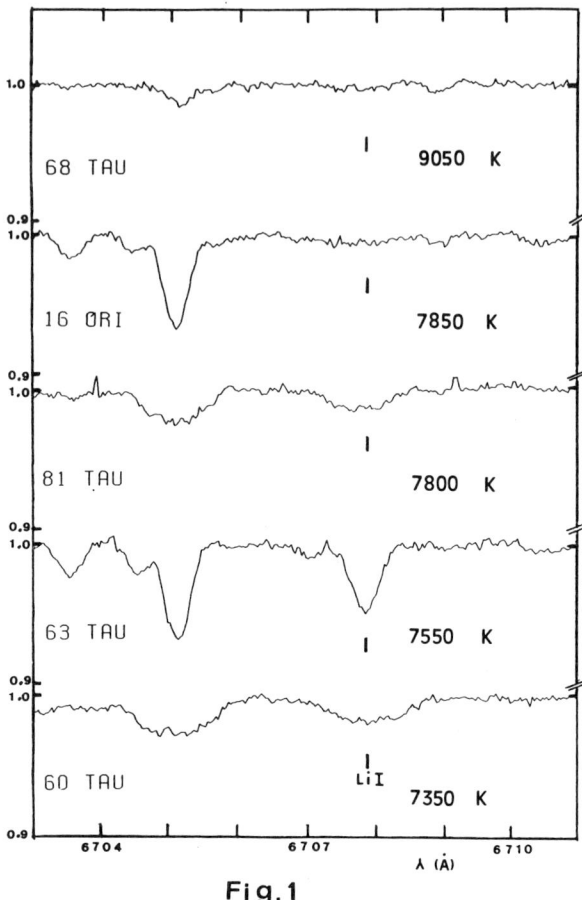

Fig. 1

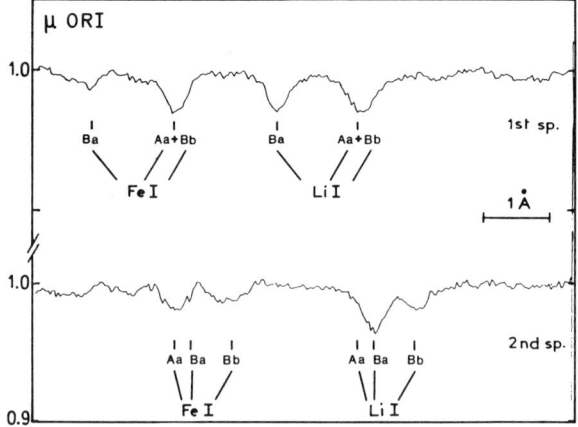

Fig. 2

2. LITHIUM IN THE Am STARS OF THE HYADES CLUSTER

Figs. 1 and 2 show a part of the spectral region observed for 6 Am stars of the Hyades cluster. Fig.3 shows the Li abundance results as a function of the temperature. The temperature T_{eff}, which is of critical importance in the abundance determination, is determined from uvby,β photometry. A set of model atmospheres (Kurucz,1979) is used to calculate the equivalent width of LiI-6707Å as a function of T_{eff} in the weak-line limit. The Li abundance follows from the measured $\lambda 6707$ Å equivalent width.

For the 6 Am stars, ranging from 7500 to 9000K, presumably having all the same age and the same original interstellar material, the Li abundance is constant - $\log N(Li)=3.0$ with $\log N(H)=12$, hereafter called the normal Li abundance -, excluded one star deficient by 0.7 dex, its temperature =7850K.

If we compare these results with those of A.Boesgaard (1987) for 5 Am stars of the Coma Berenices cluster of nearly the same age as the Hyades cluster, the same outline appears : the Li abundance is normal except near 8000K where the abundance range is 0.75 dex.

We have no total explanation of these results. A large range of the Li abundance is expected by microscopic diffusion processes, the actual abundance in the surface being very sensitive to the depth of the mixing zone. On the other hand, the stars with abnormal Li abundance are not very far from the turn-off of the clusters. We may wonder if changes in the structure of the stellar envelope induced by some weak evolution have no crucial effects for the processes acting upon the elements.

Other parameters than temperature and depth of the convective zone, age and degree of evolution, such as rotational velocity, mass loss... must be considered. More observations with the new spectroscopy are needed to disentangle the different parameters, but it must be certainly rewardful as well concerning the understanding of the Am stars and information about their stellar envelope as a better insight of the hydrodynamical processes involved where 2 adjustable parameters, the mixing length and the microturbulence are employed.

3. MULTIPLE SPECTROSCOPIC BINARITY

Multiple spectroscopic binarity may induce mistakes when assuming a line to some component.

Thus, the careful investigation of the close triple system μ Ori of the Hyades cluster by Fekel(1980), using Mc Donald 2.7m, the coudé spectrograph and a Reticon array, allowed us a correct connection between the lines observed and the 3 components in Fig.2 and avoided us to find μ Ori Aa to be an Li- overabundant Am star by a factor of 4, taking advantage of the accuracy of the wavelength and the line profile of our spectra.

Multiple spectroscopic binarity stops any study of abundance determination if the system is not well-known and this frequently occurs among

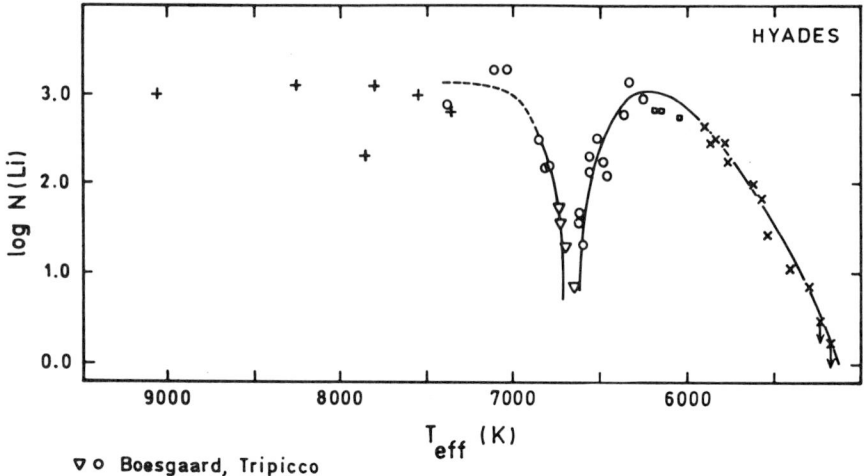

∇ ○ Boesgaard, Tripicco
× Cayrel, Cayrel, Campbell, Däppen
+ This work

Fig. 3

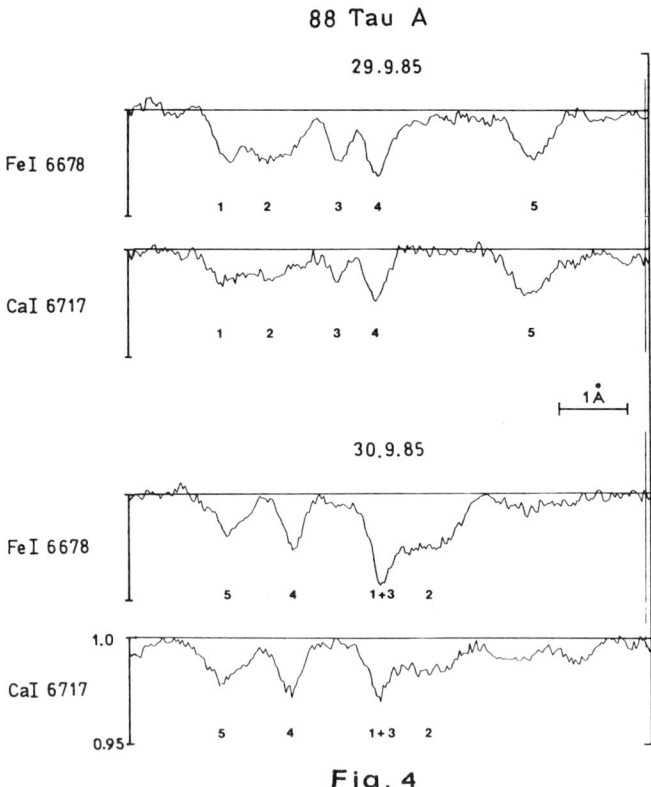

Fig. 4

the Am stars.

Thus, 88 Tau A (V=4.25) was known to be a SB2 Am star and lately resolved by speckle. In the lithium wavelength region, Fig.4, 5 systems of lines clearly appear. From the first observation to the second one, we were able to recognize each component but any abundance determination has to wait for a comprehensive study of this peculiar system.

REFERENCES
Boesgaard,A.M.: 1987, to be published in the Ap.J.
Boesgaard,A.M., Tripicco,M.: 1986, Ap.J. 302, L49
Cayrel,R., Cayrel de Strobel,G.,Campbell,B.,Däppen,W.: 1984,Ap.J.283,205
Fekel,F.C.: 1980, Publ.A.S.P. 92, 785
Kurucz,R.L.: 1979, Ap.J. Suppl. 40, 1

DISCUSSION

GERBALDI A remark : The λ 6708 region has been observed with high resolution at ESO in order to detect and analyse the abundance of Li in a sample of cool CP2 stars. Very puzzling results were obtained : see in particular : Faraggiana, Gerbaldi, Castelli, Floquet, 1986, Astron. Astrophys, <u>158</u> 200. More observations made at Observatoire de Haute-Provence (Gerbaldi, Faraggiana, 1987, Colloquium "l'Histoire et l'Avenir de l'OHP") reveal a far more complicated situation. Three stars having the same T_{eff}, the same log g ,the same spectroscopic peculiarities (Sr, Cr, Eu lines enhanced) have a blend at 6707.8 Å -the position of the LiI doublet- strongly different from one star to another, and close to it we notice unknown features.
Moreover 70% of our sample of observed cool CP2 stars do not present any feature at all at the wavelength of Li.
So we ask the question : can this feature at 6707.8 Å be attributed only to the Li or is it due to the Li and something else or even no Li at all?

MEGESSIER My question concerns the effective temperatures. In the case of Ap stars, it is known that, due to the UV flux deficiency, the T_{eff} determined from the visible flux are overestimated. What is the situation for the Am stars?

BURKHART Contrary to the Ap stars, the Am stars are only slightly abnormal and the difficulties for the T_{eff} determination exist, but are not so big.

BOESGAARD It is interesting that the Hyades seem to have another Li "quirk" at ≃8000K.
Was your temperature scale (calibrated by Moon and Dworetsky from Hβ) 100K hotter or cooler than the Boesgaard-Tripicco scale for the F stars? What is the full range of v sin i values for the Hyades Am stars and what was the v sin i range for those stars that you observed for Li?

BURKHART We have only one Hyades star in common with Boesgaard and Tripicco : our temperature is 50K cooler. The Boesgaard temperature scale (calibrated by Böh-Vitense from (B-V) for the 5 Coma Am stars is not different from the Moon scale.
The highest v sin i is 80km/s for the Hyades Am stars and we observe only up to 25km/s, i.e. a third of the full range.

SPECTROSCOPIC STUDIES OF 89 HER AND HD 161796

Sunetra Giridhar
Indian Institute of Astrophysics
Bangalore 560034, India
A.A.Ferro and L.E.Parrao
National University of Mexico
Dept. of Astronomy, Mexico City

ABSTRACT. We have studied the spectra of UU Her stars 89 Her and HD 161796 in the red spectral region. The atmospheric parameters are derived using model atmospheres and spectrum synthesis. LiI line at 6707.8 A° is identified and measured in 89 Her leading to log (Li/H) = 2.3. The variation of the H α line profile for these stars is also discussed.

INTRODUCTION

89 Her and HD 161796 are members of a group of semi-regular variables that are seen at large galactic latitudes. General behaviour of these stars is summarized by Sasselov (1984). Infrared excess has been reported for these two stars by Parthasarathy and Pottasch (1986). We felt the need for detailed spectroscopic investigation of these objects; our preliminary results are reported here.

OBSERVATIONS

The high resolution spectra covering a spectral range 5600-7000A° were taken by one of us (AAF) at DDO using 1.88 m reflector during 1981. Spectroscopic reductions were done at IIA, Bangalore, using RESPECT software (cf Prabhu et al. 1987).

RESULTS AND DISCUSSION

The spectra of these objects resemble those of F type supergiants with strong FeII and SiII lines. The most interesting feature in the spectrum of 89 Her is LiI line at 6707.8 A°. Figure 1 shows the spectral region around LiI line. Figures 2 and 3 show H α line profile variation for HD 161796 and 89 Her. P Cygni nature of the profile is obvious in the case of 89 Her with emission component seen in longer wavelength wing of the line suggesting a scattering envelope. Also the deep absorption core indicating a large optical depth might have been caused by extended atmosphere or circumstellar shell. HD161796 does not display emission component but a sharp

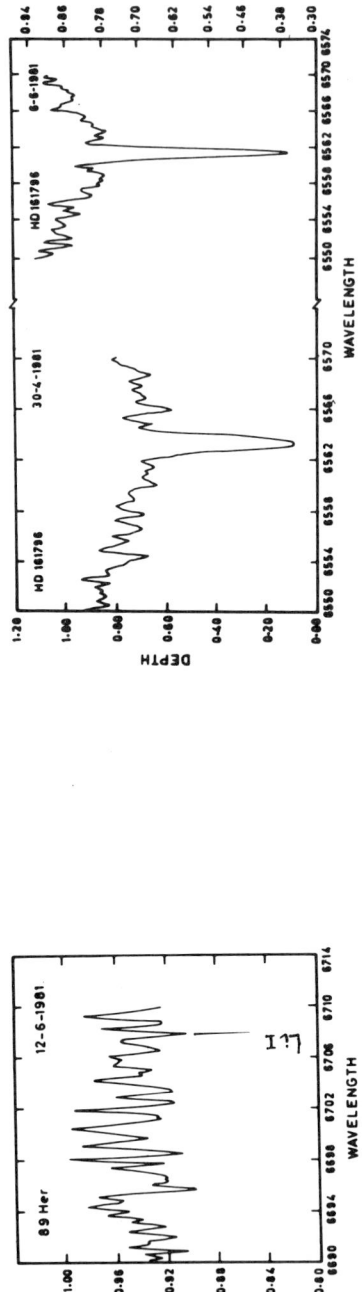

Fig.1: The spectrum of 89 Her around Li I line.   Fig. 2: The Hα line profile variation for HD 161796.

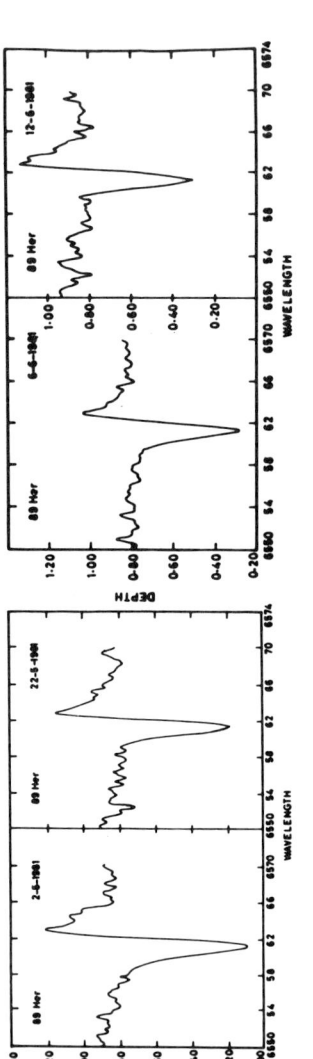

Fig. 3: The Hα line profile variation for 89 Her.

narrow absorption superposed on a broad absorption suggests a shell structure.

Our spectroscopic method of deriving atmospheric parameters is described in detail elsewhere (Giridhar 1984). We derive T_{eff} = 6000 K, log g = 0.5 and V_t = 4.5 km s^{-1} for 89 Her and T_{eff}= 5600K, log g = 1.5 V_t = 4.0 km s^{-1} for HD 161796. It was easier to measure equivalent width of LiI line in the spectrum of 89 Her because the neighbouring lines FeI 6707.441 and VI 6708.1 are very weak in such low gravity stars. In the case of HD 161796, due to relatively lower temperature and higher gravities the LiI line is hopelessly blended making equivalent width measurement impossible.

TABLE 1. DERIVED ABUNDANCES FOR DIFFERENT ELEMENTS

Element	No. of lines used	89 Her	HD 161796
Li/H	1	2.3	
Al/H	2	6.8	6.6
Si/H	4	7.57 ± 0.16	7.4 ± 0.2
S/H	2	7.12	7.3
Ca/H	4	6.2 ± 0.18	6.1 ± 0.25
Fe/H	26	7.57 ± 0.15	7.48 ± 0.16

As is obvious from the table 1 we find almost solar abundance for Si, S, Ca and Fe. Al shows marked overabundances in 89 Her but a marginal one for HD 161796.

It would be of interest to derive Li abundances for other members of UU Her stars and also to look for a correlation between Li/H and $^{12}C/^{13}C$ to understand the evolutionary status of these objects.

REFERENCES

Parthasarathy,M., Pottasch,S.R. 1986, Astr. Astrophys. 154, L16.
Prabhu,T.P., Anupama,G.C., Giridhar,S. 1987, Bull. Astr. Soc. India, 15, 98.
Giridhar,S. 1984, J. Astrophys. Astr. 4, 75.
Sasselov,D. 1984, Astrophys. Space Sci. 102, 161.

DISCUSSION

GRATTON The abundance of Al seems quite high. May you comment?

GIRIDHAR The derived abundance is indeed quite high. But the present estimate is based on only two lines. It would be nicer including more number of lines to improve the accuracy of abundance estimates.

ACCURATE SPECTROSCOPY OF INTERMEDIATE AND LATE SPECTRAL TYPES

H. Holweger
Institut für Theoretische Physik
und Sternwarte der Universität Kiel
Olshausenstrasse 40
D-2300 Kiel
Federal Republic of Germany

ABSTRACT. Elemental abundances in the Sun and in meteorites are examined with respect to their qualification as a standard of population I composition. Some problems that arise in connection with spectroscopic determinations of abundances in the Sun, and in other late-type stars, are discussed with emphasis on photospheric models and departures from LTE.

1. SOLAR AND METEORITIC ABUNDANCES

We are in the fortunate position of having solar matter in condensed form at our disposal which can be subjected to extremely accurate analysis. That there is indeed some type of meteoritic matter whose chemical composition closely matches that of the Sun is by no means trivial, yet it is by now well-established that carbonaceous chondrites contain elements with very different cosmochemical properties in solar proportions. The present situation is illustrated in Fig. 1, which shows elements whose solar abundances are believed to be accurate to within about $\pm$ 50 percent, as judged from available spectral lines and f-values. A closer comparison, at the $\pm$ 20 percent level, has been accomplished for six elements with favourable spectroscopic properties (Holweger 1979). As it turned out, the abundance ratios in the Sun agreed with those in C1 chondrites in all cases. No other type of meteorite was found to fit the solar composition so closely.

This close relationship between solar and meteoritic matter does not preclude that solar-system matter as a whole has suffered chemical fractionation on the long way from the sites of nucleosynthesis into the solar nebula. A sensitive test for cosmochemical fractionation was devised by Suess (1947). He noticed that the abundance of odd-mass nuclides belonging to elements of different volatility and geochemical type tended to be a smooth function of mass number, reflecting nuclear rather than chemical processes. Anders and Ebihara (1982), repeating the test with modern abundance data, confirmed this smoothness. Any significant cosmochemical fractionation - exceeding 15% or so - would have appeared as perceptible irregularities in the abundance curve.

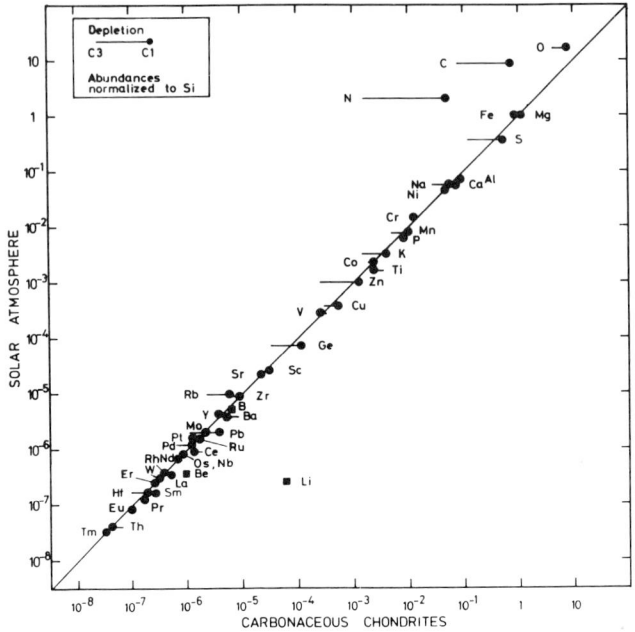

Figure 1: Comparison of solar and meteoritic abundances (carbonaceous chondrites of type 1 and 3).

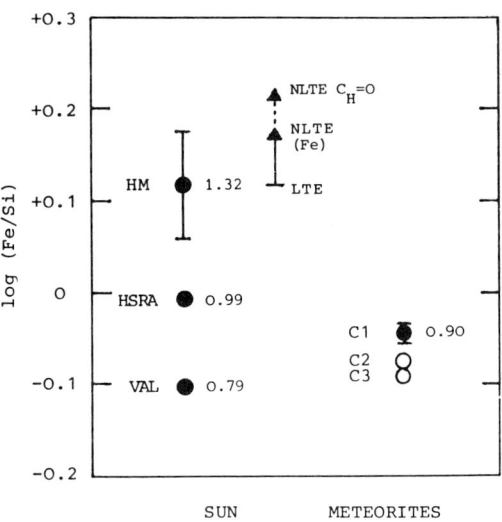

Figure 2: The Fe/Si ratio in the Sun and in carbonaceous chondrites of types 1, 2 and 3.
Solar abundances (HM model):
log N(Fe) = 7.673 ± 0.017 (Blackwell et al. 1984);
log N(Si) = 7.55 ± 0.05 (Becker et al. 1980).
Meteorites: Anders and Ebihara (1982, C1 chondrites), Mason (1971, C2 and C3). For further details see text.

Thus, the elemental (and isotopic) composition of C1 chondrites may be assumed to be typical for the condensable fraction of population I matter. Surprisingly enough, the Sun may be slightly non-standard. If we compare the solar and meteoritic ratio of the two key elements representing the siderophile and lithophile geochemical groups - iron and silicon - then a puzzling discrepancy appears (Fig. 2): the solar ratio Fe/Si = 1.32 is well above the value of 0.90 found in Cl chondrites. This poses a number of questions.

Taken at face value, this result is not implausible as soon as we view the Sun and meteorites as parts of the inner and outer solar system. It is well known that the global Fe/Si ratios of the terrestrial planets - the fractional mass of their iron core - increase towards the Sun. So the Sun itself would conform to this trend. We must bear in mind the possibility of a Fe/Si fractionation in the early solar system. Such processes might also have affected other stars and may contribute to the spread of Fe/Si ratios observed among population I F- and G-type dwarfs (Andersen et al., this conference). Gustafsson (this conference) noticed that the mean Fe/Si ratio in this sample is below the solar value, closer to that found in Cl chondrites.

However, we must also consider the possibility that the disagreement between the solar and meteoritic Fe/Si ratio is fictitious and due to our imperfect understanding of photospheric structure and spectrum line formation. This will be discussed in the next section.

2. MODELLING THE ATMOSPHERES OF SOLAR-TYPE STARS

2.1. One-Component Solar Models

The error bar quoted in Fig. 2 is the standard deviation of abundance values derived from individual lines and does not include any systematic error. The iron abundance is based on low-excitation Fe I lines (E.P. = 0-2.6 eV), whereas the Si lines are of the high-excitation type (4.9-6.0 eV). Thus, the inferred Fe/Si ratio is quite sensitive to errors in the temperature structure of the photospheric model employed.

This model-dependence is illustrated in Fig. 2. The value of 1.32 for Fe/Si, discussed in the previous section, was obtained using the HM model (Holweger and Müller 1974). If the same set of lines is analysed using the HSRA (Gingerich et al. 1971) or the VAL model (Vernazza et al. 1976), considerably lower Fe/Si values result, the logarithmic difference VAL minus HM amounting to -0.22 dex. The situation would be much more satisfactory if accurate f-values for high-excitation Fe I lines, or Fe II lines, were available: the model dependence, VAL-HM, would then reduce to -0.07 dex for 4.5 eV Fe I lines, and to +0.04 dex for 2.8 eV Fe II lines, respectively.

Deviations from LTE constitute another problem, though probably a less serious one. Recent NLTE studies (section 3.2) lead to abundance corrections of about 0.06 dex for low-excitation lines of Fe I (Fig. 2). This value would increase to 0.1 dex if neutral-atom collisions were neglected (denoted by '$C_H = 0$' in Fig. 2). No data are available for Si I. If the NLTE corrections are similar then the effect on the ratio of both elements will be smaller than indicated in Fig. 2.

How good are the <u>solar models</u>? It is tempting to remove the Fe/Si problem by simply choosing the VAL model as the appropriate one. However, there is a more realistic, more sensitive test for solar models than the comparison of solar and meteoritic abundances: to compare solar abundances derived from infrared molecular lines belonging to the same molecular band but differing in excitation potential. Vibration-rotation lines of OH and CO, and pure rotation lines of OH, have been studied in this way (Goldman et al. 1983, Grevesse et al. 1984, Sauval et al. 1984, Harris et al. 1987). The three main advantages of these lines are (i) high temperature sensitivity, (ii) likely to be found in LTE, (iii) very accurate relative f-values. In all cases, the HM model passed the examination but the other models did not.

At the same time the OH lines near 11 microns bring to light the limitations of <u>one-component model atmospheres</u>. All these models, including the HM, failed to reproduce the center-to-limb behaviour, at least close to the limb (Deming et al. 1984). The layers probed in this way are around $\tau_{5000} = 10^{-3}$. Deming et al. interpreted this as evidence for horizontal temperature inhomogeneities (thermal bifurcation) in the uppermost photosphere.

In view of this we will also have to investigate the effect of thermal inhomogeneities in the <u>lower photosphere</u>, associated with convective motions, on abundance determinations. While this is a task for the future, I wish to mention another problem that may occur in abundance determinations of solar-type stars. It is related to the <u>vertical</u> thermal structure of the photosphere.

2.2. Photospheric Abundances and Chromospheric Activity

It is well known that some G-type dwarfs possess chromospheres whose activity is much higher than that of the Sun. There is evidence that this "activity" is also perceptible in the photospheric line spectrum and may lead to fictitious underabundances if ignored. I have encountered this problem when analysing spectra of two dwarfs with very different chromospheric activity: 61 Vir (G6 V, quiet chromosphere) and Xi Boo A (very active chromosphere). The study was based on high-S/N spectra recorded with the ESO Coudé Echelle Spectrometer. The same set of 16 Fe I and 2 Fe II lines was used in both stars, giving greater weight to magnetically insensitive lines. Equivalent widths ranged from 7 to 245 mÅ. A model-atmosphere analysis was carried out using scaled solar models whose effective temperature and gravity were determined from colours and from the Fe I/II balance, respectively. The microturbulence parameter was adjusted such as to make abundances derived from medium strong and weak lines agree. The following parameters T_{eff}/log g/micro(km/s) resulted:

61 Vir: 5560/4.4/1.05; Xi Boo A: 5390/4.2/1.55.

In the case of <u>61 Vir</u>, the star with the inactive chromosphere, abundances derived from the various lines yield a highly consistent picture: weak and strong lines agree and the spread is quite small, attesting to the quality of the CES spectra. The iron abundance derived from the nine weakest and most Zeeman-insensitive lines is

[Fe/H] = +0.01 ± 0.02 (s.d.), i.e. perfectly solar.

The situation is quite different in the case of the active-chromosphere star Xi Boo A, although the spectra are of the same high quality and the analysis is done in exactly the same manner. Among the weaker lines, the scatter is twice as large and the iron abundance becomes [Fe/H] = -0.19 ± 0.04. This apparent underabundance is somewhat surprising in the case of an obviously quite young star. Much more disturbing is the fact that the three strongest lines (including one with Landé factor zero) lie about 0.2 dex above the weaker ones. Increasing log g from 4.2 to 4.55 will shift these pressure-sensitive lines downwards to match the weak ones, but then the Fe II lines will move upwards. To restore the ionisation equilibrium, T_{eff} has to be increased. If T_{eff} is treated as a free parameter, a scaled solar model characterised by 5690/4.8/1.05 meets both constraints set by Fe I/II and strong/weak lines. The iron abundance then becomes [Fe/H] = 0.00 ± 0.05. However, such a high effective temperature - 300 K hotter than derived from the colours - appears inacceptable. There is a simple way out of the dilemma: to restrict the temperature enhancement to the line-forming layers, thus keeping T_{eff} essentially unchanged. This also requires some iterative adjustment of log g, since the strong/weak line balance depends on the photospheric temperature gradient, and of T_{eff} in order to keep Fe I/II in balance. One possible model is characterised by a temperature enhancement of 200 K at optical depths smaller than 0.1, the other parameters being 5480/4.7/1.55. The iron abundance then becomes [Fe/H] = -0.03 ± 0.05, i.e. essentially solar.

Of course, this modified model of Xi Boo A is by no means uniquely defined but the trend is obvious: if excess photospheric heating is taken into account the underabundance tends to disappear.

Xi Boo A is not the only active late-type candidate of this kind. Cayrel et al. (1985) noticed anomalously low iron abundances in two Hyades dwarfs with high chromospheric activity. They suspected this to be due to chromospheric filling-in of photospheric lines. Another prominent object is ϵ Eri. Just like Xi Boo A, ϵ Eri is found to be metal deficient by 0.2 dex, when analysed with an unmodified scaled solar model (Steenbock and Holweger 1981). Indeed, in Fig. 1 of that paper, the strongest line is seen to deviate from the weaker ones as in Xi Boo A. There seems little doubt that the apparent underabundance of ϵ Eri disappears if excess photospheric heating is accounted for in the photospheric model. Interestingly, not only the thermal structure appears to be modified in these active stars but also the hydrodynamics: both exhibit anomalously high microturbulent velocities.

3. DEPARTURES FROM LTE IN STARS OF INTERMEDIATE AND LATE SPECTRAL TYPE

3.1. Observational Evidence

Departures from LTE that affect abundances were found by Ruland et al. (1980) in a detailed analysis of high-quality photographic spectra of the red giant Pollux (K0 III). When LTE was assumed, a typical pattern of inconsistencies appeared among iron-group elements. Abundances derived from individual lines of low excitation potential were found to be lower than those obtained from high-excitation lines by 0.3-0.5 dex. This dichotomy was largest for lines on the flat portion of the curve-of-growth. Similar trends were observed by Brown et al. (1983) and by other authors in G and K giants in general. Steenbock (1983) and Magain (this conference) have found NLTE effects in halo giants.

In contrast to these low-gravity objects, main-sequence stars are expected to show less conspicuous departures from LTE. Abundance anomalies due to NLTE have been suspected in the AOV standard Vega (Gigas 1986, and references therein) and Procyon (F5 IV-V; Steffen 1985). In the Sun (G2 V), high-precision spectroscopy and f-values have made it possible to discern NLTE effects in Ti I and Cr I at the 10-20% level (Blackwell et al. 1987).

3.2. Statistical-Equilibrium Calculations for Cool Stars

Recent NLTE calculations have taken into account the thermalising effect of inelastic collisions with hydrogen atoms and made efforts towards a realistic representation of the atomic term structure and the ionising UV radiation field. Here, I would like to report some unpublished results for iron, obtained by W. Steenbock in Kiel, a continuation of earlier work (Steenbock and Holweger 1984, Steenbock 1985, Watanabe and Steenbock 1986). The Fe I/II/III model atom comprises 79+20+1 levels and 52+23 line transitions (see Gigas 1986). The results are expressed in terms of the NLTE abundance corrections $\Delta \log \epsilon$ to be added to the (logarithmic) abundances derived from observed equivalent widths of individual lines under the assumption of LTE.

Pollux. Fig. 3 shows the results for a typical red giant with T_{eff} = 4840 K, log g = 2.2, and solar metallicity such as appropriate for Pollux (K0 III). Most $\Delta \log \epsilon$ values are positive, indicating a general overionisation of Fe I. When compared with the above-mentioned empirical results (see Fig. 4 in Ruland et al. 1980), an agreement is noted which gives some confidence in these fairly involved computations. In particular, the dependence of $\Delta \log \epsilon$ on excitation potential and line strength is reproduced. The relatively large NLTE corrections for lines of intermediate strength can be understood: these lines are almost box-shaped, with dark cores and practically no damping wings. Thus, their equivalent width is formed in the uppermost photosphere where departures from LTE are larger.

The thermalising effect of inelastic collisions with hydrogen atoms is important in Pollux. The $\Delta \log \epsilon$ values shown in Fig. 3 are derived by applying a scaling factor S_H = 0.2 to Drawin's approximation for the cross-sections, in accordance with Watanabe and Steenbock (1986). If we choose S_H = 1 instead, the calculated NLTE effects become too small. If we neglect neutral-atom collisions, on the other hand, the spread in the $\Delta \log \epsilon$ values increases from about 0.3 dex to 0.6 dex. Not shown in Fig. 3 are the NLTE corrections for Fe II lines, simply because they are so small (see also Steenbock 1985).

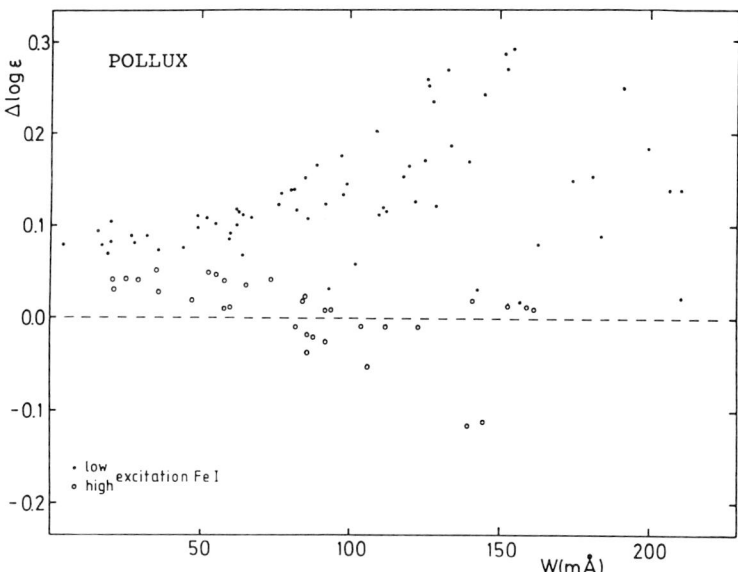

Figure 3: NLTE abundance corrections for Fe I lines in the red giant Pollux (K0 III) (Steenbock 1987).

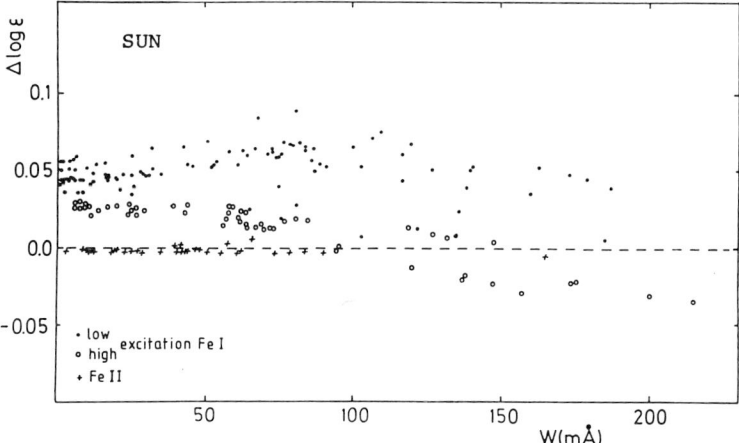

Figure 4: NLTE abundance corrections for the Sun (G2 V) (Steenbock 1987).

Arcturus (K2 III). Similar calculations have been performed for T_{eff} = 4260 K, log g = 0.9 and one-third solar metallicity. These parameters may represent a more evolved object than Pollux. The results are qualitatively similar. Surprisingly, despite the lower gravity, the NLTE corrections turned out to be a factor of two smaller than in

Pollux, at least for lines weaker than about 60 mÅ. This is probably due to the combined effect of reduced temperature and metallicity on the UV radiation field. Again, Fe II lines are in LTE.

Main-Sequence Stars. NLTE abundance corrections for the Sun (G2 V) are shown in Fig. 4. The pattern is similar to that of Pollux, but the NLTE effects are much smaller because of the increased frequency of collisions with hydrogen atoms and electrons. If the neutral-atom collisions were neglected, the $\Delta \log \epsilon$ values would increase by a factor of two. The results displayed in Fig. 4 are valid for the HM model. If the HSRA is used instead, somewhat larger departures from LTE result, but the overall pattern remains the same.

In the Sun, NLTE corrections for iron are small, amounting to at most 0.06 dex for low-excitation Fe I lines. The $\Delta \log \epsilon$ values for high-excitation lines are about half as large, while Fe II lines are in LTE. This weak sensitivity to NLTE effects is another attractive property of the latter two types of lines in addition to the weak model-dependence of their LTE abundances mentioned in section 2.1. Other iron-group elements in the Sun may be expected to exhibit similar departures from LTE. Indeed, NLTE effects of this magnitude are apparent in the high-precision LTE analysis of solar Ti I and Cr I by Blackwell et al. (1987).

Similar NLTE calculations have been carried out for two hotter main-sequence stars, Procyon (F5 IV-V) and Vega (A0 V). The results for Vega have reported elsewhere (Gigas 1986, and this conference). For Procyon, a flux-constant model with T_{eff} = 6500 K, log g = 4.0 and solar metallicity was adopted. The resulting pattern is quite similar to that in Fig. 4, with the important exception that the departures from LTE are significantly larger than in the Sun. The $\Delta \log \epsilon$ values for low-excitation Fe I lines weaker than 50 mÅ are typically 0.12 dex, while high-excitation lines deviate only by 0.07 dex, and Fe II is in LTE. Again, the largest departures occur in the equivalent width range 100-150 mÅ, with $\Delta \log \epsilon$ values of up to 0.20 dex.

3.3. NLTE and Galactic Chemical Evolution

The dependence of NLTE abundance corrections on effective temperature, gravity and metallicity will have to be studied in a more systematic manner, and for a variety of elements. Increasingly precise stellar spectra combined with differential model-atmosphere analyses permit the investigation of subtle abundance trends in galactic stellar samples which are important as constraints on models of galactic structure and evolution. Recent illustrations of the high accuracy that can be obtained are the studies of galactic disk stars by Tomkin et al. (1985) and by Nissen et al. (1985) with the supplementary data reported by Andersen et al. (this conference).

One of the basic assumptions made is LTE. This may be critical as soon as one deals with small abundance trends as a function of metallicity or, explicitly or implicitly, as a function of effective temperature or gravity. For example, one important subject of these studies is the possible variation of the abundance ratio (odd-Z elements)/Fe with Fe/H. The odd-Z elements are determined from lines of Na I and Al I, species whose ionisation potential is 5.1 and 6.0 eV, respectively - much lower than that of Fe I (7.9 eV). Thus, departures from LTE in the ionisation balance are likely to be larger for the odd-Z elements than for iron. Ignoring NLTE effects may introduce fictitious dependences of (odd-Z/Fe on Fe/H or T_{eff}. Real trends may be either exaggerated or underestimated in this way.

REFERENCES

Anders, E., Ebihara, M. 1982, Geochim. Cosmochim. Acta **46**, 2363.
Becker, U., Zimmermann, P., Holweger, H. 1980, Geochim. Cosmochim. Acta **44**, 2145.
Blackwell, D.E., Booth, A.J., Petford, A.D. 1984, A.A. **132**, 236.
Blackwell, D.E., Booth, A.J., Menon, S.L.R., Petford, A.D. 1987, A.A. **180**, 229.
Brown, J.A., Tomkin, J., Lambert, D.L. 1983, Ap.J. **265**, L93.
Cayrel, R., Cayrel de Strobel, G., Campbell, B. 1985, A.A. **146**, 249.
Deming, D., Hillman, J.J., Kostiuk, T., Mumma, M.J. 1984, Solar Phys. **94**, 57.
Gigas, D. 1986, A.A. **165**, 170.
Gingerich, O., Noyes, R.W., Kalkofen, W., Cuny, Y. 1971, Solar Phys. **18**, 347.
Goldman, A., Murcray, D.G., Lambert, D.L., Dominy, J.F. 1983, Mon.Not.R.astr.Soc. **203**, 767.
Grevesse, N., Sauval, A.J., Dishoeck, E.F. 1984, A.A. **141**, 10.
Harris, M.J., Lambert, D.L., Goldman, A. 1987, Mon.Not.R.astr.Soc. **224**, 237.
Holweger, H., Müller, E.A. 1974, Solar Phys. **39**, 19.
Holweger, H. 1979, XXIInd Liège Internat. Astrophys. Symposium, Université de Liège, p. 17
Mason, B. (Ed.) 1971, Handbook of Elemental Abundances in Meteorites, Gordon and Breach, New York.
Nissen, P.E., Edvardsson, B., Gustafsson, B. 1985, in Proc. ESO Workshop on Production and Distribution of C, N, O Elements, I.J. Danziger, F. Matteucci, K. Kjär (eds.), Garching, p. 131.
Ruland, F., Holweger, H., Griffin, R., Biehl, D. 1980, A.A. **92**, 70.
Sauval, A.J., Grevesse, N., Brault, J.W., Stokes, G.M., Zander, R. 1984, Ap.J. **282**, 330.
Steenbock, W., Holweger, H. 1981, A.A. **99**, 192.
Steenbock, W. 1983, A.A. **126**, 325.
Steenbock, W., Holweger, H. 1984, A.A. **130**, 319.
Steenbock, W. 1985, in Cool Stars with Excesses of Heavy Elements, M. Jaschek, P.C. Keenan (eds.), D. Reidel, p. 231.
Steenbock, W. 1987, private communication.
Steffen, M. 1985, A.A. Suppl. Ser. **59**, 403.
Suess, H.E. 1947, Z. Naturforsch. **2a**, 311.
Tomkin, J., Lambert, D.L., Balachandran, S. 1985, Ap.J. **290**, 289.
Vernazza, J.E., Avrett, E.H., Loeser, R. 1976, Ap.J. Suppl. **30**, 1.
Watanabe, T., Steenbock, W. 1986, A.A. **165**, 163.

DISCUSSION

BUDGE   You showed graphs of abundance determinations for a chromospherically active dwarf using a conventional model and a model with a hot upper photosphere. It appears from these graphs that the abundances derived from weak lines were affected more by the hot upper photosphere than the abundances from the strong lines. Intuitively it seems to me that the weak lines would be less affected than the strong lines.

HOLWEGER   Weak lines are not necessarily formed in deeper layers. For example, weak molecular lines, or Li I 6708, are formed in the upper photosphere.
The evidence for a hot upper photosphere of Xi Boo A is indirect. When using a scaled solar model, abundances derived from strong and weak lines are found to disagree. This cannot be removed by changing the upper photospheric temperature because the height of formation of strong and weak lines is not sufficiently different. It can be removed by an increase of gravity, i.e. pressure broadening. This also affects the ionization balance, which can be restored by making the line-forming layers hotter.

RUTTEN   I have two comments. First, it is important to realize that most high-excitation lines observed in optical spectra are also most probably lines with log gf > −2. They provide the most probable depopulation path for all levels at about their upper-level excitation energy, and their departure coefficients drop below unity where the lines start feeling the surface. This excitation deficiency affects their source function, and may cancel the effect of collective overionization, which reduces the populations and opacities but not the source functions. So, a difference between low and high excitation lines can result from this cancelling effect : two NLTE mechanisms are acting together and making high-excitation lines appear seemingly in LTE.
My second comment is on the standard solar models. You don't have to compare the HOLMUL model any more with the HSRA or VAL models. These latter models represent a progression in the sophisticated NLTE solar continuum modeling at Harvard that has now culminated in a model of which the photospheric part is nearly the same as HOLMUL, and in which the metal overionizations in the temperature minimum region are much smaller than for the older, cooler models. Simply because the flatter gradient keeps B_ν closer to J_ν in the ultraviolet. This came about because Avrett now includes more of Kurucz's ultraviolet line list in detail, so shifting the computed height of formation of the observed continua outward, and flattening the temperature gradient.
I should add here that even though these models possess chromospheres in contrast to HOLMUL, the actual model assumed for the ultraviolet line haze within the fitting procedure is without one. Avrett specifies line source functions for the haze lines that are set to scattering near the temperature minimum in an ad-hoc parametrization, effectively adopting the HOLMUL model as a line excitation temperature. So, although the HOLMUL model is never mentioned in the Harvard work, it has in effect been retrieved and verified by it —within the constraints of plane-parallel modeling.

ABUNDANCES OF LIGHT METALS IN FIELD STARS WITH METALLICITY
RANGE -1.2 < [Fe/H] < + 0.3

C. Abia[1], R. Rebolo[1], J.E. Beckman[1], L. Crivellari[2], B. Vila[3]
1. Instituto de Astrofisica de Canarias, Tenerife, Spain.
2. Osservatorio Astronomico di Trieste, Via G B Tiepolo 11,
I - 34131, Trieste, Italy.
3. Nuffield Radioastronomy Laboratories, Jodrell Bank,
Macclesfield SR11 9DL, United Kingdom.

ABSTRACT. High resolution, high S:N spectra are used to determine the abundances of Fe, Ni, Ca, Al and Si in 25 field dwarfs with $-1.2 \leqslant$[Fe/H]< +0.3. We find overabundances for Al, Ca and Si in stars with -1.2<[Fe/H] < -0.5 and solar [Ni/Fe] over the whole studied range.

1. INTRODUCTION

Detailed studies of abundances of light elements 8 < Z < 20 in stars can provide a wealth of information about the chemical history of the Galaxy. These elements are synthesized in stars with different masses (Arnett, 1971; Woosley and Weaver, 1982) and their relative abundances and isotope ratios vary with the mean metallicity of the stars. An excellent review of the recent observational results is given by Lambert (1987).

2. OBSERVATIONS

The metallicity range +0.3 >[Fe/H]> -0.6 was observed with the ESO "CAT+CES+Reticon" combination, which gave $\lambda / \Delta\lambda \sim 10^5$, with S:N ~ 200 for stars with m_v ~ 5 in exposures of 1 hour. Reduction is described in Crivellari et al (1987). Stars with -0.5 > [Fe/H] > -1.2 were observed with the INT+IDS+CCD combination on La Palma giving $\lambda / \Delta\lambda$ ~ 2x10 with S:N ~150 for m_v =8 in exposures of 20 min. Reduction is described in Rebolo et al. (1987). Spectra come in 40 A (CES) and 125 A (IDS) ranges around the λ 6707 A Li doublet. Fig. 1 gives a sample CES spectrum for HD 209100, with S:N=180, indicating general spectral quality.

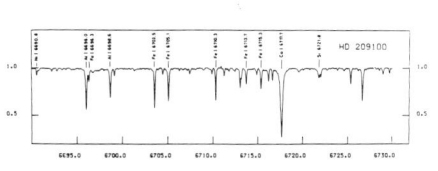

Fig. 1: Spectrum of HD 209100 in range λ6690 A - λ 6730 A with S:N ~ 180; an example of the material used for this work.

3. ANALYSIS

We derived abundances using spectral syntheses with Kurucz (1979) models for Teff >5500 K, or Gustafsson et al. (1975) models for Teff < 5500 K (all these have [Fe/H] >-0.6). Both curves of groth and direct matching of synthetic to observed spectra were used, with differences always < 0.03 dex. The solar abundances using the model types on our own lunar spectrum never differ by more than 0.06 dex. Our Teff's were from b-y or R-I and V-K indices, calibrated by Carney (1983) or Olsen (1984) for the metal rich stars. Our log g's come from b-y, c_1 calibrations from the same authors. A microturbulence "default" of 1.5 km s^{-1} was used where not previously published. Log gf values were from Gurtovenko and Kostik (1981). The analysis is differential referring all abundances to our derived solar values. We estimate the errors in our ratios as ±0.1 dex in [Al/Fe], ±0.12 dex in [Ca/Fe], ±0.1 dex in [Si/Fe] and ±0.14 dex in [Ni/Fe].

In Fig. 2 we plot [Al/Fe], [Si/Fe], [Ca/Fe] and [Ni/Fe] against [Fe/H].

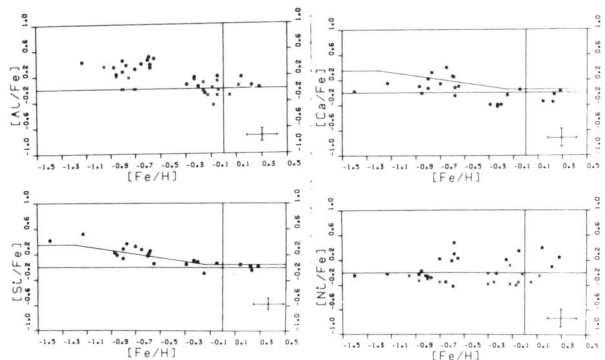

Fig. 2: Plots of [Al/Fe], [Si/Fe], [Ca/Fe] and [Ni/Fe] against metallicity [Fe/H]. For Si and Ca we include trend lines for α-elements from Lambert (1987). Key: • This work × Edvardsson et al. (1984).

4. DISCUSSION

The trends for [Si/Fe] and [Ca/Fe] illustrated here confirm the overabundances reported for α elements below [Fe/H] =-0.5, notably for [O/Fe] (Clegg et al., 1981; Edvardsson et al., 1984). This is broadly explicable by the earlier ejection of massive star products (M > 10 M_0) via SNII, compared with those from lower mass stars. The decrease in slope of this tren with increasing Z, marginally confirmed here, may be due to the exact stellar mass range where each Z range originates.

Al is produced by C combustion, favoured by neutron excess and therefore by Pop I (Pardo et al., 1984). This would explain the [Al/Fe] deficiency at very low [Fe/H], i.e. below [Fe/H] =-1 (François, 1986). Over excess for [Al/Fe] between [Fe/H] =-0.5 and -1.0 agreeing with Edvardsson et al. (1984) and Tomkin et al. (1985) is not presently explicable by galactic chemical evolution models.

Ni follows Fe right down to [Fe/H] =-1.5 as would be expected from a similar origin. Luck and Bond (1985) show an apparent overabundance at [Fe/H] =-1.8 but Gratton and Sneden (1987) have recently found [Ni/Fe]~ 0 for [Fe/H]< -1.0 with no evidence for any abundance excess.

Abundance measurements in field stars show reasonable consistency from author to author with presently achievable high S:N, the principal sources of error are the use of models and model parameters.

References

Arnett, W.D.: 1971, Astrophys. J. **166**, 153.
Carney, B.W.: 1983, Astron. J. **88**, 623.
Clegg, R., Lambert, D., Tomkin, J.: 1981, Astrophys. J. **250**, 262.
Crivellari, L., Beckman, J.E., Foing, B.H., Vladilo, G.: 1981, Astron. Astrophys. **174**, 127.
Edvardsson, B., Gustafsson, B., Nissen, P.E.: 1984, The Messenger **38**, 83.
François, P.: 1986, Astron. Astrophys. **160**, 264.
Gratton, R., Sneden, C.: 1987, Astron. Astrophys. **178**, 179.
Gustafsson, B., Bell, R.A., Eriksson, K., Nordlund, A.: 1975, Astron. Astrophys. **51**, 71.
Gurtovenko, E.A., Kostik, R.I.: 1981, Astron. Astrophys. Suppl.Ser. **46**, 239.
Kurucz, R.L.: 1979, Astrophys. J. Suppl. Ser. **40**, 1.
Lambert, D.: 1987, in ESO Workshop "Stellar Evolution and Dynamics of the outer halo of the Galaxy", In press.
Luck, R.E., Bond, H.E.: 1985, Astrophys. J. **292**, 599.
Olsen, E.H.: 1984, Astron. Astrophys. Suppl. Ser. **40**, 1.
Pardo, R.C., Couch, R.G., Arnett, W.D.: 1974, Astrophys. J. **191**, 711.
Rebolo, R., Molaro, P., Beckman, J.e.: 1987, Submitted. Astron. Astrophys.
Tomkin, J., Lambert, D., Balachandran, S.: 1985, Astrophys. J. **290**, 289.
Woosley, S.E., Weaver, T.A.: 1982, An Essay on Nuclear Astrophys.
    Ed. C.A. Barnes, D.D. Clayton, D.N. Schramm, Cambridge University Press, p. 377.

DISCUSSION

HOLWEGER The dependence of [Al/H] or [Fe/H] derived assuming LTE may be different from that you will find when including NLTE effects. Departures from LTE, if present, are likely to be stronger in Al because of the lower ionization potential. These NLTE effects will depend on [Fe/H] because the ionizing UV field depends on metallicity.

ABIA Our Al lines have excitation potentials slightly higher than 3 e.v., so we expect departures from LTE are not very important.

GUSTAFSSON How strong are your Al lines ?

REBOLO The equivalent widths of our Al lines at 6696.03 Å and 6698.66 Å are in no case higher than 80 mÅ. For instance in a star with [Fe/H] = −0.5 these lines have ≈ 40 mÅ and ≈ 20 mÅ respectively.

A PHOTOSPHERIC SOLAR IRON ABUNDANCE FROM WEAK FE II LINES

U. Pauls[*1], N. Grevesse[°] and M. C. E. Huber[*1]
* Institut für Astronomie
Eidgenössische Technische Hochschule
ETH-Zentrum
CH – 8092 Zürich, Switzerland
° Institut d'Astrophysique
Université de Liège
B – 4200 Ougrée-Liège, Belgium

ABSTRACT. The high resolution and the high light–gathering power of a Fourier–transform spectrometer (FTS) afford the observation of very weak lines in laboratory spectra. Thus it became possible to determine an accurate solar iron abundance from Fe II lines that are weak in the solar spectrum: we measured the branching fractions of a few such lines whose upper levels lifetimes are known.

1. INTRODUCTION

The history of the solar iron abundance has been described by numerous authors (see e.g. Withbroe 1971; Grevesse 1984a,b). The solar iron abundance has repeatedly been determined over several decades. Still, uncertainties and discrepancies remain and we do not consider the problem to be solved.

To be suitable for a reliable abundance determination, Fe II lines have to be weak enough to lie on the linear part of the solar curve of growth. This requirement leads to an abundance that is independent of damping and microturbulence. But owing to the large photospheric number density of Fe^+, weak Fraunhofer lines are inherently very faint and thus difficult to produce and measure in the laboratory.

2. HIGH SIGNAL-TO-NOISE MEASUREMENTS IN THE LABORATORY

The spectrometer ideally suited for our purposes was the solar Fourier–transform spectrometer (FTS) on Kitt Peak (cf. Brault 1985). A FTS does not scan the single spectral elements in sequence but observes the entire spectrum at once; in our case this means that (small) variations in the population of the upper levels and hence in the intensity of the emission lines do not affect the measurement of branching fractions. A second point in

---

[1] U.P. and M.C.E.H. were guest observers at the National Solar Observatory, Kitt Peak, National Optical Astronomy Observatories, operated by the Association of Universities for Research in Astronomy, Inc., under contract with the U.S. National Science Foundation.

favour of a FTS is its resolution; by use of path differences ranging from one to several decimeters we did always resolve the line profiles and consequently had to deal with line blends inherent in the spectrum of the light source only. Furthermore the light-gathering power of a FTS helps to perform measurements on weak lines with a laboratory light-source. Nevertheless, further steps were necessary to improve the signal-to-noise ratio (S/N) to get access to lines feeble enough for a reliable abundance determination.

As a FTS observes the whole spectrum at the same time, the noise in the recorded spectrum is proportional to the square root of the total amount of incoming photons. Consequently a drastic improvement in S/N could be reached by use of narrow-band filters (FWHM of 1.5 and 2.0 nm) — we actually achieved a factor of 10^2 compared with broadband scans that extended from 0.2 to 1 μm. The effect of such a filter is clearly illustrated in figure 1.

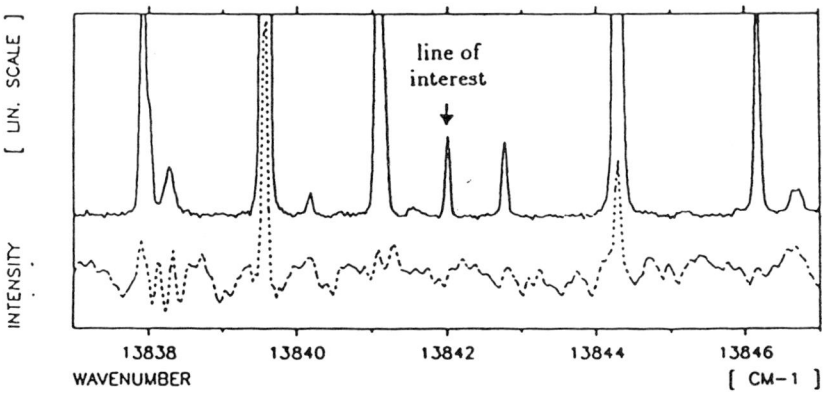

Fig. 1: Comparison of two recordings of the same spectrum by a Fourier-transform spectrometer (FTS). In one case (lower trace) the spectrum was recorded over a wide wavelength range (0.2 to 1 μm), while in the other case (upper trace) the bandwidth of the light admitted to the FTS was restricted by a narrow-band filter (full width at half maximum: FWHM = 1.5 nm). The scale of the upper trace is expanded by a factor of 10; accordingly, the actual increase in S/N is of the order of 100.

The laboratory spectrum was generated in a hollow-cathode discharge run with neon as carrier gas and observed with a FTS and a 3-m grating spectrometer.

On the FTS we recorded spectra favouring the weak lines (at source conditions of 10 mbar neon and 1 A). The relative radiometric sensitivity of the FTS was determined with the aid of known branching ratios of argon lines that were generated in seperate runs where argon replaced (or was added to) neon as carrier gas (Adams & Whaling 1981).

The strong Fe II lines below 300 nm were measured in optically thin conditions (3 mbar neon and 0.3 A) on a 3-m McPherson grating spectrometer, that had been calibrated with an argon mini-arc.

We derived the transition probabilities of three suitable Fe II lines ($\lambda\lambda$ 720 - 750 nm) by measuring their branching fractions and converting them into Einstein A-values by means of the lifetime of their upper levels (Hannaford & Lowe 1983). The transition probabilities are of the order of 10^3 s^{-1}.

Our value for the solar photospheric iron abundance is based on the model by Holweger and Müller (1974).

3. CONCLUSIONS

By judicious choice of weak Fe II lines it is possible to obtain a solar photospheric iron abundance whose accuracy is essentially dependent on the transition probabilities only. These, however, are very small and correspondingly difficult to measure. A Fourier-transform spectrometer and narrow-band filters are essential to reach the required S/N to detect and measure such lines in the laboratory.

A final value for the solar iron abundance will be released after control measurements have confirmed our preliminary results.

ACKNOWLEDGEMENT.
We thank J.W.Brault, R.Hubbard and J.Wagner for their kind and expert support during the FTS runs. Part of this research was supported by the Schweizerischer Nationalfonds.

REFERENCES.
Adams DL & Whaling W 1981 *J.Opt.Soc.Am.* **71** 1036-1038
Brault JW 1985 in *High Resolution in Astronomy* 15th Advanced Course of the Swiss Society of Astronomy and Astrophysics, Saas-Fee (Sauverny: Obs. de Genève) pp 1-61
Grevesse N 1984a *Phys. Scripta* **T8** 49-58
Grevesse N 1984b in *Frontiers of Astronomy and Astrophysics* Proceedings of the 7th European Regional Astronomy Meeting (Firenze: Italian Astron.Soc.) pp 72-81
Hannaford P & Lowe R 1983 *J.Phys.B: At.Mol.Phys.* **16** L43-L46
Holweger H & Müller EA 1974 *Solar Phys.* **39** 19-30
Withbroe GL 1971 *Nat.Bur.Stand.(U.S.)Spec.Publ.* **353** 127-148

DETAILED ANALYSES OF FOUR SOLAR ANALOGS ANALYSED
ON HIGH S/N CFH AND ESO SPECTRA

C. Bentolila and G. Cayrel de Strobel
Observatoire de Paris, Section de Meudon
F-92195, Meudon Cedex, France

ABSTRACT. Four solar type G stars claimed to be photometrically very similar to the Sun have been analyzed in detail on high resolution, high S/N spectra. Their atmospheric parameters : effective temperature, spectroscopic gravity, microturbulence and iron abundance, [Fe/H], have been determined.

1. INTRODUCTION

The search for solar analogs has been very active during these last ten years. At the present time, this research bears a special interest between astronomers interested in discovering planetary systems. A very large list of solar analogs is given in Hardorp (1978). His search has been done by means of photoelectrical observations. Other authors have also looked for stars having the same physical and chemical characteristics than the Sun. Cayrel de Strobel et al. (1981) and M.N. Perrin and M. Spite (1981) tryed to complete the search using quantitative spectral analyses of some of these objects. They were not completely successful, in the sense that at least one physical parameter was different between the Sun and the solar analogs.
Here we present a new analysis of four solar analogs based on high S/N spectra.

2. THE SELECTION OF CANDIDATES FOR SOLAR ANALOGS

Table I contains some catalog data of the four stars we have selected for this study. The first three, HD44594, HD76151 and HD20630 have been taken from Table I in Hardorp's (1978) paper: "The Sun among the stars". The star HD44594 is contained in section 1 of this table headed : "spectra indistinguishable from solar". HD76151 and HD20630 are contained in Section 3 which contains stars headed : "Spectra very close to solar". Our attention has been called on the star HD81809 by D. Mihalas (private communication). In his letter of May 6, 1981 to one of the authors, Mihalas said that : "O.C. Wilson's CaII data reveal that HD81809 has an activity cycle that looks identical to the Sun's" : (1) same mean Ca index; (2) same amplitude of Ca index; (3) same period. After such statements : HD81809 seemed to us an attractive solar twin candidate, and we added it to our list.

Table I

	Catalogue Data			
	HD 44594	HD 76151	HD 20630	HD 81809
V	6.60	6.00	4.83	5.38
B-V	0.66	0.67	0.68	0.64
Sp	G3V	G3V	G5V	G2V
π (0".001)	41	90	106	37
R_V (kms^{-1})	+60	+28	+20	+54
U (kms^{-1})	+29	-32	-23	-35
V (kms^{-1})	-73	-19	-4	-46
W (kms^{-1})	-5	-5	-5	+9

3. OBSERVATIONS AND DATA REDUCTIONS

The four stars have been observed with the CFH Telescope using the coudé spectrograph and a cooled Reticon array of 1872 photodiodes. The average S/N ratio was about 250, and the resolution about 0.2 Å (35000). HD44594, which is a southern hemisphere star, has been observed also at ESO, using the Coudé Echelle Spectrograph fed by the 1.4m Coudé Auxiliary Telescope.
In this paper we discuss only results for iron lines falling in the spectral intervals centered at Hα (6563 Å) and at ≈6750. The equivalent widths of the chosen Fe-lines have been obtained with a profile fitting technique that account for blends, following Cayrel et al. (1985).

4. DETAILED SPECTRAL ANALYSES OF THE STARS

The method is the same as the one followed by Cayrel et al. (1981). We remember that no abundance can be determined unless the three physical parameters T_{eff}, log g and ζ_t have been obtained. The effective temperatures have been derived from the wings of the Hα lines, the gravities from the ionisation equilibrium, the microturbulence was relatively unimportant: the analyzed lines were all sufficiently weak, so that a common curve of growth could be used.
The iron abundance of each program star has been obtained by comparing the observed FeI equivalent widths with those of appropriate models

interpolated in a grid of model atmospheres (Gustafsson, 1981). The results from the detailed spectral analyses are contained in Table II.

Table II

Atmospheric parameters of the stars

	HD 44594	HD 76151	HD 20630	HD 81809
T_{eff} (K)	5770 ± 30	5710 ± 40	5630 ± 50	5630 ± 50
logg (cms^{-2})	4.50	4.50	4.50	3.75
ξ_t (kms^{-1})	1.0	1.0	1.0	1.0
[Fe/H]$_\odot^*$	+0.13 ± .06	+0.13 ±.06	+0.03 ±.05	−0.28 ± .09

5. DISCUSSION OF THE RESULTS

From Table I we see that the B-V's of the stars are higher than 0.64 with the exception of HD81809, and that, again, at the exception of HD81809 the spectral types of the stars are more advanced than that of the Sun. The space velocities of the stars with the exception of HD20630 seem to indicate that they are belonging to the old disk population.
Keeping in mind that Gustaffson's solar model has been computed with T_{eff}=5770 K, logg=4.50, [A/A$_\odot$]=0.0 and ξ_t=1.0 kms^{-1}, we see from Table II that the best candidate for a solar twin is HD44594: it has the same effective temperature, gravity and microturbulence as the Sun, the star has however a slight overabundance of Fe, which could be responsible for its spectral type, one unit more advanced than that of the Sun and for its relatively high B-V value.
HD76151 is cooler and slightly overabundant in iron.
HD20630 is very much cooler than the Sun but its iron abundance is normal.
The detailed analysis of HD81809 has revealed that its T_{eff}, logg and [Fe/H] values differ strongly from those of the Sun. Following the BS Catalog the star is also a spectroscopic binary. It is difficult to understand why its activity cycle is "identical" to the Sun.

6. CONCLUSION

First of all, let us recall that the CFH and ESO observations of the four solar analogs, we have analysed, have produced results being by almost an order of magnitude more accurate than those of Cayrel de Strobel et al. (1981) and Perrin and M. Spite (1981). Nevertheless our conclusion is the same as in Cayrel et al. (1981): none of the four solar analogs we have presented here is a "real solar twin".

References
- Cayrel, G., Knowles, N., Hernandez, G., and Bentolila, C.: 1981, Astron. Astrophys. 94, 1
- Cayrel, R., Cayrel de Strobel, G., and Campbell, B.: 1985, Astron. Astrophys. 146, 249
- Gustafsson, B.: 1981 (private communication)
- Hardorp, J.: 1978, Astron. Astrophys. 63, 383
- Mihalas, D.: 1981 (private communication)
- Perrin, M.N., Spite, M.: 1981, Astron. Astrophys. 94, 207.

WELL DETERMINED ATMOSPHERIC PARAMETERS FROM HIGH S/N RETICON SPECTRA OF FOUR G AND K DWARFS WITHIN 10 pc OF THE SUN

M.-N. Perrin[1], G. Cayrel de Strobel[2] and M. Dennefeld[3]
[1] Observatoire de Paris, Paris, France
[2] Observatoire de Paris, Section de Meudon, Meudon, France
[3] Institut d'Astrophysique, Paris, France

ABSTRACT. The detailed analysis of high resolution, high signal-to-noise spectra of four G and K dwarfs lying within 10 pc of the Sun is presented. This analysis is part of a project aiming at building up a homogeneous set of reliable atmospheric parameters for all F, G and K stars nearer than 10 pc according to the Gliese catalogue.

High resolution ($\Delta\lambda \simeq 0.13$Å), high signal-to-noise (S/N > 200) spectra in four spectral intervals centered at λ 6165, 6560, 6715 and 8520 Å have been obtained for four G and K dwarfs lying within 10 parsecs of the Sun, HD 115617, HD 125072, HD 156274 and HD 156384, using the coudé Echelle Spectrograph (CES) fed by the 1.4m coudé Auxiliary Telescope (CAT) at ESO. The stars were chosen at random among those of the 69 non-degenerate stars of the Gliese catalogue (1969) with T_{eff} > 4000 K and π > 0."100, which have not yet been analysed in detail. The following results were derived.

The detailed analysis of the spectra in the λ 6165, 6560 and 6715Å intervals, based on atmospheric models of Gustafsson (Gustafsson et al. 1975, Gustafsson, 1981), has provided the well determined atmospheric parameters of the stars, given in Table I.

Table I. Relevant data for the program stars

	HD 115617	HD 156274	HD 125072	HD 156384
Sp.T.	G6V	G8V	K3V	K3V
V (mag.)	4.74	5.53	6.66	6.10
π''	0.113	0.133	0.106	0.140
T_{eff} (K)	5585	5295	4965	4930
$\sigma_{T_{\text{eff}}}$ (K)	±30	±60	±60	±100
log g	4.5	4.5	4.5	4.5
$\sigma_{\log g}$	±0.2	±0.2	±0.2	±0.2
ξ_t (kms^{-1})	1.0	0.5	0.5	0.5
$[\text{Fe/H}]^*_\odot$	−0.02	−0.35	+0.26	−0.59
$\sigma_{[\text{Fe/H}]^*_\odot}$	±0.03	±0.07	±0.09	±0.08

The range of metallicity obtained for a very small randomly selected sample is surprisingly large : $[\text{Fe/H}]^*_\odot$ from −0.59 to +0.26 dex.

A check for chromospheric activity was provided by the analysis of the two lines of the Ca II infrared triplet, lying in the $\lambda\,8520\,\text{Å}$ interval. For the three stars observed in this interval, HD 115617, HD 125072 and HD 156274, the depth of the two Ca II lines, as compared with the filled in lines of the very active solar-type Hyades dwarf VB 64 (Cayrel et al. 1983), is indicative of low chromospheric activity.

From the analysis of the lithium feature in the $\lambda\,6715\,\text{Å}$ region, it is derived that no lithium feature is apparent in the four stars.

More detailed results will be published in Astronomy and Astrophysics.

REFERENCES

Cayrel,R., Cayrel de Strobel,G., Campbell,B. Mein,N., Mein,P., Dumont,S.: 1983, *Astron. Astrophys.* **123**, 89
Gliese,W.: 1969, *Veröff. Astron. Recheninst. Heidelberg.* **22**
Gustafsson,B.: 1981 (private communication)
Gustafsson,B., Bell,R.A., Eriksson,K., Nordlund,Å.: 1975, *Astron. Astrophys.* **42**, 407

SOLAR FLUX ATLAS FROM 296 TO 1300 nm

Ingemar Furenlid
Georgia State University
Atlanta, Georgia 30303, USA

ABSTRACT: The solar flux atlas is presented and some basic information on the observations is given.

The "Solar Flux Atlas from 296 to 1300 nm" by Kurucz, Furenlid, Brault and Testerman (1984) is a high S/N spectrum of integrated sunlight, produced mainly to serve as a reference in stellar spectroscopy. The Fourier transform spectrometer of the McMath solar telescope at Kitt Peak offered an almost ideal instrument for such a purpose; a relatively short observing time could give a large wavelength coverage at extremely high S/N and spectral resolution. Observations for the atlas were carried out in the later part of 1980 and the first half of 1981. The efforts of Dr. Kurucz were crucial in bringing the atlas to completion and in obtaining the funding commitment needed to produce the printed version.

The S/N at continuum level of the full resolution atlas is probably not less than 2000 at any point longward of 303 nm and is typically somewhat less than 3000. The resolution ranges from around 350 000 in the UV to around 500 000 in the near IR. Being a FTS spectrum, the instrumental profile is essentially perfect with no stray light present; an uncertainty in the zero point at the 0.2% level in UV and visual mimics a corresponding possible level of scattered light. Another concern in very high S/N work is the absorption by ozone, which extends through the visible part of the spectrum.

The flux spectrum has been ratioed to the adopted continuum before being plotted. The continuum has been determined by global fits to high points in the original FTS scans, each fit covering from around 30 nm in the UV to around 300 nm in the IR. The continuum determination is thus entirely empirical and yields strictly speaking a quasi-continuum. Deviations from the true solar flux continuum are of course particularly marked in the UV and blue part of the visible.

When the atlas is used as a reference in stellar work it is important to strive for consistency between solar and stellar spectra, for instance in continuum location. The absence of stray light in the atlas has no significant impact on equivalent widths for most applications but must be considered in profile work.

The printed version of the atlas may well be used as a note pad for recording various data concerning individual spectral lines; at a price of $13.00 this seems a perfectly reasonable suggestion!

The Solar Flux Atlas can be purchased for US $13.00 from:
        National Solar Observatory
        Sunspot, New Mexico 88349, U.S.A.

A tape version can be obtained by sending a 2400-foot tape to:
        Dr. Robert Kurucz
        Center for Astrophysics
        60 Garden Street
        Cambridge, Mass., 02138, U.S.A.

A High S/N Spectroscopic Study of Alpha Centauri A

Ingemar Furenlid and Tom Meylan
Georgia State University
Atlanta, Georgia 30303, USA

ABSTRACT. A fine analysis of Alpha Cen A relative to the Sun has been carried out. Over 500 absorption lines have been used in deriving a consistent solution involving abundances, Teff, log g, microturbulence and blanketing. Light elements, up to zinc, are overabundant by around 0.15 dex, while heavy elements have solar abundances.

1. INTRODUCTION

Alpha Centauri A is apparently bright and very similar to the Sun. These properties make it a rewarding object for a differential analysis with the Sun as the comparison star. The term "high S/N" in the title stems on a primary level from the basic observational data, but "high S/N" implies more than precise spectroscopic data in a full analysis of a stellar spectrum. In this work we have spent considerable efforts to carry reductions, measurements and analysis to a level where these different operations are commensurate with each other in "S/N".

2. DATA

The work reported here was carried out essentially as two separate, but coupled, model atmosphere projects with one part devoted to the Sun and one to Alpha Cen A.

The solar spectrum data were provided by the "Solar Flux Atlas from 296 to 1300 nm" (Kurucz et al., 1984). The resolution of the atlas spectrum was degraded by a Gaussian to a value of 80 000 to match the stellar data; the resulting effective S/N is around 10 000. The equivalent widths of over 500 well-behaved lines of 26 elements were measured, where well-behaved lines are defined as those which have at least one half of the profile free of visible or otherwise known blends.

All measured solar lines of iron, 197 lines of neutral iron and 19 lines of singly ionized iron, were used as input parameters for the Sun in the usual model atmosphere iteration scheme. The stellar models used in iterating until a consistent solution was found were taken from the grid by Kurucz (1979). The iteration process was terminated when no dependence could be discerned of the iron abundance on excitation

potential, line strength, height of line formation or level
of ionization. The converged values of Teff, log g and microturbulence
(5800, 4.4, 1.0) were adopted for the Sun. The abundance of iron, as
well as of all other elements, were then set equal to the standard set
of solar abundances listed in the grid of models and "solar gf-values"
determined for all lines measured using the adopted physical parameters.

The spectra of Alpha Cen A were obtained at the CAT of the ESO
facility in Chile, using a Reticon as a detector. Around 100 frames
cover most of the visible spectrum with integration times chosen to
give S/N values in the reduced data of 500 - 600. Reductions were
carried out in standard fashion, the resolution degraded to 80 000,
dispersion solutions determined for each frame and continuum normalized
values plotted for every 0.02 A. The solar spectrum was treated in
exactly the same way so that the two spectra at this stage could be
superimposed point by point. Great care was taken in placing the
stellar continuum, and the solar flux atlas was consulted throughout
this operation to ensure that the continuum was consistent between two
stars.

3. ANALYSIS

The analysis of Alpha Cen A can be traced back to a "first look" at
the spectrum, consisting of measured equivalent widths for around 25
iron lines. A preliminary abundance determination based on these lines
only, led to an overabundance of iron in Alpha Cen A of around 65 %
(Furenlid and Meylan, 1984). This limited set of data converged on
physical parameters(except for the iron abundance) essentially equal
for the Sun and Alpha Cen A. This result led to the use of metal-enhanced models with a metal abundance relative to the Sun of 0.2 dex
in the first pass of the full analysis of Alpha Cen A.

Using the measured equivalent widths for Alpha Cen A and the solar
gf-values it was found that the first pass of iterations with <u>metal-enhanced</u> models gave an overabundance of all the important electron
donor elements of around 0.1 dex. A second pass of iterations was
therefore made with models metal enhanced by 0.1 dex, which then
resulted in stellar metal abundances up by around 0.15 dex. The
iterations done by metal-enhanced models used sub-grids kindly computed
for us by Dr. R. Kurucz. The final result is that using models with an
overabundance of metals relative to the Sun of 0.1 dex a consistent
solution for Alpha Cen A was found for a Teff of 5710°K, a log g of 4.0,
a microturbulence of 1.0 km/sec and an average overabundance of light
elements of 0.15 dex (formally 0.13 dex). The internal error in the
abundance determination can be estimated by a comparison of abundances
from neutral and singly ionized species of the same elements as shown
in <u>fig. 1</u>; element, ionization stage and number of measured lines are
marked over each point. The maximum difference is 0.03 dex, adopted
as an error estimate. Significant abundance variations between different light elements were obtained. The average overabundance of the
three major atmospheric electron donors (Mg, Si and Fe) is around 0.15
dex, the importance of these elements of course being their effect on
the continuous (H-) opacity. The average overabundance of 0.15 dex

reaches only to Z = 30 (zinc), whereas elements with Z larger than 30 have the same abundance as in the Sun as can be seen in fig. 2. We conclude that heavy elements occur in Alpha Cen A with nearly solar abundances but that elements up to and including the iron group are definitely overabundant by around 0.15 dex.

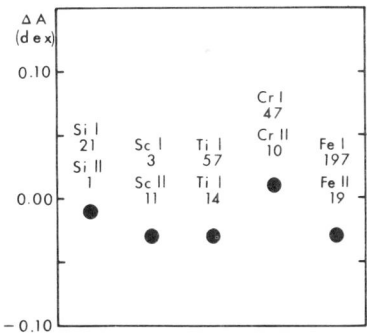

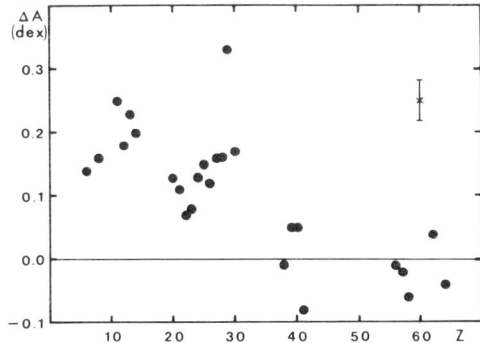

Fig. 1. The abundance difference, ΔA, between ionized and neutral species of five elements. Element, ionization stage and number of lines used in the abundance determination are marked over each plotted point.

Fig. 2. The abundance difference, ΔA, between Sun and Alpha Cen A plotted versus atomic number, z. The estimated internal error in an abundance value is marked in the upper right part of the figure.

4. DISCUSSION

The first striking result of this analysis is the strong dependence of the derived physical parameters on the metal abundance of the models. It always seems possible to find a combination of physical parameters that yields a consistent solution within acceptable margins. Using a grid of solar abundance models for Alpha Cen A underestimates blanketing and continuous opacity in the star but leads nevertheless to a consistent set of parameters; Teff will be nearly solar, thereby compensating in a broad way for the smaller continuous opacity in the model. Making the overall metal abundances part of the iteration scheme by including them in the models, allows blanketing and opacity to be correctly evaluated and has a notable effect on the final parameters, particularly the effective temperature, which is now substantially lower (90°K) than the Sun's; the overabundance is also significantly reduced.

The second striking result is the remarkable dichotomy in the abundance pattern; light elements up to zinc are overabundant, those heavier than zinc have close to solar abundances. The overabundance of lighter elements, which are the abundant and therefore important ones, is interesting in the context of a recent discussion of interior models of Alpha Cen A and B by Demarque et al. (1986). In order to find an equal age for both components of the Alpha Cen system it is necessary to assume on overabundance relative to the Sun of around 0.15 dex for both stars, leading us to the prediction that Alpha Cen B is also overabundant by around 0.15 dex.

5. REFERENCES

Demarque, P., Guenther, D. B. and van Altena, W. F. 1986, Ap. J., 300, 773.
Furenlid, I. and Meylan, T. 1984, ESO Messenger, 37, 10.
Kurucz, R. L., Furenlid, I., Brault. J. and Testerman, L. 1984, Solar Flux Atlas from 296 to 1300 nm, National Solar Observatory, Sunspot, N.M. 88349, U.S.A.
Kurucz, R. L. 1979, Ap. J. Suppl., 40, 1.
Kurucz, R. L. 1985, priv. comm.

DISCUSSION

COTTRELL Could I make a request that we adopt a uniform method for describing the enhancement or depletion of the elements rather than percentage, factors relative to the Sun or logarithmic scale factors (dex).

FURENLID Good point ; the dex-scale makes most sense and will be used in the written version.

A HIGH S/N SPECTROSCOPIC SURVEY OF CHEMICAL ABUNDANCES IN DISK POPULATION F0-G2 DWARFS

J. Andersen, Copenhagen University Observatory, Denmark,
    and Center for Astrophysics, Cambridge, Mass.

B. Edvardsson, Uppsala Astronomical Observatory, Sweden

B. Gustafsson, Uppsala Astronomical Observatory,
    and Stockholm Observatory, Sweden

P. E. Nissen, Institute of Astronomy, University of Aarhus,
    DK-8000 Aarhus C, Denmark.

In collaboration with D. Lambert and J. Tomkin high resolution, high S/N spectra have been obtained for about 250 disk population F0-G2 dwarfs. The aim of this large programme is to study various abundance ratios as a function of [Fe/H] and stellar age. A detailed description of the project and some preliminary results were given by Nissen et al. (1985).

The stars were selected by E. H. Olsen from uvby-β photometric catalogues containing all F0-G2 stars brighter than V = $8^m\!.3$. By means of the δm_1 index these stars were divided into 9 metallicity groups with [Fe/H] ranging from -1.0 to 0.3. Furthermore, the δc_1 index was used to select the somewhat evolved stars, such that ages could be estimated. Of these stars the 30 brightest in each metallicity group were put on the observing programme.

The northern programme stars have been observed with the McDonald Observatory 2.7 m telescope, Coudé spectrograph and Reticon detector, and the southern stars with the ESO 1.4 m CAT telescope and Coudé Echelle Spectrometer. For 10 stars that have been observed from both observatories the corresponding two sets of equivalent widths agree within ±3 mÅ.

Until now equivalent widths have been measured for 57 southern stars, and a preliminary LTE model-atmosphere analysis of the data has been carried out as described by Nissen et al. (1985). The parameters, T_{eff}, g and overall metal abundance of the model atmosphere corresponding to a given star, are estimated from the uvby-β photometry. The analysis is differential with respect to the Sun in the sense that gf-values of the lines are derived from solar equivalent widths as measured in the spectrum of reflected sunlight.

In the figures some results of the abundance survey are shown. The following trends are seen:

i) The O/Fe ratio increases with decreasing iron abundance according to the formula [O/Fe] = -0.5 [Fe/H].
ii) The elements Mg and Si are overabundant with respect to iron for [Fe/H] < -0.6. A similar effect is present for Ca and Ti. [Si/Fe] shows a significant offset from zero at [Fe/H] = 0.0.
iii) Aluminium shows an enhanced odd-even effect with respect to magnesium for [Fe/H] < -0.3. A similar well-defined effect is not found for sodium, but the scatter in [Na/Fe] tends to increase for [Fe/H] < -0.3.

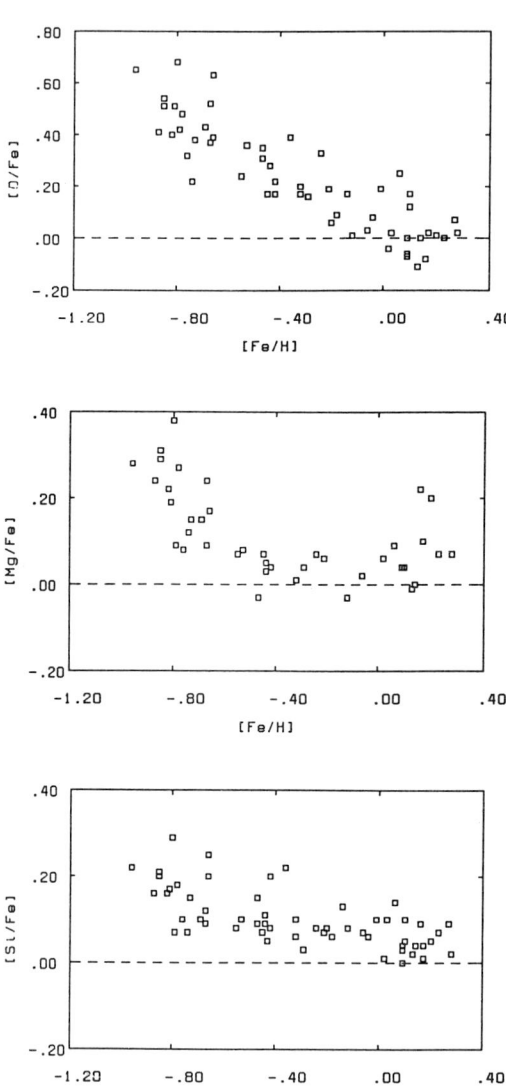

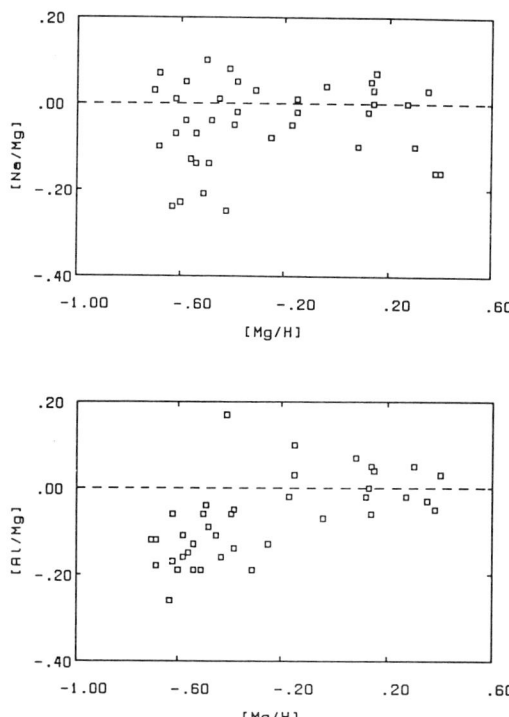

These trends should not be taken too seriously until the LTE analysis has been replaced by non-LTE computations, which are presently being carried out, and the effects of convection and inhomogeneities have been scrutinized.

The most important conclusion at present is:

All abundance ratios with respect to iron have a surprisingly small scatter at a given [Fe/H]. The rms dispersion is ±0.10 for [O/Fe], ±0.04 to ±0.06 for the α-elements, and ±0.05 to ±0.10 for [Al/Fe] and [Na/Fe]. This is the kind of scatter one would expect from the errors in the analysis and in the equivalent widths. Thus, there is no evidence for any cosmic scatter (except maybe in the case of Na/Fe for the metal-poor stars). In particular the small scatter in the N_α/N_{Fe} ratio (less than 10%) at a given [Fe/H] is remarkable. Either the α-elements and iron are produced in constant proportions at a given time by the different supernovae contributing to the nucleosynthesis or the ejecta of supernovae are very well mixed with the interstellar gas before star formation takes place.

REFERENCE

Nissen P.E., Edvardsson B., Gustafsson B., *Proc. ESO workshop on "Production and Distribution of C,N,O Elements"*, Eds. E.J. Danziger, F. Matteucci and K. Kjär, p.131.

IRON LINES AND SURFACE GRAVITY DETERMINATION FOR τ CETI

S. Arribas and C. Martinez Roger,
Instituto de Astrofisica de Canarias, 38200 - La Laguna,
Tenerife,
Spain.

ABSTRACT. We present a spectroscopic analysis for the gravity determination of τ Ceti on the basis of a new and high quality set of data. The iron lines are studied by using the Oxford oscillator strengths. The results are compared with those obtained with "solar" oscillator strengths. Non-LTE effects seem to be important in this star. The FeI at λ 526.955 nm line is used to apply the Blackwell and Willis (1977) method to determine the surface gravity. The derived log g is 4.70±0.1.

1. OBSERVATIONS AND ANALYSIS

The observations were made in October 1985 at ESO using CAT+CES and a Reticon detector with a resolving power of 80000. Reductions were carried out using the ESO IHAP system. Where possible the lines were analysed using the Oxford oscillator strength (Blackwell et al., 1986 a, and references therein). For high excitation lines where no laboratory measurements of adequate precision are available, we have determined "solar" oscillator strengths from the solar flux spectrum, in a similar way to Smith et al. (1986). The sources for the damping constant values are the works by Simmons and Blackwell (1982) and Gurtovenko and Kondrashova (1980). For the damped line used to determine the surface gravity, γ_H/N was obtained by adjusting its profile in the solar flux spectrum (Kurucz et al., 1984) to the synthetic one generated with the Holweger and Müller (1974) solar model.

In order to obtain the value for Teff we have applied the Infrared Flux Method (Blackwell et al., 1986 b and references therein), obtaining a value of 5250 K with a probable error of ± 50 K.

The equivalent widths were interpreted in terms of log (Agf) values by using an LTE code. For the entire analysis we have used model atmospheres generated by the MARCS suite of programs. The MARCS atmospheres are line-blanketed and flux constant atmospheres of the same type as those published by Bell et al. (1976).

The analysis to determine the microturbulence is analogous to the one used by Smith et al. (1986 and references therein). Thus, the microturbulence (ξ) and the abundance (A) of an element are obtained simultaneously from the function $A=A(\xi)$ for a set of lines with a wide range in equivalent widths.

In Figures 1 a, b and 2 a, b we present the behaviour for the FeI lines in
τ Ceti and in the Sun, when their respective MARCS models were used.
Figures b indicate the scatter in abundance as a function of ξ. We can
observe that minima are sited at 1.12 and 1.19 Km s^{-1} for τ Ceti and the
Sun respectively. In Figures 1 a, b the lowest excitation lines have not been
considered for reasons discussed below.

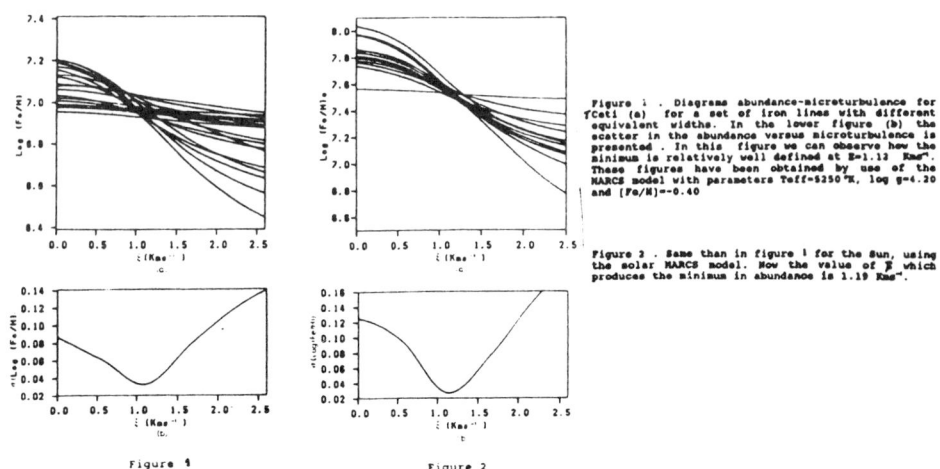

Figure 1. Diagrams abundance-microturbulence for τCeti (a) for a set of iron lines with different equivalent widths. In the lower figure (b) the scatter in the abundance versus microturbulence is presented. In this figure we can observe how the minimum is relatively well defined at ξ=1.12 Kms^{-1}. These figures have been obtained by use of the MARCS model with parameters Teff=5250 K, log g=4.30 and [Fe/H]=-0.40

Figure 2. Same than in figure 1 for the Sun, using the solar MARCS model. Now the value of ξ which produces the minimum in abundance is 1.19 Kms^{-1}.

2. FeI LINES

When the iron lines with accurate oscillator strength (low excitation) were
analysed a clear correlation between abundance and excitation was found.
Lines with excitation higher than 3 eV were analysed by use of "solar"
oscillator strengths. In this case, the absolute abundance is not directly
significant but the abundance relative to the Sun is reliable. We observed
also an important difference in the relative abundance value when comparing
with the lower excitation lines. These effects are unlikely to be explained
by errors in the atomic parameters, equivalent widths or effective temperature
taking into account the estimated errors for these magnitudes. An explanation for this could be non-LTE effects in the solar photosphere. Athay and
Lites (1972) pointed out that such effects could be important for the FeI and
FeII in the solar photosphere. However, Rutten and Kostik (1982) have found
that the Holweger and Müller (1974) model corrects these effects as well as
the Bell et al. (1976) model. As also in this case reasonable changes in the
model parameters do not explain the effect, we must think of non-LTE in
excitation in τ Ceti. For all these reasons the value for the iron abundance
remains more uncertain than expected from the quality of the atomic and
observational data used here. Thus we consider

$$\log (Fe/H) = 7.00 \pm 0.07 \quad \text{and}$$
$$[Fe/H] = -0.53 \pm 0.07$$

3. GRAVITY DETERMINATION

For the surface gravity determination we applied the Blackwell and Willis (1977) method to the FeI λ 526.955 line, which is strongly broadened by collisions. In this case, abundance is obtained from lines with similar excitation, in order to minimize non-LTE effects. When the MARCS solar models are used, the best fit is found for log g=3.94. Other authors have also found important differences in the log g values when using different solar models (Ruland et al., 1980). This shows the difficulty in absolute g determinations. For τ Ceti we found the best adjustement with the MARCS model with log g=4.20 (Figure 3). When the log g value obtained for τ Ceti si normalized to the proper solar one, we obtain log g=4.70 ± 0.10.

The log g value obtained disagrees with the one obtained by Smith and Drake (1987) from the calcium line λ 616.217 nm (log g=4.50± 0.1). Partially, this can be explained by the difference in the effective temperature assumed. On the other hand it is also noticeable that Bell et al. (1985) obtained a difference of 0.1 dex in the log g value for Arcturus when this was obtained from the CaI λ 616.217 nm and from the FeI λ 526.955 lines in the same sense as obtained here.

Acknowledgements

We are grateful to the Department of Astrophysics, University of Oxford for the use of its facilities. We also thank Dr. G. Smith who gave us some results in advance to publication.

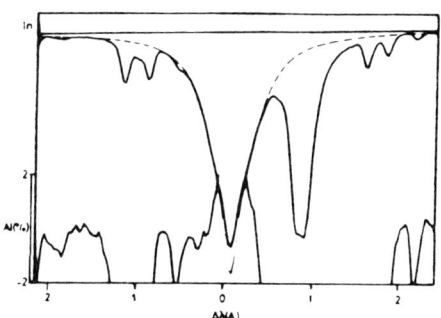

Figure 3. Best fit for the observed spectrum for τ Ceti, assuming Teff=5250°K with the MARCS models. The log g value was 4.30. Continuum line represents the observed line and discontinuum one indicates the synthetic profile. The lower scale indicates the difference in percentage between both types of profiles.

References

Athay, R.G., and Lites, B.W.: 1972, Astrophys. J. **176**, 809.
Bell, R.A., Edwardsson, B., Gustafsson, B.: 1985, MNRAS **212**, 497.
Bell, R.A., Ericsson, K., Gustafsson, B., Nourld, A.: 1976, Astron. Astrophys. Suppl. Ser. **34**, 229.
Blackwell, D.E., Booth, A.J., Petford, A.D., Leggett, S.K., Mountain, C.M., Selby, M.J.: 1986 b, MNRAS **221**, 427.
Blackwell, D.E., Booth, A.J., Haddock, A.D., Petford, A.D., Leggett, S.K.:

          1986 a, MNRAS **220**, 549.
Blackwell, D.E., Willis, R.B.: 1977, MNRAS **180**, 169.
Gurtovenko, E.A., Kondrasova, N.N.: 1980, Solar Phys. **18**, 347.
          Astron. Astrophys. **42**, 407.
Holweger, H., Müller, E.A.: 1974, Solar Phys. **39**, 19.
Kurucz, R.L., Furelid, I., Brault, J., Testerman, L.: 1984, Solar Flux Atlas
          from 269 to 1300 nm. National Solar Observatory Atlas No. 1.
Ruland, F., Holveger, H., Griffin, R.R., Biehl, D.: 1980, Astron. Astrophys.
          92, 70.
Rutten, R.J., Kostik, R.I.: 1982, Astron. Astrophys. **115**, 104.
Smith, G., Drake, J.J.: 1987, (In Press).
Smith, G., Edwardsson, B., Frisk, V.: 1986, Astron. Astrophys. **165**, 126.
Simmon, G.J., Blackwell, D.E.: 1982, Astron. Astrophys **112**, 209.

THE IRON ABUNDANCES, [FE/H] IN THE FOUR NEAREST OPEN CLUSTERS :
PLEIADES, URSA MAJOR STREAM, COMA BERENICES AND HYADES

R. Cayrel[1], G. Cayrel de Strobel[2] and B. Campbell[2]
[1] Observatoire de Paris, 61 avenue de l'Observatoire
F-75014 PARIS
[2] Observatoire de Paris, Section de Meudon
F-92195, Meudon Cedex, France

ABSTRACT. Iron abundances, [Fe/H] have been studied in G and K dwarfs of the four nearest clusters. With the exception of the Pleiades stars which are all fainter than the 10th magnitude, the observational material consists of high resolution, high S/N spectra taken at the coudé Reticon spectrograph of the Canada, France Hawaii Telescope. The Hyades is the only cluster with a significative overmetallicity with respect to the Sun.

1. INTRODUCTION

The variation of metallicity as a function of space and time is a central problem in the knowledge of the chemical evolution of our Galaxy and of other galaxies. We are studying this problem using the stellar content of the solar neighborhood in which objects of all ages can be found. We know that our nearest neighbours, the field stars are chiefly old disk population stars, sometime quite older than the Sun. Their metallicities vary from factors by 1 to 3, more, to factors by 1 to 5, less, than that of the Sun. If we want to study the metallicities of young disk population stars we have to turn towards stars belonging to nearby open clusters. Indeed, in the understanding of the chemical composition of our Galaxy the nearby open clusters play an important role. Cluster members are often used as reference stars for photometric and/or low resolution spectroscopic abundance techniques. Their abundances have to be reliably determined, if we want to use these stars as milestones for the study of the chemical evolution of a galaxy.
Here, we present an analysis of the determination of iron abundance in some G and K stars belonging to the Pleiades, Ursa Major stream, Coma Berenices and Hyades. We have already published iron and metal abundances of 12 G and K dwarfs of the Hyades (Cayrel et al. 1985). The same reduction procedure and spectral detailed analysis as in this research have been used in studying the stars of the other three clusters.

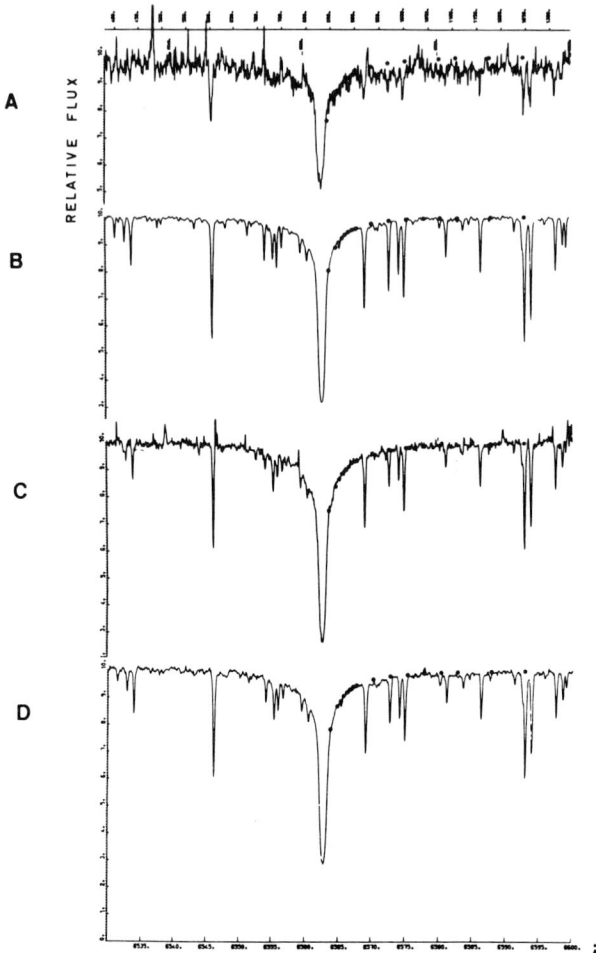

Fig. 1. CFH Reticon spectra of the Hα region of:
A HII 1776, G5V, in the Pleiades,
B HD 41593, K0V, in UMa,
C TR 132, G5V, in Coma,
D VB 92, G8V, in the Hyades,
dots represent the best fitting computed profiles.

2. OBSERVATIONS, DATA REDUCTIONS, AND DETAILED SPECTRAL ANALYSES
   OF THE CLUSTER STARS.

All the stars contained in Table 1 have been observed with the CFH

Telescope using the Coudé spectrograph and a cooled Reticon array of 1872 pixels. In this table are also included five specimen of the twelve Hyades stars previously analyzed. (Cayrel et al. 1985). As we can see from Table 1, the signal/noise ratio of the Pleiades stars (all fainter than the 10th magnitude) is much smaller than that of the stars of the other clusters. The most nasty feature, during long exposures, is the large number of cosmic ray spikes which damage the spectrum. Special attention has been payed in the reduction of the equivalent widths of the stars of the Pleiades.

Table 1

Some basic parameters for the stars of the four clusters

PLEIADES					
Age	~ 80 10^6 yrs				
Distance	~ 135 pc				
Star	S/N	V	Sp	$T_{eff}K$	[Fe/H]
HII 1101	110	10.26	F9V	6100	+0.09±0.12
1776	44	10.91	G5V	5600	+0.02±0.19
2462	44	11.55	-	5300	+0.16±0.12
296	38	11.46	G8V	5100	+0.26±0.17

URSA MAJOR					
Age	~ 160 10^6 yrs				
Distance	from ~ 8 to ~ 20 pc				
Star	S/N	V	Sp	$T_{eff}K$	[Fe/H]
γLepA	700	3.60	F6V	6200	−0.14±0.04
γLepB	400	6.15	K2V	4950	+0.02±0.06
π'UMa	500	5.64	G0V	5850	−0.01±0.06
HD115043	370	6.85	G1V	5830	−0.03±0.04
HD41593	600	7.23	K0V	5350	+0.08±0.04

COMA BERENICES					
Age	~ 500 10^6 yrs				
Distance	~ 85 pc				
Star	S/N	V	Sp	$T_{eff}K$	[Fe/H]
TR 162	150	8.61	G0V	6250	−0.07±0.14
85	150	9.33	G1V	5850	−0.06±0.09
132	140	9.91	G5V	5700	+0.02±0.11
213	90	10.51	K0V	5300	0.00±0.14

HYADES					
Age	~ 650 10^6 yrs				
Distance	~ 44 pc				
Star	S/N	V	Sp	$T_{eff}K$	[Fe/H]
VB 73	300	7.85	G1V	5900	+0.143±.035
52	440	7.80	G1V	5840	+0.028±.035
64	200	8.12	G2V	5770	+0.138±.035
92	300	8.66	G8V	5540	+0.157±.050
46	200	9.11	K1V	5170	+0.070±.100

The effective temperatures have been determined from the wings of the Hα lines (Fig. 1). The iron abundance of each star has been obtained by comparing the observed equivalent widths of iron lines with those of appropriate models interpolated in a grid of model atmospheres (Gustafsson 1981). The results from the detailed spectral analyses are contained in Table 1.

3. DISCUSSION OF THE RESULTS.

The list of dwarfs analyzed in each cluster is presented in Table 1. In this table the clusters do rank in order of increasing cluster age. The mean metallicities of the first three clusters does not seem to deviate from the solar one. Among the stars of a same cluster slight difference in [Fe/H] are found. The authors of this paper have considered this problem and have proposed that such differences may be caused by stellar activity (Cayrel et al. 1985). The Hyades are the only cluster with a significative overmetallicity with respect to the Sun.

4. CONCLUSION

We have studied the iron abundances, [Fe/H] in G and K dwarfs of the four nearest clusters. The observational material was excellent except for the stars of the Pleiades. Significant differences in the [Fe/H] values have been found between the stars of a single cluster in the Hyades, Ursa Major and Pleiades. The Hyades are the only cluster showing a significative enhancement in [Fe/H] with respect to the Sun, confirming that determined from Strömgren's photometry.

REFERENCES.

Cayrel, R., Cayrel de Strobel, G., and Campbell, B.: 1985 Astron. Astrophys. 146, 249
Gustafsson, B.: 1981 (private communication).

DISCUSSION

BASRI   How worried are you by the effects of chromospheric activity on the wings of Hα used to make T_{eff} determinations, especially for very active stars.

R. CAYREL   For stars not more active than the Hyades we are confident that keeping away of the (−2 Å, +2 Å) region around the center of Hα our T_{eff} is not contaminated by activity. For the Pleiades the activity is so strong that we do not believe that Hα is still a reliable T_{eff} indicator.

BESSELL   Comment: It is always advantageous to use continuum colors or fluxes such as R-I, V-K and IR flux method, for effective temperature derivation in addition to hydrogen line profiles.

DUNCAN   Comment: Paul Butter, Geoff Marcy, Ross Cohen, and I have completed a study of 4 rapidly rotating and 4 slowly rotating K dwarfs in the Pleiades [Ap.J. LeH., in press]. We find that all of the rapid rotators, usually thought to be older, have much more Li than the slow rotators. HII 296 may show a similar effect.

CHEMICAL COMPOSITION OF THE FOUR HYADES GIANTS

N. Arimoto and G. Cayrel de Strobel
Observatoire de Paris, Section de Meudon
F-92195, Meudon Cedex, France

ABSTRACT. Differential detailed analyses with respect to ϵ Vir of the four Hyades giants have been carried out on high resolution high S/N CFHT Reticon spectra. The preliminary results are given below.

1. INTRODUCTION

Differential detailed analyses of 12 dwarfs of the Hyades cluster have been carried out by Cayrel et al. (1985). This paper deals with a similar analysis, but for the 4 giants of the cluster. Its aim is to compare the metal abundance found in two samples of stars of the same cluster being in a different stage of evolution.

2. OBSERVATION, REDUCTION AND METHOD OF ANALYSIS

High resolution, high S/N spectra in three spectral intervals at $\lambda 5890$, 6560, and 6750A have been obtained for the Hyades giants γ Tau, δ Tau, ϵ Tau, and θ^1 Tau, using the coudé spectrograph of the Canada-France-Hawaii Telescope (CFHT). The average S/N ratio obtained for these spectra was about 300:1 and their resolution $\simeq 0.2Å$. Equivalent widths of metal lines have been obtained with a profile fitting technique that accounts for blends. Special care has been given in determining the continuum, always a delicate question in these late type stars.
The spectra have been interpreted with theoretical line computations using the grid of model atmospheres computed by Bell and Gustafsson (1978). The G9III star ϵ Vir has been used as comparison star. The spectra of ϵ Vir have been obtained with the same instrumentation and the same S/N ratio.

3. RESULTS

The effective temperatures of the 4 giants have been derived from the photometric (T_{eff}, V-K) calibration for giants by Ridgway et al. (1980). The spectroscopic gravity has been obtained from the ionization equilibrium and the microturbulent velocity from the curves of growth. No significant difference in gravity and microturbulence has been found between the 4 giants. We have found log g = 2.7 and ξ_t = 1.7 km.s^{-1} for the 4 stars. Their iron abundance is also very similar, the mean value is :

[Fe/H] $\frac{\text{Hyades giants}}{\odot}$ = + 0.11 ± 0.03.

ϵ Vir

However, in analysing ϵ Vir with respect to the sun, we have found slightly, but significantly, enhanced iron abundance , i.e. :

[Fe/H] $\frac{\epsilon \text{ Vir}}{\odot}$ = + 0.09 ± 0.03.

In taking T_{eff}(ϵ Vir) = 4990, log g = 2.7 and ξ_t = 1.7 km.s^{-1}, the iron abundance of the giants with respect to the sun is then :

[Fe/H] $\frac{\text{Hyades giants}}{\odot}$ = + 0.20 ± 0.03.

All the other elements analyzed in the 4 giants have abundances varying in lockstep with iron, except sodium. The mean abundance value for sodium is :

[Na/H] $\frac{\text{Hyades giants}}{\odot}$ = +0.5 ± 0.1.

4. DISCUSSION

Table 1 gives chromospheric and coronal emissions, X-ray luminosity, effective temperature and Fe and Na abundances for the 4-Hyades giants. Ca II K and Mg II (h+k) are taken from Baliunas et al. (1983), logL$_X$ from Stern et al. (1981), T_{eff}, [Fe/H], and [Na/H] come from the present study.

Table 1 Chromospheric and coronal spectra CaII K, MgII(h+k), X-ray luminosity, effective temperature, abundance.

Star	Ca II K	MgII(h+k) (10^3ergs/cm²/s)	LogL$_X$ (ergs/s)	T_{eff} (K)	[Fe/H]$_\odot$	[Na/H]$_\odot$
θ^1 Tau	0.60	587	30.0	5050	0.19±0.02	0.5
γ Tau	0.48	566	29.4	4990	0.20±0.02	0.5
δ Tau	0.26	345	28.9	5000	0.21±0.02	0.5
ϵ Tau	0.18	334	—	4950	0.23 :	—

From columns 2 and 3 of Table 1 we see that θ^1 Tau and γ Tau have a more active chromosphere than δ Tau and ϵ Tau. No significant effect of chromospheric activity on the derived abundances is visible, in contrast to what was found by Cayrel et al. (1985) in a sample of 12 dwarfs of the Hyades. Surprisingly, the iron abundance found in the giants is slightly larger (+0.08 dex) than the one found in the dwarfs, Cayrel et al.

(1985). As no iron enrichment is possible from core nuclear burning and mixing, this difference can be only attributed to spurious non LTE or chromospheric activity effects.

Finally the evolutionary stage of the 4-Hyades giants is discussed. Figure 1 shows the (B-V,V) diagram of the Hyades from Johnson et al. (1962). Superimposed are evolutionary tracks by Iben (1965, 1967) for stars of 2.25 and 3.0 $M_\odot$; Y=0.29, Z=0.02. Two major nuclear burning phases, the hydrogen-core and the helium-core burnings, are indicated by thick lines. Fig. 1 demonstrates that the positions of the 4-Hyades giants in the (B-V,V) diagram correspond to the core-helium burning phase of stars with 2.5≲m/$M_\odot$≲3.0. Thus, the estimated mass is well beyond the critical mass m≃2.25$M_\odot$ for the electron degeneracy in the core during the first ascent of the red giant branch.

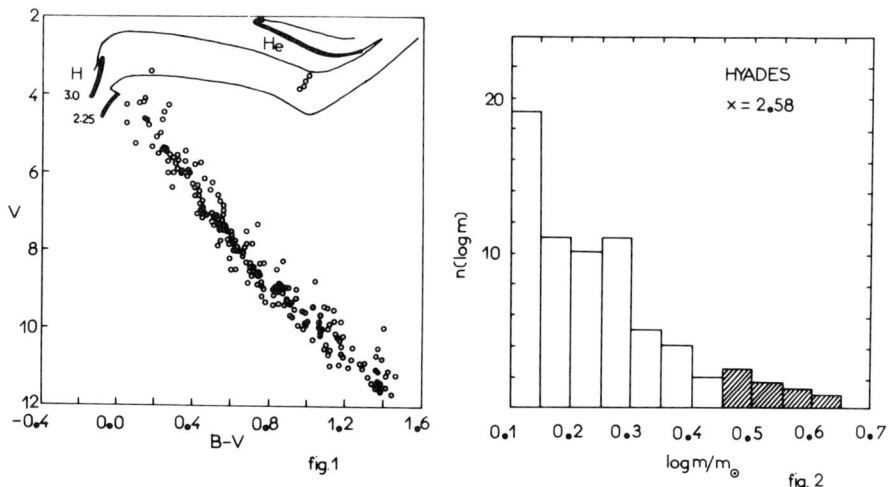

Fig.1. The (B-V,V) diagram of the Hyades. Superimposed are evolutionary tracks by Iben (1965,1967) for stars of 2.25 and 3.0 $M_\odot$; Y=0.29, Z=0.02.

Fig.2. The mass distribution of the Hyades dwarfs (non-hatched) from Tarrab (1982) and the predicted mass distribution for the giants (hatched).

It is well known that only 4 yellow giants have been found in the open cluster of the Hyades. It was therefore interesting to estimate the expected number of giants in the cluster by using its initial mass function (IMF). Tarrab (1982) found that the observed mass function of the Hyades is well approximated by a power-law of mass, $\phi(m) \propto m^{-\mu}$, with $\mu=2.58$ [for the Salpter's (1955) IMF, $\mu=1.35$]. Figure 2 shows the mass distribution of the Hyades dwarfs (non-hatched bins) in the mass range of $1.3 \leq m/M_\odot \leq 2.8$ (Tarrab,1982) and the predicted mass distribution for the Hyades giants (hatched bins) of $2.8 < m/M_\odot \leq 4.5$ by using the same slope of $\mu=2.58$. From Fig.2, we find that the expected number of giants in the mass range of $m/M_\odot \leq 3.0$ is about 2. The difference between the predicted and the observed number of the Hyades giants is consistent with the Hyades IMF so far as all the 4-giants are in the core-helium burning phase.

5. CONCLUSION

The iron abundance is very similar in the 4-Hyades giants, its mean value is [Fe/H]=+0.20±0.03. This significantly higher metal abundance of the Hyades giants with respect to the dwarfs and the constancy of their values in all the giants could be a non LTE effect or a consequence of a different effect from chromospheric activity between giants and dwarfs. With the help of the IMF ($\mu=2.58$) and stellar evolutionary models, it seems that all the 4-giants are in the core-helium burning phase.

REFERENCES

Baliunas,S.L., Hartmann,L., Dupree,A.K.: 1983, Astrophys.J. **271**, 672.
Bell,R.A., Gustafsson,B.: 1978, Astron. Astrophys. Suppl. **34**, 229.
Cayrel de Strobel,G., Chauve Godard,J., Hernandez,G., Vaziaga,M.J.: 1970, Astron. Astrophys. **7**, 408.
Cayrel,R., Cayrel de Strobel,G., Campbell,B.: 1985, Astron. Astrophys. **146**, 249.
Iben,I.Jr.: 1965, Astrophys.J. **142**, 1447.
Iben,I.Jr.: 1967, Astrophys.J. **147**, 650.
Johnson,H.L., Mitchell,R.I., Iriarte,B.: 1962, Astrophys.J. **136**, 75.
Lambert,D.L., Ries,L.M.: 1981, Astrophys.J. **248**, 228.
Ridgway,S.T., Joyce,R.R., White,N.M., Wing,R.F.: 1980, Astrophys.J. **235**, 126.
Salpeter,E.E.: 1955, Astrophys.J. **121**, 161.
Stern,R.A., Zolcinski,M.-C., Antiochos,S.K., Underwood,J.H.: 1981, Astrophys.J. **249**, 647.
Tarrab,I.: 1982, Astron. Astrophys. **109**, 285.

DISCUSSION

GRIFFIN How did you estimate the surface gravity?

ARIMOTO With the help of the ionization equilibrium.

GRATTON Two questions and a comment. 1) What is the mass you infer from the gravity you use? 2) What is the influence of an atmospheric structure which might be different from dwarfs and giants on your abundances? 3) It would be interesting to have more than the Na D lines to derive Na abundances. Chris Sneden and me found an analogous result for Na abundances derived from D lines in metal rich stars.

CAYREL DE STROBEL 1) In the paper by Cayrel et al. (1970) using the same log g=2.7 we have found 2 $M_\odot$ for the giants. 2) Dr. Gustafsson will comment on this question, in this invited paper, (see in particular the best part of the discussion after Gustafsson's paper). 3) In the three Reticon spectral intervals we have studied no weak sodium lines are present, so we have had at our disposal only the strong D lines.

ON TECHNETIUM ABUNDANCE IN LATE-TYPE STARS

T.A. Kipper
Tartu Astrophysical Observatory
202444 Tõravere
Estonian SSR, USSR

ABSTRACT. Technetium abundances in the atmospheres of S- and C-stars are estimated by comparing the synthetic spectra and high dispersion observed spectra for the TcI λ 5924.47 region. It was found that reliable estimates are possible only for the pure S-stars and SC-stars. For the late carbon stars only the upper limit of the Tc abundance can be found. For TX Psc (C6.5) it was found to be $\lg \varepsilon_{Tc} = 0.5$.

1. INTRODUCTION

Three of the most intense resonance lines of TcI occur in the $\lambda\lambda$ 4000 - 4300 spectral interval. It is quite difficult to obtain high-dispersion spectrograms for that region, particularly for carbon-rich stars. Even with a very efficient spectrograph coupled to the DAO 1.2-meter telescope 6 hours were needed to Dominy and Wallerstein (1986) to obtain 2.4 Å/mm spectra for the bright mira χ Cyg ($m_B \sim 5.0$). Therefore most observers use a comparatively modest dispersion, 9 - 18 Å/mm (Kipper and Kipper, 1984).

Some years ago Smith and Wallerstein (1984) analysed the spectra of some stars with a dispersion of 5 - 6.7 Å/mm near the TcI intercombination line λ 5924.47. Their observing list contained a barium star, S- and SC-stars. In the spectra of the observed stars the CN red-system bands are relatively weak and it was possible to identify and measure the TcI line. The use of the yellow spectral region has some serious advantages: first, late-type stars are brighter in this region, and, secondly, this region can be observed by most solid-state detectors.

I decided to study the possibilities of using the yellow spectral region for Tc abundance determination for even later-type S- and C-stars.

2. OBSERVATIONS

The observing list contained two late-type carbon stars V460 Cyg (C6,2), TX Psc (C6,5) and a S-star χ Cyg (S7,1). Observations were made in 1986 with a CCD-detector attached to the coude spectrograph of the 2.6-meter telescope of the Crimean Astrophysical Observatory. The spectra have a dispersion of 3 Å/mm and cover a spectral region of 30 Å.

3. ANALYSIS OF SPECTROGRAMS

The spectra were reduced using the subroutines from the semiautomatic spectrophotometric reduction system developed at the Tartu Astrophysical Observatory (Kipper, 1986). Due to a very high signal to noise ratio the relative intensities found from various exposures do not differ more than 0.5%, when the traces left by cosmic particles have been removed.

The dispersion curve was constructed and the instrumental broadening was found using the narrow telluric lines in the spectrum of the Moon. Using 21 such lines this curve was obtained with the accuracy of 0.008 Å. The line positions were determined as the points, which halve the equivalent widths of the lines.

In the spectrum of χ Cyg it was not possible to identify the TcI λ 5924.47 line though Dominy and Wallerstein (1986) were able to measure the TcI blue line λ 4088.7. They estimated the Tc abundance $\lg \varepsilon_{Tc}$ = 1.7. Using the photographic spectra obtained with the 1.5-meter telescope of the Tartu Astrophysical Observatory with the dispersion of 9.6 Å/mm and the synthetic spectra for the model (3000/0.0/1.0) (Johnson, 1982) I estimated the Tc abundance for χ Cyg $\lg \varepsilon_{Tc}$ = 2.0 from the TcI λ 4262.26 line. In the spectrum of χ Cyg in the region containing the TcI λ 5924.47 line the very strong TiO bands are observed, which is consistent with the low ZrO abundance index for this star. It is impossible to calculate a yellow region synthetic spectrum for χ Cyg for the lack of molecular data for all the TiO γ'-system bands having lines in the relevant region (L. Hänni, private communication). Therefore for S-stars one can hope to identify and measure the TcI yellow line only if the star has a high ZrO index (pure S-stars) or for SC-stars as this was done by Smith and Wallerstein (1983).

In the spectrum of TX Psc there is a feature, which can be identified with the TcI line. Technetium lines in the blue spectra of this star are very strong (Little-Marenin and Little, 1979).

In order to estimate Tc abundances the synthetic spectrum approach was used. Johnson et al. (1982) when modelling the spectra of TX Psc and V460 Cyg using carbon-rich models (Johnson, 1982) found for both stars T_{eff} = 3000 K, the ratio of carbon and oxygen abundances C/O $\sim$ 1.02, the carbon isotopic abundance ratio $^{12}C/^{13}C$ $\sim$ 25 and the microturbulent velocity ξ_t $\sim$ 4 - 6 km/sec.

These data together with the line list compiled and kindly supplied to us by R. Bell were used to calculate the synthetic spectra. The wavelenghts of the CN red-system lines observed in the solar spectrum (Sotirovski, 1972) were corrected. The oscillator strength for the TcI λ 5924.47 line calculated by Garstang (1981) was used.

As Bell's line list was compiled for oxygen-rich stars and did not contain all the lines observed in late carbon stars it was impossible to reach a good fit for the whole 30 Å region. Nevertheless it was possible to estimate Tc abundance. For comparing the synthetic spectra with the observations the computed spectra were smeared with a Gaussian profile with the half-width of 0.3 Å. This profile contains the apparatus profile 0.25 Å and macroturbulence (8 km/sec). The width of the instrumental profile was determined from the observed width of narrow telluric lines in the spectrum of the Moon.

4. RESULTS

The best fit of the synthetic spectrum with the observed ones for both stars was obtained using the microturbulent velocity of ξ_t = 2 - 3 km/sec and macroturbulence in addition to the instrumental broadening. As the pertinent wavelength region contains the isotopic lines of ^{13}CN the carbon isotopic abundance ratio was also estimated as ^{12}C/^{13}C ∿ 50. This estimate depends mainly on the visibility of the $Q_2(9)$ λ5923.44 line of the (7 - 2) ^{13}CN band. As an illustration, the figure compares the observed spectra with the synthetic spectrum computed for the model (T_{eff} = 3000 K, lg$g$ = 0.0, C/O = 1.02). The positions of the TcI line and some of the ^{12}CN (7 - 2) band Q_1 branch lines are marked with vertical strokes. One of those strong lines closely coincides with the TcI line and therefore there is no hope to find accurate Tc abundances for these carbon stars. One can only estimate the upper limit for technetium abundance, which in case of TX Psc turns out to be lgεTc = 0.5.

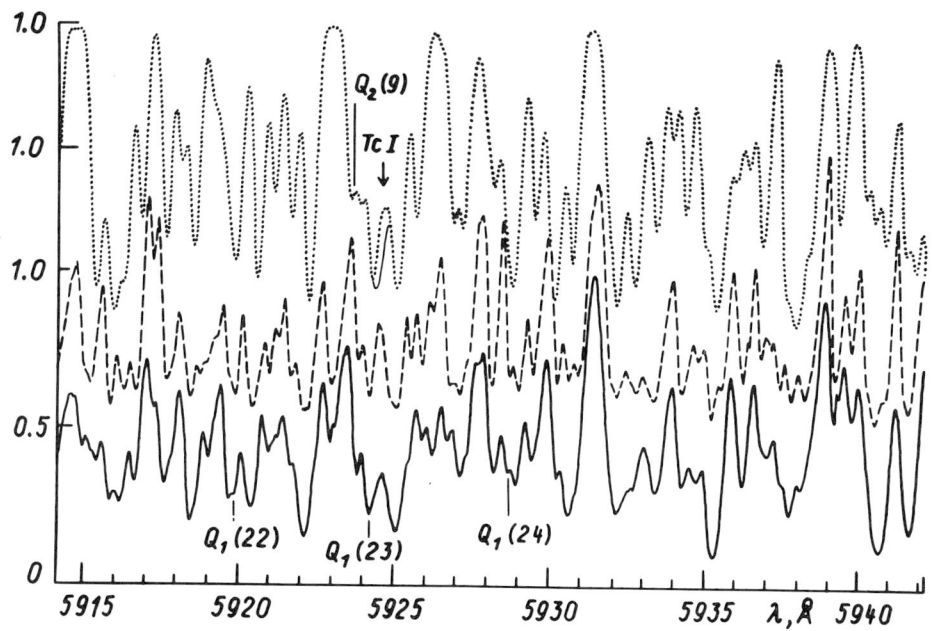

Fig. 1. The observed spectra of TX Psc (full line) and V460 Cyg (dashed line) around 5925 Å as compared with a corresponding synthetic spectrum (dotted line). The TcI line at λ 5924.48, some of the ^{12}CN (7 - 2) Q_1-branch lines and a ^{13}CN line Q_2 (9) are indicated.

REFERENCES

Dominy, J., Wallerstein, G.: 1986, Astrophys. J. **310,** 371.
Garstang, R.H.: 1981, Publ. Astron. Soc. Pacific **93,** 641.
Gray, D.F., Toner, C.G.: 1986, Astrophys. J. **310,** 277.
Johnson, H.R., O'Brien, G.T., Climenhaga,J.L.: 1982, Astrophys. J. **254,** 175.
Johnson, H.R.: 1982, Astrophys. J. **260,** 254.
Kipper, T.A., Kipper, M.A.: 1984, Sov. Astron. Lett. **10,** 363.
Kipper, T.A.: 1986, Estonian Acad. Sci., Preprint A-2.
Little-Marenin, L.R., Little, S.J.: 1979, Astron. J. **84,** 1374.
Smith, V.V., Wallerstein, G.: 1983, Astrophys. J. **273,** 742.
Sotirovski, P.: 1972, Astron. Astrophys. Suppl. **6,** 85.

LITHIUM ABUNDANCES, DIFFUSION AND MACROSCOPIC MOTIONS IN STELLAR CLUSTERS

Sylvie Vauclair
Observatoire Midi-Pyrénées
14, avenue Edouard Belin
31400 Toulouse
France

The "lithium gap" observed in the Hyades and other galactic clusters by Ann Boesgaard and her collaborators (Boesgaard and Tripicco 1986, Boesgaard 1987, Boesgaard, Budge and Burck 1987) gives a challenge to theoreticians. Indeed a good fit between the theoretical results and the observations will give a clue for our understanding of the stellar internal structure and evolution.

A theoretical explanation of the "lithium gap" by gravitational and radiative diffusion has been proposed by Michaud 1986. In G type stars, the convection zone is too deep for gravitational settling to take place : the density at the bottom of the convection zone is so large that the diffusion time scale exceeds the age of the star. Increasing the effective temperature leads to a decrease of the convection zone, and consequently to a decrease of the diffusion time scale (fig. 1). In F stars it becomes smaller than the stellar age, leading qualitatively to a lithium abundance decrease as observed. When the convection zone is shallow enough, the radiative acceleration on lithium becomes important as lithium is in the hydrogenic form of li III (while it is a bare nucleus, li IV, deeper in the star - (fig. 2). This radiative acceleration may prevent lithium settling for hotter F stars.

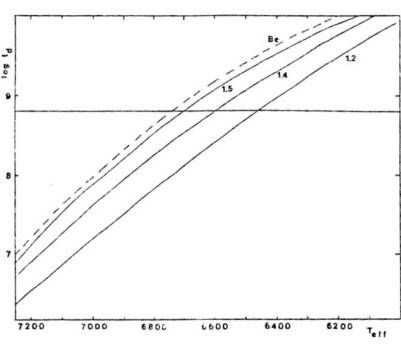

Fig. 1 : Time scales for gravitational settling at the bottom of convection zones in main sequence stars.
Solid lines = lithium, for α = 1.2, 1.4 and 1.5
Dasked line : beryllium, for α = 1.5

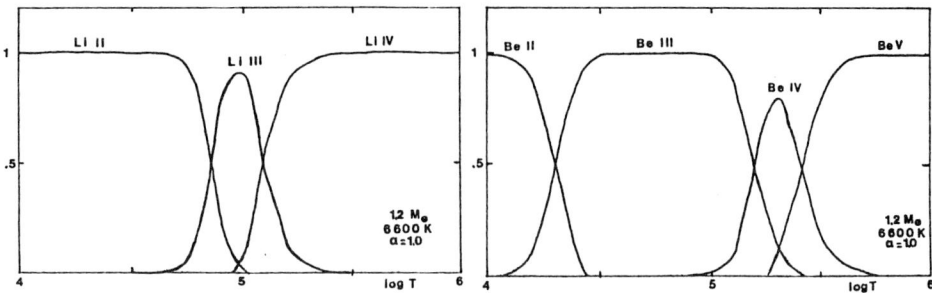

Fig. 2 : ionisation stages for lithium and beryllium in a 1.2 M_o star, with α = 1.0

This is a very attractive explanation, which leads to a minimum of the lithium abundance nearly at the place where it is observed in effective temperature. However it suffers from some difficulties : the theory predicts an increase of the lithium abundance larger than normal in the hottest F stars, which is not observed, and the predicted minimum lithium abundance is one or two orders of magnitude higher than the minimum observed in the Hyades. Also, this theory does not take any turbulence into account, while turbulence is definitly needed to explain the lithium abundance decrease in G stars (by mixing and nuclear destruction, see for example Cayrel et al 1984, Baglin, Morel, Schatzman 1985).

In the present paper we have focused on the problem of the computation of radiative accelerations, which are of prime importance in this theory. The radiative acceleration on a given element, through a bound-bound transition, may be written :

$$g_R = \frac{1}{m} \frac{N_{i,n}}{N} \int_0^\infty \sigma_{i,n}(v) \frac{\phi_v \, dv}{c} \qquad (1)$$

where m is the mass of the considered element, $N_{i,n}/N$ the fraction of the element in the lower level of the line, $\sigma_{i,n}(v)$ the transition section and $\phi_v \, dv$ the available photon flux.

With the diffusion approximation, a lorentz profile for the line, and after integration over v, g_R becomes :

$$g_R = \frac{1}{mN} \frac{8\pi^2 k^3}{3h^2 c^3} T^2 \left(-\frac{dT}{dr}\right) \frac{z^4 e^z}{(e^z - 1)^2} \frac{\Delta/2}{\sqrt{\frac{\kappa_c}{\kappa_L}\left(\frac{\kappa_c}{\kappa_L} + 1\right)}} \qquad (2)$$

with

$$z = \frac{h\nu}{kT} \quad \text{and} \quad \kappa_L = N_n \frac{\pi e^2}{m_e c} \frac{f}{\pi} \frac{2}{\Delta}$$

where f is the oscillator strength of the line and $\Delta/2$ the half width.

κ_c is the monochromatic opacity due to all the opacity sources except the considered line.

For an unsaturated line ($\kappa_L \ll \kappa_c$), (2) may be transformed into :

$$g_R = \frac{1.6 \times 10^{-4}}{A} \frac{N_{i,n}}{N} f \frac{z^4 e^z}{(e^z-1)^2} \frac{T_e^4}{T} \frac{R^2}{r^2} \frac{\bar{\kappa}}{\kappa_c} \quad (3)$$

where $\bar{\kappa}$ is the Rosseland mean opacity

T_e the effective temperature
T the local temperature
R the stellar radius
r the local radius

The ratio $\bar{\kappa}/\kappa_c$ which appears in g_R represents the fact that the radiative acceleration through one line strongly depends on the other sources of opacity at the same frequency. Up to now, the radiative accelerations have been computed with the approximation $\kappa_c \approx \bar{\kappa}$. However if, for example, a line of an abundant element sits at the same place as the Ly α lithium resonance line, the radiative acceleration on lithium may be strongly decreased.

		Lithium					Beryllium		
	ion	λ	level (eV)	f		ion	λ	level (eV)	f
Lyα 135.0Å	O V	134.473	19.7		Lyα 75.93Å	Mg VI	75.834	0.0	0.01
	"	135.175	19.7			"	75.890	0.0	0.01
	"	135.523	0.0	0.016		Mg VII	75.975	5.1	
	Ne V	134.48	0.0			Al VII	75.846	0.0	
	"	134.84	7.9	0.01		Al VIII	75.894	0.2	
	Fe VII	134.940	3.6			"	75.985	0.5	
	"	135.488	3.6			Fe V	75.685	0.0	
						"	76.006	0.0	
Lyβ 113.94Å	Mg V	113.703	0.0	0.12	Lyβ 64.06Å	Mg VII	64.12.	5.08	
	"	113.934	0.22	0.01		"	64.38	5.08	
	"	113.990	0.0	0.03		Mg VIII	64.24	0.0	
	"	114.059	0.0	0.17		"	64.38	0.4	
	"	114.183	0.22	0.02		Si VIII	63.88	0.0	
	"	114.199	0.22	0.06		"	63.90	8.6	
	Ne VI	113.95	0.16			"	64.28	8.6	
	"	114.07	0.0			Al VIII	63.93	0.0	
	"	114.13	0.0			"	63.96	0.2	
						"	64.004	0.55	
						"	64.086	0.55	
						Fe XIII	64.139		
						Fe XIV	63.96		

Sources :
Kelly 1973 and 1982, Kurucz and Peytremann 1975, Wiese 1985

A table of the important lines which may blend the lithium and beryllium resonance lines is given above. This table is not exhaustive as this part of the spectrum is not well known. Also the atomic parameters of these lines are very uncertain. As a quantitative result cannot be given without more precise atomic parameters, let us discuss here what would happen if the radiative acceleration on lithium was strongly overestimated : then the theoretical minimum would be smaller and could possibly match the observations. The red part of the "lithium gap" could be well explained with an α parameter (ratio of the mixing length to the presure scale height) of 1.2 (fig. 3).

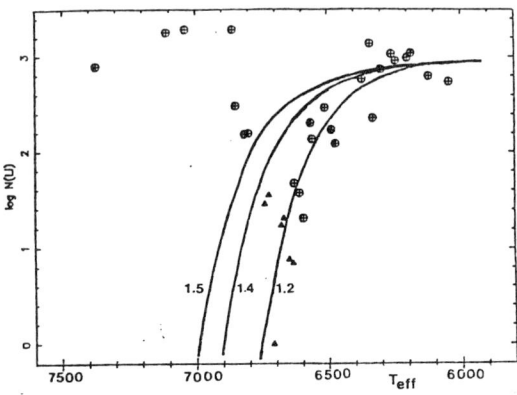

Fig. 3 : Lithium depletion by gravitational settling alone, compared with Boesgaard and Tripicco 1986 observational data

However the radiative acceleration alone would not be sufficient in this case to account for the increase of the lithium abundance back to normal in the hotter F stars.

Macroscopic motions could be invoked to explain this feature : either turbulence induced by rotation (as the rotation velocity in stars increase from G to F stars, Benz et al 1985 ; Boesgaard 1987) or mass loss.

Observational tests of these scenarii can be found in the observation of beryllium : if the lithium feature is entirely due to gravitational and radiative diffusion as proposed by Michaud 1986, beryllium should also show a minimum, less deep than the lithium one and at a smaller effective temperature (as beryllium is in the hydrogenic form deeper than lithium see fig. **1**).

On the other hand, the lithium and beryllium minima would be at the same place if macroscopic motions were the reason for their abundance increase back to normal in F stars.

More spectroscopic observations, and more atomic parameters for the far UV line are needed before reaching any precise conclusion.

Acknowledgements

Thanks are due to Marie-Christine Artru for many fruitful discussions and help in my search for atomic parameters.

References

Baglin,A., Morel,P., Schatzman,E. 1985, Astron. Astrophys. 149, 309
Benz,W., Mayor,M. Mermilliod,J.C. 1984, Astron. Astrophys. 138, 93
Boesgaard,A.M., Tripicco,M.J. 1986, ApJ 302, L49
Boesgaard,A.M., Budge,K.G., Burck,E.E. 1987, preprint
Boesgaard,A.M. 1987, P.A.S.P., in press
Boesgaard,A.M. 1987, ApJ, in press
Cayrel,R., Cayrel de Strobel,G., Campbell,B., Däppen,W. 1984, ApJ 283, 205
Kelly,R.L. 1973 and 1982 "Atomic and Ionic spectrum line below 2000 Å", office of fusion energy, ORNL 5922
Kurucz,R.L. Peytremann,E. 1975 "Table of semi empirical g f values" Smithsonian Astrophysical Observatory Special report 362
Michaud,G. 1986, ApJ 302, 650
Wiese,W.L. 1985, "Atomic data for fusion" ORNL 6089

DISCUSSION

BOESGAARD A comment: These are very important calculations for the interpretation of the Li dip in the F stars. I would like to say that thanks to the high technology at this meeting I can give you a new constraint for the theoretical calculations. I got a computer message this morning from my student at Caltech on the Li dip in the Pleiades cluster. It is not nearly as deep apparently as that in the Hyades. It goes down to perhaps log N(Li) = 2.5. This is consistent with the fact that the Pleiades are an order of magnitude younger than the Hyades so diffusion has not had as long a time to operate.

ROTATION, ACTIVITY AND LITHIUM DEPLETION IN THE HYADES LATE MAIN SEQUENCE

R. Rebolo and J.E. Beckman,
Instituto de Astrofisica de Canarias,
38200 - La Laguna,
Tenerife,
Spain.

ABSTRACT. Selecting stars with previously measured or estimated rotational periods we measured Li equivalent widths in 8 stars of the Hyades late main sequence. Combining our data with literature values, we derive a homogeneous set of Li abundances. Plotting log N(Li) against Rossby number we can separate the effects of temperature and rotation. We find for stars later than F8 increased depletion with longer period, at constant surface temperature. Discounting, as improbable, spurious abundances due to surface activity effects, we give a tentative view of the implications of these results for stellar modelling.

1. INTRODUCTION

A star's present Li abundance depends on its initial value, Li or and on the mechanisms which transport Li down to its nuclear burning depth. The well-known monotonic relation log N(Li) v. Teff between F8 and K0 in clusters is due to the increase in the convection zone depth with decreasing stellar mass. Use of Li to determine age or to probe interiors is inhibited by our limited knowledge of the processes transporting Li across the "gap" between the base of the convection zone and the nuclear burning zone. Among recent models which agree quite well with the log N(Li) v. Teff observations in the Hyades is that of Baglin et al. (1985) using a turbulent diffusion mechanism proposed by Zahn (1984); (see also Baglin et al. (1987), this volume). Here we use the Hyades, where many rotation periods are measured to probe rotation effects on Li depletion.

2. LITHIUM DATA

We obtained spectra of 8 Hyades MS stars, at the Cassegrain ID spectrograph of the 2.5m Isaac Newton Telescope, La Palma, with spectral resolution $\lambda/\Delta\lambda$ ~2×10^4, and S:N ratios between 150 and 200 at $\lambda 6708$ A, using a CCD detector. To these we added literature Li abundances (Cayrel et al., 1984) and equivalent widths, (Duncan and Jones 1983; Zappala, 1972). We analyzed data using the same Teff and log g calibrations, and the same model atmospheres as Cayrel et al. (1984). Relative Li abundance errors are less than 0.10 dex.

3. LITHIUM v. PERIOD AND ROSSBY NUMBER

Periods were either from rotational modulation of the Strömgren b,y colours (Lockwood et al., 1984) with errors ±3%, or from H and K fluxes by Duncan et al. (1984) using the relation of Noyes et al. (1984) between these and the period, which has errors of ±20%. In Fig. 1a we plot log N(Li) v. period P, and in Fig. 1b we show log N(Li) v. Teff. The two curves are strikingly similar, with a little more scatter in the former. This result is not surprising, given the role of deeper convection zones in destroying Li, and also in facilitating magnetic braking.

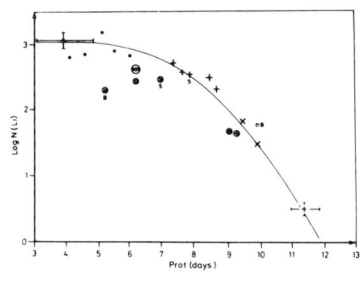

Fig 1a: Plot of Li abundance v. period for Hyades main sequence stars.
Key:
• Li data from Duncan and Jones (1983) or Zappala (1972). Periods predicted from Ca fluxes.
+ Li data from Cayrel (1984). Period measurements from Lockwood et al. (1984).
× Li from Cayrel. Periods estimated from Ca Fluxes.
⊖ Li data from this work. Periods from Lockwood et al. (1984). ⊕ Li from this work. Periods from Ca fluxes.

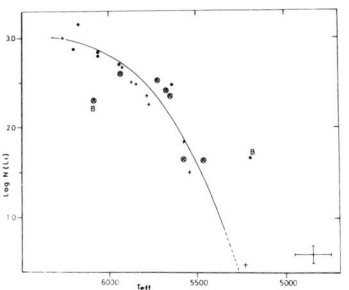

Fig 1b: Li abundance v. temperature for the same stars as in Fig 1a.

In Fig. 2 we plot log N(Li) v. Rossby number $N_R = P/\tau_c$ where τ_c is the convective zone turnover time (see Noyes et al., 1984). Note that this plot is a dispersed family of curves, in contrast to the monotonic dependence of Ca H and K emission on N_R in Noyes et al. (1984). Note also that along each line of constant B-V, Li is more depleted at longer period. This feature implies a rotation effect independent of Teff or mass.

A note of caution: activity effects accompanying rotation could just possibly produce the curves in Fig. 2. The presence of sunspots could raise the apparent Li abundance (Giampapa, 1984); plage cover, on the other hand, would tend to reduce it. We believe such effects to be insignificant, because no modulation of Li eq. width has been detected even in stars rotating more rapidly that this sample (see comments after Soderblom in this volume), and because the colour "anomaly" in B-V against V-K (see Campbell, 1984) is very small for our stars.

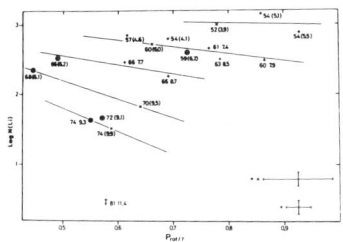

Fig. 2: Li abundance versus Rossby number for the Hyades stars. Trend lines indicate objects of similar (B-V) values. Integer numbers are (B-V)x100. Decimal numbers are periods (brackets estimated from Ca fluxes; without brackets measured directly). Symbols are as in Fig 1.

4. ROTATION AND LI DEPLETION

There are three possible causes for the different periods P exhibited by the Hyades main sequence stars. These are: (a) They started with different initial periods P_0, or (b) they started with identical values of P_0 but have different ages, or (c) they have been differentially braked by different "frozen in" initial magentic fields. We can discount (c) because the timescale for poloidal field regeneration is very short, on the order of the stellar magnetic cycle time, i.e. a few years (Durney et al., 1981) which implies that any frozen-in fields would not last longer than a few cycles at most. Whilst we cannot rule out (b) it is difficult to sustain numerically. For stars with (B-V)=0.66, there is an apparent dispersion in period of 40% which would correspond to an age spread of 80% if the braking were according to the empirical law of Skumanich (1972). This large age range is not easily compatible with the Hyades H-R diagram. However, given the uncertainty in the estimates of P, especially the indirect estimates, we would not be justified in ruling out differential age effects in the cluster.

Option (a) is a very probable cause of the present dispersion in P. It is not easy, however, to reconcile this as the principal cause with the dependence of Li abundance on Rossby number shown in Fig. 2. If a slower rotation now means that the star has always rotated more slowly, then slower rotators deplete Li more rapidly (for equal mass) than more rapid rotators. This can be consistent with models where the cause of Li depletion is rotation-driven turbulent diffusion (e.g. Schatzman, 1977; Baglin et al., 1985) only by assuming that the rotation also tends to inhibit the depth of the subphotospheric convective zone (or of the overshooting) especially for small values of P. We must therefore suspend final physical judgement until better data, especially on the periods, but also more Li abundances, are available for the Hyades.

References

Baglin, A., Morel, P.J., Schatzman, E.: 1985, Astron. Astrophys **149**, 309.
Campbell, B.: 1984, Astrophys. J. **283**, 209.
Cayrel, R., Cayrel de Strobel, G., Campbell, B.: 1984, Astrophys. J. **283**, 205.
Duncan, D.K., Baliunas, S.L., Noyes, R.W., Vaughan, A.H., Frazer, M.J., Canning, H.: 1984, PASP, **96**, 707.
Duncan, D.K., Jones, B.: 1983, Astrophys. J. **271**, 663.
Durney, B.R., Mihalas, D., Robinson, R.: 1981, PASP **93**, 537.
Giampapa, M.S.: 1984, Astrophys. J. **277**, 235.

Lockwood, G.W., Thompson, D., Radick, R.R., Osborn, W.H., Baggett, W.E., Duncan, D.K., Hartmann, L.W.: 1984, PASP **96**, 714.
Noyes, R.W., Hartmann, L.W., Baliunas, S.L., Duncan, D.K., Vaughan, A.H.: 1984, Astrophys. J. **279**, 763.
Schatzman, E.: 1977, Astron. Astrophys. **56**, 211.
Skumanich, A.: 1972, Astrophys. J. **171**, 565.
Vaughan, A.H., Baliunas, S.L., Middelkoop, R., Hartmann, L.W., Mihalas, D., Noyes, R.W., Preston, G.W.: 1981, Astrophys. J. **250**, 276.
Zahn, J.P.: 1984, In Physics of Early-Type Stars, Eds. B. Hauck, A. Maeder, Geneva Obs. Press.
Zappala, R.R.: 1972, Astrophys. J. **172**, 57.

DISCUSSION

MARCY Couldn't an age spread in the Hyades explain the decrease in Lithium with rotation ?

BECKMAN According to our semi-empirical understanding of stellar braking mechanisms the observed rotational period difference, up to 40%, between stars of similar mass would imply an age difference of well over 50% in age; such differences are not entirely excluded, as we explain in the text of the paper.

BIMODALITY AND LITHIUM ABUNDANCE ON THE UPPER MAIN SEQUENCE OF THE OPEN CLUSTER NGC 752

J E Beckman and R Rebolo,
Instituto de Astrofisica de Canarias,
38200 - La Laguna,
Tenerife,
Spain.

ABSTRACT. Spectra of resolution $\lambda/\Delta\lambda \sim 2\times10^4$ and good S:N ratio are presented in the range containing the ^{7}Li doublet at 6707 Å for 9 main sequence or slightly evolved stars in NGC 752 (age $\sim 2\times10^9$ years). We investigate the suggested main sequence bimodality using spectroscopic indications of binarity and high rotational velocity, as well as the Li abundance to supplement previous photometry.

1. INTRODUCTION

NGC 752 is a moderately old open cluster, with age estimated by Demarque (1980) as 1.7×10^9 years. Several photometric studies most recently Twarog (1983), have shown an apparently double main sequence. ^{7}Li is a probe of time-dependent processes; its abundance is one tool to explore cluster evolution. Hobbs and Pilachowski (1986, HP) have published main sequence Li abundances showing warmer stars with values up to log N(Li)=3.2, a "gap" at mid-F (cf. Boesgaard and Tripicco, 1986 for the Hyades), and a fall-off from late F through G. All the stars in HP lie on the "blue" main sequence, shown in Fig. 1. On this HR diagram using data from Eggen (1963) we also show the 9 stars in the present study.

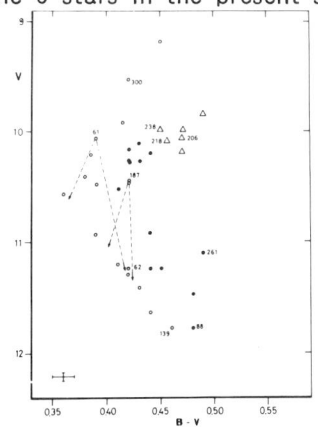

Fig. 1: HR diagram for NGC 752.
Key:
○ Blue MS.
● Red MS
△ Slightly evolved group
  Suggested decomposition of binaries into components

2. OBSERVATIONS

Data are from the cassegrain focus spectrograph of the 2.5m Isaac Newton Telescope (La Palma; 28-30 July 1986 and 16/17 October 1986). Slit limited spectral resolution was ~0.4 Å, 2 pixels of the 70% q efficient CCD camera at $\lambda 6707$ Å. Standard reduction included flat-fielding, calibration via emission lamp, and cosmic ray spike removal. For stars in range m_v =10-12, S:N for exposures of 30-60 minutes was 100-150; the brightest object gave S:N=200.

Abundances were from curves of growth derived from Kurucz (1979) models, carefully abstracting the Fe line at 6707.44 A, with log g=4.0, microturbulence 1.5 km s^{-1}, and T_{eff} from literature photometry (eg. Eggen, 1963; Twarog, 1983). For more detail see Rebolo and Beckman (1987).

3. POSSIBLE CAUSES OF BIMODALITY

(a) Binarity. A binary will, in general, appear displaced upwards and redwards in the HR diagram. If the two stars are of equal mass, the displacement is 0.75 mag brighter. Pilachowski et al. (1986) measured the parameters of the known spectroscopic binary H 300 which is on the upper red MS (see Fig. 1).

(b) Rotation. It is known (eg. Collins and Sonneborn, 1977) that rotation reddens a star's flux, and could therefore contribute to produce a shifted MS.

(c) Age. It is possible that the red MS is older than the blue. The HR diagram separation would correspond to an age difference of ~3×10^8 years (Ciardullo and Demarque, 1979) not excluded by the overall cluster age. A test could be to observe Li later than G2, using depletion as a chronometer but for stars with m_v>12.5 this is not easy. For hotter stars a bimodal metallicity could indicate an age split.

4. OUR Li RESULTS

These are summarized in Table 1.

Table 1.

Star (1)	B-V	V	T_{eff} (2, 3)	W(mÅ) (4)	LogN(Li)	MS	Bin (5)	Rot (7)	"gap" (10)
H61	0.39	10.06	6810	54	3.05	B	Yes[6]	---	---
H62	0.42	11.24	6740	63	3.1	B	---	---*	---
H88	0.48	11.78	6240	<25	<2.2	R	---	Yes	Yes
H139	0.46	11.79	6420	<40	<2.7	B	---	---	Yes
H187	0.42	10.45	6660	31/18 (9)	3.1/2.8	R	Yes	---	---
H206	0.47	10.06	6440	74	3.05 (8)	N	---	Yes	---
H218	0.455	10.09	6510	84	3.1 (8)	N	---	Yes	---
H238	0.45	9.98	6360	82	3.0 (8)	N	---	Yes	---
H261	0.49	11.15	6410	42	2.6	R	---	---	---

(1) Heinemann's (1926) identification.
(2) (3) Hearnshaw's (1974) calibration using Hβ photometry of Crawford and Barnes (1970).
(4) Fe I line at 6707.441 A included. B "blue" MS; R "red" MS; N neither MS.
(5) Spectroscopic indication of binarity.
(6) Component estimates: T_{eff} = 7000 K, log N(Li) = 3.3; T_{eff} = 6700 K, log N(Li) = 2.7.
(7) Position in Fig. 1 possibly affected by rotation.
(8) Sub-giants with little Li depletion.
* Low v sin i, but rotational reddening needed to explain Li "gap" position.
(9) Two component estimates. T_{eff} = 6900 K, log N(Li) = 3.1; Teff = 6700 K, log N(Li) = 2.8.
(10) Li abundance possibly reduced as star falls in the "F gap" for Li (Boesgaard and Trippico, 1986).

The short format precludes details of our physical inferences, which will be presented elsewhere (Rebolo and Beckman, 1987).

5. CONCLUSIONS

(a) The initial undepleted Li abundance in NGC 752 was log N(Li) $\simeq$ 3.1 (cf. HP, 1986).
(b) A group of somewhat evolved stars shows no Li depletion.
(c) We identify two new spectroscopic binaries (H61, H187).
(d) We identify two clear rapid rotators (H206, H218).

The presence of two near parallel MS branches in NGC 752 appears to be something of artefact due to the small sample (some 35) of stars which make up the cluster. Future studies with a greater fraction of the cluster stars would help to decide this issue.

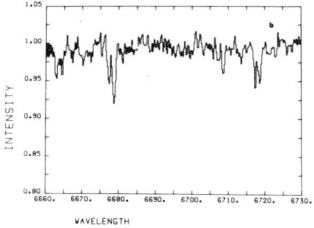 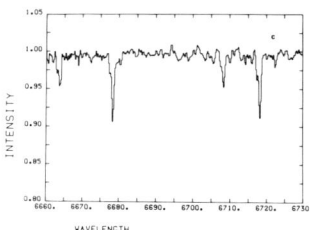

Fig. 2: b Spectrum of H187 showing clear evidence of binarity in line splittings. c Spectrum of H61 showing evidence for binarity with weak secondary components to the lines.

References
Boesgaard, A.M., Tripicco, M.J.: 1986, Astrophys. J. **302**, L49.
Ciardullo, R.D., Demarque, P.: 1979, Dudley Obs. Rep. 14, 317.
Collins, G.W., Sonneborn, G.H.: 1977, Astrophys. J. Suppl. Ser. **34**, 41.
Crawford, D.L., Barnes, J.V.: 1970, Astron. J. **75**, 946.
Demarque, P.: 1980, IAU Symp. No. 85, "Star Clusters" Ed. J.E. Hesser (Dordrecht, Reidel), 281.

Eggen, O.J.: 1963, Astrophys. J. **138**, 356.
Heinemann, R.: 1926, Astr. Nach. **227**, 213.
Hearnshaw, J.B.: 1974, Astron. Astrophys. **34**, 263.
Hobbs, L.M., Pilachowski, C.: 1986, Astrophys. J. **309**, L17.
Michaud, G.: 1986, Astrophys. J. **302**, 650.
Pilachowski, C., Willmarth, D.W., Halbedel, E., Mathieu, R.D., Hobbs, L.M., Milkey, R.W., Saha, A.: 1986, PASP **98**, 1321.
Rebolo, R., Beckman, J.: 1987, (In preparation).
Twarog, B.A.: 1983, Astrophys. J. **267**, 207.

CHEMICAL COMPOSITIONS OF POPULATION II MID- AND LATE-TYPE STARS

Howard E. Bond
Space Telescope Science Institute
3700 San Martin Drive
Baltimore, Maryland 21218 USA

R. Earle Luck
Department of Astronomy
Case Western Reserve University
Cleveland, Ohio 44106 USA

ABSTRACT. The abundance patterns in Population II red giants provide information about galactic chemical evolution and nucleosynthesis sites for various chemical elements. These patterns were first revealed by moderate-S/N photographic spectra and are now being refined and extended to additional chemical elements and fainter stars with high-S/N data. The well-established features include overabundances of O and Ca, underabundances of s-process elements, and solar ratios of heavy r-process elements. An excess of nickel found by us on moderate-S/N spectra is less certain, and should be checked with high-S/N data. New information on oxygen abundances in Population II red giants is becoming available from high-S/N observations of weak [O I] lines, and consistently indicates an oxygen overabundance due to the preferential sampling in Population II stars of ejecta from massive stars of previous generations. Recent studies of highly evolved, post-AGB Population II stars are revealing remarkable overabundances of light elements (C, N, O, and S) and deficiencies of the heaviest elements; these are probably signatures of internal nucleosynthesis followed by mixing up to, or exposure by mass loss at, the stellar surface.

1. WHY DETERMINE CHEMICAL ABUNDANCES IN POPULATION II STARS?

The chemical compositions of the oldest stars in our galaxy are of interest for several reasons:

(1). The wide range of metallicities ($-4.5 < [Fe/H] < -1$) that is observed among halo stars provides "snapshots" of the process of chemical enrichment during the early stages of formation of our galaxy.

(2). Population II stars sample material ejected by stars with short lifetimes (*i.e.*, high masses), and thus provide information on the sites of nucleosynthesis of various chemical elements. Thus, for instance, chemical elements that are preferentially produced by high-mass stars would be expected to show positive [element/Fe] ratios at low [Fe/H]; examples to be discussed below are O, Ca, and possibly Ni. Secondary elements should have [element/Fe] < 0; *e.g.*, the *s*-process elements such as Sr, Y, Zr, and Ba, which require pre-existing Fe for their synthesis. On the other hand, primary elements should have [element/Fe] = 0; an example would be heavy *r*-process elements, if they are directly synthesized in Type II supernovae without the need for pre-existing iron-group seeds.

(3). Population II stars with peculiar abundances can provide information on internal nucleosynthesis and the exposure of processed material at the stellar surfaces via mixing and/or mass loss, and thus provide constraints on the evolution of low-mass stars. For example, the CN cycle depletes C and enhances N, but helium burning makes C and O, so that very distinct abundance signatures are produced.

General discussions of this subject have been given by several recent authors, including Luck and Bond (1985a), Spite and Spite (1985), Truran and Thielemann (1986), Lambert (1987), and Truran (1988). This brief paper will concentrate on the importance of high-S/N spectroscopy to current and future developments. Moreover, we will concentrate on abundance analyses of *field* Population II stars, for which the highest-S/N data can be obtained, leaving discussion of the fainter globular-cluster stars to other speakers at this Symposium.

2. ABUNDANCE PATTERNS IN METAL-DEFICIENT FIELD RED GIANTS

2.1. Population II Abundances from Moderate-S/N Data

The capabilities and limitations of classical photographic spectroscopy for abundance analyses of 8th- to 11th-magnitude stars are typified by our results for field metal-deficient red giants (Luck and Bond 1985a). We used photographic (IIIa-J) echelle image-tube spectrograms obtained with the KPNO and CTIO 4-m telescopes, and carried out LTE model-atmosphere analyses. The spectra had S/N $\simeq$ 35, leading to measured equivalent widths that were uncertain by about a factor of 2 at ~ 25 mÅ (*cf.* Luck and Bond 1985a, Fig. 1). From such material it is possible to determine abundances for species like Fe, which are represented by numerous absorption lines of moderate strength, to an accuracy of about ± 0.2 dex; most of this error is actually contributed by uncertainties in the atmospheric parameters, $T_{\rm eff}$ and $\log g$. For species represented by fewer or weaker lines, the contribution from equivalent-width errors becomes more important, and systematic errors due to continuum placement and line blending (both in the program star and in our reference star, the Sun) could also be significant.

The chief results of this and similar studies are the following:

(1). Oxygen shows a general overabundance of [O/Fe] $\simeq +0.4$ for all stars with [Fe/H] < 1 (see below), and the [Ca/Fe] ratio behaves similarly. These results suggest that both oxygen and calcium are preferentially synthesized in massive Type II supernovae, in agreement with theoretical expectation.

(2). Sr, Y, Zr, and Ba are underabundant at low [Fe/H]; this supports the conclusion that these elements have, at least in part, a secondary (*s*-process) origin in Population II stars.

(3). However, the elements heavier than barium do *not* appear to be secondary elements. For example, we found that [Nd/Fe] $\simeq 0$ at all [Fe/H]. This indicates that these are primary *r*-process elements, possibly synthesized in Type II supernovae as suggested by Truran (1981). The heavy-element patterns in individual Population II stars have been analyzed by Sneden and Parthasarathy

(1983) and Sneden and Pilachowski (1985), and these patterns also support an r-process origin for the heaviest elements. The lines of the s- and r-process elements are often very weak in Population II stars, and it is clear that the high S/N now available from modern detectors will provide much new information (cf. Sneden, Pilachowski, and Krishnaswami 1988).

(4). Our most unexpected result was that nickel is overabundant in field red giants with [Fe/H] < −1.8 (Luck and Bond 1983, 1985a). The nickel overabundance rises to [Ni/Fe] ≃ +0.5-0.8 at [Fe/H] = −3. We suggested that this effect could indicate that the mass cut (between ejected and retained material) may have been deeper in massive Population III Type II supernovae, leading to ejection of more highly neutronized matter. Some support for this view comes from our observation that the Ni abundance appears to be anticorrelated with the Ti abundance for Population II red giants, as would be expected from the mass-cut argument. In the next subsection, we discuss the reality of the nickel overabundances in more detail.

2.2. Is Nickel Overabundant at Low [Fe/H]?

There are reasons for doubting the high [Ni/Fe] ratios in Population II red giants that were just mentioned. First of all, our studies were the first in which a significant number of Ni I lines were measured in extremely metal-poor stars (because there are numerous Ni I lines around 5100 Å, a region covered by our echellograms but not by conventional direct IIa-O plates); nevertheless, these Ni I lines are still quite weak. Fig. 1 shows histograms of the Ni I equivalent-width distributions in four of our metal-poor red giants covering a wide range of [Fe/H]; the equivalent widths have been published by Luck and Bond (1985b).

At moderate iron deficiency, e.g., in HD 26297 ([Fe/H] = −1.4), the strengths of the Ni I lines are such that a rather accurate Ni abundance can be derived (comparable in accuracy to Fe, in fact). However, as we move below [Fe/H] ≃ −2, the Ni I lines become quite weak, and for [Fe/H] ≤ −2.6 virtually all of the Ni I lines are below the equivalent width at which the uncertainty is at least a factor of two. At such low metallicities, it thus becomes possible that we are preferentially retaining only the Ni I lines for which the noise, line blending, or continuum placement has created a spuriously large equivalent width.

A further argument against the reality of our nickel overabundances is that several other authors have not confirmed it. Bessell and Norris (see discussion after this paper) found normal [Ni/Fe] in several of our stars, on the basis of low-excitation Ni I lines near 3400 Å. Peterson (1988), working in the same spectral region as we did, also did not find Ni overabundances in several of our stars.

On the other hand, we can mention several reasons for believing that the Luck and Bond nickel results are correct. (1) Other elements with weak lines (e.g., Cr I) did not show overabundances in the Luck and Bond (1985a) analyses; (2) our observed anticorrelation of Ni and Ti (over the whole range −3 < [Fe/H] < −1) is as expected from the mass-cut argument given above; (3) the change in slope of the [element/Fe] vs. [Fe/H] ratios occurs at the same [Fe/H] for Ni and the s-process elements, suggesting a common and real cause for both.

This is clearly an issue that will be resolved only with high-S/N spectra obtained in the region near 5100 Å; we, as well as several other groups, are now obtaining such data. Meanwhile, we warn that a nucleosynthetic explanation can probably be found for any abundance trend, even if the trend does not really exist!

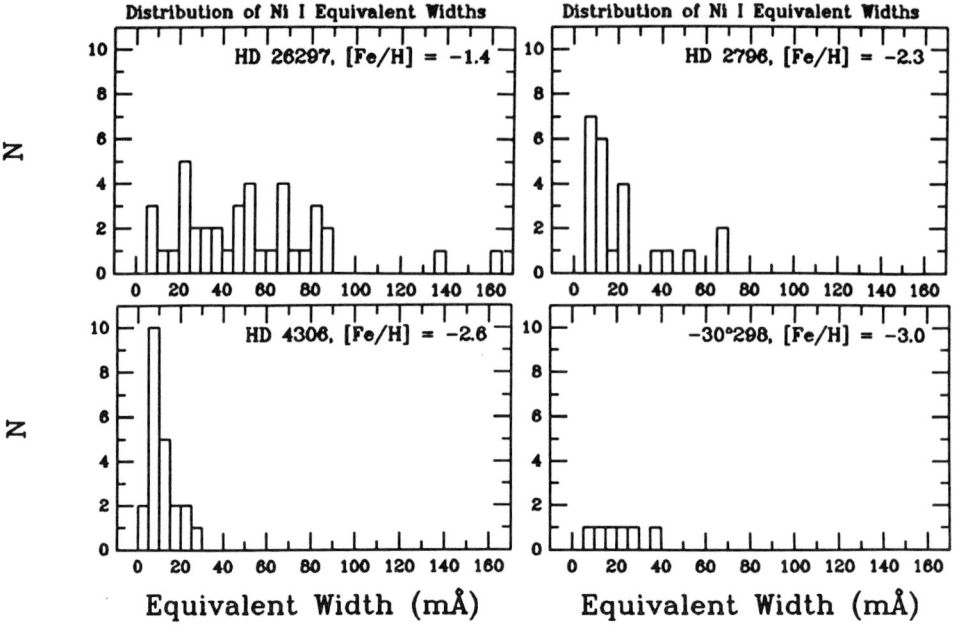

Figure 1. Ni I equivalent-width distributions in four metal-deficient red giants. Note that in the most metal-poor stars, nearly all of the Ni I lines are very weak.

2.3. Oxygen Overabundances in Metal-Poor Field Stars

As discussed above, a clear prediction of chemical-evolution theory is that [O/Fe] > 0 is expected in Population II stars because they preferentially sample the ejecta of massive objects.

In fact, an overabundance of oxygen now appears to be a well-established feature in Population II stars. In Fig. 2, we plot [O/Fe] vs. [Fe/H] for field (sub)dwarfs and giants. All oxygen determinations of which we are aware for stars with [Fe/H] < −1 are included. Most of the measurements for [Fe/H] < −1 refer to the [O I] line at 6300 Å, which is extremely weak in the most metal-deficient stars.

Different symbols are used in Fig. 2 to distinguish measurements made with photographic (moderate-S/N) or electronic (high-S/N) detectors. The figure clearly shows that the scatter in [O/Fe] is smaller for the high-S/N data. Whether or not any residual cosmic scatter remains is a significant question, since a scatter in [O/Fe] among globular clusters might be related to horizontal-branch anomalies. Pilachowski, Sneden, and Wallerstein (1983) did report a larger scatter in oxygen abundances among globular clusters than is seen in our Fig. 2, but since that work was based on photographic spectra it would be important to obtain higher-S/N data. So far, the high-S/N observations for field red giants have failed to reveal any analogs of the oxygen-*deficient* globular-cluster stars claimed by Pilachowski *et al.*

The high [O/Fe], if it is as pervasive among globular-cluster stars as it is

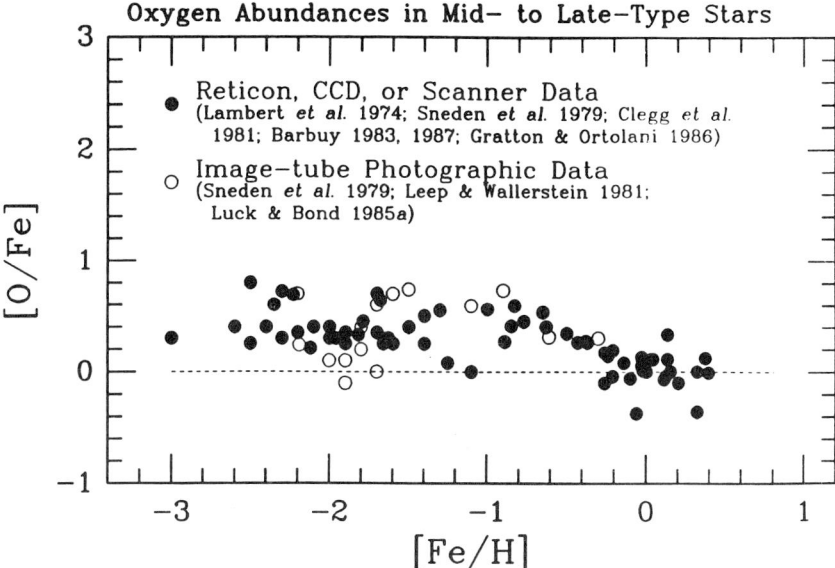

Figure 2. [O/Fe] vs. [Fe/H] for field stars, taken from the sources noted in the legend. Stars with [Fe/H] < −1 show a scatter around a constant [O/Fe] ≃ +0.4. *Open circles* refer to data obtained with photographic techniques, while *filled circles* plot higher-S/N data obtained with digital detectors.

among the field stars in Fig. 2, is also of fundamental significance because it acts to decrease the ages of globular clusters as derived from color-magnitude diagrams (*e.g.*, Fahlman, Richer, and VandenBerg 1985).

The availability of high-S/N detectors on large telescopes is making possible significant advances in this area; for example, the authors are now determining oxygen abundances in an additional three dozen field Population II giants on the basis of excellent CCD spectra obtained with the 4-m reflector at CTIO (Luck and Bond, in preparation).

3. PECULIAR CNO, S, AND s-PROCESS ABUNDANCES IN POST-AGB STARS

We will conclude this article by discussing some recent findings of remarkable abundance patterns in stars that are in their final evolution off the asymptotic giant branch (AGB) toward the realm of the nuclei of planetary nebulae.

3.1. The Population II Supergiant HD 46703

HD 46703 is a 9th-magnitude F-type star that appears to be a post-AGB object of the type just described. Our initial study of its chemical composition (Luck and Bond 1984) showed a halo metallicity ([Fe/H] = −1.6) and overabundances of both carbon and oxygen by factors of more than ten. More recently, we used a CCD detector at the 2.1-m KPNO reflector in a follow-up high-S/N study of the

N I lines near 8700 Å, and got a double surprise (Bond and Luck 1987).

Nitrogen is overabundant in HD 46703 by more than a factor of 50; the excesses of C, N, and O point to substantial mixing to the stellar surface (or exposure there by extreme mass loss) of CN-cycled material, along with additional material from the CO core. Our second surprise was serendipitous; by chance, several lines of S I happen to lie near 8700 Å, and their extraordinary strengths imply that sulfur is overabundant by nearly a factor of 20. We suggested that this is ^{32}S, synthesized by successive α-captures on ^{12}C during a transitory high-temperature episode in the stellar core and subsequently exposed at the surface. If so, an unexpected new feature of the evolution of low-mass stars is indicated.

BD +39°4926 is an A-type field star that is in a very similar evolutionary stage to that of HD 46703. On the basis of photographic spectra, Kodaira, Greenstein, and Oke (1970) reported pronounced overabundances of C, O, and, tentatively, S. It would be extremely interesting to repeat this analysis using high-S/N digital spectra to see whether the extraordinary abundance pattern of HD 46703 exists in a second object.

3.2. The s-Process Elements in Post-AGB Stars

A further remarkable property of HD 46703 is that the s-process elements are underabundant by ~ 0.7 dex (Luck and Bond 1984). Underabundances of this amount are encountered in Population II red giants, but only at [Fe/H] $\simeq -2.5$ (Luck and Bond 1985a); in HD 46703, the iron abundance is nearly an order of magnitude higher.

In fact, there are several studies that indicate that unusually low s-process abundances may be a pervasive phenomenon among low-mass post-AGB stars. These include abundance analyses of high-latitude A- and F-type supergiants (Luck, Lambert, and Bond 1983, and in preparation), W Virginis stars (Barker et al. 1971; Anderson and Kraft 1971), and RV Tauri variables (Luck 1981; Luck and Bond, in preparation).

One tentative explanation for this phenomenon would be to suppose that the stars were initially of very low iron content, and that they had the low s-process abundances typical of such objects. Exposure of processed, hydrogen-deficient layers has now produced higher observed [Fe/H] ratios, while leaving the [s/Fe] ratios at their original low values. Alternatively, one will have to conclude that plane-parallel, LTE model-atmosphere analyses are inadequate for highly evolved objects that have extended atmospheres and may even be surrounded by cool proto-planetary nebulae. In any event, the heavy-element abundances are based on very few, and often very weak, absorption lines. High-S/N data would help place this phenomenon on a firmer basis.

H.E.B. thanks the Space Telescope Science Institute Associates Program and the International Astronomical Union for travel support. R.E.L. acknowledges support from the National Science Foundation (grant AST 86-02028) and from the STScI visitor program.

REFERENCES

Anderson, K.S., and Kraft, R.P. 1971, *Ap. J.*, **167**, 119.
Barbuy, B. 1983, *Astr. Ap.*, **123**, 1.
_____. 1987, preprint.

Barker, T., et al. 1971, Ap. J., **165**, 67.
Bond, H.E., and Luck, R.E. 1987, Ap. J., **312**, 203.
Clegg, R., Lambert, D.L., and Tomkin, J. 1981, Ap. J., **250**, 262.
Fahlman, G.G., Richer, H.B., and VandenBerg, D.A. 1985, Ap. J. Suppl., **58**, 225.
Gratton, R.G., and Ortolani, S. 1986, Astr. Ap., **169**, 201.
Kodaira, K., Greenstein, J.L., and Oke, J.B. 1970, Ap. J., **159**, 485.
Lambert, D.L. 1987, in *ESO Workshop on Stellar Evolution and Dynamics in the Outer Halo of the Galaxy*, in press.
Lambert, D.L., Sneden, C., and Ries, L. 1974, Ap. J., **188**, 97.
Leep, E., and Wallerstein, G. 1981, M.N.R.A.S., **196**, 543.
Luck, R.E. 1981, Pub. A.S.P., **93**, 211.
Luck, R.E., and Bond, H.E. 1983, Ap. J. (Letters), **271**, L75.
_____. 1984, Ap. J., **279**, 729.
_____. 1985a, Ap. J., **292**, 559.
_____. 1985b, Ap. J. Suppl., **59**, 249.
Luck, R.E., Lambert, D.L., and Bond, H.E. 1983, Pub. A.S.P., **95**, 413.
Peterson, R.C. 1988, this Symposium.
Pilachowski, C.A., Sneden, C., and Wallerstein, G. 1983, Ap. J. Suppl., **52**, 241.
Spite, M., and Spite, F. 1985, Ann. Rev. Astr. Ap., **23**, 225.
Sneden, C., Lambert, D., and Whitaker, R. 1979, Ap. J., **234**, 964.
Sneden, C., and Parthasarathy, M. 1983, Ap. J., **267**, 757.
Sneden, C., and Pilachowski, C.A. 1985, Ap. J. (Letters), **288**, L55.
Sneden, C., Pilachowski, C.A., and Krishnaswami, K. 1988, this Symposium.
Truran, J.W. 1981, Astr. Ap., **97**, 391.
_____. 1988, this Symposium.
Truran, J.W., and Thielemann, F.-K. 1986, in *Stellar Populations*, eds. C.A. Norman, A. Renzini, and M. Tosi (Cambridge: Cambridge University Press), p. 149.

DISCUSSION

BESSELL The Ni I lines that Bessell and Norris have measured use low-excitation lines near λ3400 Å. These lines indicate no deficiencies relative to Fe in the stars with [Fe/H] < −2.5, but Dr. Holweger's paper this morning showed large difference for Fe I between abundances for low and high excitation.

BOND The NiI lines we used are near 5100 Å and have excitation of ≈ 1.6−3.6 eV.

HOLWEGER It is difficult to assess the non-LTE effects for low-and-high excitation lines without detailed calculations. Empirical evidence indicates that low-excitation lines give larger abundances than those of high excitation in Pollux (T_{eff} ≈ 4840K, [M/H] ≈ 0) but that the situation is reversed in the halo giant HD 122563. In any case I would strongly discourage anybody to use excitation temperatures in the analysis of red giants.

SNEDEN Why do you recommend 5100 Å for followup Ni I observations ?

BOND Simply because Ni I lines are particularly numerous in this region.

PETERSON The N I and S I in HD 46703 are very high-excitation lines and are sensitive to the temperature gradient at large optical depths. The abundances derived from such lines should be checked, by deriving abundances for species represented by both very high-excitation lines and normal (i.e. Fe I like) lines, e.g. Mg II and Mg I.

BOND It is not sufficient to change just the overall effective temperature −the dependance of [S/Fe] on the assumed T_{eff} is too weak to remove the huge sulfur overabundance. We have not checked Mg I-II, but both low-excitation [C I] and high-excitation C I give virtually the same carbon abundance.

R. CAYREL You convinced us that no Pop.III has been seen, already six years ago.
Is it not true that we have not even seen a second generation star, because s-process elements are always present at some level, even in the most iron-poor stars ?

BOND If Sr, Y, Zr, and Ba are s-process elements (or have an s-process component) in our extreme Pop.II stars, then there must have been at least 2 previous unseen generations.

NON-LTE EFFECTS AND ABUNDANCE ANALYSES OF HALO STARS (*)

Pierre Magain
European Southern Observatory
Casilla 19001
Santiago 19
Chile

ABSTRACT. The possible existence of departures from LTE affecting the abundance analyses of halo dwarfs and subgiants is analysed from the observational point of view, and illustrated by the case of the intermediate halo subgiant HD 76932. High resolution and high S/N Reticon and CCD spectra have been obtained with the ESO Coude Echelle Spectrometer. A detailed model atmosphere analysis has been carried out, which reveals a number of inconsistencies. In particular, the iron abundance derived from the neutral lines shows a very clear excitation potential dependence. Similar effects appear for oxygen and calcium and, possibly, for magnesium, chromium and ionized iron. Some overionization also seems to be present in a number of elements. The impact of these effects on the derived abundances may be rather large (some 0.2 to 0.6 dex). In particular, doubts might be raised about the reality of the oxygen overabundance and the odd-even effect in Na, Mg and Al.

1. INTRODUCTION

The use of solid-state detectors has allowed dramatic improvements in the quality of the spectroscopic data available for abundance determinations in Pop II stars. New, more precise analyses have shown clear variations of some relative abundances with overall metallicity, which have important implications for the chemical evolution of the Galaxy. All these analyses, however, are based on the assumptions of local thermodynamic equilibrium (LTE) and plane parallel model atmosphere, which we will refer to as the "classical" assumptions. Although the quality of the presently available data allow these assumptions to be tested, little effort has been done in that direction. The aim of this paper is to present the main results of such a test on the bright halo subgiant HD 76932.

(*) Based on observations collected at E.S.O. (La Silla).

2. OBSERVATIONS AND METHOD OF ANALYSIS

The observations were carried out at ESO (La Silla), with the Coude Echelle Spectrometer (CES) fed by the 1.4 m Coude Auxiliary Telescope. A total of 25 spectral regions were observed, either with the long camera and Reticon or with the short camera and CCD. The resolving power was set to 80000 and the S/N was generally higher than 200. The spectra were reduced with the IHAP facility at La Silla.

The Strömgren b-y and β indices as well as the Johnson V-K and spectrophotometric data all agree to indicate an effective temperature of 5815 (±15) K. The surface gravity, as determined by comparing the wings of strong lines to weak lines originating from the same atomic levels (Blackwell and Willis, 1977), is log(g) = 4.1. A microturbulent velocity of 1.2 km/s is deduced from Fe I and Ca I lines with accurate oscillator strengths.

The temperature structure of the model was chosen as:

$$T(\tau) = T_*(\tau) / T_\odot(\tau) \times T_{HM}(\tau) \qquad (1)$$

where τ is the optical depth at 500 nm, $T_*(\tau)$ and $T_\odot(\tau)$ are the temperature distributions of the theoretical models for the star and sun, and $T_{HM}(\tau)$ the temperature distribution of the solar Holweger - Müller (1974,HM) model. The theoretical models are from Magain (1983) and are computed with the MARCS program (Gustafsson et al., 1975). Eq. (1) means that the whole model grid is changed so that the grid solar model agrees with the HM model. This model was found to give the best fit of the strong line profiles. It must be pointed out, however, that none of the conclusions of this paper would be changed by the use of a purely theoretical model.

For each line, an element abundance is deduced from the measured equivalent width, under the assumption of LTE. Whenever available, accurate laboratory gf values are used, especially those of the Oxford group (Blackwell and collaborators). Otherwise, they are derived from the solar spectrum, but only if the solar line is found to be reasonably unblended and has a sufficiently well defined local continuum. The solar analysis is carried out with the HM model.

3. RESULTS

The Fe abundance is plotted as a function of the line excitation potential EP in Fig. 1, for both Fe I and Fe II. It appears that:
(1) all Fe I lines with EP < 2.6 eV indicate the same abundance, with a very small scatter (σ = 0.029 dex, which reduces to 0.023 if one discrepant line is excluded);
(2) the Fe I lines with EP > 2.8 eV give an abundance which increases with increasing excitation potential;
(3) the Fe II lines, all having EP > 2.8 eV, essentially show the same effect as the high excitation Fe I lines.

Similar effects appear for O, Mg, Ca and Cr, all showing a line abundance increasing with EP. The only clear exception is Ti (Fig. 2),

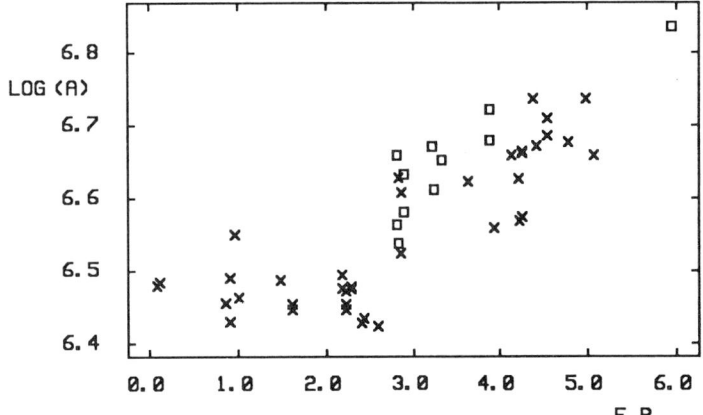

Figure 1. The computed Fe abundance is plotted versus the excitation potential for neutral lines (crosses) and ion lines (squares).

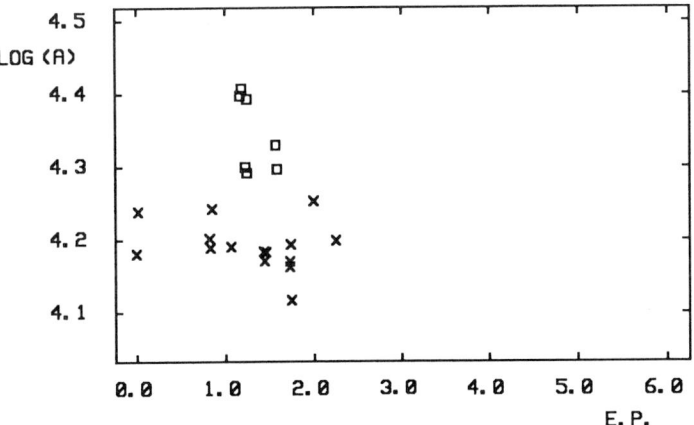

Figure 2. Same as Fig. 1 for Ti I (crosses) and Ti II (squares).

for which the deduced abundance is independent of EP, but which shows a rather strong overionization effect.

If these effects are the signature of model problems (such as temperature inhomogeneities), one might expect similar trends in all elements. On the other hand, if they are due to departures from LTE, a more erratic behaviour from element to element would be expected. The data show that the trend is remarkably similar for most elements, the slope of the log(A) - EP relation being close to 0.06 eV^{-1}. This would point to model problems, were it not for Ti I and the low excitation Fe I lines, which show a drastically different behaviour (slope $\approx$ -0.01 eV^{-1}). The conclusion is thus unclear: the data show definite departures from the classical assumptions, but do not allow to decide which one has to be relaxed. Maybe both are wrong!

4. CONSEQUENCES FOR ABUNDANCE DETERMINATIONS

Once it has been established that there are significant variations in the deduced abundances with excitation potential or with ionization stage, it is of interest to investigate the consequences of these variations for the abundance determinations.

The abundances deduced from different sets of lines are listed in Table I for the elements considered. For most of them, the uncertainties are quite large, ranging from 0.18 dex (Fe) to 0.59 dex (O). The discrepancies in the element abundances deduced from different lines are, at least in some cases, of the same order as the variations in relative abundances interpreted as signatures of galactic evolution. It cannot thus be excluded a priori that some of these trends just reflect errors in the spectroscopic analyses, and have nothing to do with galactic evolution. Let us illustrate this point by two specific cases.

Table I. Element abundances as given by different sets of lines

Element	EP	[el/H]	Element	EP	[el/H]
O I	0.0	-0.96	Ti I	0.0-2.3	-0.89 (±0.02)
	9.1	-0.37 (±0.03)	Ti II	1.1-1.6	-0.70 (±0.04)
Mg I	0.0	-1.00	Cr I	0.9-3.2	-1.15 (±0.05)
	4.3-6.0	-0.66 (±0.11)	Cr II	3.8-4.1	-0.96 (±0.06)
Ca I	0.0	-0.94	Fe I	0.0-2.6	-1.21 (±0.01)
	2.5	-0.79 (±0.03)		2.8-5.1	-1.04 (±0.03)
Ca II	7.5	-0.40	Fe II	2.8-6.0	-1.03 (±0.05)

Note: the listed uncertainty is twice the standard deviation of the mean value.

One very well established effect is the overabundance of oxygen relative to iron in metal-poor stars. In Fig. 3, [O/H] is plotted as a function of [Fe/H] according to the classical paper of Sneden et al. (1979). The oxygen abundances are deduced from the high excitation infrared triplet around 777 nm. The point corresponding to HD 76932, with the oxygen abundance deduced from the same lines, agrees very well with the general trend. Now, if the oxygen abundance was determined from the forbidden 630 nm line, nearly nothing would be left of the overabundance. (Imagine all the points with [Fe/H] < -1 moving down by a similar amount). Since there are some reasons to think that the forbidden line is a good abundance indicator, while the infrared triplet might possibly be affected by departures from LTE, it may not be unreasonable to question the reality of the oxygen overabundance.

Another interesting consequence of our results concerns the odd-even effect in the carbon burning products. Although the detailed behaviour of [Na/Mg] and [Al/Mg] in Pop II stars is subject to some controversy (see, e. g., Magain, 1987), it is generally agreed that Na and Al are overdeficient with respect to Mg, at least by some 0.5 dex.

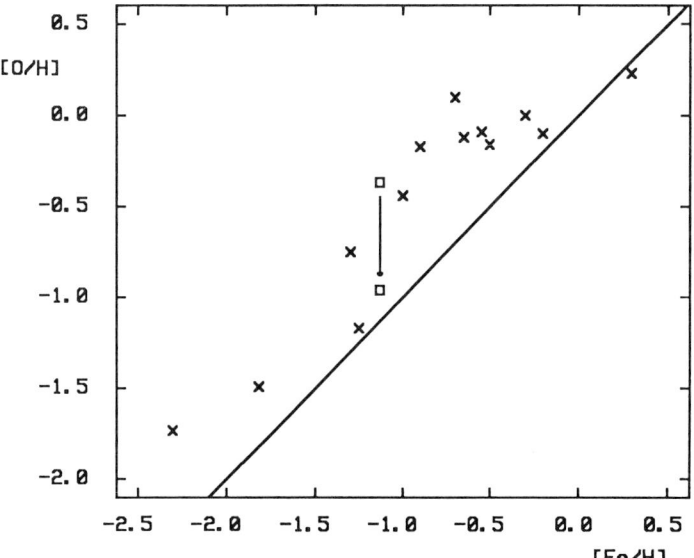

Figure 3. The oxygen abundance in HD 76932 as deduced from the infrared triplet (upper square) and from the forbidden line (lower square) is compared with the results of Sneden et al. (crosses).

The HD 76932 line abundances of Na, Mg and Al relative to H are plotted in Fig. 4 as a function of the line EP. The points for Na and Al fall on the curve defined by the Mg lines. This means that if the abundances of these elements were deduced from lines of the same excitation potential, it is likely that the odd elements overdeficiency would disappear. The reality of the enhanced odd-even effect in Pop II stars could thus also be questioned.

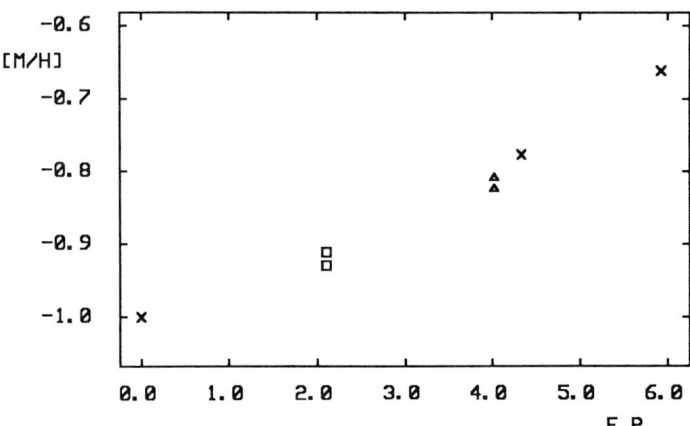

Figure 4. The deficiencies in Na (squares), Mg (crosses) and Al (triangles) are plotted versus the line excitation potential.

6. CONCLUSIONS

It is by no means our intention to conclude that the non-zero relative abundances found in Pop II stars just reflect departures from LTE or other flaws in the classical spectroscopic analyses. After all, HD 76932 might not be representative, and it cannot be excluded that the effects found in this analysis are just due to some peculiarity of that star, like binarity (but is it a peculiarity ?). However, as far as we know, no evidence for any peculiarity of HD 76932 has been reported. Nevertheless, it is urgently needed to analyse other stars, in order to confirm these results and to investigate the variations of these effects with temperature, gravity and metal abundance.

Although it might turn out that the classical LTE analyses are not always as bad as would appear from this paper, we have shown, at least, that some non-LTE effects may be present in the atmospheres of Pop II stars, and may have important consequences as far as the relative abundances are concerned. The solution of this problem is in our hands: we have the instruments to investigate these effects and the theoreticians are beginning to have the appropriate non-LTE codes. It is just a matter of will: do we intend to continue to provide the galactic evolution theorists with data we cannot reasonably guarantee their reliability, or will we concentrate part of our efforts on checking the validity of our assumptions?

REFERENCES

Blackwell, D. E., Willis, R. B.: 1977, M. N. R. A. S. 180, 169
Gustafsson, B., Bell, R. A., Eriksson, K., Nordlund, Å.: 1975, Astron. Astrophys. 42, 407
Holweger, H., Müller, E. A.: 1974, Solar Phys. 39, 19
Magain, P.: 1983, Astron. Astrophys. 122, 225
Magain, P.: 1987, Astron. Astrophys. 179, 176
Sneden, C., Lambert, D. L., Whitaker, R. W.: 1979, Ap. J. 234, 964

DISCUSSION

BECKMAN How would the effects you discuss affect the derivation of the Li, Be and B abundances for HD 76932 ?

MAGAIN It is impossible to give a definite answer. But, since the lines used are all resonance lines, if these elements follow the general trend of increasing abundance with excitation potential, one might expect an eventual underestimate of these abundances. All this is, however, highly speculative.

PETERSON How did you determine T_{eff} ? Wouldn't choosing a higher T_{eff} alleviate this dependence of abundance on excitation potential ?

MAGAIN The effective temperature was determined from 3 colour indices plus spectrophotometric observations, all giving the same value within 20K. To reproduce the trend shown by most (but not all) elements, one would need to increase T_{eff} by some 300 to 400K. Such a high temperature would be completely incompatible with the colours as well as with the Hα wings. Moreover, it would introduce other problems (i.e. in the TiI excitation equilibrium).

GUSTAFSSON For 8 metal poor dwarfs with $T_{eff} \approx 6000K$ and $-0.8 \geqslant$ [Fe/H] $\geqslant -1.0$ which we have analysed in our survey project (Andersen et al., this volume), we have around 20 Fe I lines with W_λ < 50 mA$^\circ$, spanning a range in χ_l from 2eV to 5eV. From these lines we find similar slopes, positive for all the stars, with a mean of $\delta[Fe/H]/\delta\chi = 0.05$ eV^{-1} and a scatter of ±0.01 eV^{-1}. Also our effective temperatures are determined from photometry.

MAGAIN I am pleased by this confirmation of my findings, which indicates that HD 76932 does not show a peculiar behaviour in that metallicity range.

BESSELL
1. There are significant differences between the T–τ relations of solar and low metal abundance models. What models do you use ?
2. What temperature calibration do you use ? (Your analysis of HD 140283 used an effective temperature and gravity different from that used by Peterson, Bessell and Norris, and Spite.Spite.)
3. Bessell and Norris use low excitation lines of NH and OH in the UV to derive abundances for O and N and obtain the same trends of O/Fe, N/Fe as seen in other data using the OI lines.

MAGAIN
1. I use a kind of scaled HM model, but taking into account the differences in T_{eff}, log g and [Fe/H], so this is a model which is adapted to metal-poor stars. But I have checked that my conclusions do not change if I use a theoretical stellar model instead.
2. I use my own calibrations of V–K and b–y (Astron. Astrophys. 181, 323), which are based on the IR flux method. These calibrations essentially agree with those of Carney (differences of less than a few tens of K). The present analysis is completely independent from my previous work on HD 140283.
3. Around [Fe/H]=–1, I would not give too much weight to determinations based on the UV spectral region, because of problems in locating the continuum. However, I fully agree that it would be very interesting to

have comparisons with abundance determinations from molecules. A caution about the use of gravities determined from ionization equilibria : departures from LTE in the latter could cause large systematic errors in abundances derived from molecules.

RUTTEN Since you have lines from different wavelength regions, it may be interesting to plot abundance values against upper-level excitation energy instead of lower-level excitation energy, and to see whether the spread changes. If so, this is a NLTE diagnostic because ionization departures affect both levels and therefore only the opacities, while excitation departures due to photon losses (which I would suspect here) affect only the upper-level population and the line source function.

THE EFFECT OF NON-LTE AND ATMOSPHERIC PERTURBATIONS ON RELATIVE
ABUNDANCE DETERMINATIONS IN METAL-POOR GIANTS

Ruth C. Peterson
Whipple Observatory, Smithsonian Institution
Steward Observatory, Room 458
University of Arizona
Tucson, AZ 85721

ABSTRACT. This paper reports an abundance analysis of spectra of 10 extremely metal-poor giants, including two in the globular cluster M92, in collaboration with Robert Kurucz and Eugene Avrett of the Center for Astrophysics and Bruce Carney of the University of North Carolina. An intercomparison of equivalent widths indicates that Luck and Bond have seriously overestimated the strengths of very weak lines and so deduced a spuriously high nickel abundance in the most metal-poor stars. A review of line-formation effects in the Sun strongly suggests that the rather high metallicities found for some extremely metal-poor stars by Pilachowski and coworkers is due to errors in their solar gf values, which unfortunately are unpublished. In our own analyses, stars with asymmetric H-alpha line profiles tend to have an excitation temperature at odds with the effective temperature found from infrared photometry. The discrepancy is removed by lowering the surface temperature by 100 K or by increasing the microturbulent velocity, either by a constant 0.5 km/s or by an amount which increases toward the surface. These changes have the most impact on those abundances deduced from strong lines or low-excitation lines of the neutral species; this includes sodium and barium as well as all the Fe I lines with very accurate gf values. Despite these difficulties, our findings reveal a low abundance for M92, and intrinsic scatter in the relationship between the relative abundances of sodium and oxygen and the overall iron abundance. In one field star, oxygen is apparently overabundant by an order of magnitude or more, while in the two M92 giants, the sodium abundance appears to differ by nearly this much.

1. INTRODUCTION

One of the major motivations for spectroscopic determinations of elemental abundances in metal-poor stars is to establish or constrain the mechanism(s) of nucleosynthesis during the early history of the Galaxy from trends in relative abundances (reviewed by Spite and Spite 1985). To establish a consistent set of analyses from which such trends may be reevaluated, my colleagues and I have undertaken a series of abundance analyses of metal-deficient giants. The following results are extracted from a series of papers to be submitted to the Astrophysical Journal.

2. OBSERVATIONAL DATA, EQUIVALENT WIDTHS, AND THE NICKEL ABUNDANCE

At Kitt Peak and Cerro Tololo Interamerican Observatories, we obtained 4m echelle spectra (resolution 15 km/s) with the long-focus camera and Carnegie image tube and baked IIIaJ plates. All stars were observed in the red, from 4860 to 6750 A, and four in the blue also. Two spectra were obtained for several stars, with a combined S/N of ~50. Spectra were digitized on the Lockheed PDS microdensitometer with a 6 micron slit and the curve-following procedures of Peterson and Title (1976).

A detailed intercomparison of our equivalent widths (EW) with modern ones published for the same stars is generally encouraging. Below 100 mA there is excellent agreement with the CCD values of Sneden and Parthasarathy (1983): the mutual error is ± 10 mA, which translates into an abundance error of ± 0.17 dex throughout this range. The strongest line detected by them that we missed was 12 mA. Above 100 mA, our values are larger and the scatter increases; this might be due to subjective differences in the measurement of line wings, or to scattered light in the CCD spectra. It has no effect the abundance determinations, however, because it is absorbed in the determination of the microturbulence, discussed below.

At all EW values, our measurements are on the same scale as those of Luck and Bond (1985b), though the scatter is much larger. However, these authors often list EW values of 25 mA for lines which are not detectable in our spectra. This assumes critical importance given their result (Luck and Bond 1985a) that nickel is overabundant with respect to iron in the most metal-poor stars. Unfortunately, the element nickel is represented only by rather weak lines in the spectral region redward of 4000 A. The overabundance was detected only where virtually all the Ni I lines they measured have EW < 25 mA. Since our analysis does not show an overabundance of nickel in these stars, we attribute their result to the overestimation of very weak features.

3. METHOD AND ASSUMPTIONS OF ABUNDANCE DETERMINATIONS

We determine abundances by computing a theoretical strength for every line individually using Kurucz's program WIDTH. A model stellar photosphere is provided as input, along with the line transition probability (gf value) and a trial abundance.

Unfortunately the theoretical EW depends on assumptions of the model. The stellar effective temperature is a critical parameter for most of the elements under study here (see Pilachowski, Sneden, and Wallerstein 1983). A higher temperature tends to ionize more atoms, which produces more continuous opacity and depopulates lower ionization and excitation states. The gradient of temperature

towards surface layers also affects the strongest lines of such
species: a steeper gradient can dramatically enhance lines by
increasing level populations, especially for the neutral species of
an element which is predominantly once-ionized.

Strong lines of all types are subject to the choice of
microturbulent velocity v_t and its variation with depth, for this
desaturates moderately strong lines by inducing an artificial
Doppler velocity over and above thermal. In K giants such as these,
gas densities are so low that broadening by van der Waals damping
mechanisms is ineffective below 100 mÅ, so v_t is very influential
for EW ~ 50 - 100 mÅ (referred to 5000 Å). In the Sun, though, the
higher gas density produces damping effects at EW ~ 50 mÅ.

Since an error in the line transition probability translates
directly into an error in the abundance of the opposite sign, it is
important to use values without systematic errors. Very accurate gf
values are now available for certain Fe I lines from the Oxford
group, as summarized by Blackwell, Petford, and Simmons (1982).
Their comparisons with Fe I gf-value determinations of other groups
frequently show sizable errors in the latter, often dependent on
transition strength or excitation level.

Because the Oxford furnace measurements are limited to rather
strong, low-excitation lines, the analysis of the iron abundance in
the Sun depends on both the solar model and on the choice of v_t and
damping constants. Blackwell and Shallis (1979) show that log(Fe/H)
from Fe I lines with Oxford gf values decreases by 0.2 dex in going
from the solar model of Holweger and Müller (1974) to that of
Vernazza, Avrett, and Loeser (1976). We would expect this given the
steeper temperature gradient of the latter model. Their analysis
also demonstrates that a smaller value of v_t is deduced when using a
model with a steeper temperature gradient. Note also that the best-
fit v_t value also depends on the source of solar EW values. Those
from the solar intensity spectrum at the center of the solar disk
(e.g. Delbouille, Roland, and Neven 1973) require smaller v_t values
that those from the integrated light of the solar disk (e.g. by
Beckers, Bridges, and Gilliam 1979).

It appears that inappropriate solar parameters may be largely
responsible for rather high abundances found for the most metal-poor
stars by Pilachowski and coworkers. Table 1 of Buonanno, Corsi, and
Fusi Pecci (1985) illustrates the problem: for the three extremely
metal-poor clusters M15, M92, and NGC 5466, the Pilachowski results
are consistently the highest of the dozens listed, 0.2 - 0.4 dex
above the mean. Most (but not all) analyses by Pilachowski and
Sneden and their coworkers rely on somewhat dubious solar gf values.
Pilachowski, Wallerstein, and Leep (1980) chose v_t for the disk of
the Sun but used solar intensity EW values; Sneden and Parthasarathy
(1983) used Fe I damping constants that are about four times too
large for the low-excitation lines (see Simmons and Blackwell 1982,
Table 3). Because the strongest solar lines are the most affected,
the systematic errors introduced are largest for the most extremely
metal-poor stars. Unfortunately, none of the papers lists the gf
values, so this potential source of error cannot be checked.

Our own Fe I analysis is based primarily on lines with Oxford gf values. In stars with low interstellar reddening, or in clusters such as M92 where reddening is well determined, the model effective temperature T_{eff} is established to 100 K from the infrared color V-K (Cohen, Frogel, and Persson 1978). Yet in several such stars, the dependence on lower excitation potential of the iron abundance deduced solely with Oxford gf values indicated a lower T_{eff} unless v_t was made artificially large. This applied equally to models from the Kurucz (1979) and the Gustafsson et al. (1975) grids. The discrepancy appeared only for stars with a blueshifted H-alpha core.

We have begun to explicitly calculate non-LTE effects as a possible cause. The sodium calculation is straightforward, and suggests that only the very core of the NaD lines are enhanced, so the abundance effect is small. The Fe I calculation is much more difficult because many transitions are involved. In particular, the non-LTE excitation of the lower levels of various transitions in the visible is strongly influenced by which ultraviolet excitations are included, and what is assumed for the stellar UV flux.

The discussion above suggests a steeper surface temperature gradient or an increase in v_t could also remove the discrepancy. WIDTH calculations show that a model with the surface temperature lowered by 100 K does so, as do models with v_t increased by a constant 0.5 km/s or progressively enhanced by up to 2 km/s at shallow levels. There is no physical basis for these alterations, however. The Fe I abundance deduced from Oxford lines does depend rather sensitively on whether such a change is invoked, but not on which prescription is followed. So the overall iron abundances in our work should be considered uncertain by at least 0.2 dex, the amount of change when v_t is raised by 0.5 km/s. The abundances of species such as oxygen and nickel that are represented only by weak lines are independent of v_t, so a comparable uncertainty pertains to their relative abundance. And no matter what approach is taken, the strongest Ti I lines can give abundances differing by about 0.3 dex from those of the Ti II lines; the sodium abundances deduced from the NaD lines should be considered uncertain by this amount in those cases.

4. CONCLUSIONS

Despite these difficulties, we feel that our iron abundances are an improvement over previous values. In general our results are lower; for M92, for example, we find [Fe/H] = -2.5 ±0.3, which is within 0.2 dex of the abundances of the extremely metal-poor field stars.

It seems to us that genuine cases exist of radical overabundances of oxygen and sodium. In most stars the oxygen line at 6300.3 A, the strongest accessible, is too weak to be detected. In star XII-8 in M92, for example, our upper limit of 14 mA places an upper limit of -2.0 ±0.2 on the oxygen abundance relative to the Sun. However, in one field star, HD 184711, this line appears at a strength of 63 mA, and the weaker member of the doublet is also

detected at 15 mÅ. The implied oxygen-to-iron ratio is nearly
thirty times solar. Also, the NaD lines are extremely strong in M92
III-13, and a weaker line is also detected; their analysis yields a
sodium-to-iron ratio 0.5 dex above solar. This is significantly
larger than the remainder of our sodium-abundance ratios, notably
that in XII-8. Thus we find evidence for intrinsic scatter in
whatever relationship may exist for the relative abundances of these
elements versus overall stellar metallicity.

ACKNOWLEDGEMENTS

It is a pleasure to acknowledge the capable assistance of both KPNO and
CTIO staff at the telescope and on the KPNO VAX computer, and I am
grateful for generous allocations of time on both. My research on
metal-poor stars is supported by NSF grant AST 85-21487.

REFERENCES

Beckers, J. M., Bridges, C. A., and Gilliam, L B. 1976, A High
    Resolution Spectral Atlas of the Solar Irradiance From 380 to 700
    Nanometers, AFGL-TR-76-0126.
Blackwell, D. E., Petford, A. D., and Simmons, G. J. 1982, M.N.R.A.S.,
    201, 595.
Blackwell, D. E., and Shallis, M. J. 1979, M.N.R.A.S., 186, 673.
Buonanno, R., Corsi, C. E., and Fusi Pecci, F. 1985, Astron.
    Astrophys., 145, 97.
Cohen, J. G., Frogel, J. A., and Persson, S. E. 1978, Ap. J., 222,
    165.
Delbouille, L., Roland, G., and Neven, L. 1973, Photometric Atlas of
    the Solar Spectrum from $\lambda 3000$ to $\lambda 10000$ (Liège: Institut
    d'Astrophysique).
Gustafsson, B., Bell, R. A., Eriksson, K., and Nordlund, A. 1975,
    Astron. Astrophys., 42, 407.
Holweger, H., and Müller, E. A. 1974, Solar Phys., 39, 19.
Kurucz, R. L. 1979, Ap. J. Suppl., 40, 1.
Luck, R. E., and Bond, H. E. 1985a, Ap. J., 292, 559.
Luck, R. E., and Bond, H. E. 1985b, Ap. J. Suppl., 49, 249.
Peterson, R. C., and Title, A. M. 1976, Appl. Opt., 14, 2527.
Pilachowski, C. A., Sneden, C., and Wallerstein, G. 1983, Ap. J.
    Suppl., 52, 541.
Pilachowski, C. A., Wallerstein, G., and Leep, E. M. 1980, Ap. J.,
    236, 508.
Simmons, G. J., and Blackwell, D. E. 1982, Astron. Astrophys., 112,
    209.
Sneden, C., and Parthasarathy, M. 1983, Ap. J., 267, 757.
Spite, M., and Spite, F. 1985, Ann. Rev. Astron. Astrophys., 23, 225.
Vernazza, J. E., Avrett, E. H., and Loeser, R. 1976, Ap. J. Suppl.,
    30, 1.

DISCUSSION

JUDGE With respect to your third suggestion for the solution of the colour vs. excitation temperature discrepancy by the desaturation of photospheric lines by velocity fields associated with mass flow, I get the feeling that you would need very large mass loss rates indeed to produce the effect in your photospheric lines. Significant mass flows indicated by asymmetry in Hα almost certainly occur in the chromospheres at much lower column masses than where your photospheric lines are formed, if are taken recent theoretical work (e.g. of Cram and Mullan, Mallick) and observations (e.g. Zarro and Rodgers) of population I giants into account. Current estimates of mass loss in K III population I stars are $\approx 10^{-10} - 10^{-9}$ M_O yr^{-1}. If I am wrong, here we have a potential diagnostic of mass loss rates !

PETERSON The mass loss rate depends on the column density and thus on the radius at which the flow becomes strong. This is presently unknown but some constraints may be established from calculations such as those of Dupree, Hartmann and Avrett (Ap J. Lett, 1984). However, Hα is less suitable than Ca II K, but that is a very difficult line to observe.
The desaturation mechanism I am suggesting is a crude representation since the dependence of flow velocity on optical depth is virtually unknown. I will model it only as a dependence of microturbulent velocity on optical depth. A small increase in microturbulent velocity is probably all that is needed.

BUDGE You mention that you could solve some of yours problems if you could find a reason to lower the temperature of the outer photospheres of your models. As I recall, ATLAS assumes plane-parallel geometry. Since these are giant stars, and especially since you find evidence at mass outflow in these stars, I wonder it perhaps the effects of spherical geometry might not be important and result in such a lowering of the boundary temperature ?

PETERSON Plane-parallel atmospheres are probably valid since the extent of the photosphere is small with respect to the stellar radius. Recall that these are giants, not supergiants, and the brightest have $M_{bol} \approx -3$ and $T_{eff}=4000K$. Furthermore, several stars analyzed, e.g. the fainter one in M92, lie rather far from the red-giant-branch tip. This star has very nearly the same T_{eff} and log g and [Fe/H] as HD 122563, for which T_{exc} and T(V-K) are in excellent agreement.

G. CAYREL Are you calibrating gf values also in the near infrared $\approx$ 6750 region ?

PETERSON Our data end at approximately this wavelength, as the S/N becomes low. There is no reason that this could not be done in the future from additional data.

GRATTON The derivation of accurate gf's for high excitation lines, where we have to rely on solar gf's, is a potential hazard for metal poor stars. In fact in this case we have to use lines which are rather strong in the Sun. The derivation of solar gf's for quite strong Fe lines depends very much in the way we handle damping. Therefore, we must be careful when considering trends with line strength or line excitation in metal poor stars.

PETERSON Important guidelines are available from intercomparaisons of laboratory oscillator strengths, especially by the Oxford group. Our T_{exc} is based solely on Fe I lines with Oxford gf values. The high-excitation Fe I lines are run separately to check the consistency of the scale. Since the high-excitation Fe I lines have strengths which change more slowly with temperature, we hope to bootstrap gf values over ranges of stellar metallicities by including hot, rather metal-rich stars as well as cool, very metal-poor ones. Thus we are trying to ensure that iron abundances are on a consistent scale over the range $-2.5 \leqslant [Fe/H] \leqslant -1.0$.

DISTRIBUTION OF R- AND S-PROCESS ELEMENTS IN THE GALACTIC HALO

C. Sneden[1], C. A. Pilachowski[2], K. K. Gilroy[1], and J. J. Cowan[3]

1. Department of Astronomy and McDonald Observatory, University of Texas, Austin, Texas 78712, USA
2. Kitt Peak National Observatory, Box 26732, Tucson, Arizona 85726, USA
3. Department of Physics and Astronomy, University of Oklahoma, 440 W. Brooks, Norman, Oklahoma 73019, USA

ABSTRACT. Current observational results for the abundances of the very heavy elements (Z>30) in Population II halo stars are reviewed. New high resolution, low noise spectra of many of these extremely metal-poor stars reveal general consistency in their overall abundance patterns. Below Galactic metallicities of [Fe/H] $\approx$ -2, all of the very heavy elements were manufactured almost exclusively in r-process synthesis events. However, there is considerable star-to-star scatter in the overall level of very heavy element abundances, indicating the influence of local supernovas on element production in the very early, unmixed Galactic halo. The s-process appears to contribute substantially to stellar abundances only in stars more metal-rich than [Fe/H] $\approx$ -2.

1. INTRODUCTION

In the past 4 decades high resolution spectroscopy has been one of principal tools used to probe the nature of the chemical evolution of the Galaxy. Analyses of the overall metallicities of halo stars have figured prominently in this work, beginning with the study of the classical metal deficient dwarf-subgiant HD 140283 by Chamberlain and Aller (1951). Solutions to many questions concerning the origin of the elements have been sought in these studies. In this paper we concentrate on recent abundance results for the very heavy elements (Z > 30) in the metal-poor halo stars. These abundances can be used to determine the relative contributions of slow neutron captures (the so-called s-process) and rapid neutron captures (the r-process) to the buildup of the very heavy elements. An assessment of these r- and s-process abundance contributions is now possible.

2. OBSERVED VERY HEAVY ELEMENT ABUNDANCES

It is not easy to determine reliable abundances of very heavy elements in metal-poor stars. For a couple of these elements (Sr and Ba), only a few very strong lines of the singly ionized species are detected easily in the spectra of very metal-poor stars, and analyses of these lines are notoriuosly difficult. For the other very heavy elements which are routinely detectable in very metal-poor stars (defined here as Y, Zr, La, Ce, Pr, Nd, Sm, Eu, Gd, and Dy), many transitions of singly ionized species do exist, but almost all of them are extremely weak and thus undetectable with high noise photographic spectra. Here we summarize the recent observations of these elements in metal-poor stars, which have improved both the quantity and quality of available data.

It has been understood for a long time that the abundance ratios [M/Fe] of the very heavy elements are distinctly non-solar in some very metal-poor stars. For instance, Ba was recognized to be often underabundant with respect to iron, while Eu was claimed to be nearly normal. Recent studies of many other metal-poor stars provide much more extensive information about trends in very heavy element abundance ratios in the halo. For example, in figure 1 we have plotted the ratio [Eu/Fe] versus [Fe/H]. Three principal conclusions may be drawn from this figure. 1) *Eu often is overabundant in metal-poor stars relative to Fe*, especially in the range -1.5 ≤ [Fe/H] < -2.5. Eu may be made in bulk only in the r-process, and so these relative overabundances clearly demonstrate the existence of substantial r-process synthesis (presumably by high-mass supernovae) early in the history of the galaxy. 2) *[Eu/Fe] displays a considerable star-to-star scatter*, perhaps increasing with decreasing [Fe/H]. This result supports the common ideas that the r-process elements and the Fe-peak elements probably were synthesized by stars of different masses, and that the Galaxy was largely unmixed during the early formation of the halo. 3) *[Eu/Fe] may begin to decline below [Fe/H] ≈ -2.5*. The data here are intriguing but scant, and suggest that the first Galactic nucleosynthesis events may not have been able to manufacture this element (and those of similar Z?) very efficiently.

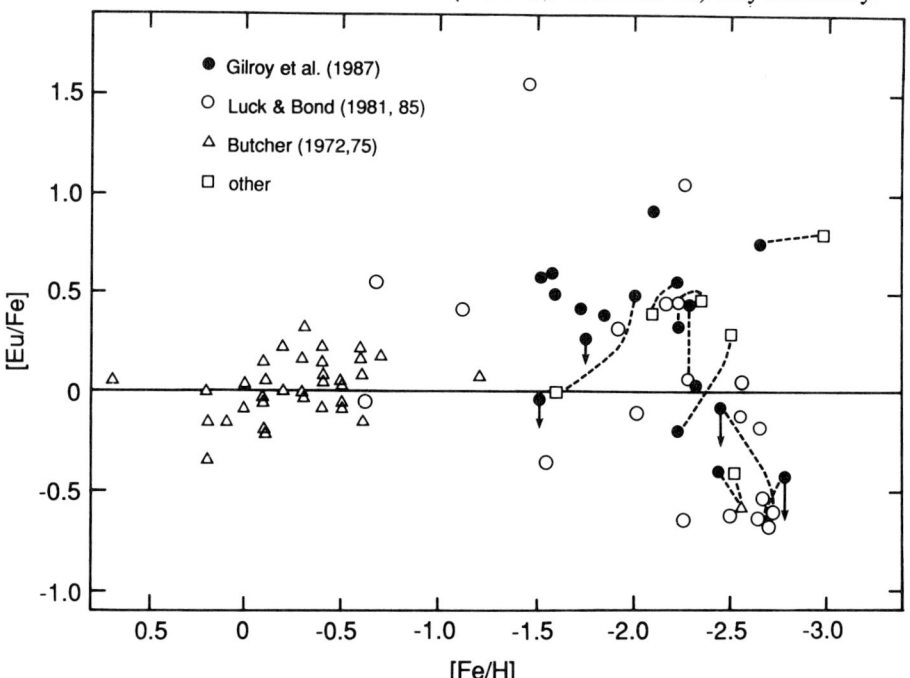

Figure 1. A correlation of Eu/Fe ratios with stellar metallicities. The points labeled "other" are from Spite and Spite (1978), and Griffin *et al.* (1982).

A comparison of Ba with Eu abundances can show the relative contributions of the r- and s-process efficiencies, since Ba may be manufactured both processes. Indeed, it is manufactured so easily during s-process events that often it is believed mistakenly to be exclusively an s-process element. In figure 2 we have correlated the [Ba/Eu] ratios with [Fe/H] values in the same manner as in figure 1. [Ba/Eu] ratios decline steadily from [Fe/H] ≈ -1 to -2. This trend usually is attributed, correctly, to a lack of significant s-process nucleosynthesis in the early Galaxy. The Ba abundances are alleged to decline nearly linearly below [Fe/H] ≈ -1.5 (see summaries by Spite and Spite, 1985; Luck and Bond, 1985). However, the evidence for a continuation of this decline at the very lowest metallicities is not strong. Moreover, the [Ba/Eu] ratios seem constant below [Fe/H] ≈ -2. Finally, the Ba content in the ultra-low metallicity star CD-38° 245 ([Fe/H] ≈ -4.5) is not extremely abnormal: [Ba/Fe] ≈ -0.5 (Bessell and Norris, 1984).

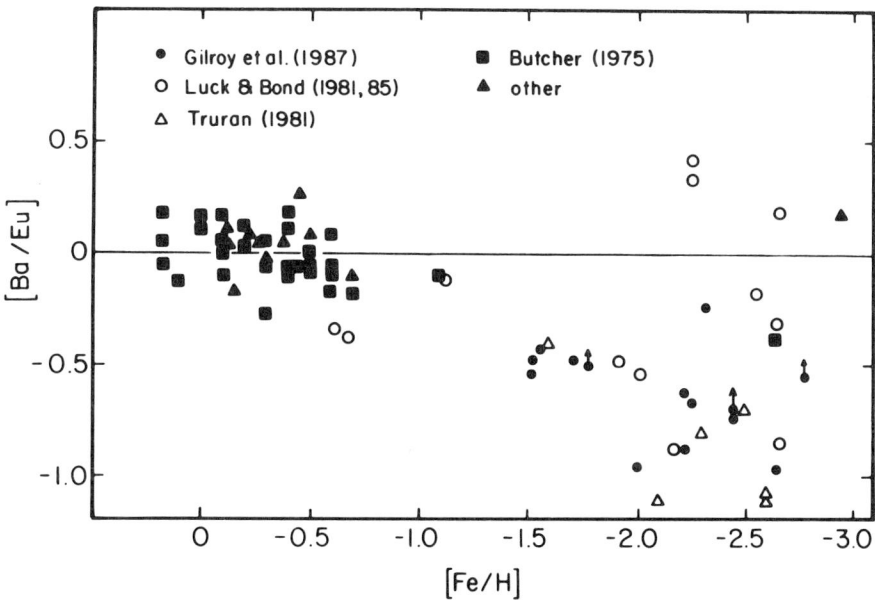

Figure 2. Ba/Eu ratios with stellar metallicities. The points labeled "other" include data from Helfer and Wallerstein (1968), Mackle *et al.* (1975), and Ruland *et al.* (1980).

The Ba/Eu behavior can be explained simply by asserting that *both Ba and Eu are r-process elements in extremely metal-poor stars; no appeal to the s-process for either element is necessary.* Truran (1981), from consideration of the few very heavy element abundances available for halo stars at that time, first suggested that Ba is an r-process element in these stars. His idea has been amplified observationally by Sneden and Pilachowski (1985), and by Gilroy *et al.* (1987), who present, for the first time, a set of r-process predictions for these elements, based on the work of Cowan, Cameron, and Truran (1985, and references therein). One feature of these calculations is the substantial amount of Ba produced in typical r-process model calculations: typically Ba/Eu ≈ 1, a result not too far from the observed ratio in extreme Population II stars.

Other abundance correlations would show similar effects. For instance, in a

[Ba/Nd] versus [Fe/H] plot, a decline in the [Ba/Nd] ratio is seen from [Fe/H] ≈ -1 to -2 (albeit with a large amount of scatter). This again undoubtedly signals the disappearance of the s-process part of the abundances of these elements. The apparent constancy of this ratio below [Fe/H] ≈ -2, especially in the data of Gilroy et al. (1987), reaffirms the idea that these two elements are made in the same r-process syntheses that produced Eu. Note also that the Ba/Nd ratio is one for which considerable globular cluster data exist (Cohen, 1980, and references therein: Gratton, 1982). The abundances for the globular cluster stars were obtained almost exclusively from low signal-to-noise image tube spectra, and the resulting abundance ratios should be interpreted with caution. Still, decent agreement is seen with the abundance ratios of the field stars. No very low Ba/Nd ratios in globular clusters have been detected, which could be another indication that the globular clusters formed after the most metal-poor filed stars, and after the appearance of s-process synthesis events in the halo. Indeed, stars of the globular cluster M55 ([Fe/H] ≈ -1.2) have detailed very heavy element abundance patterns which are nearly solar (Pilachowski, Sneden, and Green, 1984).

3. COMPARISONS WITH NEUTRON CAPTURE THEORIES

Enough observational data on a lot of very heavy elements now exists to be able attempt a comparison with r- and s-process theories, and the theoretical predictions recently have become available in element-by-element forms. In figure 3 we have plotted the average abundance results for stars of the Gilroy *et al.* (1987) study which have [Fe/H] < -1.8, along with an s-process prediction of Malaney (1987), and an r-process prediction of Cowan and collaborators employed by Gilroy *et al*. From the previous discussion it should be no surprise that the s-process predictions do not fit the observed Population II star data very well. The r-process curve fits the elements from Ba through Dy better, matching all elements well except for Eu, which is overproduced in the theoretical models. However, for a couple of the most metal-poor stars Gilroy *et al.* (1987) determined only upper limits for Eu. If Eu begins to decline below [Fe/H] ≈ -2.5, then the observed data possibly should have a lower abundance cutoff as well as an upper cutoff for a cleaner comparison to r-process theory.

Some comments on both the theory and observation should be given here. 1) *No single curve can represent adequately the abundance patterns produced by the r-process.* The r-process is a dynamic event which, unlike the s-process, cannot be constrained in terms of beta-decay rates. Therefore the curve displayed here is only one of a family of possible r-process curves with different assumptions about seed nuclei and neutron fluxes. Such curves can look very different that the one displayed here. 2) *The r-process synthesis events which preceded the formation of the extreme Population II stars have the same character as those which built the r-process fractions of the solar system abundances.* The observed abundances of these metal-poor stars and the solar system r-process abundances (Cameron, 1982) match fairly well. The theoretical r-process curve which we display here also is the one which matches best the solar system abundances (Cowan et al., 1985). 3) *The relative abundances of the Sr,Y,Zr element group versus the Ba-Dy group are not good diagnostics for the existence of the r- or s-process.* This is simply because either process can produce a variety of different ratios of very heavy elements which are widely separated in Z. 4) *Ba, La, Eu, and Dy are critical elements for comparison with nucleosynthesis theories*. These four elements both seem most sensitive to the different modes of synthesis, and can be observed most easily in metal-poor stars.

In summary, available abundances of very heavy elements in Population II stars indicate the prevalence of r-process abundance patterns, and these abundances are only loosely correlated with overall stellar metallicities. The importance of continued work on the determination of reliable very heavy element abundances for extremely metal-poor stars ([Fe/H] < -2.5) cannot be overemphasized. It remains of great interest to discover whether an evolution of these abundances can be seen in the most metal-poor stars.

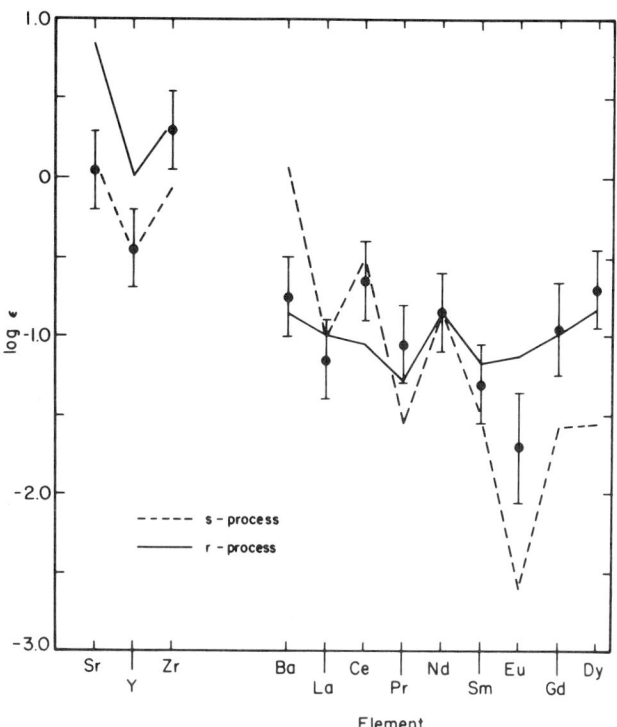

Figure 3. A comparison of neutron capture predictions with observations of very heavy elements in very metal-poor stars. The observational data are from Gilroy et al. (1987).

REFERENCES

Bessell, M. S., and Norris, J. 1984, *Astrophys. J.*, **285**, 622.
Butcher, H. R. 1972, *Astrophys. J.*, **176**, 711.
Butcher, H. R. 1972, *Astrophys. J.*, **199**, 710.
Cameron, A. G. W. 1982, *Astrophys. Space Sci.*, **82**, 123.
Chamberlain, J., and Aller, L. H. 1951, *Astrophys. J.*, **114**, 52.
Cohen, J. G. 1980, *Astrophys. J.*, **241**, 981.
Cowan, J. J., Cameron, A. G. W., and Truran, J. W. 1985, *Astrophys. J.*, **294**, 656.
Gilroy, K. K., Sneden, C., Pilachowski, C. A., and Cowan, J. J. 1987, *Astrophys. J.*, submitted.
Gratton, R. 1982, *Astron. Astrophys.*, **115**, 171.
Griffin, R., Griffin, R., Gustafsson, B., and Vieira, T. 1982, *Mon. Not. Roy. Ast.*

Soc., **198**, 637.
Helfer, H. L., and Wallerstein, G. 1968, *Astrophys. J. Suppl.*, **16**, 1.
Luck, R. E., and Bond, H. E. 1981, *Astrophys. J.*, **244**, 919.
Luck, R. E., and Bond, H. E. 1985, *Astrophys. J.*, **292**, 559.
Mackle, R., Holweger, H., Griffin, R., and Griffin, R. 1975, *Astron. Astrophys.*, **38**, 239.
Malaney, R. A. 1987, *preprint.*
Pilachowski, C. A., Sneden, C., and Green, E. M. 1984, *Publ. Ast. Soc. Pac.*, **96**, 932.
Ruland, F., Holweger, H., Griffin, R., Griffin, R., and Biehl, D. 1975, *Astron. Astrophys.*, **92**, 70.
Sneden, C., and Pilachowski, C. A. 1985, *Astrophys. J. (Lett.)*, **288**, L55.
Spite, M., and Spite, F. 1978, *Astron. Astrophys.*, **67**, 23.
Spite, M., and Spite, F. 1985, *Ann. Rev. Astron. Astrophys.*, **23**, 225.
Truran, J. W. 1981, *Astron. Astrophys.*, **97**, 391.

DISCUSSION

PETERSON   Have you compared r/s -process abundances anomalies versus those of light elements ?

SNEDEN   No, but it should be done if enough light element data exist.

MAGAIN   Do you think that undetected duplicity could explain some of the differences in relative abundances that you find in your study ?

SNEDEN   Probably not, for in the example of HD 108317 vs. HD 128279, the spectral lines, except for the very heavy element lines, seem quite similar in the 2 stars.

BOND   You contrasted HD 108317 and HD 128279 as having different contents of very heavy elements at the same [Fe/H], implying a scatter at fixed iron content. But couldn't HD 108317 be self-polluted as in barium or CH stars ?

SNEDEN   The low Ba/La ratios of both stars seem to suggest the r-process for both stars' very heavy elements, which in turn implies a primordial origin for these element abundances for both stars.

LITHIUM ABUNDANCES IN A NEW SAMPLE OF METAL DEFICIENT DWARFS

R. Rebolo[1], P. Molaro[2], J. E. Beckman[1]
1. Instituto de Astrofisica de Canarias, 38200 - La Laguna, Tenerife, Spain.
2. Osservatorio Astronomico di Trieste, via G.B. Tiepolo 11, I - 34131, Trieste, Italy.

ABSTRACT. We have measured the Li abundance in a sample of 37 field dwarfs of low metallicity ($-0.6 \geqslant$ [Fe/H] $\geqslant -3.2$). Stars with [Fe/H]<-1.4 ("EMD" stars) form a much more homogenous population w.r.t. ^{7}Li than do the stars with somewhat higher metallicities. We argue in detail that the warmer EMD stars (6000K$< T_{eff} <$ 6300K) show little Li depletion, and that their asymptotic Li abundance log N(Li) = 2.2 (± 0.15) is the best available approximation to a primordial value.

1. INTRODUCTION

Since the pioneer study of Spite and Spite, SS (1982) the ^{7}Li abundance has been recognized as a practical tool for probing cosmological models. The basic dilemma at present facing interpreters to Li abundances is whether the primordial abundance was log N(Li) = 3 and that those stars with lower abundances show only the effects of depletion, or whether the primordial value was log N(Li) = 2, and subsequent galactic enrichment has increased the abundance in relatively younger stars, to values near 3, coincident with reduction by convective depletion in cooler stars, or by diffusive depletion in warmer stars. The measurements reported here were made to throw more light on this question (see also Duncan, this volume).

2. OBSERVATIONS

All our spectra were obtained with a resolution $\lambda/\Delta\lambda \sim 2 \times 10^4$ with the Intermediate Dispersion Spectrograph of the 2.5m Isaac Newton Telescope on La Palma (Canary Islands). Signal to noise ratios of order 100 were achievable for stars of magnitude m_v = 10 in periods of order 30 minutes. 70% of the stellar spectra have S/N >100. In the best cases lower limits of 4mÅ could be set to the imprecision in the equivalent widths of the Li doublet.

For full details of the analysis see Rebolo et al. (1987). The input parameters for the models were set as follows: (a) T_{eff} by averaging V-K, R-I and Strömgren b-y (see Carney, 1983). Our estimated error is ΔT_{eff} = ±80 K; (b) log g, from c_1 and b-y indices (Relyea and Kurucz, 1978; Carney,

1983), or U-B, B-V where necessary (Relyea and Kurucz, 1978), with an upper limit to Δ(log g) of ±0.5 dex; (c) metallicity from Cayrel de Strobel et al. (1985) or Carney (1983). To derive the values of [Li/H] and [Fe/H] we used models by Kurucz (1979) for $T_{eff} \geqslant 5500K$ or by Gustaffson for stars with lower T_{eff}, in a synthesis programme by Castelli, computing our own curves of growth, from which abundances were read off for observed equivalent widths. The quadratic sum of all the observational and model errors is estimated at ±0.11 dex, to which we must add possible systematic NLTE effects, which would tend to raise the quoted values by of order 0.1 dex but which cannot really be quantified due to theoretical uncertainty (see Steenbock and Holweger, 1984).

3. RESULTS AND INTERPRETATION

In Fig. 1 we have plotted all of our data for $T_{eff} \geqslant 5500K$ together with a wide selection of literature Li abundances against metallicity. The clear break between the abundances below [Fe/H] =-1.4, where the dispersion is small, and above this value caused us to classify our observed stars into "extremely metal deficient" (EMD), with [Fe/H] <-1.4 and "highly metal deficient" (HMD) stars. The upper envelope of Fig. 1 can be interpreted as indicating galactic enrichment, and the scatter below it as due to the effects of depletion in stars with different metallicities and histories (see below).

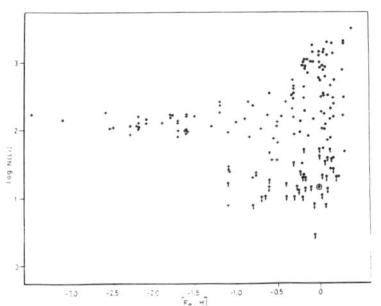

Figure 1:
Lithium abundance against metallicity for stars in the literature with $T_{eff} \geqslant 5500K$. Typical error bars log N(Li) are 0.2 dex and for [Fe/H] are 0.15 dex. Temperature ranges: • $T_{eff} \geqslant 6000$ K, + 6000K > T_{eff} >5500K.

In Fig. 2 we plot Li abundances v. T_{eff} for EMD stars measured here, and taken from the literature (see caption for details), and compared this with the equivalent curve for the Hyades. For T_{eff} <5500K the depletion in the Hyades is greater than for EMD stars, even though the latter are more than ten times older. If only equivalent processes produce depletion in the two groups, the EMD stars with $T_{eff} \geqslant 6000K$ should be undepleted. Further evidence of low depletion rates for EMD stars is presented in Fig. 3 where differential depletion as a function of metallicity shows up in the comparison of log (N(Li) v. T_{eff} for EMD stars, HMD stars and M67. These data can be globally explained if we assume that the initial Li for each group was greater if it formed at a later epoch, but that Li depletion increases with metallicity. There are, however, a number of stars with low metallicity and low Li abundance - a fact which merits further research. Further support for a value around 2 comes from Fig. 4 where we plot log N (Li) against mass; the asymptotic value at high mass is ~2.2.

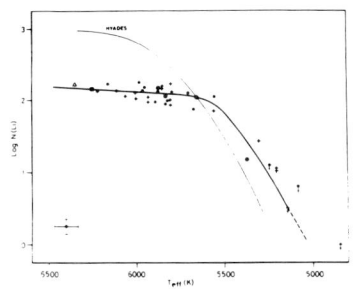

Fig. 2: Lithium abundances against T_{eff} for EMD stars ([Fe/H] < -1.4). Sources of data: + Spite and Spite (1982, 86, 87), Spite et al., (1984). Rebolo et al., (1987 a, c). Boesgaard (1985). Also shown is an observational log N(Li)-T_{eff} curve for the Hyades.

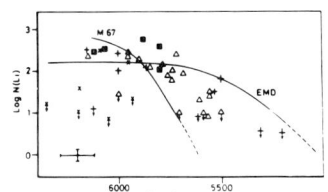

Fig. 3: Lithium abundances of our HMD stars plus literature values for -0.3>[Fe/H] >-1.3 in the effective temperature range 5000K - 6300K. Sources of data: ■ Duncan (1981); + Spite and Spite (1982); x Boesgaard and Tripicco (1986); △Rebolo et al. (1987). Also included for comparison are the curves for EMD stars and for M67 (Hobbs and Pilachowski, 1986; Spite et al., 1987).

4. CONCLUSIONS

Neglecting possible diffusion effects which according to Michaud et al. (1984) would not readily produce the "plateau" in log N(Li) v. T_{eff} illustrated in Figs. 2 and 4. We would expect very low Li depletion for EMD stars between 6000K and 6300K. This would put log N(Li)$_p$ (p primordial) at 2.20 (±0.15). In Rebolo et al. (1987 c) we show that spallative production of Li before the formation of the EMD stars is highly improbable, so that we are justified, following SS, in using this value as primordial. Supporting arguments are the invariance of Li with metallicity between [Fe/H]=-1.5 and-3.5 and invariance with site of origin of the high velocity EMD objects. This value for log N(Li)$_p$ together with the new theoretical η-Li curve of Kawano et al. (1987) in the frame of a canonic Big Bang model, sets limits of $1.5 \times 10^{-10} < \eta < 8 \times 10^{-10}$ for the universal baryon:photon ratio, and taken with the limits to primordial ^{4}He (Pagel, 1987) and the accepted range of values for the neutron half-life, limits the possible number of light neutrino types, N_ν, to 3 or less and the fraction of the density contribution by baryons to $0.004 < \Omega < 0.10$.

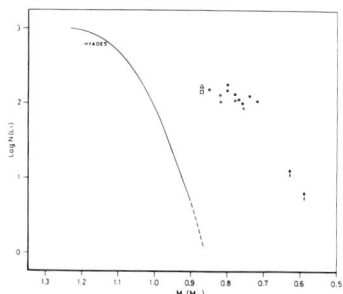

Fig. 4: Lithium abundance as a function of mass for stars with [Fe/H]<-2.0. Masses estimated from VandenBergh (1983). Also included in the equivalent curve for the Hyades. Sources of data:
+ Spites and collaborators.
▫ Boesgaard (1985)
• Rebolo et al. (1987 a, c).

References

Boesgaard, A.M.: 1985, PASP, **97**, 784.
Boesgaard, A.M., Tripicco, M.J.: 1986, Ap. J. **303**, 724.
Carney, B.W.: 1983, Astron. J. **88**, 623.
Hobbs, L.M., Pilachowski, C.: 1986, Ap. J. **311**, L37.
Kawano, L., Steigman, G., Schramm, D.N.: 1987, Ap. J. (In Press).
Kurucz, R.L.: 1979, Ap. J. Supp. Ser. **40**, 1.
Michaud, G., Fontaine, G., Beaudet, G.: 1984, Ap. J. **282**, 206.
Pagel, B.E.J.: 1987 in "Advances in Nuclear Astrophysics", Vangioni-Flam et al. (Ed.) p. 53.
Rebolo, R., Molaro, P., Abia, C., Beckman, J.: 1987a, A & A (In Press).
Rebolo, R., Beckman, J., Molaro, P.: 1987b, A & A **172**, L17.
Rebolo, R., Molaro, P., Beckman, J.: 1987c, A & A (In Press).
Relyea, L.J., Kurucz, R.L.: 1978, Ap. J. Suppl. Ser. **37**, 45.
Spite, F., Spite, M.: 1982, A & A **115**, 357.
Spite, M., Maillard, J.P., Spite, F.: 1984, A & A **141**, 56.
Spite, F., Spite, M.: 1986, A & A **163**, 140.
Spite, F., Spite, M., Peterson, R.C., Chafee, F.M., Jr.: 1987, A & A **171**, L8.
Steenbock, W., Holweger, H.: 1984, A & A **130**, 319.
VandenBergh, D.H.: 1983, Ap. J. Suppl. Ser. **51**, 29.

DISCUSSION

R. CAYREL The shape of the depletion curve of Li in the subdwarfs can be reproduced by a combination of convection (negligible depletion rate) overshooting, turbulent mixing and nuclear burning. The problem is that the parameters which describe the Hyades depletion curve (cf. Baglin paper in this Symposium) have to be drastically changed to account for the subdwarf depletion curve.

REBOLO The simplest explanation for the curve Li versus effective temperature (or mass) for extreme metal deficient stars is that depletion has been negligible in the hotter stars. This assumption is not contradictory with the observations in more metallic stars (higly metal deficient stars and open clusters).

BORON ABUNDANCE IN THE POPII HD 76932

P. Molaro
Osservatorio astronomico di Trieste,
Via G.B. Tiepolo 11, 34131 I,
Italia

ABSTRACT. Aware of the importance of B observations in cool stars we have attempted to search for BI 2496 lines in high resolution IUE spectra of HD 76932, with [Fe/H]=-1.1, and HD 216385, with [Fe/H]=-0.6. Line blending and the impossibility of establishing the continuum level are the major obstacles in such an investigation. Molaro (1987) showed that, in the case of extremely metal-deficient stars, the consequent line-blocking reduction allows us to derive significant upper limits for the B abundance. In HD 140283, [Fe/H]=-3.0, Molaro (1987) derived [B/H]< 2×10^{-11}. In HD 76932 a preliminary analysis gives a B abundance of [B/H]<2×10^{-10}, i.e. less than the present "cosmic" value. On the other hand no useful limits can be placed for HD 216385. The presence of an unidentified line, not resolved with IUE, prevents B analysis in cool stars with similar or greater metallicity. Better data with higher spectral resolution and greater photometric accuracy are required to confirm these results. The simultaneous presence of Li in the atmospheres of these stars rules out stellar depletion for B, providing evidence that the B abundance in the Galaxy was lower in the past than it is today. These observations also imply that spallation proccesses were not able to produce significant quantities of B, or other light elements at the beginning of the Galaxy. In particular, spallation processes cannot be adduced as a possible source for the Li observed in the Pop II stars.

1. INTRODUCTION

Boron is one of most elusive elements in nature. A compilation of the more recent B determinations in various media is given in Table 1. Boron cannot be manufactured by stars and is destroyed in the hot stellar interiors. Together with Be and Li, it is likely the product of spallation processes between energetic galactic cosmic rays and interstellar gas (Reeves et al. 1970). The presence of B in stellar atmospheres is useful to probe the internal structure (cfr. A. Boesgaard these proceedings), and from the evolutionary abundance curve it is possible to infer information on the behaviour of GCR and chemical

evolution in the Galaxy from the time of its origin.
The most important problem in the abundance derivation is the severe blending of the BI 2496 A resonance doublet. Other difficulties are: the uncertainty in placing the ultraviolet continuum, the lack of experimental oscillator strengths, the many missing lines in the ultraviolet region, and the limited S:N today available in ultraviolet data. Here we show that some of the above-mentioned problems can be alleviated by analyzing high resolution IUE spectra of very metal-deficient stars.

Table 1

source	[B/H]	references
stars	$2(\pm 2)^{-10}$	Boesgaard and Heacox 1978
meteorites	2^{-9}	Weller et al 1977
	3^{-10}	Curtis et al. 1980
sun	$4(\pm 2)^{-10}$	Kohl et al. 1977
	$<1.8^{-10}$	Hall and Engvold 1975
interstellar	$1.5(\pm 0.7)^{-10}$	Meneguzzi and York 1980
	$<7.6^{-11}$	Morton 1975

2. OBSERVATIONS AND ANALYSIS

Following the analysis of Molaro (1987) on HD 140283 we have searched the B lines in other two metal-weak line stars: HD 76932, [Fe/H]=-1.1 and HD 216385, [Fe/H]=-0.6 (LWR 7182 and LWR 5308 respectively). The analysis has been performed using the Synth procedure to compute a synthetic spectrum, in the LTE aproximation, from the blanketed Kurucz model atmosphere (Castelli 1985). Details on the line list, log gf, broadening sources and wavelenght calibration can be found in Molaro (1987).
For HD 76932, [Fe/H]=-1.1, the synthetic spectrum has been computed from a model with Teff = 5860 K, gravity log g=3.5 and a microturbolence of 1 Km/sec. In this case the van Der Waals constants of the strong lines, such as FeI 2491.155 A, have been increased by up to a factor of 5 over classical values in order to obtain better fits of strong lines. The synthetic spectrum has been integrated over a large range to take into account the contribution on the wing of the FeI 2491.155 A strong line.
Fig. 1 shows a preliminary analysis of the synthetic spectrum around BI 2496 A. The log gf values of the three strong Fe lines near the BI doublet are reduced similarly to what has been done for HD 140283. This reduction is quite ad hoc and precludes the use of the 2497.725 for B analysis. Unlike the case of HD 140283, from Fig 1 is possible to see that an absorption is present in correspondence to 2496.7 A. This can be ascribed to B, but also to an unidentified line located at 2496.88 A, which is present in the solar spectrum and noted by Kohl et al 1977. There is no way to account here for the unidentified line and we must wait for better resolution data in order to solve the problem. At first glance we attribute this contribution to the B line. The absorption is consistent with an abundance of [B/H]=2×10^{-10} that we consider as an

upper limit to the possible presence of B.
HD 216385, with [Fe/H]=-0.6, can be considered representative of moderate metal-deficent stars. The synhtetic spectrum is performed with Teff=6070 logg=4.0. Fig. 2 gives the sintetic spectrum with [B/H]=1x10^{-12} and [B/H]=4x10^{-10}. As can be seen from the figure, due to the presence of the above mentioned unidentified line, we cannot infer any B abundance or upper limits with the present data. This case is illustrative of what happens as the metallicity of the star increases and also of the difficulties that one faces in trying to derive B abundances in cool stars with solar abundances.

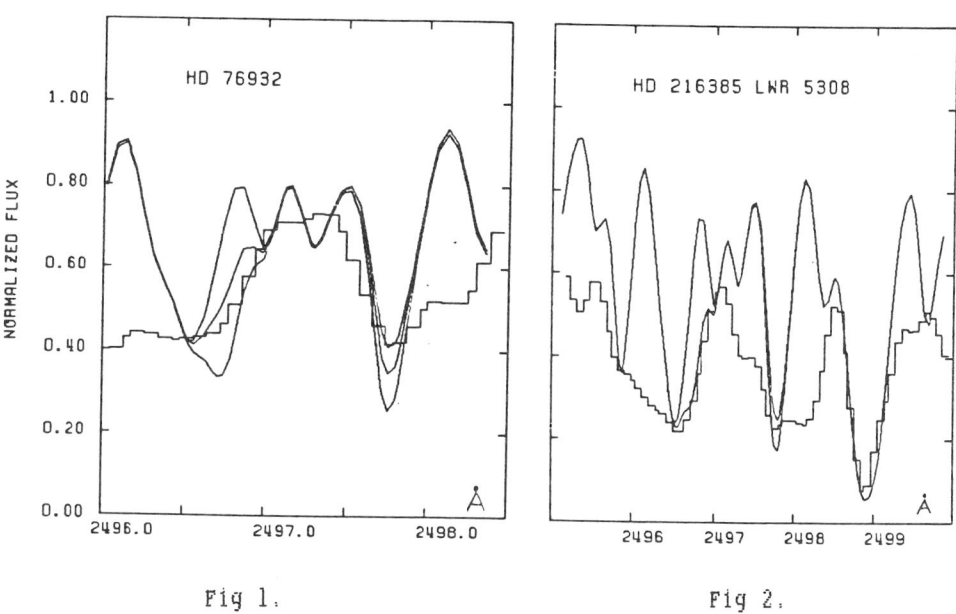

Fig 1.   Fig 2.

Fig 1. Observed (histogram) and synthetic spectrum for HD 76932 around Bi 2496 A. The models are for [B/H] abundances of 10^{-12}, 2x10^{-11} and 2x10^{-10}.
Fig 2. As Fig 1 for HD 216385. The models are for [B/H] of 10^{-12} and 4x10^{-10}.

3. DISCUSSION

The upper limit for HD 76932, together with the previous result for HD 140283, provide some evidence of a low B abundance in the atmospheres of stars older than the sun. In both these two PopII stars, Li has been observed at the "typical" value for the pop II stars (i.e [Li/H]=1x10^{-10}, Spite and Spite 1982, Beckman et al 1986). Since B is destroyed by thermonuclear reactions at a temperature of 5 10^6 K, compared with the 2.5 10^6 of Li, the survival of Li in the atmosphere of these old stars rules out stellar depletion of B. This in turn implies that the gas of the nebula from which these stars have been formed was also free of B. These results seem consistent with the view that the light elements are produced by spallation processes in the interstellar medium with a gradual increase of this element, starting from zero values at the beginning of the Galaxy. The absence of B also has implications for the origin of the Lithium observed in the popII. Differently from Li which, apart from spallation, can be synthesized through a number of other processes, B can be produced only through spallation processes. Thus, the difference in the nucleosynthesis mechanisms enables us use the B observations to verifie the fraction of Li that can be produced by spallation processes. Taking the most stringent upper limit the observed Li/B ratio is more than 25 times that predicted by the spallation theory of high energy GCR ruling out a significant synthesis of the Li observed in the population II stars.

4. REFERENCES

Beckman , J.E., Rebolo, R., Molaro P.:1986, Proc. Second IAP Rencontre on nuclear astrophysics, in press.
Boesgaard, A.M. and Heacox, W. D.: 1978, Astrophys. J. 226,888.
Castelli, F.:1985 "A procedure to compute a stellar synthetic spectrum", Publ. OAT %N. 984.
Curtis, D., Gladney, E., and Jurney E.: 1980, Geochim. Cosmochim. Acta, 1945.
Hall, D.N.B., and Engvold, O. :1975 Astrophys. J. 197,513.
Kohl, J. L., Parkinson W. H., Withbroe G.L.: 1977 Astropys. J. 212, L101.
Meneguzzi M., and York D. G.: 1980, Astrophys. J. %235, L111.
Molaro,P.:1987 Astron. Astrophys. in press
Morton. D.C. 1975, Astrophys.J., 197,85.
Reeves, H,, Fowler, W.A., Hoyle F.: 1970, Nature 226, 727.
Spite,F., Spite,M.: 1982 Astron. Astrophys. 115, 357.
Weller M.R., Furst, M., Tombrello, T. A., and Burnett D.S.:1977 Astrophys.J. 214,L39.

DISCUSSION

C. Sneden: How did you set your continua, and how did you treat the u. v. opacities?
P. Molaro: The setting of the continuum is one of the most critical points. I assumed that the highest points of the order 93 of the IUE echelle spectrum are points of true continuum and the results depend from this assumption. For the continuum all the opacities of the most abundant elements are considered, in particular the bound-free absorption of the the ^3P level of MgI with the ionization threshold at 2513 A. The line opacities are mainly from Kurucz and Peytremann (1975).

OXYGEN IN HALO GIANTS

B. Barbuy
Universidade de Sao Paulo
Depto Astronomia, C.P. 30627
Sao Paulo - 01051
Brazil

ABSTRACT: High resolution CCD spectra were obtained in the region $\lambda\lambda 6290-6330$ Å containing the |OI| $\lambda 6300.3$ Å line for a sample of halo giants. Synthetic spectra were computed to derive their oxygen abundances. A mean value [O/Fe] $\simeq$ +0.35 is obtained.

I. Introduction

The oxygen-to-iron ratio in the halo stars is a boundary condition of major importance in contexts such as (1) the chemical evolution of the Galaxy, and (2) the mass distribution of the (first) stars preceding the halo stars, a subject connected to theories of star formation at low metallicities.

Oxygen overabundances in the halo have been found by different authors - see Sneden (1985) for a review - the available data showing however a non-negligible scatter.

In the present work, the oxygen abundance was derived for a sample of 17 halo giants, where a relatively low scatter is found.

II. Observations and theoretical spectra

The spectra were obtained at the 1.4 m Coudé Auxiliary Telescope, using the Coudé Echelle Spectrometer (CES) and a 1024x512 element RCA CCD detector, at the European Southern Observatory (ESO), La Silla, Chile.

The list of stars, selected essentially from Bond (1980) is presented in table I.

The synthetic spectra calculations (cf. Barbuy, 1981), include a molecular dissociative equilibrium, where carbon and oxygen are compounds of several molecules. In particular, the oxygen association in CO molecules affects non-negligibly the strength of atomic oxygen lines in cool giants.

The derivation of stellar parameters is based on photometric indices and measured equivalent widths, as well as on detailed analyses given in the literature. The model atmospheres employed are interpolated in the grids by Bell et al (1976).

III. Results

The iron-to-hydrogen ratio adopted and the resulting oxygen-to-iron ratios are given in table I.

The [O/Fe] ratios as a function of metallicity [Fe/H] are shown in figure 1, where data for disk stars by Nissen et al (1985) are also plotted.

HD nº	[Fe/H]	[O/Fe]	HD nº	[Fe/H]	[O/Fe]
2796	−2.4	+0.4	103036	−1.7	+0.7
6268	−2.5	+0.25	122956	−1.9	+0.25
21581	−1.65	+0.25	165195	−2.1	+0.4
26297	−1.4	+0.25	184711	−2.3	+0.3
23798	−2.2	+0.35	187111	−2.0	+0.4
29574	−1.4	+0.5	204543	−1.63	+0.3
36702	−2.0	+0.3	206739	−1.6	+0.25
44007	−1.7	+0.35	222434	−1.9	+0.35
83212	−1.5	+0.4			

Table I - Program stars, metallicities adopted and oxygen-to-iron ratios obtained.

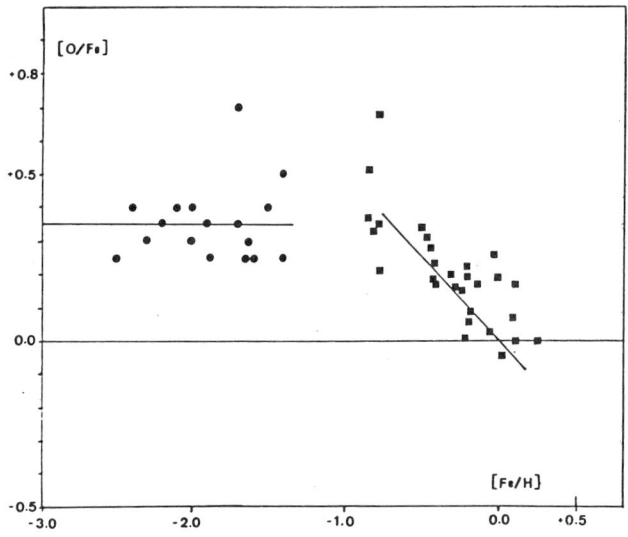

Figure 1 - Oxygen-to-iron ratio versus iron to hydrogen ratios for: ●: typical halo stars (present work), ■: typical disk stars (Nissen et al, 1985).

IV. Discussion

In the present work, a constant [O/Fe] ratio is found for typical halo stars in the metallicity range $-2.5 <$ [Fe/H] < -1.4. A mean value [O/Fe] $\simeq +0.35$ is obtained, showing an overabundance lower than the previously reported values of +0.6 dex (e.g., Clegg et al, 1981).

This set of data combined to results for typical disk stars by Nissen et al (1985) are consistent with the following scenario for the chemical enrichment in oxygen and iron in the Galaxy:

The O/Fe ratio in halo stars represents the signature of nucleosynthesis in massive stars, where oxygen is produced in excess with respect to iron, relative to the solar value (Arnett, 1978). The oxygen excesses increase with stellar mass and it could be expected that [O/Fe] would show a slight decrease from the more metal-poor to the intermediately metal-poor halo stars, corresponding to a gradual enrichment by stars of decreasing mass and longer timescales of evolution, among the massive stars: this decrease is not seen in the present determinations.

The enrichment in iron is provided by the intermediate (and low) mass stars ($M < 9$ $M_\odot$), and their effect starts to be seen at metallicities around [Fe/H] $\simeq -1.0$. The gradual enrichment in iron causes a progressive decreases of the O/Fe ratio in the disk, which is clearly seen from the data by Nissen et al (1985).

A next important step to be done is a study of O/Fe for a sample of intermediate metallicity stars, characterized by kinematics and metallicities ($-1.2 <$ [Fe/H] < -0.8) of the old disk and thick disk populations.

Acknowledgements: The present observations were obtained at the European Southern Observatory (La Silla, Chile) and I wish to express my gratitude to those who have made this possible. I am grateful to the International Astronomical Union for a travel grant. Travel costs to Chile were sponsored by the foundation FAPESP (Sao Paulo).

References:

Arnett, W.D.: 1978, Astrophys. J. 219, 1008
Barbuy, B.: 1981: Astron. Ap. 101, 365
Bell, R.A., Eriksson, K., Gustafsson, B., Nordlund, A.: 1976, Astron. Ap. Suppl. 23, 37
Bond, H.E.: 1980, Astrophys. J. Suppl. 44, 517
Clegg, R.E.S., Lambert, D.L., Tomkin, J.: 1981, Astrophys. J. 250, 262
Nissen, P.E., Edvardsson, B., Gustafsson, B.: 1985, in "Production and Distribution of C,N,O elements", ESO Conference and Proceedings no 21, eds I.J. Danziger, F. Matteucci, K. Kjar, p. 131
Sneden, C.: 1985, in "Production and Distribution of C,N,O elements", ESO Conference and Proceedings no 21, eds I.J. Danziger, F. Matteucci, K. Kjar, p. 1

DISCUSSION

      R. CAYREL   We are in one of the rare cases, with the 6300 [OI], where one has not to worry about departure from LTE because 99% of O happens to be in the ground level of OI. Could you do the same type of analysis with subdwarfs near the turn off instead of giants ?

      BARBUY   It is a forbidden line forming preferentially in low density media, and appearing therefore stronger in giants. It practically disappears for gravities log g ⩾ 3.5.
For subdwarfs the OI triplet at λ7771-5 Å is convenient (showing however pronounced non-LTE effects).

      JUDGE   A general comment about Oxygen in the UV region in cool giants. Space telescope should enable the detection of the OI λ1300 chromospheric emission lines in Pop. II giants which IUE has shown to be so strong in population I giant stars. The enhanced relative abundance of OI reported here and elsewhere during this Symposium in Population II stars should have observable effects in UV spectra. Eventually these lines may provide valuable constraints on the heating of chromospheres, related to the subphotospheric convection zone structure, as a function of metallicity.

The chemical composition of K dwarfs.

Olof Morell
Astronomical Observatory
Box 515
S-751 20 UPPSALA
Sweden

ABSTRACT. A project with the purpose of determining the chemical composition of a sample of K dwarfs is presented, the project is yet not completed so this is a status report.

1. INTRODUCTION:

High precision analysis of cool stars has until recently been an impossible task due to the severe line crowding in the spectra. However, the possibility to use high resolution spectrographs, such as the CAT/CES at ESO, Chile, in combination with spectral synthesizing has now made it possible.
    For the understanding of the chemical composition and evolution of the galactic disc it is important to have a coverage of many types of stars. F and G-type stars are now under close investigation by Andersen et al., and this project aims at extending this sample towards the cooler end of the main sequence. For this purpose all K dwarfs in the Brigth Star Catalogue observable from Chile, except the close binaries, were selected, in total 28, and we intend to observe all those in five wavelength regions acording to table 1.

2. OBSERVATIONS.

In two observing runs, September 1985 and April 1986, twenty stars have been observed, almost all of them in all five wavelength regions. The equipment used was the CAT/CES at La Silla, the only available detector then was a RETICON array. Twenty nights have been used and the observing times range from a few minutes per star and wavelength region, to three hours. The V magnitudes for the faintest stars are approximately 6.4.

TABLE 1.

Wavelength region	Species
5070 - 5110 Å	C_2 Swan band, (2,2) and (3,3) bands.
6135 - 6185 Å	Na, Si, Ca, Fe, Zr.
6290 - 6340 Å	O, Ca, Mg, Sc, Ti, V, Fe, Ni.
6670 - 6730 Å	Ca, Al, Si, Ti, Cr, Fe, Co, Ni.
7955 - 8020 Å	CN red (2,0) band will yield nitrogen abundances when combined with the carbon abundances from C_2.

3. CONCLUSION.

Data are partially reduced, mainly with the ESO IHAP system, and the analysis has begun, model atmospheres calculated with the MARCS programme (Gustafsson et al. 1975) are being used. Some preliminary results have been obtained but they are interesting enough to need further confimation before being published.

Also more photometry should be obtained, two observing runs have been destroyed by bad weather.

REFERENCES.

Andersen, J., Edvardsson, B., Gustafsson, B., Nissen, P.E., IAU 132, 'A high S/N spectroscopic survey of chemical abundances in disc population F-type dwarfs'.
Gustafsson, B., Bell, R.A., Eriksson, K., Nordlund, Å., Astron. Astrophys., **42**,407, 1975.

THE SULPHUR ABUNDANCE IN METAL DEFICIENT DWARFS *

Patrick FRANCOIS
- European Southern Observatory, D-8046 Garching bei München
- Observatoire de Paris, Section de Meudon,
  F-92195 Meudon Principal Cedex, France

* based on observations collected at the European Southern
  Observatory, Chile and at the Canada-France-Hawaii telescope.

INTRODUCTION

The study of the chemical composition of metal deficient dwarfs is very interesting because thse stars have been formed at different epochs and their atmosphere is representative of the chemical composition of the gas at the time of their formation. The study of faint lines in dwarf stars has made it possible to obtain reliable information on the early evolutionary phase of the Galaxy (for a review, see Spite et al. 1985).
    However, some elements have not been yet well studied in these stars. In particular, the determination of the [S/Fe] ratio in the halo dwarfs is not well known (François 1987). In this paper, the Sulphur abundance has been determined in 11 dwarfs, in which 5 are halo stars.

OBSERVATIONS

The observations were carried out partly with the CES spectrometer fed by the 1.4m telescope (CAT) of ESO and partly at the coudé focus of the 3.6m CFH telescope in Hawaii. In both cases, the detector was a RETICON silicon array with 1872 photodiodes (width: 15 μm).
    The resolution measured on stellar lines is about 0.14 Å at the CES and 0.26 Å at the CFH.

ANALYSIS

The main characteristics of the atmospheres of the stars are known and are given in the "Catalogue of [Fe/H] determinations" (Cayrel et al. 1985) or in the paper of Laird (1985). With these parameters, we have interpolated the models in the grid of Gustafsson's model for dwarf stars (Gustafsson 1981). The main assumptions and procedures used for computing the lines are described in François (1986).

In this study, we have used weak lines of Sulphur and Iron. Thus, systematic errors are due to uncertainties in the atmospheric effective temperatures T_{eff}. An error of 100K in T_{eff} leads to a relative abundance change of 0.1 dex. The errors arising from the uncertainty in the microturbulent velocity are negligible. Departures from LTE have been studied by different authors (Müller et al. 1969, Rutten 1982, Lites et al. 1974, Steenbock 1984) and it seems that abundance ratio determinations are not greatly affected when differential analysis is used.

RESULTS AND DISCUSSION

In figure 1, we have plotted the [S/Fe] abundance ratios found in our sample of stars as a function of metallicity. We have also plotted the [S/Fe] ratios found by François (1987). Our results confirm previous results found by François (1987) and Clegg et al. (1981).

An important result is that the [S/Fe] ratio is constant and higher than solar in the halo as it has been found for the other α-elements already studied. The mean value of the [S/Fe] ratio in the halo is about 0.60 dex. In figure 3, we have also plotted the prediction of the [S/Fe] ratio as a function of [Fe/H] by Matteucci (1987b). Her model takes into account the nucleosynthesis of both type II (Woosley and Weaver 1986, Matteucci 1987a) and type I (Nomoto et al. 1984, Nomoto et al. 1986) supernovae and also of the lifetime of the precursors. Her predictions show a good agreement with other α-elements (O, Si, Mg) but the prediction of the [S/Fe] ratio in the halo seems to be slightly lower. This is not surprising because the final stages of the nucleosynthesis of massive stars are not well known, in particular the quantity of matter of the envelope which remains bound

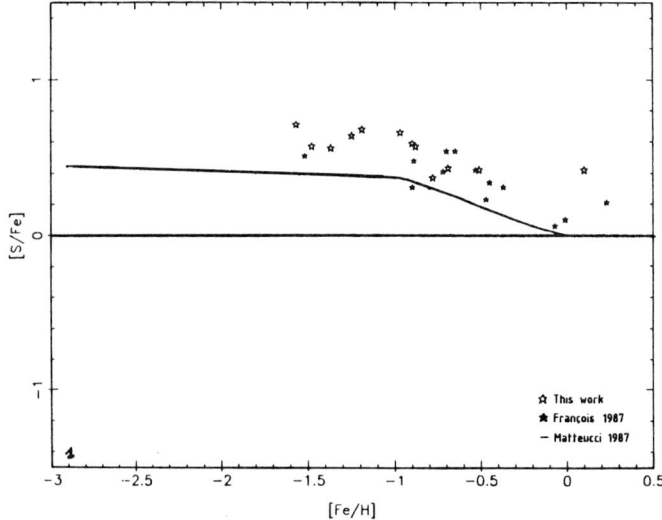

to the core after the explosion of the SN II. Our results may be used to put constraints on this parameter (Matteucci 1987b).

However, a scenario for the formation of the halo population proposed by Cayrel (1986) would lead also to a constant ratio. Both models show that massive stars have played an important role during the early phase of the Galaxy.

CONCLUSION

High resolution spectra of metal deficient dwarfs have been obtained. The high S/N ratio of the spectra has allowed us to measure very faint lines of Sulphur and to determine the chemical composition abundance of this element in old stars. These results have been used to trace the chemical evolution of the Galaxy. An important conclusion is that these results confirm the role played by massive stars during the beginning of the life of the Galaxy, in particular the fact that massive stars have mostly contributed to the nucleosynthesis of S but not to that of Iron, which should be mainly produced by low and intermediate mass stars (type I SNe).

ACKNOWLEDGMENTS

I am indebted to F. Matteucci for communicating results before publication.

REFERENCES

Branch, D., Nomoto, K.: 1986, Astron. Astrophys., **164**, L13.
Cayrel, R.: 1986, Astron. Astrophys., **168**, 81.
Cayrel de Strobel, G., Bentolila, C., Hauck, B., Duquennoy, A.: 1985, Astron. Astrophys. Suppl. Ser. **59**, 149.
Clegg, R.E.S., Lambert, D.L., Tomkin, J.: 1981, Astrophys. J., **250**, 262.
François, P.: 1986, Astron. Astrophys., **160**, 264.
François, P.: 1987, Astron. Astrophys., **176**, 294.
Gustafsson, B.: 1981 (private communication).
Laird, J.B.: 1985, Astrophys. J. Suppl. Ser., **57**, 389.
Lites, B.W., Cowley, C.R.: 1974, Astron. Astrophys. **31**, 361.
Matteucci, F.: 1987a, ESO Workshop, "Stellar evolution and dynamics of the outer halo of the Galaxy", April 7-9 1987, (in press).
Matteucci, F.: 1987b, in preparation.
Müller, E.A., Mutschlecner, J.P.: 1969, Astrophys. J., **84**, 1.
Nomoto, K., Thielemann, F.K., Yokoi, K.: 1984, Astrophys. J., **286**, 644
Rutten, R.J., Kostik, R.I.: 1982, Astron. Astrophys. **115**, 104.
Spite, M., Spite, F., 1985: Ann. Rev. of Astron. and Astrophys., **23**, 225.
Steenbock, W., Holweger, H.: 1984, Astron. Astrophys. **130**, 319.
Woosley, S.E., Weaver, T.A.: 1986, Proc. IAU Coll. 89 "Radiation Hydrodynamics in Stars and Compact Objects", ed. D. Mihalas and K.H. Winkler, p 91.

RECENT ADVANCES IN HIGH DISPERSION SPECTROSCOPY OF GLOBULAR CLUSTER STARS

Raffaele G. Gratton
Osservatorio Astronomico di Roma
Via del Parco Mellini 84
00136 Roma
Italy

ABSTRACT. The use CCD detectors has allowed a major progress in abundance derivations for globular cluster stars in the last years. Abundances deduced from high dispersion spectra now correlates well with other abundance indicators. I discuss some problems concerning the derivation of accurate metal abundances for globular clusters using high dispersion spectra from both the old photographic and the most recent CCD data. The discrepant low abundances found by Cohen (1980), from photographic material for M71 giants, are found to be due to the use of too high microturbulences.

1. INTRODUCTION

The introduction of Echelle spectrographs equipped with image tubes at the half of '70s allowed the systematic observation of globular cluster giants at a rather high spectral resolution (~0.2÷0.3 A), though at a low S/N level (~25÷30). Results obtained with these instruments were conflicting with the usual adopted metallicities for metal rich globular clusters (see e.g. Pilachowski et al., 1983a, and Figure 1a). However, the introduction of CCD detectors at the beginning of '80s improved by a factor of two the S/N ratio of observations. Furthermore, accurate flat-fielding was achieved, which is specially important in echelle spectra, whose compact format is obtained at the expense of a fast variation of the blaze function over the single orders. Several papers on globular cluster giants were published in the last years using high dispersion spectroscopy with CCD's detectors (Cohen, 1983; D'Odorico et al., 1985; Gratton et al., 1986, 1987; Gratton, 1987a, 1987b; Leep et al., 1986, 1987; Wallerstein et al., 1987; Pilachowski and Sneden, 1988). Results from these studies are comfortably close each other, and abundances derived from these spectra correlates well with other indicators (see Figure 1b). However, we have to consider in more detail errors that are likely to be present in current as well as old abundance derivation for globular clusters.

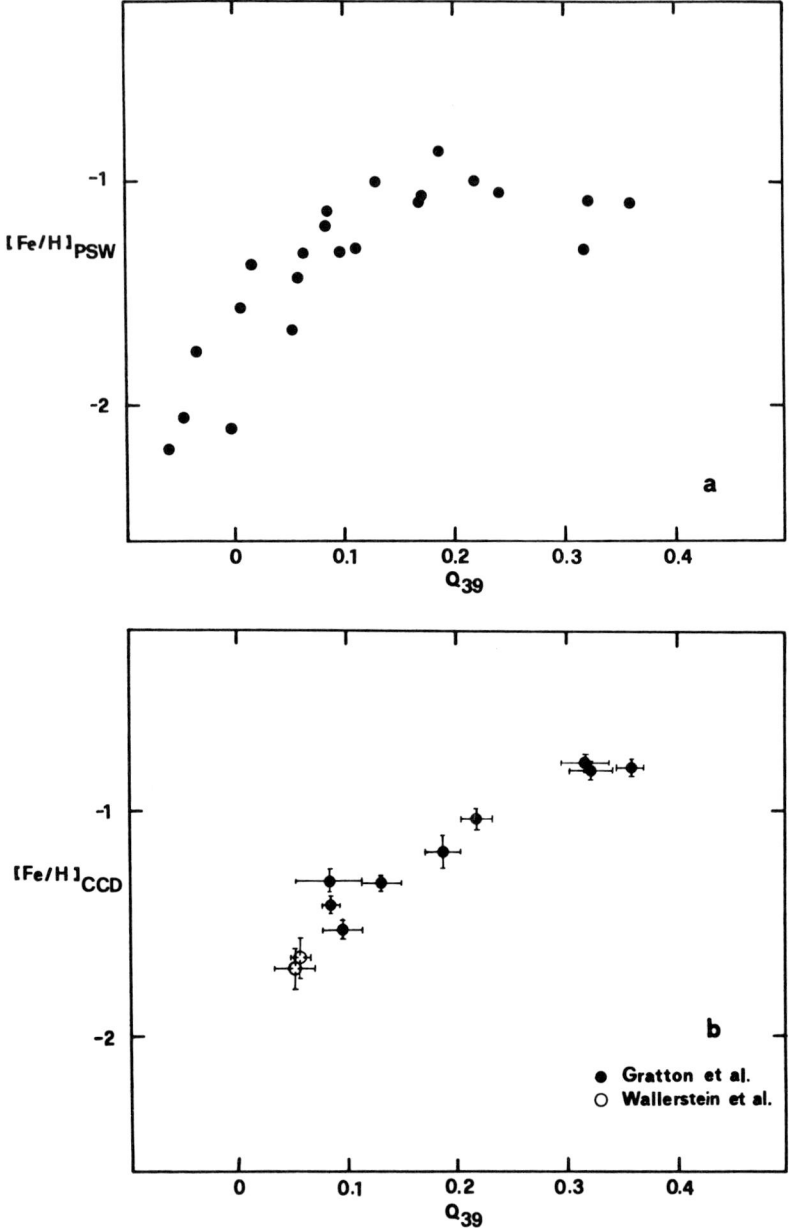

Figure 1. Metal abundances from high dispersion spectra against the Q_{39} index by Zinn (1980) and Zinn and West (1985). Panel (a) shows photographic data (Pilacwhoski et al., 1983a); Panel (b) displays CCD data (Gratton et al, 1986; Gratton, 1987; Wallerstein et al., 1987).

2. SYSTEMATIC ERRORS

I will focus my discussion on the problem of the scale of metal abundance for globular clusters, which is a basic parameter required in the deduction of ages for globular clusters. High dispersion results may be used to address other questions too, like spread in abundance of various elements within globular clusters, or general trends for abundance ratios between different elements. Metal abundance must be determined with an accuracy of 0.15 dex to estimate the age of a cluster like M71 with an accuracy of 2 Gyrs, without use of the controversial absolute magnitude of the horizontal branch. Several issues may preclude the construction of such an accurate scale of abundances for globular clusters (see Pilachowski et al., 1983b; Peterson, 1983; Gustafsson, 1983). An incomplete list includes:
- systematic errors in equivalent widths (EWs) due to scattered light in the spectrographs, continuum tracing, or undetected blends;
- use of erroneous atmospheric parameters (mainly temperatures and microturbulences);
- inadequate consideration of the solar metal abundance;
- deviations from LTE.

I will now consider some of these points.

2.1. Systematic Errors in the Equivalent Widths

Systematic errors in the EWs provided by the old photographic material were discussed comparing EWs derived from spectra analogous to those used for globular cluster stars, with the best available material for well studied stars (Gratton, 1982; Pilachowski et al., 1982). These comparisons showed a rather good agreement among the various sets of EWs. However they did not show conclusively that there were not systematic errors in the photographic EWs for giants in metal rich globular clusters, since spectra of these stars are richer than those of comparison stars. Furthermore, Geisler (1986) suggested that a haze of faint TiO lines significantly lower the continuum in the most metal rich and cool giants of globular clusters. He concluded that a similar effect causes systematic errors of 0.2 dex or less on the inferred metal abundances.

A careful discussion of the problem of continuum tracing in CCD spectra was made by Gratton et al. (1986). These authors applied several criteria in order to lessen problems related to continuum tracing, including observation of stars warmer than 4200 K, use of higher S/N spectra, accurate flat fielding and reduction for the blaze function. A comparison with higher resolution spectra for Arcturus provided evidences for a small systematic overestimate of their EWs (likely due to undetected blends). However, they also derived abundances by a synthetic spectrum analysis, which avoid any continuum tracing, and showed that abundances deduced by this technique agreed with those inferred from line analysis for the stars under scrutiny.

The discrepancy between various estimates of metal abundance for metal rich globular clusters suggests that the photographic spectra yielded erroneously low metal abundances. At present, most of the researchers accept that these errors were due to erroneous continuum tracing, while a minority think it is simply the effect of a large scatter due to poor observational material. However, both these explanations are not substantiated by a careful discussion.

2.2. Atmospheric Parameters and gf's

There is an appreciable systematic difference between the metal abundance scales by Pilachowski and coworkers and by Cohen. Pilachowski et al. (1980) showed that this difference (~0.2 dex) is due to their use of a more severe criterion of line selection, which ruled out a large fraction of the stronger lines. They justified this selection by the possible presence of systematic errors in the adopted solar gf's, due to improper consideration of damping in the Sun. However, a reanalysis of Cohen (1980) EWs for M71, using high quality gf's (Gurtovenko and Kostyk, 1981) shows that a large difference (~0.3 dex) between abundances obtained with and without the application of these selection criteria suggested by Pilachowski et al. still exists. From a comparison with curves of growth, it is clear that the microturbulences used in Cohen analysis (~2 Kms^{-1}) are too large. Observations suggests a value of ~1 Kms$^-$. Repeating the analysis of Cohen's EWs with a similar low value of the microturbulence, a value of the abundance as high as [Fe/H]=-0.87±0.06 is derived for M71, whatever selection criteria is applied. This abundance is in good agreement with the most recent determinations from hotter stars. I conclude that the low metal abundance found for M71 by Cohen mainly depended on the use of microturbulences which were too low, and that the use of adequate values for this parameter not only better fits her data, but also resolves the discrepancy among the different abundance scales for globular clusters.

It is important to note that investigators who analyzed globular cluster stars (and I include myself in this list), generally derived microturbulences by avoiding any trend of the abundances (for Fe I) with EWs. This method works well if lines on the damping portion of the curve of growth are dropped. However, if some strong line is present, an absence of slope may be obtained even with an erroneous microturbulence. I therefore exhort to verify microturbulences derived using this technique by means of curves of growth.

2.3. Solar Model

The most recent papers on metal rich globular cluster stars compared the metal abundances derived for stars in metal rich globular clusters with that deduced for Arcturus by similar material. Arcturus has temperature, surface gravity and metal abundance close to the globular cluster giants considered in those papers. In this case, therefore, the method of differential abundances reduces incertitudes concerning atmospheric parameters, deviations from LTE, etc. It is very welcomed that these studies agree that M71 is about 0.25 dex more metal poor than Arcturus, and that they define a very similar abundance scale. Differences between individual values are well within the error bars, as far as the same temperature for Arcturus is adopted. However, as argued by Leep et al.

(1987), there is an incertitude of about 0.2 dex in the correct abundance scale for globular clusters, which is a consequence of an analogous incertitude in the metal abundance of Arcturus. This incertitude depends on the present knowledge about the scale of temperatures for giants, and on the failure of present theoretical models to describe accurately the atmospheric structure of the Sun and of stars.

REFERENCES

Cohen, J.G.: 1980, Astrophys. J. **241**, 981
Cohen, J.G.: 1983, Astrophys. J. **270**, 654
D'Odorico, S., Gratton, R.G., Ponz, D.: 1985, Astron. Astrophys. **142**, 232
Geisler, D.: 1986, Astrophys. J. **304**, L41
Gratton, R.G.: 1982, Astrophys. J. **257**, 640
Gratton, R.G.: 1987a, Astron. Astrophys. **179**, 181
Gratton, R.G.: 1987b, Astron. Astrophys. **177**, 177
Gratton, R.G., Quarta, M.L., Ortolani, S.: 1986, Astron. Astrophys. **169**, 208
Gratton, R.G., Quarta, M.L., Ortolani, S.: 1987, Astron. Astrophys. Suppl. **68**, 21
Gurtovenko, E.A., Kostyk, R.I.: 1981, Astron. Astrophys. Suppl. **46**, 239
Gustafsson, B.: 1983, Publ. Astron. Soc. Pacific **95**, 101
Leep, E.M., Wallerstein, G., Oke, J.B.: 1986, Astron. J. **91**, 1117
Leep, E.M., Oke, J.B., Wallerstein, G.: 1987, Astron. J. **93**, 338
Peterson, R.C.: 1983, Publ. Astron. Soc. Pacific **95**, 98
Pilachowski, C.A., Sneden, C.: 1988, in Globular Cluster Systems in Galaxies, D. Philip and J. Grindlay eds., Reidel, Dordrecht
Pilachowski, C.A., Wallerstein, G., Leep, E.M.: 1980, Astrophys. J. **236**, 508
Pilachowski, C.A., Sneden, C., Dominy, J.F., Cottrell, P., May, D.C.: 1982, Publ. Astron. Soc. Pacific **94**, 1029
Pilachowski,, C.A., Sneden, C., Wallerstein, G.: 1983a, Astrophys. J. Suppl. **52**, 241
Pilachowski, C.A., Olszewski, E.W., Odell, A.: 1983b, Publ. Astron. Soc. Pacific **95**, 713
Wallerstein, G., Leep, E.M., Oke, J.B.: 1987, Astron. J., in press
Zinn, R.: 1980, Astrophys. J. Suppl. **42**, 19
Zinn, R., West, M.J.: 1984, Astrophys. J. Suppl. **55**, 45

DISCUSSION

MAGAIN What is your estimate of the present uncertainty on the globular clusters abundances and its consequences on the determination of globular cluster ages and the business of constraining the Hubble constant ?

GRATTON I think that the difference in metal abundance between M71 and α Boo is quite well established ([Fe/H]=-0.25 dex). However the metal abundance of α Boo is not well determined. I think that the error bar is approximately ±0.2 dex at least. This error bar implies an uncertainty of at least 2.5 GYRs.

LINE-BLANKETING IN THEORETICAL MODEL ATMOSPHERES FOR F, G, AND
K-TYPE STARS

Roland Buser
Astronomical Institute, University of Basel

Robert L. Kurucz
Harvard-Smithsonian Center For Astrophysics, Cambridge

ABSTRACT

We have computed improved flux distributions from Gustafsson et al.'s extensive grids of theoretical model atmospheres for late-type (F-K) giant and dwarf stars. In order to enhance the realism of the computed fluxes throughout the optical and the ultraviolet wavelength ranges, the massive list of atomic opacity sources used for the hotter Kurucz models was also employed in calculating the cooler spectra. For both the giant and the dwarf models, the resulting synthetic UBVRI photometry provides excellent matches to the spectroscopic calibrations of the observed metallicity and temperature scales based on the ultraviolet excesses, δ (U-B), and the BVRI colors, respectively. We conclude that atomic line-blanketing due to known, rather than "missing", opacity sources accounts for the largest fraction of the systematic UV discrepancy existing between observations and earlier model calculations.

1. INTRODUCTION

Synthetic spectra calculated from theoretical model atmospheres have become the natural representations of our knowledge of stellar physics applied in the evaluation and interpretation of both spectroscopic and photometric observations of the stars. In principle, the (same) synthetic spectra should simultaneously explain both the high-resolution spectroscopic and the low-resolution photometric data of the (same) stars. In practice, however, synthetic spectra have been used mainly in spectroscopic analyses of relatively narrow wavelength ranges suitable for the determination of the physical parameters of individual stars, and these spectroscopic results then also serve to calibrate the observed photometric data of larger stellar samples in terms of temperature and/or metal abundance (for a review, see, e.g., Buser 1985).
    Direct calibration of photometric systems by synthetic

photometry from theoretical models has however been deemed somewhat premature, because most currently popular grids cannot yet reproduce, within the desired ranges of parameter values, the observed colors to sufficient systematic accuracy (for reviews, see, e.g., Gustafsson 1986 and refs. therein; Kurucz 1986). One of the principal reasons lies in the difficulty of adequately calculating, over the wide wavelength range covered by photometric observations, the effects of atomic and molecular line blanketing on the emergent flux spectra necessary for synthetic photometry. For example, Gustafsson and Bell (1979) have shown a large systematic "ultraviolet discrepancy" to exist between observed and computed colors of G-K-type giants, which they concluded to be due to uv line opacity missing in the model flux calculations by Bell and Gustafsson (1978; hereafter BG).

In order to make these models useful for synthetic photometry purposes throughout the ultraviolet and optical wavelength ranges (cf. Buser 1986), we computed new flux spectra employing the much larger list of atomic opacity sources compiled by Kurucz and Peytremann (1975). Our successful re-evaluation of the UBV colors of giant stars (for a preliminary report, cf. Buser and Kurucz 1985) led us to extend the same calculations to Eriksson et al.'s (1979) unpublished grid for late-type dwarfs and to near-infrared colors as well. For a full account of our work, the reader is referred to Buser and Kurucz (1987); here, we will be mainly concerned with the improved metallicity and temperature calibrations of UBVRI photometry provided by our new theoretical fluxes.

2. METALLICITIES FROM $\delta(U-B)$

As expected, the impact on the synthetic fluxes and colors of a more than tenfold increase in opacity sources from atomic lines is a strong function of the metallicity. As the metallicity is increased from [M/H]=-3 to the solar value, the improved U-B colors tend to become increasingly redder, by up to 0.4 mag, than those computed by BG with fewer lines. The amplitudes of these differences also depend on the temperature and the surface gravity, with the highest values for the lowest temperature and highest gravity models, and gradually decreasing to essentially zero amplitudes at slightly (0.1 mag) bluer colors for the hottest low-gravity models. In fact, our new theoretical colors differ from those calculated by BG in very much the same way as the observed colors, whence we conclude that the largest fraction of the "uv discrepancy" is indeed accounted for by the more comprehensive empirical line list, although there may be faint lines, due to atomic transitions which have not been observed in the laboratory and which are still missing from the present calculations (Kurucz 1986).

Since the blanketing effects on the B-V colors are much smaller than for the shorter wavelengths, the theoretical two-

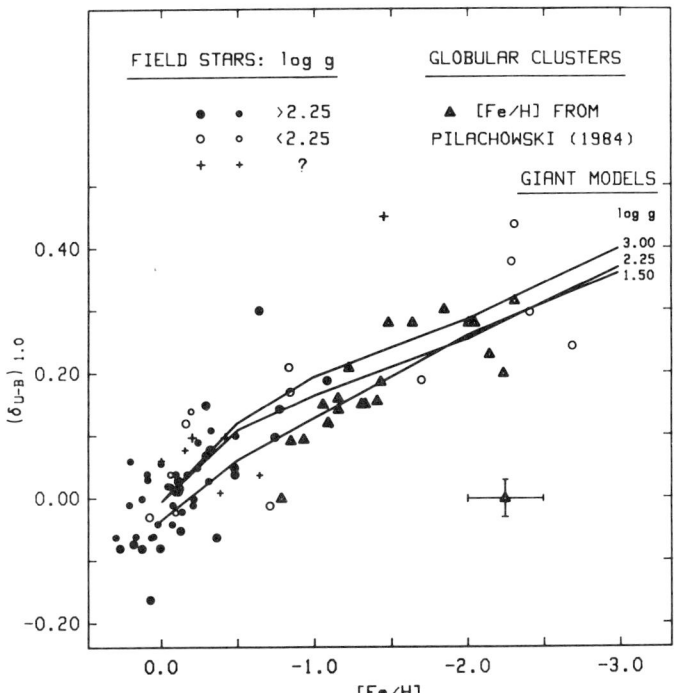

Fig.1. The observed ultraviolet excess for field giants and globular clusters plotted as a function of iron abundance; the theoretical relations from models with different surface gravities are superimposed.

---

color diagrams obtained from the present models now also provide excellent matches to the observed dwarf and giant sequences for normal population I stars. This suggests that the differential as well as the absolute photometric properties of our model grids are accurate enough to permit direct application in calibrating the ultraviolet excesses, $\delta(U-B)$, in terms of metal abundances.

Fig. 1 compares the observed $\delta(U-B)$, read at $(B-V)_o = 1.0$, as a function of [Fe/H], with the theoretical results for giant stars. The spectroscopic abundance data for the individual stars are from Cayrel de Strobel and Bentolila (1983). Although we have taken giants only that have well known (large symbols) or probably negligible (small symbols) E(B-V) values, the scatter is still appreciable and the data for metal-poor field stars are rare. The observed relation is however sufficiently well defined to provide a significant test of the superimposed models. It is especially gratifying to see the intermediately metal-poor globular clusters (triangles) to fit well with the (appropriate) theoretical prediction. This appears to provide independent evidence in favor of Pilachowski's (1984)

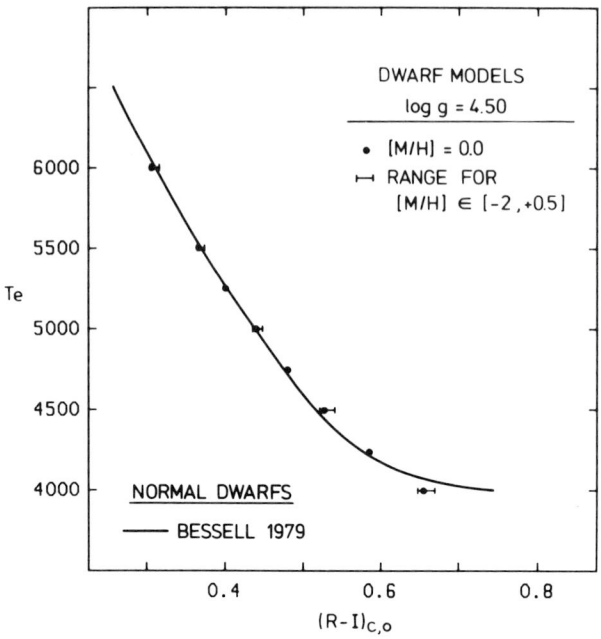

Fig. 2. The empirical temperature calibration of the Cousins R-I color index for normal dwarf stars; the synthetic relation for models with different metallicities is superimposed.

---

recalibrated cluster metallicities, even though the most metal-rich cluster in the present sample (M 71 at [Fe/H]=-0.79) appears to indicate the kind of distortion of the metallicity scale which has plagued the high-resolution spectroscopic abundance determinations due to the difficulties in continuum placement in line-crowded wavelength bands.

An interesting feature of the present theoretical calibration is the usefulness of $\delta(U-B)_{1.0}$ as an abundance parameter continuing down to the lowest metallicities. The early empirical calibration by Wallerstein and Helfer (1966) suggests a gradually flattening relation through the few scattered data with [Fe/H]<-1, whereas the present relation averaged over the different gravities is quite linear for -0.5 > [Fe/H] > -3.0. This uniform increase in the uv excess is due to the fact that for abundances decreasing below [Fe/H]<-1, the slower decrease in U-B goes along with an increase in B-V, such that the deblanketing vectors in the two-color diagram tend to curve away from the metal-normal line and thus make the measured uv excess larger. The reddening of B-V at small abundances is caused by the enhanced importance of Rayleigh scattering in the cooler models.

The success of our synthetic calibration of photometric metallicity parameters is fully shared by the results for the improved dwarf model spectra. Unlike the recent calculations by VandenBerg and Bell (1985), our more completely blanketed models perfectly match both the zero-point and the scale of the $\delta(U-B)_{0.6}$ vs. [Fe/H] calibrations based on spectroscopic abundances of solar-neighborhood dwarfs (Carney 1979, Cameron 1985). Moreover, they demonstrate that the "normalization" of the observed uv excess to its "maximum sensitivity" at $(B-V)o = 0.6$ (Sandage 1969) is obsolete now because, for all gravities, the iso-abundance lines in the two-color diagram tend to be more widely spaced rather than converging for increasing $(B-V)o$. A similar result was already found by Bessell and Wickramasinghe (1979).

3. TEMPERATURES FROM R-I

The improved accounting for atomic opacities in the ultraviolet (where most of the additional line-blocking occurs) should also improve our flux spectra at longer wavelengths, and hence the determination of effective temperatures from suitable visual or near-infrared colors. Indeed, we find our synthetic V-R, V-I, and R-I colors in very good agreement with recent empirical temperature calibrations.

An example is given in Fig. 2, which compares the dwarf models with the observed R-I vs. Teff relation published by Bessell (1979). Note that R-I provides an excellent temperature parameter almost completely independent of metallicity. On the other hand, while V-I may be preferable because of its larger baseline, due to the V band it is also sensitive to metal abundance. This in turn makes it more susceptible to remaining errors in our models, such as the neglect of molecular lines.

ACKNOWLEDGEMENTS

R.B. acknowledges partial support through the Swiss National Science Foundation as well as the National Aeronautics and Space Administration during his tenure as a consultant at the Space Telescope Science Institute.

REFERENCES

Bell, R.A. and Gustafsson, B. 1978, Astron. Astrophys. Suppl. 34, 229 (BG).
Bessell, M.S. 1979, Publ. Astron. Soc. Pacific 91, 589.
Bessell, M.S. and Wickramasinghe, D.T. 1979, Astrophys. J. 227, 232.
Buser, R. 1985, in: La Composition Chimique des Etoiles dans le Voisinage Solaire, eds. A. Florsch, C. Jaschek, M. Jaschek, Comptes Rendus sur les Journees de Strasbourg, 7eme reunion, Obs. de Strasbourg, p. 71.
Buser, R. 1986, in: Highlights of Astronomy, Vol. 7, ed.

J.P. Swings, D. Reidel Publ. Comp., Dordrecht/Boston/
Lancaster/Tokyo, p. 799.
Buser, R. and Kurucz, R.L. 1985, in: Calibration of Fundamental
Stellar Quantities, eds. D.S. Hayes, L.E. Pasinetti,
A.G. Davis Philip, IAU Symp. No. 111, D. Reidel Publ.
Comp., Dordrecht/Boston/Lancaster/Tokyo, p. 513.
Buser, R. and Kurucz, R.L. 1987, Astron. Astrophys. (in press).
Cameron, L.M. 1985, Astron. Astrophys. 146, 59.
Carney, B.W. 1979, Astrophys. J. 233, 211.
Cayrel de Strobel, G. and Bentolila, C. 1983, Astron. Astrophys.
119, 1.
Eriksson, K., Bell, R.A., Gustafsson, B. and Nordlund, A. 1979,
Trans. IAU 17A, Part 2, D. Reidel Publ. Comp.,
Dordrecht, p. 200.
Gustafsson, B. 1986, in: Highlights of Astronomy, Vol. 7, ed.
J.P. Swings, D. Reidel Publ. Comp., Dordrecht/Boston/
Lancaster/Tokyo, p. 805.
Gustafsson, B. and Bell, R.A. 1979, Astron. Astrophys. 74, 313.
Kurucz, R.L. 1986, in: Highlights of Astronomy, Vol. 7, ed.
J.P. Swings, D. Reidel Publ. Comp., Dordrecht/Boston/
Lancaster/Tokyo, p. 827.
Kurucz, R.L. and Peytremann, E. 1975, Smithsonian Astrophys.
Obs. Spec. Rep. No. 362.
Pilachowski, C.A. 1984, Astrophys. J. 281, 614.
Sandage, A. 1969, Astrophys. J. 158, 1115.
VandenBerg, D.A. and Bell, R.A. 1985, Astrophys. J. Suppl. 58,
561.
Wallerstein, G. and Helfer, H.L. 1966, Astron. J. 71, 350.

DISCUSSION

GUSTAFSSON: Naturally, I am very pleased with the improved
agreement between observed ultraviolet colours and calculated
ones, and I congratulate you to these results. Our colour
calculations were based on a much shorter line list (as regards
atomic lines; about 100000 such lines were included), and we
expected that a more complete list would more or less solve
this problem. Also, the choice of the U passband (which was
highly uncertain) is important, however. Would you comment
further on that?

BUSER: In fact, before calculating synthetic colors from
theoretical model atmosphere fluxes, you must first evaluate
the passbands by calculating synthetic colors from observed
spectrophotometric data and comparing to the observed colors
of the same stars. This is what I did for the Johnson UBV
system ten years ago, when I found the U passband to be
different from anything that was offered in the literature
at the time.

ACCURATE ABUNDANCES IN SOME GIANT STARS OF THE GLOBULAR CLUSTER ω CENTAURI

[1]F. Spite, [1]M. Spite, [2]P. François
[1]Observatoire de Paris, section de Meudon
 92195 Meudon Cedex (France)
[2]E.S.O., Karl Schwarzschild Str. 2
 8046 Garching bei München (FRG)

ABSTRACT. The globular cluster ω Cen is known as a very peculiar cluster : some stars show anomalous abundances. Accurate abundance determinations of various elements have been made in six giants of ω Cen, using spectra with high signal to noise ratios. The results suggest some constraints on the origin of the anomalies, but also show the great complexity of the problem, which requires more observations for approaching a solution. A solution would help to understand the evolution of other objects : globular clusters in the Magellanic Clouds, and dwarf galaxies.

1. INTRODUCTION

Some stars of the cluster ω Cen display enhanced abundances of some elements. This peculiarity is found also in very few other clusters, generally in an attenuated form. Several interpretations have been proposed in the literature. An obvious way of trying to trace the processes responsible for the observed peculiarities of ω Cen, is the determination of the abundances of several elements in several stars.

2. OBSERVATIONS AND REDUCTIONS

Six stars of the list of Persson et al. (1980) were observed with the CASPEC spectrograph (D'Odorico et al. 1983) at the 3.6m telescope of ESO.The resolving power of the spectrograph is 20000 and the signal to noise ratio obtained is generally larger than 200. The method used for abundance determination is described in a forthcoming paper in A & A. About 300 lines of different elements were measured for each star; the quality of the data enables to check the accuracy of the temperature determination (the main cause of error). The agreement with the photometric determinations is excellent.
According to the CO index of Persson et al. (1980) two of the 6 observed stars are in the CO-rich class, three are in the CO-normal class; the sixth one has no measured CO index and shows some emission at the H_α line. One of the CO-rich star is also N-rich, following Cohen and Bell (1986).

3. RESULTS

Here are the most important results .
The values of the metallicity [Fe/H] of the stars are found to vary between -1.58 and -1.95. These values suggest a rather small variation of metallicity in ω Cen which does not account for the spread in B-V.
In figures 1 to 3 the pattern of the chemical elements in the stars of ω Cen is compared to the pattern of the elements in a typical halo field star with about the same metallicity ([Fe/H] ≤ -1.5) . This standard pattern has been established from the results of Spite and Spite (1978), Luck and Bond (1985), Barbuy

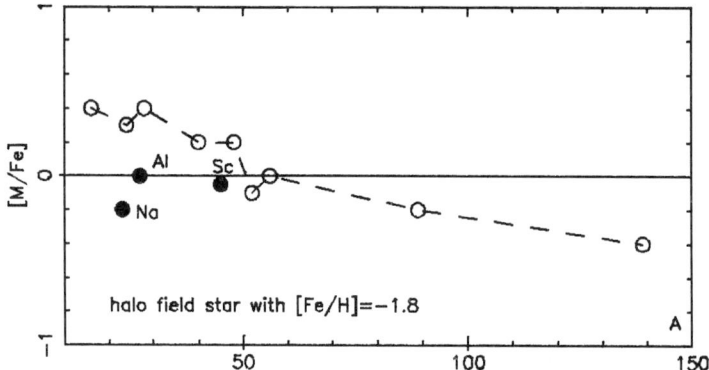

Figure 1 Pattern of the element abundances for a typical field halo star with the same metallicity as the stars of ω Cen. The light odd elements ^{23}Na ^{27}Al ^{45}Sc ^{51}V are represented by filled circles. The abscissa is the atomic mass of the element and the ordinate the abundance of this element in the star relative to the Sun in a logarithmic scale.

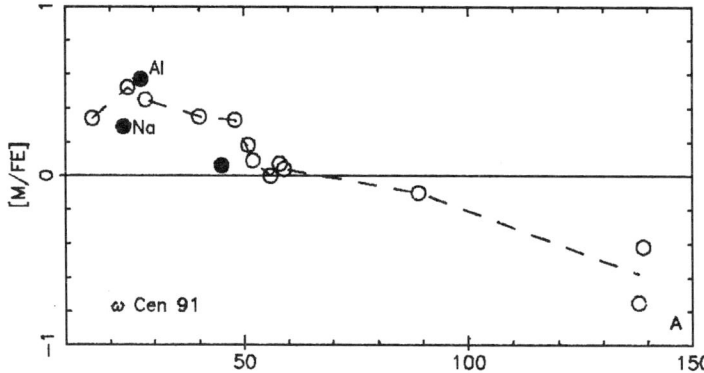

Figure 2 Pattern of the element in a typical star of the CO normal group in ω Cen. The meaning of the symbols is the same as in fig.1.

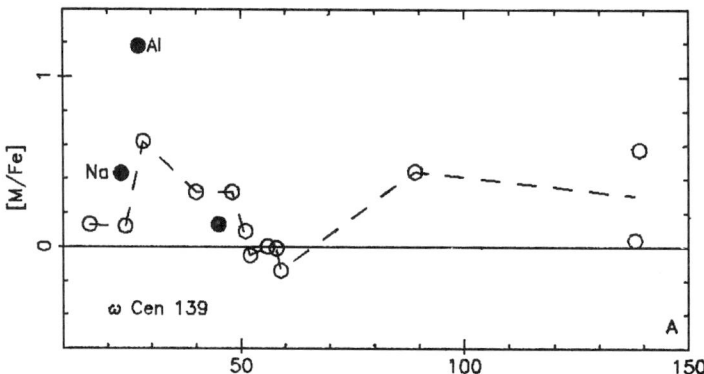

Figure 3 Pattern of the elements in a N-rich CO-rich star of ω Cen.

(1985), François(1986a and b), Gratton and Sneden (1987).
The comparison shows that all the observed stars have enhanced abundances of the "odd"(neutron-rich) metals Na and Al. The CO-rich stars have moreover enhanced abundances of the heavy (s-process ?) elements Ba and La. The star which is both CO-rich and N-rich, has Aluminum *strongly* enhanced. This peculiarity is also found in the only other CO- and N-rich star observed by Cohen (1981).

4. DISCUSSION

According to the present nucleosynthesis theories, the spread in the iron abundance found in the giants of ω Cen has to be attributed to inhomogeneities in the primordial matter which formed the stars. The abundance anomalies found in ω Cen for the elements other than iron could all be explained also by such inhomogeneities. However, this explanation may seem to be *ad hoc,* only shifting the problem to an earlier phase. Moreover, the enhancement of ^{14}N in some giants suggests a mixing to the surface layers of the products of the CNO cycle (such as ^{14}N) built in the deep layers of the star. It has also been proposed (Chieffi and D'Antona 1983) to consider the accretion on the surface of some giants of ω Cen of elements built and ejected by highly evolved intermediate mass stars. Another explanation is the nucleosynthesis of the elements, which appear enhanced, in the "thermal pulses" expected in AGB stars (Iben and Renzini 1984). This last explanation could fit the abundances observed in the CO-rich stars, in which light neutron-rich metals (Na,Al) and heavy (*s* -process?) elements are enhanced. However, this process cannot explain the abundance pattern of the CO-poor stars. Moreover, it is difficult to admit that all the CO-rich stars are on the AGB. This objection is avoided in another hypothesis : the enhanced elements would be formed and mixed at the surface by an excentered core helium flash (which takes place at the tip of the Giant Branch). However, Deupree and Wallace (1987) find that such a process cannot explain the observed enhancement of Na and Al, especially because these enhancements would imply an enhancement of Mg (which is not observed : see fig. 3).
Finally, it could be that the primordial abundance variations were the good hypothesis, which do not exclude other secondary processes. These variations can be produced by a self pollution of the cluster. The self pollution bears of course some resemblance to the well known scenario of globular cluster formation by R.Cayrel (1986).

5. CONCLUSION

The complexity of the stellar abundances in ωCen and the complexity of the processes possibly at work, make that it is necessary to observe stars in each subclass (CO-rich/poor, N-rich/poor, Fe-rich/poor) in order to disentangle the signatures of the processes. Such a study should bring some useful information about slightly different subjects for example the nucleosynthesis in dwarf galaxies such as Leo I (Fox and Pritchet 1987) which also shows a spread of the abundances in the stars of the giant branch.

REFERENCES

Barbuy,B. :1985 in ESO Conferences and workshops Proc. N° 21, I.J.Danziger F.Matteucci and K.Kjär eds., "Production and Distribution of C,N,O elements", p.49
Cayrel, R.: 1986, Astron. Astrophys. **168**, 81
Chieffi, A. , D'Antona, F. : 1983, Astron. Astrophys. **126**, 372
Cohen, J.: 1981, Astrophys. J. **247**, 869
Cohen, J.G., Bell,R.A.: 1986, Astrophys. J. **305**, 698
Deupree R.G., Wallace R. K.:1987, Astrophys. J. **317**,724
D'Odorico S., Enard,D., Lizon,J.L., Ljung, B., Nees, W., Ponz,D., Raffi, G.: 1983, The Messenger, 33,2
Fox, M.F., Pritchet,C.J. : 1987 Astron. J. **93**, 1381

François, P. : 1986 a, Astron. Astrophys. **160**, 264
François, P. : 1986 b, Astron. Astrophys. **165**, 183
François,P., Spite, F., Spite,M. :1987 , Astron. Astrophys. (in press)
Gratton,R.G. , Sneden,C. : 1987, Astron. Astrophys. **178** , 179
Iben I. Jr., Renzini,A. : 1984, Physics Reports (Rev. Sec. of Phys. Letters) **105**, n°6
Luck,R.E., Bond,H.E. : 1985, Astrophys. J. **292**, 559
Persson, S.E., Frogel, F.A., Cohen, J.G., Aaronson, M., Matthews,K. : 1980 Astrophys. J. **235**, 452
Spite, M., Spite, F. : 1978, Astron. Astrophys. **67**, 23

DISCUSSION

CAYREL R. Is it possible to study, in ωCen, stars less bright than giants (may be subgiants) in which the interpretation is not complicated by endogenic production of elements (N, "s" elements) ?

SPITE F. Low resolution spectra were obtained by Hesser et al. (1985) for fainter stars of ωCen (reaching V=17.8) including subgiants near the main sequence. The spread in B-V and in abundances found in the giant branch , seems to be similar in the subgiant branch. This may be an argument in favor of the primordial origin of the anomalies. It would be very interesting to obtain high resolution spectra of low luminosity stars of ω Cen: this would provide (if the primordial origin is confirmed) the detailed yield of Pop.III supernovae.

BESSEL Part of the width of the B-V giant branch is caused by the strong CN 4215 band seen in some of the stars in addition to the metallicity variation in B-V.

SPITE F. I wanted only to stress that the width of the giant branch cannot be attributed only to a global variation of the metallicity, as some older papers seem to suggest. Iron abundance variations can only be primordial. Nitrogen abundance variations could possibly be attributed to the mixing of deep layers with the surface.

TRURAN The trends you have discussed with respect to Na and Al seem in fact to be characteristic of other globular clusters as well (although they are quite different from those for field halo stars). Do you believe that stellar processes in situ can generally explain these trends ?

SPITE F. No . The CO-poor stars (CO-normal) stars are against such an interpretation : Al and Na (which are neutron-rich elements) are enhanced but not the heavy metals Y and Ba ; the CO-rich giants are a more favorable case, since both Al+Na and Y+Ba are enhanced.

THE CHEMICAL COMPOSITION OF GIANTS IN THE GLOBULAR CLUSTER ω CEN

Verne V. Smith
Department of Astronomy
The University of Texas
Austin, TX 78712 USA

ABSTRACT: High-resolution, high S/N spectra have been obtained for 22 giant stars in the globular cluster ω Cen. These stars span a wide range of temperature and position on the giant branch and include some of the S-type and Barium star members. Spectra were obtained on the CTIO 4m telescope with the echelle spectrograph and CCD detector. Initial abundance results for Iron, Calcium, and Oxygen are presented for three stars: one of the most luminous cluster red-giants, a giant on the blue edge of the ω Cen giant branch, and an extreme Barium star member.

I. INTRODUCTION

The globular cluster ω Centauri is unique among globular clusters in showing a large spread in (B-V) color at fixed V along the giant branch. Studies using a variety of techniques show that this spread in color reflects a spread in the metal abundances: (1) low-resolution spectroscopy of RR Lyraes by Freeman and Rodgers (1975), (2) spectrophotometry and IR photometry of red giants by Rodgers *et al*. (1979) and Persson *et al*. (1980), and (3) synthesis of low-resolution spectra by Dickens and Bell (1976). This abundance spread raises obvious questions; are the abundance variations primordial or perhaps quasi-primordial (i.e., created early in the life of the cluster), or are the variations the result of internal nucleosynthesis within the current cluster members? To answer these questions, one must obtain abundances of many elements, such as a sample of light elements (O, Mg, Al, Si), Fe-peak elements and some of the heavier elements: Y, Zr, Ba and Eu, for example. The Dickens and Bell (1976) study used low-resolution spectrum synthesis of some 30 stars and noted large variations in the abundances of Nitrogen and s-process elements.

Thorough element-by-element analyses provide more detailed information, but have been restricted by the difficulties in obtaining high-resolution spectra of the faint cluster stars. True, high-resolution studies of selected stars in ω Cen have been carried out by Mallia and Pagel (1981) using 0.5 Å resolution spectra from the AAT RGO-Cassegrain spectrograph, and Cohen (1981) using 0.3 Å resolution from the CTIO 4m echelle spectrograph with image-tube photographic plates. Both Mallia and Pagel (1981) and Cohen (1981) found variations in [Fe/H] (-1.2 to -1.9 and -1.5 to -1.9, respectively). In addition, Gratton (1982) observed a sample of ω Cen giants with the CTIO 4m and image-tube plus echelle combination and obtained [Fe/H]'s ranging from -1.8 to -1.2. In all studies, variations comparable to that seen in Iron are observed for elements spanning the range from Carbon to Neodymium. Variations in [Fe/H] suggest primordial variations as the low-mass stars currently evolving into giants (turnoff mass ~ 0.8 $M_\odot$) would not be expected to alter internally their Iron-abundances. On the other hand, low-mass stars can produce changes in their C and N abundances as well as s-process elements. The chemical evolution of ω Cen is interesting from the points of view of both stellar evolution and nucleosynthesis. In addition, as the largest globular cluster, ω Cen, may share some traits in common with some dwarf spheroidal galaxies, such as Sculptor and Ursa Minor, which also show the wide (B-V) giant-branches (Norris and Bessell, 1978).

With the advent of high quantum-efficiency CCD detectors, new studies of ω Cen via high-resolution spectroscopy are timely, as evidenced by the similar probing of ω Cen by Francois, Spite, and Spite (1987, this volume) using the ESO 3.6m telescope and CASPEC spectrograph.

2. OBSERVATIONS AND ANALYSIS

Spectra were obtained with the CTIO 4m echelle spectrograph with the air-Schmidt camera and GEC CCD detector. This particular combination yielded a resolution of about 0.34 Å (two pixels) and complete wavelength coverage from about 6200 - 7600 Å. A total of 22 stars in ω Cen were observed over four nights in May of 1986. In Figure 1 we show a color-magnitude diagram of the stars observed, with the star designations as well as V-magnitudes and (B-V) colors from Woolley (1966).

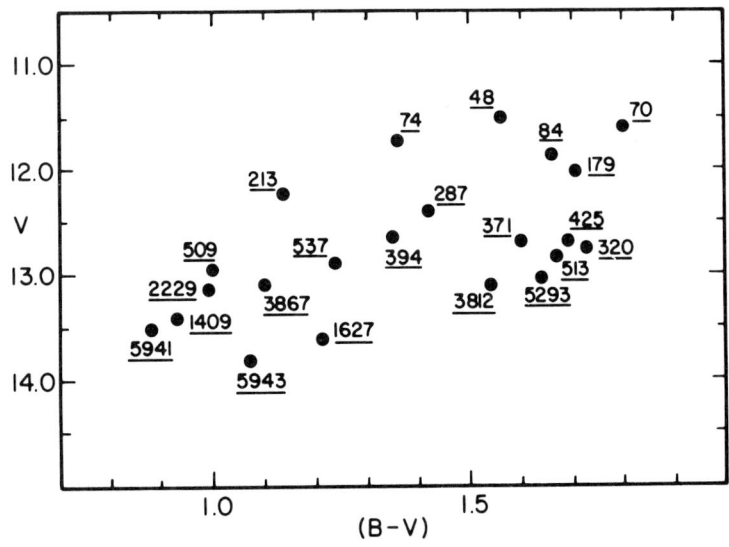

Fig. 1. The color-magnitude diagram of the stars observed in ω Cen.

The individual echelle orders were extracted and reduced using the NOAO reduction package IRAF. Final S/N of the spectra range typically from 50 to 100, depending upon V-magnitude and particular echelle order.

Analysis of the spectra utilize the grid of metal-poor model atmospheres from Bell *et al.* (1976-BEGN) and the bright red giants α Ser and γ Dra as standard stars. All abundance determinations are carried out differentially with respect to the standard stars. Our abundance analysis is just beginning, and we present some preliminary results for a few stars.

3. SOME INITIAL RESULTS

In Table 1 we present results for three ω Cen giants: ROA48, 213, and 371. ROA48 is one of the cluster's luminous red giants ($M_v \approx -2.4$), while ROA213 is a less-luminous giant near the blue edge of the wide giant branch, and ROA371 represents an example of an extreme barium star. Effective temperatures are derived from the IR-photometry of Persson *et al.* (1980), surface gravities are determined from stellar luminosities and an assumed mass of 0.8 $M_\odot$, and the microturbulent velocities are derived from demanding equal abundances from both weak and strong Fe-I lines. As our abundance analyses are just being started and our final line-selection is far from complete, we present results only for Fe, Ca, and O. An extensive analysis will follow.

Table 1

Stellar Parameters and Abundances

Star	T_{eff}(K)	Log g	ξ (km s^{-1})	[Fe/H]	[Ca/H]	[O/H]
ROA48	4050	0.5	3.5	-1.9	-1.6	-1.3
ROA213	4500	1.0	2.0	-2.0	-1.5	-1.5
ROA371	4000	1.0	2.5	-1.6	-1.4	-1.4

The internal dispersions of the derived abundances from the set of Iron and Calcium lines are about ± 0.2 dex. The observed [Fe/H]'s in these three stars are within the range found for ω Cen by previous high-resolution analyses. Note that ROA371 is slightly more metal-rich than ROA48 and ROA213 and displays greatly enhanced s-process spectral lines. No quantitative abundance analysis of the s-process elements has been carried out on our data yet; however, the rather large enhancements in the line-strengths of many s-process lines (such as Zr I, Y II, Ba II, and La II) suggest that ['S'/Fe] in ROA371 is comparable to that seen in extreme Barium stars. A complete analysis will shed some light on primordial or internal abundance variations within ω Cen.

I thank the CTIO staff for their assistance with the observing and Dr. David L. Lambert for valuable collaboration. This research is supported in part by the National

Science Foundation (grant AST 83-16635) and the Robert A. Welch Foundation of Houston, Texas.

REFERENCES

Bell, R. A., Eriksson, K., Gustafsson, B., Nordlund, A. 1976, *Astron. Astrophys. Suppl.*, **23**, 37.
Cohen, J. G. 1981, *Astrophys. J.*, **247**, 869.
Dickens, R. J., and Bell, R. A. 1976, *Astrophys. J.*, **207**, 506.
Freeman, K. C., and Rodgers, A. W. 1975, *Astrophys. J.*, **201**, L71.
Gratton, R. G. 1982, *Astron. Astrophys.*, **115**, 336.
Mallia, E. A., and Pagel, B. E. J. 1981, *Mon. Not. Roy. Astron. Soc.*, **194**, 421.
Norris, J., and Bessell, M. S. 1978, *Astrophys. J.*, **225**, L49.
Persson, S. E., Frogel, J. A., Cohen, J. G., Aaronson, M., and Matthews, K. 1980, *Astrophys. J.*, **235**, 452.
Rodgers, A. W., Freeman, K. C., Harding, P., and Smith, G. H. 1979, *Astrophys. J.*, **232**, 169.
Woolley, R. v.d.R. 1966, *Roy. Obs. Ann.*, No. 2.

DISCUSSION

F. SPITE What is the resolution of your spectra ?

V. SMITH The resolution is about 0.34 Å or a $\lambda/\Delta\lambda \approx 20000$.

NISSEN I note that you have obtained some very accurate spectra down to magnitude $V \approx 14^m$. Is that your limiting magnitude or can you go fainter by using longer exposure times without being disturbed too much by e.g. cosmic ray events on the CCD ?

V. SMITH One can certainly go fainter. Our spectra at magnitude 14 required four exposures of 30 minutes per exposure, to minimize cosmic ray problems, which were then averaged.

ABUNDANCES OF HEAVY ELEMENTS IN THE MAGELLANIC CLOUDS

Stephen C. Russell, Michael S. Bessell and Michael A. Dopita
Mount Stromlo and Siding Spring Observatories
Private Bag PO Woden ACT 2611
Australia

ABSTRACT. This paper provides the first reliable determination of the iron abundance of the Magellanic Clouds. There is clear evidence for carbon and nitrogen depletion, with respect to iron, in the Clouds which can be explained by the relative youth of the Clouds compared to our own Galaxy. Oxygen depletion relative to iron is also in evidence and it is suggested here that this is due to preferential loss of ejecta from high mass stars. Finally, our discovery of an extremely low abundance star in the SMC, coupled with the identification of a similar star by Spite et al. (1987), calls in to question whether or not the SMC is indeed a well mixed system as previously assumed.

1. INTRODUCTION

This paper presents a preliminary report on a part of a complete survey of elements in the Magellanic Clouds. Traditionally, abundances in the Clouds have been derived from spectra of HII regions for the lighter elements, and from spectra of supergiants for the heavier elements. The supergiants selected are young objects with atmospheric abundances little changed from their initial values. Unfortunately previous efforts at determining these heavy element abundances have relied on high dispersion observations of the brightest, most extreme stars (for example Przybylski 1968, 1971, 1972; Wolf 1972, 1973). These studies have resulted in abundance estimates for the same star by different authors that differ by up to 0.7 dex. The problem with these stars is that physical parameters are difficult to determine due to their temperature and a break down in the classically held assumptions of Local Thermodynamic Equilibrium (LTE) and radiative equilibrium.

2. DISCUSSION

In order to avoid the problems noted above, we chose stars with the lowest luminosity and highest gravity possible. The need for a plentiful supply of lines in the spectra while still allowing adequate continuum placement, and the requirement that molecular opacity remains unimportant, further restricted our choice to spectral classes between

F0 and F8. Since the Magellanic Clouds were both expected to be well mixed (see Pagel et al 1978), only a small number of stars were required to be observed in order to obtain the global abundances of the Clouds. This paper gives the preliminary results from 5 stars in the SMC and 6 in the LMC, the latter of which have only iron and titanium abundance measures so far.

Estimates of the effective temperatures and gravities of the program stars were made using 4-color and and Hβ photometry (Russell et al 1987). Spectra were then obtained using the 5m Anglo-Australian Telescope (AAT), and the 2.3m telescope at SSO with dispersions in the range of 3 to 5Åmm^{-1}. The equivalent widths of the lines were measured and the abundances derived with the Kurucz program WIDTH6. Fine analyses were run using models with different surface gravity and effective temperature (interpolated from the grid of Kurucz (1979)), and different values for microturbulence, until abundances were derived independent of line strength, ionization state and excitation potential.

The first interesting result from the analysis concerned the physical properties of the stars themselves. It seemed that there was a systematic difference between the microturbulent velocities measured for the stars we observed in the two Clouds. The range of velocities for the SMC were between 4 and 5 kms^{-1}, while for the LMC they were between 7 and 12 kms^{-1}. Although both ranges were well within those found for supergiants in the Galaxy, their lack of overlap with each other may suggest that metallicity affects a star's microturbulent velocity.

The results of the abundance analysis are summarized in Table 1, where our values represent deficiencies *with respect to* Canopus as measured by us using the same wavelength intervals and the same method of reduction. Our values for [Fe/H] in the SMC are somewhat higher than estimates made by other authors from photometry and low resolution observations of the calcium H and K lines, although still within their error bars, while our detailed abundances are, in general, somewhat lower than those estimated by Foy(1981) and Thévenin and Foy (1986) from medium resolution observations of G0Ia stars. These latter stars are however, of high luminosity and low gravity, and considering the relatively low resolution observations made on them, their results may be somewhat less reliable than our own. In the LMC, on the other hand, our results agree fairly well with those from other authors, especially the most relevant and accurate results of van Genderen et al (1985). These results then, represent the first reliable, direct measure of the iron abundance in the Clouds. From this basis is possible to say much concerning the abundances of other elements relative to iron as compared to the same ratios in the Galaxy.

Figure 1 show the depletion of all the elements measured in the literature (see Dopita 1987 for a summary) for the Clouds, where values from this study (Table 1) have been used whenever possible. The values for the depletion of iron as measured by us, have been marked with dashed lines since, in general, elements in our own galaxy tend to follow the deficiency of iron. The most striking feature is the underabundance of carbon and nitrogen with respect to iron in both the Clouds. The variation of the carbon to iron ratio against the iron abundance in our own Galaxy see for example (Twarog and Wheeler 1982), is close to being constant. This would indicate that both elements are derived from similar mass stars. However, comparison with the results for the Clouds shows that there is a drastic underabundance of carbon relative to iron in the Clouds. This would tend to indicate that in fact iron is produced by somewhat higher mass stars than those producing carbon, and the latter stars have not yet had time to enrich their local Interstellar Medium (ISM) to equilibrium proportions in the Clouds. Peimbert (1987) came to a similar conclusion in an effort to explain the difference between the $^{12}C/^{13}C$ ratio in the Sun and in the Galactic ISM. He suggested that a substantial fraction of the ^{12}C is produced by intermediate mass stars in the $M/M_\odot \leq 2$ range, and therefore the instantaneous recycling approximation is not valid. Similar deficiencies are obtained for nitrogen (see Twarog and Wheeler 1982), so although the bulk of both nitrogen and carbon are produced by similar mass stars, iron must come from the higher mass stars.

Oxygen production in the Galaxy, see for example (Matteucci,1986), is on the other hand, far from being a linear relation with iron. The constant [O/Fe] ratio below an iron abundance of [Fe/H] = -1.0 is interpreted as being due to production of both oxygen and iron (to a limited extent) by high mass star supernovae (Type II). However, the change in slope beyond an iron abundance of -1.0 is due to the bulk of iron being produced by the medium mass stars, with the appropriate time lag. Plotting the values for the SMC and LMC on the same diagram reveals a large underabundance of oxygen with respect to iron as compared to the Galaxy. Any chance of explaining this by invoking a different Initial Mass Function (IMF) for low metallicity environments, is largely ruled out due to the effect that this would have on, for instance, the G-dwarf problem. Indeed Larson (1986) was tempted to suggest that a double peaked IMF with a steeply declining Star Formation Rate (SFR), would provide an explanation for the G-dwarf problem while accounting for the missing dark matter with the resulting large density of white dwarfs. Unfortunately this scenario would result in yet further enrichment in oxygen relative to iron in the Magellanic Clouds.

One possible explanation is that oxygen is preferentially lost, relative to iron, from the Clouds, while being retained in the Galaxy. A possible scenario for this is suggested by observations of supergiant HI loops in the Clouds (see for example Dopita, Mathewson and Ford 1985). These are vast loops of HI gas, on scales of several kiloparsecs, observed to be expanding at velocities of the order of 35 kms^{-1}, and thought to be powered by supernovae explosions from contagious massive star formation. They eject material into the hot haloes of their parent galaxies as efficiently in the case of the Clouds as in the case of the Galaxy (Dopita 1985), but each galaxy can only retain the material according to the strength of its gravitational field. Iron, on the other hand, is produced much later by lower mass stars which are unable to provide the necessary combined energy to blast a hole in the ISM through to the hot galactic halo. Moreover, the readymade holes from the high mass stars dissipate in the order of 10^7 years, long before the stars responsible for iron production end their lives. Thus we conclude that large amounts of oxygen escape preferentially from smaller systems, which are nevertheless able to hang on to the majority of their iron.

TABLE 1

Magellanic Cloud Star Abundances

	SMC			LMC	
ELEMENT	[M/H] (4 F Ib)	No.of Lines	[M/H][1] (F, TF)	[M/H] (6 F Ia-b)	No.of Lines
Mg	− 0.61 ± 0.26	2	−0.2		
Ca	− 0.33 ± 0.24	2	−0.5		
Sc	− 1.04 ± 0.13	1	−0.3		
Ti	− 0.52 ± 0.07	20	−0.35	−0.22 ± 0.14	18 - 101
Cr	− 0.49 ± 0.15	4	−0.25		
Fe	− 0.55 ± 0.10	20	−0.4 ± 0.2	−0.18 ± 0.19	19 - 61
Zr	− 0.43 ± 0.23	1	−0.4		
Ba	− 0.80 ± 0.18	1	−0.4		
Ce	− 0.51 ± 0.22	1	−0.6		

[1] Foy (1981), Thévenin and Foy (1986)

Other estimates of the SMC metallicity: Smith, 1980: −0.8 ± 0.2; Harris, 1981,83: −0.65 ± 0.4; Pel el al., 1981: −0.7 ± 0.25

Other estimates of the LMC metallicity: Smith, 1980: −0.30 ± 0.2 Harris, 1983: −0.09 ± 0.3; van Genderen et al., 1985: −0.18 ± 0.02

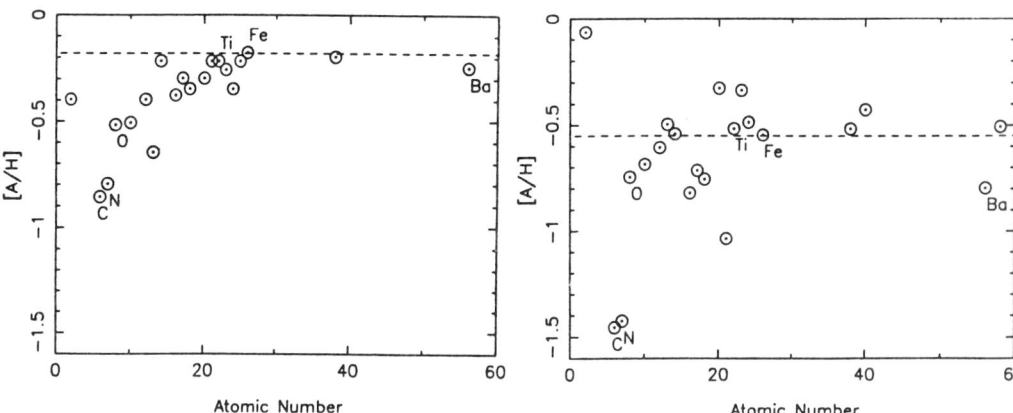

Figure 1. Abundances of the elements in the LMC and SMC relative to Canopus. The dashed lines represents the iron abundance.

The general trend in abundance for the LMC (Figure 1a) is one of rapidly decreasing overdeficiency for the lighter elements until calcium, after which there is a normal deficiency. This is similar to, but not so steep as the enrichment expected from the carbon deflagration supernovae, as put forward by Nomoto et al. (1984, model W7). The results of Thielmann and Arnett (1985) for the expected contributions to the light elements by Type II supernovae suggest that these would tend to flatten out any trend in overdeficiencies produced by other means. The observed slope in overdeficiency therefore, supports the proposed loss to the system of elements ejected from high mass supernovae.

The SMC (Figure 1b), on the other hand, shows very little slope in overdeficiency. This may reflect the youth of the system, whereby there has been too little time for enrichment due to carbon deflagrating stars to have had much effect. Barium is one of the few elements known to be produced largely by the s-process and therefore its measurement is of special interest. Unfortunately, all of our measurements of its abundance depend upon one strong line, and as such they are very sensitive to the exact microturbulent velocity. The results for barium are therefore only tentative at present, and should be treated with caution until we measure more lines. The same problem is true for Scandium which our results show as being of rather low abundance for the SMC.

We must point out that one of the stars we observed in the SMC (Azzopardi and Vigneau 1975: No.79) has a much lower abundance ([Fe/H] $\approx$ -1.0, [Ti/H] $\approx$ -1.7) than the others. The iron abundance is similar to the deficiency ([Fe/H] = -1.2) measured by Spite et al. (1986) for a star in the SMC cluster NGC 330. The latter star is in a very young globular cluster which has been suggested from other observations to be extremely metal deficient, thus to some extent adding weight to the findings of Spite et al. These stars tend to suggest that the SMC at least is not so well mixed as was previously thought.

3. CONCLUSIONS

We have measured the iron abundance in the Magellanic Clouds directly and reliably and found the following results:
    LMC: [Fe/H] = -0.18 $\pm$ 0.19
    SMC: [Fe/H] = -0.55 $\pm$ 0.10
Combining these results with those derived from the literature we are able to say that both carbon and nitrogen are deficient with respect to iron in the Clouds. This is probably due to the youth of the Clouds and the fact that the these elements are produced mainly by somewhat lower mass stars than those responsible for the bulk of the iron production. Oxygen is underabundant with respect to iron in the Clouds relative to our own galaxy. We suggest that this is due to the preferential loss of ejecta from high mass stars in low mass galaxies. Finally, we have identified a very metal deficient star, which together with one reported by Spite et al.(1986), suggest that the SMC is not so well mixed as previously thought.

4. REFERENCES

Azzopardi, M. and Vigneau, J. 1975 *Astr. Astrophys. Suppl*, **22**,285.
Clegg, R.E.S., Lambert, D.L. and Tomkin, J. 1981 *Astrophys.J.*, **250**, 262.
Dopita, M.A. 1985, in *'Birth and Evolution of Massive Stars and Stellar Groups'*, ed. by W.Boland and H. van Woerden, p. 269.
Dopita, M.A. 1987, in *'The Magellanic System'*, ed. by M.A.Dopita, Princeton U.P., in preparation.
Dopita, M.A., Mathewson, D.S. and Ford, V.L.1985 *Astrophys. J.*,**297**, 599.
Foy, R. 1981 *Astr. Astrophys.*, **103**, 135.
Harris, H.C. 1981 *Astr. J.*, **86**, 1192.
Harris, H.C. 1983 *Astr. J.*, **88**, 507.
Kurucz R.L. 1979 *Astrophys. J.*, **40**, 1.
Larson R. 1986 *Mon. Not R. Astr. Soc.*, **218**, 409.
Matteucci, F. 1986 *Publ. Astr. Soc. Pac.*, **98**, 173.
Nissen, P.E., Edvardsson, B. and Gustafsson, B. 1985 *'ESO Workshop on Production and Distribution of C,N,O Elements'*, I.J. Danziger, F. Matteucci and K. Kj (eds.) (ESO Publication), p. 131.
Nomoto, K., Thielemann, F.-K. and Yokoi, K. 1984 *Astrophys.J.*, **286**, 644.
Pagel, B.E.J., Edmunds, M.G., Fosbury, R.A.E. and Webster, B.L. 1978, *Mon. Not. R. Astr. Soc.*, **184**, 569.
Peimbert, M. 1987, in *'Star Forming Regions'*, IAU Symposium No. 115, M. Peimbert and J. Jugaku (eds.) (Dordrecht: Reidel), p. 111.
Pel, J.W., van Genderen, A.M. and Lub, J. 1981 *Astr. Astrophys.*, **99**, L1.
Przybylski, A. 1968 *Mon. Not. R. Astr. Soc.*, **139**, 313.
Przybylski, A. 1971 *Mon. Not. R. Astr. Soc.*, **152**, 197.
Przybylski, A. 1972 *Mon. Not. R. Astr. Soc.*, **159**, 155.
Russell.S.C., Bessell, M.S and Dopita, M.A.1987, *Proc.Astr.Soc.Aust.*,in press.
Smith H.A.1980 *Astr.J.*, **85**, 848.
Sneden, C., Lambert, D.L. and Whitaker, R.W.1979 *Astrophys.J.*, **234**, 964.
Spite, M., Cayrel, R., Francois, P., Richtler, T. and Spite, F.1987, preprint.
Spite, M. and Spite, F. 1978 *Astr. Astrophys.*, **67**, 23.
Thévenin, F. and Foy, R. 1986 *Astr. Astrophys.*, **155**, 145.
Thielemann, F.-K. and Arnett, W.D. 1985, in *'Nucleosynthesis,Challenges and New Developments'*, W.D.Arnett and J.W.Truran (eds.) (University of Chicago Press), p. 151.
Twarog, B.A. and Wheeler, J.C. 1982 *Astrophys. J.*, **261**, 636.
van Genderen, van Driel W.and Greidanus, H.1985 *Astr.Astrophys.*,**155**, 72.
Wolf, B. 1972 *Astr. Astrophys.*, **20**, 275.
Wolf, B. 1973 *Astr. Astrophys.*, **28**, 335.

DISCUSSION

SNEDEN How do the microturbulent velocities that were derived for LMC stars compare with those determined for stars of similar spectral type in our galaxy ?

BESSELL The range of microturbulent velocities in the LMC stars is similar to that in the F supergiants analysed by Osmer. However, the gravity of the galactic supergiants are more difficult to assess for correlation of microturbulence with log g.

HIGH SIGNAL TO NOISE ANALYSIS OF THE CHEMICAL COMPOSITION OF STARS IN THE MAGELLANIC CLOUDS

M. Spite[1], F. Spite[1], and P. François[2]
[1] Observatoire de Paris-Meudon
92195 Meudon Cedex (France)
[2] E.S.O., Karl Schwarzschild str. 2
8046 Garching bei München (F. R. G.)

ABSTRACT. A few high S/N spectra of F-G supergiants in the field of the Magellanic Clouds, and a spectrum of a star in the in the young SMC cluster NGC330 have been obtained at the échelle spectrograph of the 3.6m telescope of ESO (CASPEC). Preliminary results of the analysis of these stars and their consequences in the evolution of the clouds are presented here.

1. INTRODUCTION

In 1985 Stryker et al. from a photometric study of the metallicity of some clusters in the Small Magellanic cloud suggested that, at a fixed age, the metallicity is weaker in the SMC than in the solar vicinity of our Galaxy and that moreover, a strong and sudden enrichment of the matter has taken place in the SMC 2 billions years ago.
The installation of the échelle CASPEC spectrograph at the 3.6m telescope of ESO gave us the opportunity of studying the metallicity in the Magellanic Clouds as a function of age and place, from high Signal to Noise high resolution spectra.

2. THE YOUNG SMC CLUSTER NGC330

In 1983 Richtler and Nelles had derived from photometric measurements (in the Strömgren system), preliminary metallicities for Magellanic globular clusters; they found a very young SMC globular cluster (10^7years) which was very metal deficient : NGC330.
It was important to confirm this large deficiency. A spectrum of a star in NGC330 was obtained and in collaboration with Cayrel and Richtler we analyzed it (Spite et al. 1986).
The spectrum shows indeed a large metal deficiency. The star is rather similar to the galactic globular cluster stars, so that the metal deficiency may be considered as reliable. The result of the analysis is [Fe/H]=-1.4.
The pattern of the abundances of the elements is different from the pattern found in the galactic halo field stars, but it is similar to the pattern found in the CN rich star in ω Cen (fig.1). A larger sample of stars in NGC330 is necessary in order to have an unbiased estimation of the primordial distribution of the elements in this cluster.
This observation extends the range of the "very large metal deficiency" to very small ages. This is an argument in favor of the scenario of globular cluster formation of Cayrel (1986).

3. THE FIELD SMC STARS AND THE ENRICHMENT OF THE SMC

It is also important to check the metal abundance of the field young stars of the SMC in particular because these stars, in older publications, have been found metal-deficient by a factor of about 2, and the HII regions (following Dufour) by a factor of more than 4.

Thus we obtained spectra, with S/N ratios better than 200, for three F supergiants of the SMC. The temperatures of the stars were deduced, for the first iteration, from photometry. In the detailed analysis (using Kurucz models) the temperatures were defined by requiring the deduced abundances to be independent of the excitation potential of the line.

For each star we found a deficiency by a factor of 5 ($-0.8 <$ [Fe/H]<-0.7) in good agreement with the HII regions.

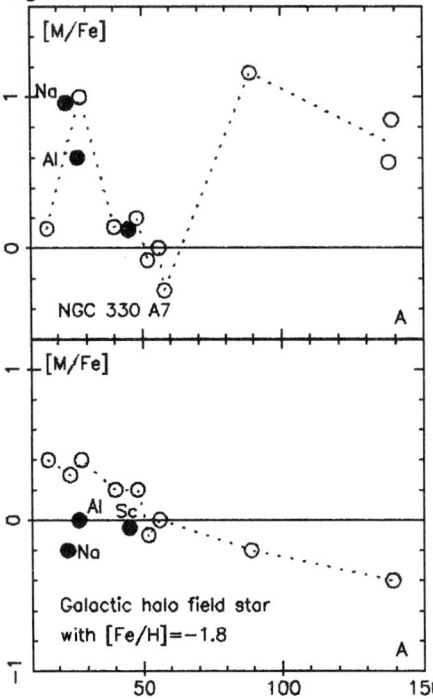

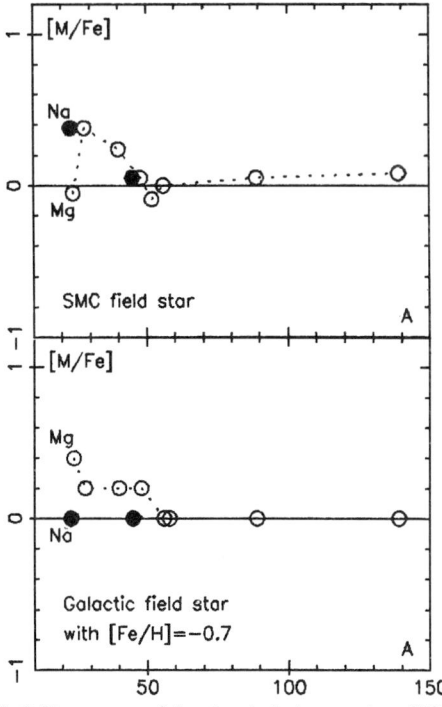

Fig.1 The pattern of the chemical elements in the star A7 of NGC 330 is compared to the pattern of the elements of a typical halo field star with the same mean metallicity.

In abscissa is plotted the atomic number of the element, and in ordinate:

[M/Fe] = log (M/Fe)$_*$ - log (M/Fe)$_\odot$

where M/Fe represents the abundance of the element relative to iron.

The pattern of the elements in the star A7 of NGC 330 looks like the pattern of the elements in the CN rich stars of ω Cen.

Fig.2 The pattern of the chemical elements in a SMC field star (mean of the three studied stars) is compared to the pattern of the elements in a typical galactic star with the same mean metallicity.

The pattern are similar, except for the light elements Na and Mg.

In the fig.1 and 2 the filled circles represent the odd light elements ^{23}Na, ^{27}Al, ^{45}Sc.

The pattern of the chemical elements is similar in the three observed stars, and is also similar to the pattern of the elements in the metal deficient galactic stars (fig.2). The main difference is in the ratio of the odd elements to the even ones (ratio Na/Mg) but this must be confirmed.

The data enable to derive a new image of the chemical enrichment of the SMC (fig.3). It seems that it

exists:
-something like a **Halo Population** with practically no enrichment (solid line) during the last ten billions years
-something like a **Disk Population** with a slight enrichment. The dashed line is from Bica et al.(1986). But from these data it is difficult to decide whether this disk was subjected to a continuous enrichment, or only to a "burst" 2 billions years ago (dotted line).

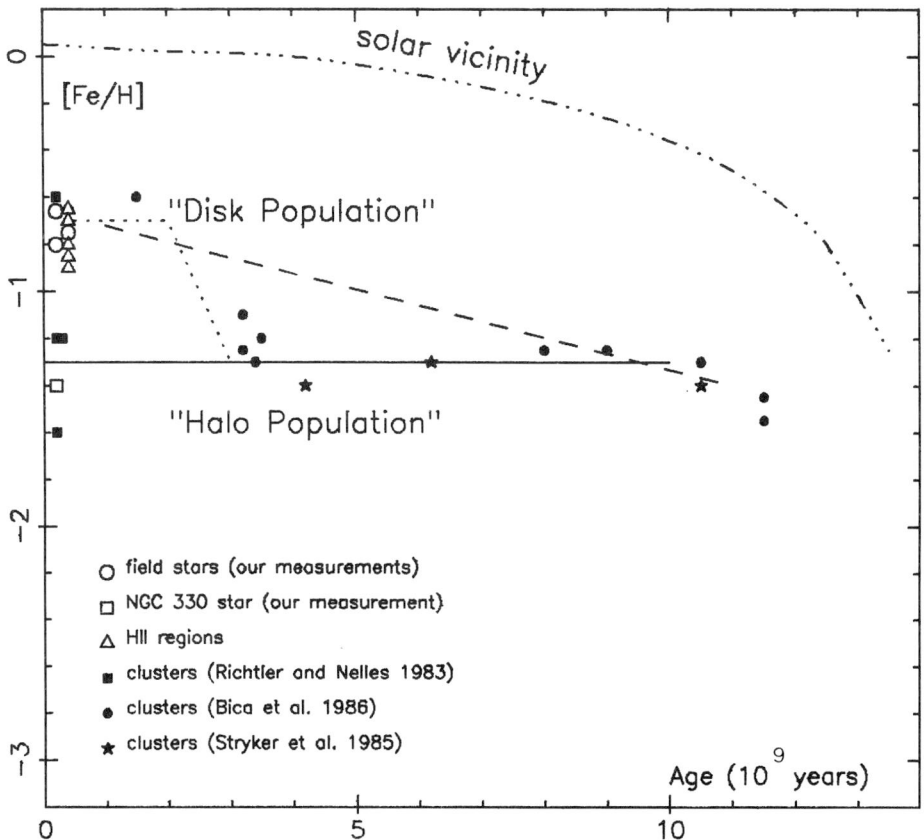

Fig.3 The chemical enrichment in the SMC. The dashed and the dotted lines represent two hypotheses of the evolution of the "disk population" in the SMC.

4. THE FIELD LMC STARS

In the LMC we observed only two field supergiants and only the reduction of G258 is almost completed. Exactly the same method and the same lines have been used in the study of the stars of the LMC and of the SMC. Thus the comparison of the metallicities must be reliable. The deficiency of G258 ([Fe/H]=-0.2) is about by a factor 1.5. It is smaller than the deficiency factor generally observed in the SMC HII regions (2.5 according to Dufour, 1983), but G258 could be located in the bar of the LMC and the metal enrichment could be particularly efficient in this region. Moreover, according to Bica et al. (1986) it seems that the LMC is not homogeneous.
The analysis of the other star could help to solve this problem.

REFERENCES

Bica,E., Dottori,H., Pastoriza,M. : 1986, *Astron.Astrophys.* **156**, 261
Cayrel,R. : 1986, *Astron.Astrophys.* **168**, 81
Dufour, R.J. : 1983, in IAU symposium108 p.353 (S. Van den Bergh and K. de Boer,eds.)
Richtler,T., Nelles,B. : *Astron.Astrophys.* **119**, 75
Spite,M., Cayrel,R., François,P., Richtler,T., Spite,F. : 1986, *Astron.Astrophys.* **168**, 197
Stryker, L.L., Da Costa,G.S., Mould,J.R. : 1986, *Astrophys.J.* **298**, 544

DISCUSSION

BOND What kind of stars did you analyse in the clusters and in the field (temperature gravity) ?

SPITE M. The parameters of the model of the cluster star are similar to the parameters of a galactic red giant cluster star : Teff=3900° log g=0.1. On the contrary the field stars are F supergiants; their temperature is about 6500° and their gravity log g=0.7.

CAYREL R. We have taken the habit of presenting relative abundances with respect of iron. However looking at the diagram showing the abundances in the star A7 of NGC330 we could perhaps more simply interpret it as a lack of iron peak elements (SN I yields) rather than an excess of all other elements.

SPITE M. This could be a good idea to keep in mind . However, if this star is really similar to the CN rich stars of ω Cen then the ratio [Mg/Fe] should be practically normal. Unfortunately we could not measure this ratio.

CAYREL G. Is the absence of the value of the abundance of Mg lacking in your diagram because the lines of Mg, are too weak in the spectrum of the star, or because you did not observed the appropriate spectral region ?

SPITE M. For the star A7 of NGC330 we covered only the spectral range $5900<\lambda<6900$. There are no measurable magnesium lines in this region. We hope to get soon a spectrum in the region 4700-5800 Å where several magnesium lines are measurable.

GRATTON It would be important to know the ratio between the abundance of α rich elements and Fe-group elements in the Magellanic clouds . We think at present that the variation of this ratio in the halo and disk of our Galaxy is due to a different time scale of production. These timescales interplay in a very different way with the timescale of stellar production (which is probably related to the overall metal abundance) in the Magellanic clouds and in our Galaxy. May you please comment on this point ?

SPITE M. For the field SMC stars, as you could see on the figure, there is a slight over abundance of the "even" light metals (Ca, Si) as in our Galaxy for stars with a metallicity of about [Fe/H]=-0.7 . However, magnesium seems to be more deficient and sodium more abundant than in the galactic similar stars, but this result must be considered as preliminary. In the field LMC stars, all the lines are stronger and therefore all these ratios become dependent of the microturbulent velocity. Thus, I cannot now comment on this point.

TRURAN The metallicity spread in the Large Magellanic cloud may be relevant to the interpretation of the blue supergiant progenitors of Supernova 1987A. Would you please comment on the observed spread in [Fe/H] in the LMC ?

SPITE M Up to now we completed the analysis of only one star of the LMC thus I cannot comment on the basis of our own measurements. Bica et al., through narrow band photometry of 41 clusters in the LMC, found an intrinsic metallicity dispersion in the LMC chemical evolution. Could E.Bica comment on this point ?

E. BICA The average spread found for the LMC at fixed age is 0.2-0.3 dex and extreme values amount to 0.5 dex, using narrow band Hβ and G band integrated photometry of stars clusters of intermediate and old ages (Bica et al. 1986). A similar metallicity range is indicated by H II region abundance studies (e.g. Pagel et al. 1978, *M.N.R.A.S.* **184**, 569).

High-dispersion spectroscopy of the Ofpe/WN9 stars R84 and S61 of the LMC

B. Wolf, O. Stahl and W. Seifert
Landessternwarte Königstuhl
6900 Heidelberg
FRG

The Ofpe/WN9 stars R84 and S61 of the LMC have been studied with high-dispersion spectroscopy in the optical and satellite-UV range. The high resolution and high S/N spectra in the optical range are particularly distinguished by strong emission lines of H, HeI and [NII]. The comparison of R84 with S61 shows that the peculiar emission line spectrum is not caused by the previously found late type companion of R84. We find that the UV spectra of both stars closely resemble those of late O-supergiants but all absorption lines are violet-shifted by about 250 km s^{-1} (R84) and about 200 km s^{-1} (S61). The absorption lines are stronger than in normal O-type stars. The UV-resonance lines indicate low terminal wind velocities of $\approx$ 900 km s^{-1} only. Unlike to normal O-type stars the AlIII-resonance lines also show pronounced P Cygni profiles with an even lower edge velocity ($v_{edge} \approx$ 400 km s^{-1}). The mass loss rates (>6.10^{-6} M$_\odot$ yr^{-1}) are comparable to rates found in normal luminous hot stars. However, the wind appears to be much more gradually accelerated similar to the wind of the galactic supergiant P Cygni. It is suggested that the Ofpe/WN9 transition type stars are the hotter counterparts of the early B-type P Cygni stars.

(A more detailed article is forthcoming in Astron. Astrophys.)

The spectral evolution of the LMC S Dor variable R127 during outburst

O. Stahl, B. Wolf
Landessternwarte Königstuhl
6900 Heidelberg
FRG

The LMC star R127 was classified as O Iafpe extr. by Walborn (1977). Later, Walborn (1982) noted the spectral similarity to late WN stars and classified R127 as Ofpe/WN9 star. Stahl et al. (1983) discovered an S Dor outburst of R127. It has now been monitored photometrically almost continuously for more than four years. In addition, several high-dispersion and high signal-to-noise spectra were taken with the ESO CASPEC and CES spectrographs. In 1986 R127 became the visually brightest star of the LMC until it was surpassed by SN1987A in Feb. 1987.

During the first phases of the outburst, the equivalent spectral type of the star was early B type. In the course of the visual brightening the spectral type became gradually later and reached middle A type. In this phase the spectrum was indistinguishable from the maximum spectrum of the prototype star S Dor, when this star was in a phase of similar brightness. Several shell absorption lines appeared in the early phase of the outburst and slowly disappeared later.

A more detailed discussion of the spectral evolution of R127 is in preparation for publication in Astronomy and Astrophysics.

References:
Stahl, O., Wolf, B., Klare, G., Cassatella, A., Krautter, J., Persi, P., Ferrari-Toniolo, M.: 1983, Astron. Astrophys. 127, 49
Walborn, N. R.: 1977, Astrophys. J. 215, 53
Walborn, N. R.: 1982, Astrophys. J. 257, 452

"2-D FRUTTI" SPECTRA OF B SUPERGIANTS IN THE MILKY WAY AND THE LMC

Edward L. Fitzpatrick
Joint Institute for Laboratory Astrophysics
University of Colorado
Boulder, Colorado 80309-0440
U.S.A.

Digital spectra of 7 B-type supergiants in the Milky Way and 15 B-type supergiants in the Large Magellanic Cloud (LMC) were obtained in December 1986 using the "2-D Frutti" detector (2-DF) and the Carnegie Image Tube Spectrograph on the 1-m telescope at the Cerro Tololo Inter-American Observatory. The 2-DF is a photon counting, 2-dimensional Shechtman-type detector, now available on both the 1-m and 4-m telescopes at CTIO. The detector/spectrograph configuration used for the December observing run yielded spectra covering the classical blue region, 3800-5000 Å, with a resolution of approximately 3 Å. The typical observing procedure was to obtain spectra for each star at several locations along the slit. The individual spectra were then averaged (to reduce the detector fixed pattern noise) resulting in S/N ratios of 50-60 in the 4300 Å region.

Spectra for 11 of the LMC supergiants, ranging from types O9.5 to B9 are shown in Figure 1. The spectra were all smoothed with a 4 point boxcar filter and normalized using the IRAF software installed on the 8600 VAX computer at JILA. The most prominent absorption lines present in the blue spectral region are listed in Table I. The spectral types listed next to the spectra were derived by the author in the usual way (from comparison with Galactic standards) from photographic spectra also obtained with the CTIO 1-m spectrograph.

The resolution and S/N ratio of the 2-DF data are clearly adequate for spectral classification work and for quantitative line strength analyses. The classification criteria developed for photographic spectroscopy may be transferred easily to the digital data. Among the more important of these criteria are: a) the changing ionization state of prominant Si lines, from Si IV in the early B-types, to Si III in the middle B's, to Si II in the later B's; b) the presence of numerous lines of O II for types B0.5 and B1; and, c) the reversal of the He I 4471/Mg II 4481 ratio, and the general weakening of the He spectrum, towards the later B-types.

In addition to their use as a classification basis, the 2-DF data also allow a detailed line-strength comparison between the LMC supergiants and their Milky Way counterparts. Figure 2 shows such a comparison for the LMC star Sk130-67 and the Galactic MK classification standard (for type B5 Ia) HD58350. The H and He spectra of the two stars are very well-matched, and their Stromgren Balmer jump indices are nearly identical ($c0 \approx 0.19$) indicating a good temperature-gravity match. Perhaps surprisingly, the metal lines are also extremely well-matched. Although the analysis is still proceeding, a general result from these observations seems to be that, at the level of classification-grade stellar spectra, little or no evidence is seen of the generally accepted factor-of-2 lower metal abundance in the LMC. We plan to enlarge the LMC data sample in the future, to search for line strength variations within the LMC, and also to observe a corresponding sample of B-type supergiants in the Small Magellanic Cloud.

This study was supported by NSF grant AST85-20728 and a generous allotment of observing time at CTIO.

TABLE I (right) - Prominant absorption lines in the classical blue spectral region.

FIGURE 1 (following page) - 2-D Frutti spectra of 11 B-supergiants in the LMC.

FIGURE 2 (below) - Comparison of B5 Ia stars in the Milky Way (HD58350) and the LMC (SK130-67).

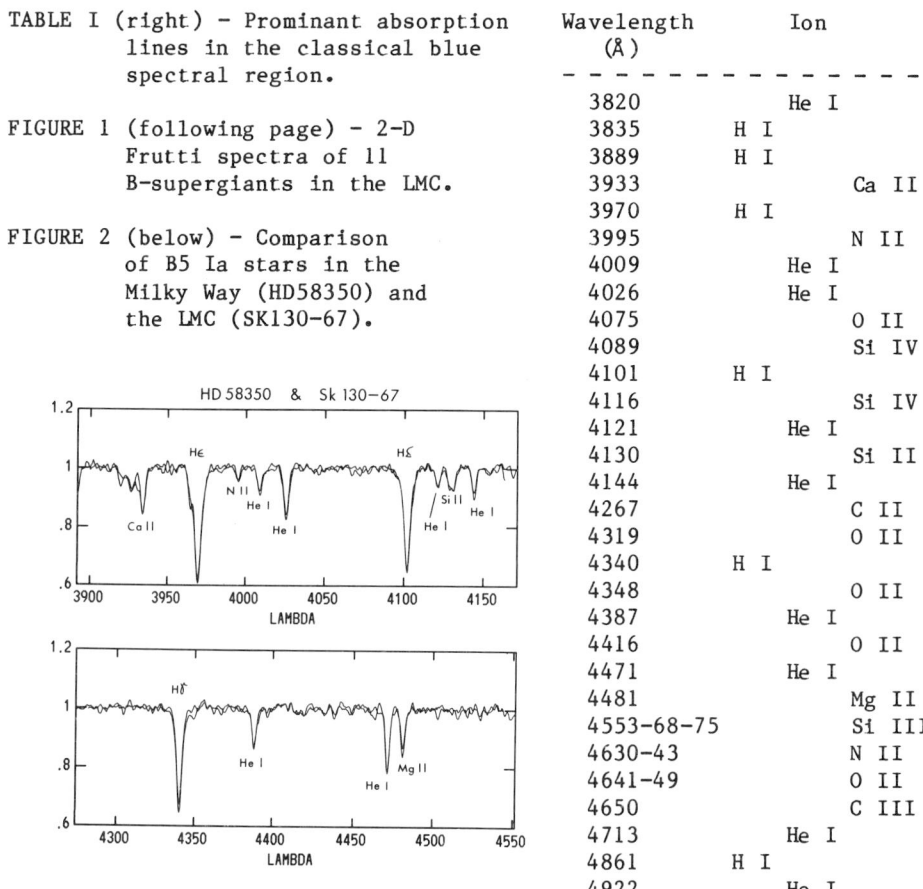

Wavelength (Å)	Ion
3820	He I
3835	H I
3889	H I
3933	Ca II
3970	H I
3995	N II
4009	He I
4026	He I
4075	O II
4089	Si IV
4101	H I
4116	Si IV
4121	He I
4130	Si II
4144	He I
4267	C II
4319	O II
4340	H I
4348	O II
4387	He I
4416	O II
4471	He I
4481	Mg II
4553-68-75	Si III
4630-43	N II
4641-49	O II
4650	C III
4713	He I
4861	H I
4922	He I

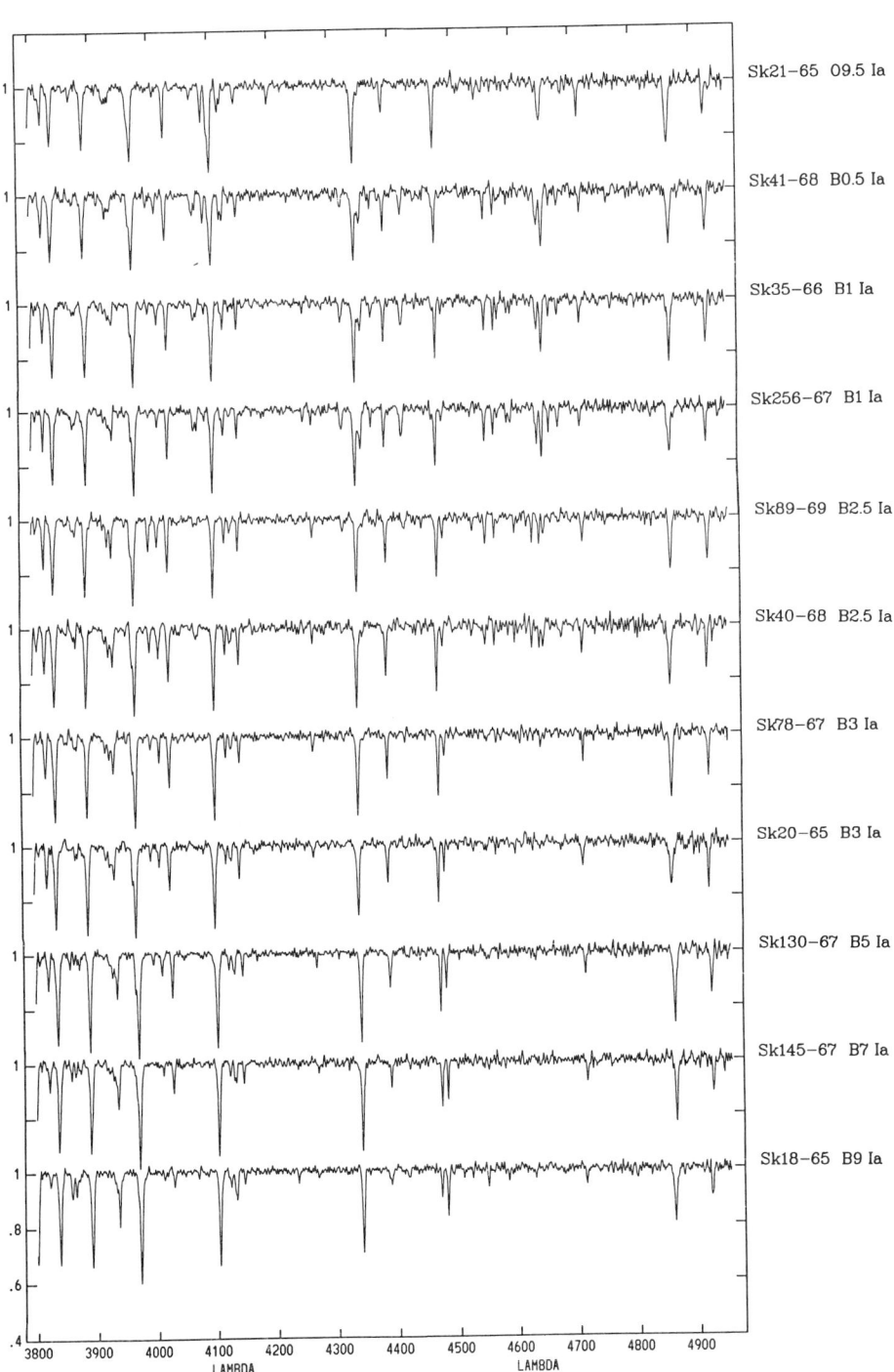

DISCUSSION

EBBETS The stars in your region of the HR diagram are likely to be a mixed bag of various evolutionary stages. Are there features that can be observed with spectra of the quality you have that could distinguish the various stages, in particular to identify those likely to become supernovae ?

FITZPATRICK Several months ago, I would have said that, by excluding the B supergiants which have shown strong photometric variability and/or emission lines, I was restricting my sample to "first-time" supergiants – i.e., stars evolving left-to-right on the HR-diagram and passing thru the B supergiant phase for the first time. If the progenitor of SN1987A turns out to be SK202-69 then these criteria are clearly not sufficient, since SK202-69 would have qualified as a "first-time" supergiant. My new answer to the question is, "I don't know". I think that quantitative abundance measurements are required for a wide range of early type supergiants, and then, if we are lucky enough, when one of them blows up we can go back, look at the spectrum, and claim that we knew all along that the star was peculiar !

CHEMICAL COMPOSITION AS A SIGNATURE OF STELLAR EVOLUTION - THE BARIUM STARS

David L. Lambert
Department of Astronomy
The University of Texas
Austin, TX 78712 USA

ABSTRACT: The hypothesis that Barium stars are the product of mass transfer across a binary system is reviewed with special attention given to the chemical composition of AGB stars (the mass-losing star in the binary) and the Ba stars (the mass-gaining star).

1. INTRODUCTION

Preparation of a spectroscopic attack on an outstanding question of stellar evolution or nucleosynthesis must review the basic questions: signal-to-noise (S/N) ratio? spectral resolution? wavelength coverage? magnitude limit? sample size? As telescopes of larger aperture and spectrometers of greater sensitivity have become available, this observer has enjoyed exploiting the less restrictive compromises. My initial intent was to illustrate how the sharpened observational attacks have led to new insights into and critical tests of theories of stellar evolution. Although some current progress is directly traceable to the acquisition of high S/N high resolution spectra, other key developments have come as the new telescopes and spectrometers have yielded modest S/N ratio and low resolution spectra of faint stars; for example, such spectra of the Magellanic Cloud's asymptotic giant branch (AGB) stars have provided novel and unanticipated results.

A review of how the panoply of stellar evolution and nucleosynthesis can be pierced by quantitative spectroscopic determinations of the stars' surface chemical composition might be organized in several ways:
· stars could be discussed according to their positions in the Hertzsprung-Russell diagram.
· abundance anomalies (i.e., an abundance differing from that expected for the star's main sequence progenitor) could be ordered by element and isotope.
· physical processes (e.g., deep convection, mixing induced by rotation, mass loss) responsible for production of the anomalies could be isolated for comment.
· principal areas of agreement and disagreement between observation and theory could be stressed.

This fourth scheme could be the basis for a fascinating review constructed around a series of observing proposals aimed at resolving the key areas of disagreement. With the editors' insistence that invited reviews not exceed a mere nine pages, the review becomes a brief essay. Through this essay on the classical Barium stars, I hope to illustrate how the 'new' spectroscopy including high S/N spectroscopy can address outstanding issues in stellar evolution.

2. THE BARIUM STARS

2.1. The Mass-Transfer Hypothesis

Barium (Ba) stars were isolated as a class by Bidelman and Keenan (1951): the low dispersion spectra of these G and K giants show enhanced absorption due to Ba II, Sr II, CH, CN, and C_2. Quantitative analyses have shown that the atmosphere of a Ba star is enriched in carbon and the heavy elements synthesized by the neutron capture s-process (see review by Lambert 1985). Enrichments range from slight in the mild Ba stars to substantial (say [s/Fe] ~ 1) in the classical Ba stars. Here, I shall refer to these stars as Ba stars. The CH stars identified by Keenan (1942) are metal-poor giants with the spectroscopic characteristics of Ba stars. Bond (1974) introduced the class of subgiant CH stars which extended detection of C and s-process enhancements to the warmer subgiants, and even the main sequence. I focus on those clues to the origin of the stars that are provided by the chemical composition of Ba and related stars. A reader to whom Ba stars are a novelty is urged to read McClure's (1984) thorough review, and several contributions in *Cool Stars with Excesses of Heavy Elements*, notably Catchpole and Feast (1985), Lambert (1985), McClure (1985), and Wood (1985).

The hypothesis under test was suggested by the radial velocity surveys conducted by McClure and colleagues at the Dominion Astrophysical Observatory. An initial survey of Ba stars and a control sample of K giants showed that the Ba stars were binary systems (McClure, Fletcher, and Nemec 1980): "all stars in the sample that have Ba II star features stronger than Ba1 on the Warner (1965) scale show velocity variations" (McClure 1985). Direct evidence that the mild Ba stars (strengths less than Ba1) belong exclusively to binary systems is not so strong but, as McClure points out, the sample is contaminated by stars that may not be Ba stars "but only stars that were suspected from objective prism plates". McClure further notes that preliminary results suggest that the giant and subgiant CH stars are also binary systems. These surveys in which an accuracy of 0.5 km s^{-1} or better is achieved exemplify, in my view, one impact of high S/N stellar spectroscopy.

For this essay, I shall assume that the origin of the Ba stars is intimately related to the fact that they belong to a binary system. In particular, I shall review the hypothesis of mass transfer. (Four alternative explanations for the occurrence of Ba stars as binary systems are listed and dismissed by McClure (1985)). Consider a wide binary in which mass transfer via Roche lobe overflow or a stellar wind cannot occur until the primary has evolved to the AGB and dredged up C and the s-process elements to its surface. Through the mass transfer from the AGB star, the companion is converted to a Ba star. The transferred mass may be mixed with the outer envelope of the companion. The companion may be a main-sequence star. On evolution to the red giant branch, the convective envelope should thoroughly erase composition gradients and, as in normal stars, dredge up material lightly exposed to the CN-cycle. When the transferred mass is much less than the mass contained within the red giant's convective envelope, the striking abundance anomalies will be diluted and the giant may appear almost normal. Some mass-gaining stars may be red giants at the time of mass transfer. I adopt the plausible assumption that all of the transferred mass is provided by the AGB star's envelope, and hence, its composition is that of the AGB star's atmosphere I suppose that the AGB star is commonly a cool carbon star. Mass transfer in a few cases may occur before the AGB star has completed its conversion from an O-rich to a C-rich star.

To an observer, the mass transfer hypothesis has four attractive features: (i) AGB stars are predicted and observed to be enriched in C and s-process elements, whereas the competing source of these elements - the He-core flash - has not been shown on either theoretical or observational grounds to be capable of mixing products of nucleosynthesis to the stellar surface, (ii) mass transfer between components of a binary will be delayed in wide systems until the more massive star has evolved to the AGB, (iii) the chemical composition of Ba stars is, in principle, predictable given the composition of the AGB stars, and (iv) the core of the mass-losing AGB star should be detectable as a white dwarf companion to the Ba star.

In the following section, I shall explore whether the mass transfer hypothesis with the derived compositions of AGB stars can account for the compositions of the Ba stars. Then, I comment briefly on the question "Do all Ba stars have a white dwarf companion?"

2.2 Chemical Compositions of AGB and Ba Stars

Recent explorations of the chemical compositions of both Ba and the MS, S, and C stars on the AGB permit simple tests of the mass-transfer hypothesis. I shall examine the following aspects of the compositions:

· Are the correlated enhancements of carbon and the s-process elements in the Ba stars consistent with those seen in AGB stars?

· Are the principal characteristics of the s-process products similar for the Ba and the AGB stars? Such characteristics include

· the identity of the neutron source

· the neutron exposure parameter τ_0

· the effective neutron density at the s-process site N(n)

Comments on the mass-transfer hypothesis and the abundances of the trace species Li, ^{13}C, ^{17}O, and ^{18}O may be found elsewhere: Li (Lambert 1985; Smith and Lambert 1986a), ^{13}C (Sneden 1983; Lambert 1985); ^{17}O and ^{18}O (Harris, Lambert, and Smith 1985, 1987).

2.2.1. Carbon and the s-process elements.
For Ba stars there is a correlation between the overabundances of carbon and s-process elements: mild Ba stars are only slightly enriched in these species, but the classical Ba stars show strong enrichments amounting to [s/Fe] ~ 1 and [C/Fe] ~ +0.3. (I use the conventional notation: [X] = log X (star) - log X (standard) where the sun is often adopted as the standard.) In Figure 1a, I show the [s/Fe] - [C/Fe] correlation compiled from analyses based on high S/N spectra and model atmospheres.

On the mass-transfer hypothesis one expects the most severe transfers of mass to create classical Ba stars with a chemical composition that is generally close to that of the AGB mass-losing star. Of course, the composition of the Ba star may be modified as the star evolves from its initial condition (say, on the main sequence) to its present status as a red giant with a deep convective envelope; i.e., the C/O ratio of the Ba giant will be less than that of the main sequence progenitor which may have a C/O ratio somewhat less than that of the AGB star. This expectation for carbon and the s-process elements is confirmed by recent analyses of the O-rich M, MS, and S and the C-rich cool carbon AGB stars (Figure 1).

After years of near-complete neglect, quantitative spectroscopy of the cool AGB stars has begun. A principal reason for current interest is the ability to obtain high resolution high S/N spectra in the infrared atmospheric windows that provide molecular lines, and hence, the abundances of the isotopes of C, N, and O, all of which are sensitive in differing ways to stellar evolution. For the O-rich stars (spectral types M, MS, S), the following molecules were considered in our CNO analysis: CO, OH, NH, and CN (Smith and Lambert 1985, 1986b, SL). For the comparable analysis of the cool carbon stars, the primary molecules were CO, CN, and C_2 (Lambert *et al.* 1986).

To obtain the s-process enhancements of the MS and S stars, we scoured near-infrared windows between molecular bandheads for suitable lines. Abundances of 'light'

(Sr, Zr, Y) and 'heavy' (Ba,Nd) s-process elements were obtained differentially with respect to the K5 giant α Tau (SL); a differential analysis reduces the systematic errors due to non-LTE effects. For the s-process abundances of cool carbon stars, I draw on Utsumi's (1985) curve-of-growth analyses based on lines in two regions (4750-4900 Å, 4400-4500 Å) not dominated by molecular lines. For his sample of 12 stars of which 10 were analyzed for CNO by Lambert et al. (1986), the mean enhancement [s/Fe] is +1.7 for the 'light' s-elements (Zr,Y) and +1.1 for the 'heavy' s-elements (Ba,La,Nd,Sm); I adopt [s/Fe] = +1.3 ±0.4 as characteristic of cool carbon stars. The chosen sample does not include the ^{13}C-rich (J-type) stars for which the s-process abundances are "nearly normal".

Figure 1 clearly shows that the C and s-process overabundances in the extreme classical Ba stars approach the levels reported for the cool carbon stars - see the line in Figure 1a showing how the composition of a GK giant is changed as an increasing fraction

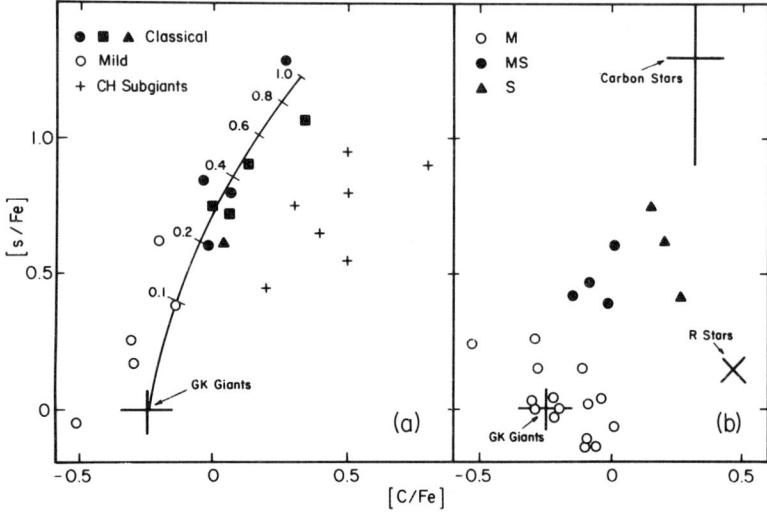

Fig. 1. - The abundance of s-process elements and carbon in (a) Ba stars and CH subgiants and (b) M, MS, S, cool and warm (R type) carbon stars. The location in this plane of the GK giants is shown in both panels. The line in panel (a) shows how the composition of a GK giant is changed as an increasing fraction of material from a cool carbon star is added. Data for mild Ba stars are from Sneden et al. (1981) and for classical Ba stars from: circles - Sneden et al. (1981), Tomkin and Lambert (1979), squares - Smith (1984), and triangle - Kovács (1983). Data for CH subgiants are from Sneden (1983) and Luck and Bond (1982). Points for GK giants and R stars are taken from Lambert and Ries (1981) and Dominy (1984), respectively. Points for M, MS, and S strs are taken from SL. The representative point for the cool carbon stars is based on Utsumi's (1985) s-process abundances (see text) and [C/O] from Lambert et al. (1986), with the assumption [O/Fe] = 0.0.

of C-star material (here, [C/Fe] = + 0.3 and [s/Fe] = + 1.2) is added. The fact that this line runs through the points representing the Ba stars would support a claim that the latter are created through mass-transfer onto a *giant* rather than a *dwarf*. Addition of material to a dwarf followed by reduction of C at the first dredge-up would lead to a Ba giant having at most [C/Fe] ~ 0.0, not +0.3. By their position in Figure 1a, the subgiant CH stars appear to be the progenitors of the Ba stars. The mass-transfer hypothesis also shows why C-rich

Ba stars appear to be rare; since the C/O ratio is close to unity in the majority of carbon stars (Lambert *et al.* 1986), a slight mixing of the transferred mass with the envelope of the companion for which C/O < 0.6 will ensure that C/O ≤ 1 for the Ba star. Nonetheless, some C-rich Ba stars could result; the class of CH-like stars may include examples (Yamashita 1972, 1975). A minority of the MS and S stars may be evolved Ba stars - see Smith and Lambert (1987) for comments on Tc-poor MS and S stars.

The N and O abundances of the Ba stars (Lambert 1985) and the MS-S stars (SL) are identical within the measurement errors to those found for G and K giants. This conclusion holds also for the O abundances of the cool carbon stars (Lambert *et al.* 1986), but not for the N abundances. The unexpectedly lower N abundances in the carbon stars may reflect a systematic error associated with the use of CN as the primary N indicator. It would be an over-reaction to consider the N abundance as grounds for rejecting the mass-transfer hypothesis. The pattern of CNO abundances in the AGB stars supports the idea that, after a first dredge-up as a K giant, the thermal pulses added (almost) pure ^{12}C (and s-process elements) to the envelope.

Although a reader of these recent papers - and especially a theoretician interested in tests of stellar evolution - should note the discussions of the major sources of error, two conclusions seem clear: (i) the enhancements of C and the s-process elements in the Ba and AGB stars are consistent with the mass-transfer hypothesis, and (ii) the mass-gaining star may be a giant at time of the transfer. Since the relative rates of C and s-process production within a star are not an invariant property of any nuclear-reaction network, but reflect the detailed structure of particular zones within the interior, one may plausibly expect mixing resulting from the He-core flash in a low-mass star and mixing from thermal pulses in an AGB star to produce different surface enhancements of C and s-process elements. Moreover, except for a few very peculiar stars (e.g., U Aqr, a RCrB star with an unusual pattern of s-process overabundances), the Ba and AGB stars are the only objects with marked overabundances of C and the s-process; i.e., there is no need to invoke operation of the s-process in multiple sites of radically different properties. The close correspondence between the compositions of Ba and AGB stars surely suggests that abundance anomalies come from sites with similar characteristics. In short, a common site - the He-shell of a AGB star - is strongly suspected, as required by the mass-transfer hypothesis.

2.2.2. The s-process.

Physical conditions at the s-process site may be estimated from the relative abundances of key sensitive elements and isotopes. Sophisticated analyses of the accurate abundance data for the solar system (i.e., meteorites) provide mean estimates of the neutron density $N(n)$, the total density and the temperature T_n together with limited information on their time-dependence during processing. Although full duplication of these analyses will never be possible for stars, acquisition and analysis of high S/N stellar spectra is yielding some interesting results.

The neutron source: Before discussing the physical conditions at the s-process sites, I shall comment on the neutron source. The two most likely neutron-producing reactions are the ^{13}C$(\alpha,n)^{16}$O and ^{22}Ne$(\alpha,n)^{25}$Mg reactions. The latter reaction is expected to operate in the thermal pulses of intermediate-mass stars and the former in the pulses of low-mass stars (see Wood 1985). In addition to composition differences resulting from the different physical conditions in low and intermediate mass stars, the two neutron sources are distinguishable through the Mg isotopic ratios. Operation of the ^{22}Ne source results in production of significant amounts of ^{25}Mg and ^{26}Mg; classical Ba stars are predicted to have ^{25}Mg/^{24}Mg and ^{26}Mg/^{24}Mg $\gtrsim$ 1 (Scalo 1978; Malaney 1987a). Production of the heavier isotopes through operation of the ^{13}C-source may occur, but is less efficient (Arnould and Jorissen 1986). Presence of MgH lines in the spectra of red giants provides the opportunity to determine the Mg isotopic ratios for both Ba and AGB stars.

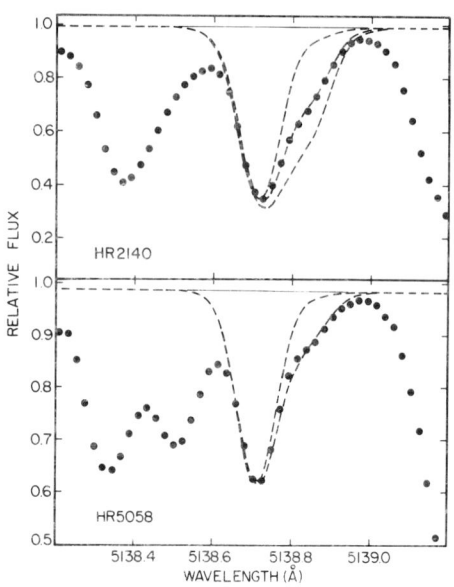

Fig. 2. - Observed and synthetic spectra near 5139 Å showing the coincident ^{24}MgH 0-0 $Q_1(22)$ and 1-1 $Q_1(14)$ lines at 5138.7 Å in HR2140, an old disk giant, and HR5058, a classical Ba giant. Synthetic spectra are shown for ^{24}Mg:^{25}Mg:^{26}Mg = (100:0:0) and (80:10:10 ≡ solar) for both stars and for (90:5:5) for HR2140 only.

None of the stars analyzed to date show evidence for the ^{22}Ne-source. Smith and Lambert (1986b) analyzed two normal M and three similar but s-process enhanced stars; all showed Mg isotopic ratios similar to the solar (the presumed initial) ratios. Earlier, Clegg, Lambert, and Bell (1979) obtained a similar result for the S star HR 1105. Isotopic analyses for two classical and two mild Ba stars also give ratios near the terrestrial values (Tomkin and Lambert 1979; McWilliam and Lambert 1987) - see Figure 2. Recently, a modest overabundance of ^{25}Mg and ^{26}Mg has been found for the cool classical Ba star HD178717 (Malaney and Lambert 1987): ^{24}Mg:^{25}Mg:^{26}Mg = 60:25:15 where 80:10:11 is the terrestrial mix. These enhancements are substantially less than those predicted for the ^{22}Ne-source (e.g., ^{26}Mg/^{24}Mg ~ 3), but may be compatible with the ^{13}C-source (Arnould and Jorissen 1986). Although the samples are still small, we may conclude that the ^{13}C-source rather than the ^{22}Ne-source appears to have been ignited in both the Ba and AGB s-process sites. This conclusion is consistent with the mass-transfer hypothesis. Identification of the source with the ^{13}C(α,n) reaction also serves as a crude effective thermometer for the s-process site, say $10^8 < T_n < 2 \times 10^8$ K.

The neutron density $N(n)$: Sensitivity of the s-process to $N(n)$ occurs when the s-process path encounters an unstable nucleus with a half-life comparable to the mean time between neutron captures. Estimates of $N(n)$ appear to be possible for the coolest O-rich AGB stars from the ZrO lines and the derived Zr isotopic abundances, and for the Ba stars from the Rb abundance. Unfortunately, a direct comparison between the two groups of stars using the same lines seems to be impossible: the ZrO lines are absent or very weak in spectra of Ba stars and the Rb resonance lines appear to be irretrievably blended in the AGB stars. The s-process path enters at ^{90}Zr and passes through ^{91}Zr and ^{92}Zr to the unstable ^{93}Zr. However, ^{93}Zr is so long-lived ($\tau_{1/2} \sim 1.5 \times 10^6$ y) that the path continues through ^{94}Zr to the unstable ^{95}Zr which with a β-decay half-life of 65 days provides the branch that is sensitive to N(n). At low densities, ^{95}Zr decays to ^{95}Nb and so avoids a n-capture. At high densities, ^{95}Zr captures a neutron to produce ^{96}Zr, the heaviest stable Zr isotope. If we define the critical density to be that at which the mean time between n-captures is equal to the β-decay half-life, ^{95}Zr gives N(n) $\sim 10^{10}$ cm^{-3}. Detection of significant amounts of ^{96}Zr among s-processed material would suggest N(n) $\gtrsim 10^{10}$ cm^{-3}.

Rubidium monitors branching in the s-process path due to unstable ^{85}Kr where the critical density is near 10^8 cm^{-3}; this branch is also sensitive to temperature because low-lying states of ^{85}Kr β-decay more quickly than the ground state. At low densities, ^{85}Rb is synthesized. ^{85}Rb is by-passed at high densities and ^{87}Rb is produced. Thanks to a large difference in the n-capture cross-sections ($\sigma(85) = 360$ mb and $\sigma(87) = 11$ mb), the Rb abundance relative to neighboring Sr, Y, and Zr increases by about an order of magnitude between the low and high density limits.

Initial determinations of the Zr isotopic ratios used photographic spectra of the ZrO γ(0-0) band near 6370 Å (Schadee and Davis 1968; Peery and Beebe 1970). Later, Zook (1978, 1985) exploited the larger isotopic wavelength shifts provided by the B-X (0-1) band near 6931 Å. Recently, we have been acquiring spectra near 6930 Å for a number of S stars using the McDonald Observatory's 2.7m telescope and coudé spectrometer equipped with a Reticon detector. Our spectra at a resolution of 0.08 Å are ratioed with the featureless spectrum of a hot star to remove telluric lines. In addition to the B-X k0-1) bandhead, the γ (1-2) R$_1$ bandhead at 6923.4 Å (^{90}ZrO) is present and may be of sufficient strength to provide a second source of the isotopic ratios.

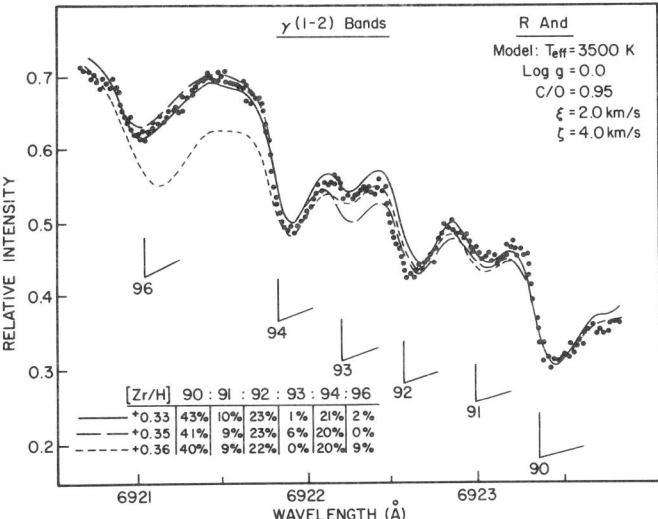

Fig. 3. - The observed spectrum of R And (filled circles) and three synthetic spectra with differing Zr-isotopic mixtures for the γ(1-2) isotopic ZrO bandheads. Locations of the bandheads are shown. The isotopic mixtures and the adopted model atmosphere are characterized by the keys on the figure.

To date, a preliminary analysis is complete for R And and R Gem (Smith 1987). In Figure 3, I show the region near the γ(1-2) head in R And. Synthetic spectra were computed from a line list of 3500 ZrO lines and a handful of atomic lines; the mean density of ZrO lines is 100 per Å! Since ^{96}Zr is the density indicator, I draw attention to the weakness of the ^{96}ZrO bandhead; the feature that appears to be the bandhead is dominated by ^{90}ZrO lines from the γ(5-6) band. This coincidence highlights the need for a complete and accurate linelist; wavelengths *and* oscillator strengths are required. For R And, the B-X (0-1) head provides a second estimate of the isotopic ratios. *Preliminary* results are summarized in Table 1; note the similar ratios for R And (both bands) and R Gem where the γ(1-2) band is too weak to be used reliably. A full error analysis is in progress. Our synthetic spectra are a substantially better fit to the observed spectra than in the comparisons offered by Zook (1985).

Table I. Isotopic Zr-Abundances by Fraction

Star and Bandhead		^{90}Zr	^{91}Zr	^{92}Zr	^{93}Zr	^{94}Zr	^{96}Zr
R And	γ(1-2)	.42	.10	.22	.02	.21	.02
R And	B-X 0-1	.32	.18	.14	.05	.24	.07
R Gem	B-X 0-1	.35	.20	.13	.02	.26	.04

The observed isotopic ratios for R And and the solar system material are compared in Figure 4 with Malaney's (1987b) predicted ratios for the low (N(n) = 10^8 cm^{-3}) and high density (N(n) = 10^{11} cm^{-3}) limits at a neutron exposure parameter $\tau = 0.5$ mb^{-1} (see below). Malaney's predictions for ^{96}Zr were adjusted to account for the recommended lower cross-section ($\sigma = 25 \pm 15$ mb, Bao and Käppeler 1987). After a simple mixing of s-processed and unprocessed material, the observed (o), primordial (p, unprocessed), and s-processed (s) isotopic fractions are related by the equation

$$f_s(i) = [(1 + E_s) f_o(i) - f_p(i)]/E_s$$

where E_s is the observed enhancement of the elemental abundance. For these S stars, we expect $E_s \sim 10$, and hence, $f_s(i) \sim f_o(i)$ for an approximately solar $f_p(i)$.

Inspection of Figure 4 suggests that N(n) $\sim 10^{10}$ cm^{-3} may offer the best fit to the Zr isotopic abundances for R And. This suggestion is based on both the apparent detection of ^{96}Zr and on the overall pattern for the more abundant isotopes. Estimates of the ^{96}Zr abundance are likely to be overestimates because the ^{96}ZrO head may be blended. The solar abundances fit the low density limit; ^{96}Zr is ascribed to the r-process. In both stars, ^{93}ZrO is observed, and this is consistent with the presence of the shorter-lived ^{99}Tc in these stars (Merrill 1952).

Rubidium and neighboring elements have been analyzed to provide N(n) estimates for three Ba stars: HR 774 (Tomkin and Lambert 1983; Käppeler 1986), ζ Cap (Smith and Lambert 1984), and HD 178717 (Malaney and Lambert 1987). The estimates are quite similar N(n) ~ 2 x 10⁷ cm⁻³, a value just above the low density limit for the ^{85}Kr-branch. It will be noted that this estimate is less than that suggested by the *preliminary* analyses of

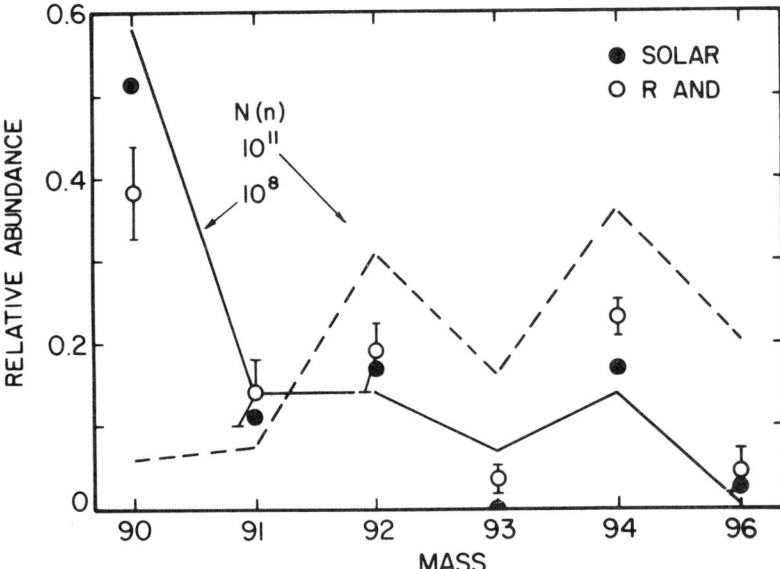

Fig. 4. - Observed and predicted Zr isotopic fractions. These fractions are normalized to unity for the sum over the stable isotopes ^{90}Zr, ^{91}Zr, ^{92}Zr, ^{94}Zr, and ^{96}Zr. Solar-system fractions are the observed values; the s-process is the dominant contributor to all isotopes except to ^{96}Zr which is ascribed to the r-process. The predicted abundances adapted from Malaney (1987b) include ^{93}Zr at the level attained 10⁵ yr after the s-process occurred.

ZrO in the S stars R And and R Gem. Before rejecting the mass-transfer hypothesis on the basis of these apparently disparate N(n) estimates, the error analyses for the S stars must be completed. Since the depth of the bandheads and the degree of overlap increases progressively from ^{96}ZrO to ^{90}ZrO, there is a possibility that systematic errors show a similar trend. In addition, ZrO and Rb should be studied in more stars; R And and R Gem may not be representative of the stars that gave their mass to create a Ba star.

The predicted neutron densities of AGB stars are typically in the range of 10^9 - 10^{12} cm⁻³ (Cosner, Iben, and Truran 1980; Malaney 1986, 1987). This is true even in low-mass AGB stars as a result of the high reaction rate of ^{13}C(α,n)^{16}O relative to ^{22}Ne(α,n)^{25}Mg. It would appear that a high N(n) is obtained for R And and R Gem. If mass transfer from an AGB star creates a Ba star, then the observed Rb/Sr in Ba stars would appear to be incompatible with current low-mass AGB models. In order to remedy this situation, temperatures at the intershell base of low-mass AGB stars are required to be higher than presently predicted (Mathews *et al.* 1986; Malaney and Boothroyd 1987). It is interesting to note that recent stellar evolution calculations indicate such a trend (Boothroyd 1987).

The neutron exposure τ_0: The neutron exposure is

$$\tau = \int N(n) \, v \, dt$$

where $N(n)$ is the neutron number density, v is the mean relative velocity of neutrons and seed nuclei, and the integral is taken over the duration of the event; τ is commonly given in the unit millibarn^{-1}. A distribution function $\exp(-\tau/\tau_0)$, is attractive not only because the exact solution for the s-processed abundances is known, but because mixing in AGB stars may simulate this distribution of exposures rather than a single burst (Ulrich 1973, Clayton and Ward 1974). The parameter τ_0 is obtained by matching predicted and derived abundances for the s-processed material; see Tomkin and Lambert (1983) for a simple recipe relating the observed and derived s-processed abundances.

Results for 7 O-rich (MS,S) AGB stars suggest that τ_0 increases with increasing enrichment of carbon (SL). This trend is confirmed by a qualitative discussion of an additional 30 stars (Smith, Lambert and McWilliam 1987). For the most extreme of the analyzed S stars, $\tau_0 \approx 0.4$-0.5, a value found for the classical Ba stars (Tomkin and Lambert 1983; Smith 1984). At least some of the mild Ba stars contain material exposed to $\tau_0 \sim 0.5$: Tomkin and Lambert (1986) showed that the element-to-element ratios in o Vir and 16 Ser are essentially identical to those in the classical Ba stars. Estimates of τ_0 should now be obtained for a larger sample of mild Ba stars. A detailed analysis of two subgiant-CH stars (Krishnaswamy and Sneden 1985; see also Luck and Bond 1982) showed moderate s-process enhancements ([light-s/Fe] ~ 0.5 to 0.8), but an abundance pattern characterized by a low neutron exposure of $\tau_0 \sim 0.2$. The subgiants are slightly metal-poor ([Fe/H] = -0.5 and -0.3), but some classical and mild Ba stars show similar deficiencies. The difference in τ_0 is one factor suggesting that Ba and subgiant-CH stars may have (slightly?) different origins - see below for a second factor.

The τ_0 estimates and the overall level of s-process overabundances (see above) support the mass-transfer hypothesis. If transfer occurred before the AGB star had undergone many thermal pulses, the transferred material would have a lower τ_0 (< 0.5) and so create a mild Ba star with modest enhancements of the 'light'-s elements (Sr, Y, Zr), but minor enhancements of the 'heavy'-s elements (Ba, La, Nd). This scenario seems unlikely to account for the subgiant CH stars because they are quite C-rich (Figure 1a); the observed (i.e., post-transfer) abundances are in the range [C/Fe] = 0.2 to 0.6 (Sneden 1983; Luck and Bond 1982) and exceed those observed ([C/Fe] ~ 0.0) in MS stars with $\tau_0 \sim 0.2$, and may even exceed the enhancements ([C/Fe] $\sim +0.3$ on the assumption that [O/Fe] ~ 0.0) reported for carbon stars.

Were it not for the apparent difference in τ_0, the locations of the subgiant CH and the Ba stars in Figure 1a would support the idea that the former are the latter's main-sequence progenitors. The first dredge-up experienced by giants decreases [C/Fe] by about 0.2 to 0.3 dex for standard models. A slightly larger cut (say 0.5 dex) would superimpose the subgiant CH stars on the Ba stars in Figure 1a. The first dredge-up would also reduce the high $^{12}C/^{13}C$ ratios of the subgiants (Sneden 1983) to the lower values reported for the Ba stars (Tomkin and Lambert 1979; Harris, Lambert, and Smith 1985). This scenario encounters a problem when Li abundances are examined. Smith and Lambert (1986a) searched unsuccessfully for the Li I 6707 Å doublet in 10 subgiant CH stars. Observed upper limits to the Li abundance in the subgiants with the predicted Li dilution factor for the first dredge-up provide predicted upper limits for classical Ba stars that are less than the observed Li abundances (Pinsonneault, Sneden, and Smith 1984) by as much as a factor of 10. Two explanations may be offered. First, the subgiant CH stars are not progenitors of the classical Ba stars. Since the predicted Li abundances overlap those of the mild Ba stars, it is possible that some of the mild Ba stars are descended from the subgiant CH stars. If, as Böhm-Vitense, Nemec, and Proffitt (1985) claim, all mild Ba stars have a hot white dwarf companion, the evolutionary ties between the two groups must be in doubt because Bond (1984) was unable to detect white dwarfs around the subgiant CH stars. Second, the Li I doublet in Ba stars may be severely blended with an unidentified 's-process' line so that the Li abundances are systematically overestimated.

A comparison of the Li abundances in cool carbon stars with those reported for Ba stars suggests that the latter were probably created as giants and have not evolved from Ba dwarfs. Torres-Peimbert and Wallerstein (1966) give a "Li-index" which Wallerstein and Conti (1969) convert to an abundance: $\log \varepsilon(Li) \sim 1.1$ for a sample of seven cool carbon stars. The "super-Li" carbon stars are excluded because Utsumi's (1985) analysis shows them not to be enriched in s-process elements. This mean Li abundance is close to the upper limit reported for the Ba stars, a coincidence consistent with production of the Ba stars as giants. If, however, the stars were created as dwarfs, the Li abundance in the giant should be at least a factor of 50 lower ($\log \varepsilon(Li) < -0.6$). Unless the Li I feature is severely blended, the observed Li abundance exceeds this prediction by such a margin that we must suppose the Ba giants to have been created as giants.

In light of the differences in τ_o and Li, it seems necessary to consider the subgiant CH and classical Ba stars as unrelated. I suggest that the subgiant CH stars are created by transfer of a small amount of mass and the abundance anomalies are effectively eased by the deep convective envelope as these stars evolve up the red giant branch. Then, two leading questions are 'where are the main-sequence progenitors of the classical Ba stars? or 'Are these Ba stars produced directly as giants?'

3. DO ALL Ba STARS HAVE A WHITE DWARF COMPANION?

After mass-transfer, the core of the former AGB star remains as a white dwarf companion to the Ba star. Radial velocity variations show that the mass of the companion is below the Chandrasekhar limit for white dwarfs: McClure (1983) estimates $m_2 \sim 0.5\ M_\odot$ for an assumed Ba star mass $M_1 \sim 1.5\ M_\odot$, and this estimate for M_2 is likely to be an overestimate (Dominy and Lambert 1983 $\equiv$ DL).

Spectroscopic detection of the white dwarf would provide more direct support for the mass-transfer hypothesis. Low dispersion IUE spectra have led to discovery of white dwarf companions around several mild and classical Ba stars including the luminous Ba star ζ Cap (Böhm-Vitense 1980). DL were unable to detect a hot white dwarf companion to the classical Ba star HR 5058 and the mild Ba star 16 Ser. For distances derived from the Ca II K-line, the ultraviolet flux limits suggest that the white dwarf companion must have cooled for in excess of 10^9 yr. Since its lifetime as a red giant is shorter than the lower limit to the cooling time, the Ba star must have been created as a main sequence star. In view of an apparent lack of main sequence Ba stars, DL suggested that the mass-transfer hypothesis be discarded.

This suggestion deserves a reconsideration. Böhm-Vitense, Nemec, and Proffitt (1984) claim that DL may have underestimated the stellar distances, and hence, overestimated the ages of the white dwarfs. Other authors have suggested that the main sequence Ba stars do exist but "perhaps we just don't recognize them" (McClure 1985). If, as Halbwachs (1985) has estimated, the Ba stars created by mass-transfer comprise 5% of all main-sequence stars of type F6 or later, they ought not be too hard to identify by means of quantitative spectroscopy. The subgiant CH stars may be too rare, too confined in spectral type, and too unlike the classical Ba stars (τ_o, Li) to be the "missing" stars.

There remains the possibility that main sequence Ba stars are not created by the mass-transfer process. I have noted that the C and Li abundances in Ba stars would support a claim that these stars were created as giants. If transfer onto either a main sequence or a giant companion occurs via Roche lobe overflow at a high rate, it appears likely that a common envelope is created around the two stars, and rapid collapse of the orbit (a "spin-down") is followed by ejection of the envelope to create a planetary nebula with the binary evolving to become a cataclysmic variable (Paczyński 1976; Meyer and Meyer-Hofmeister 1979). Discoveries of close binaries as the central objects of planetary nebulae (Bond, Liller, and Mannery 1978) support this scenario.

At lower mass-transfer rates, main-sequence Ba stars should be created. At moderate mass-transfer rates, the main-sequence star develops an extended envelope resembling a giant. This is the expected result for rates $M_{tr} > M_{ms}/\tau_{KH} \approx 3 \times 10^{-8} M_\odot yr^{-1}$ for a 1 $M_\odot$ star where M_{tr} is the mass-transfer rate, M_{ms} is the main-sequence star's mass, and τ_{KH} is its Kelvin-Helmholz time. Such transfer rates might occur as the star accretes mass lost from the AGB star through a wind or an ejection of the envelope. On the cessation of mass transfer, the envelope collapses and the secondary reassumes a main-sequence appearance, albeit now with a peculiar composition. Since the collapse back to the main sequence occurs quickly (on a Kelvin-Helmholtz time scale), this scenario cannot account for the majority of the giant Ba stars (Iben and Tutukov 1985). At low mass-transfer rates, $M_{tr} < M_{MS}/\tau_{KH}$, the main-sequence companion is simply coated with mass from the AGB star. Although the thin convective envelope of the main-sequence star may mix the outer layers and so slightly reduce the abundance anomalies, the atmosphere should have a peculiar composition.

The apparent absence of main sequence Ba stars might reflect the more efficient production of Ba giants. Large amounts of mass (M_{tr}) must be transferred to create the extreme Ba stars with carbon and s-process enhancements close to those estimated for the carbon AGB stars, say $M_{tr} \sim 0.5 M_\odot$. We speculate that the giant with its larger cross section captures substantially more material than a main-sequence companion would. Unless it is of low mass, the latter lacks a deep convective envelope, and so little mass is required to produce marked abundance anomalies; on the other hand, the hot wind off the main-sequence star may inhibit accretion. Perhaps transfer occurs mainly through the "superwind" that is thought to lead to a planetary nebula. Then, it may be no coincidence that the estimate M_{tr} is similar to the mass of planetary nebulae. Since the lifetime of a giant is less than the cooling time of a white dwarf, hot white dwarfs should accompany the Ba stars. Hopefully, ultraviolet observations with the Hubble space Telescope will provide a definitive test of this prediction.

4. CONCLUDING REMARKS

This review of the chemical compositions of Ba and AGB stars provides evidence to support the hypothesis that the former are created by mass-transfer across a binary. The strongest evidence is provided by the correlations between the carbon and s-process abundances (see Figure 1) and by the similar neutron exposures (τ_o) reported for Ba and extreme AGB stars. An apparent difference in the derived neutron densities for the s-process sites of the two groups needs to be resolved. There is observational evidence involving the carbon and lithium abundances, and the presence of white dwarfs to suggest that Ba stars are created primarily as giants and that few have evolved from a Ba dwarf. The subgiant CH stars may be not Ba dwarfs, but stars in which so small amount of mass is transferred that the abundance anomalies are erased by the giant's convective envelope.

My research is supported in part by the National Science Foundation (grant AST 86-14423) and the Robert A. Welch Foundation.

REFERENCES

Arnould, M. and Jorrisen, A. 1986, in *Advances in Nuclear Astrophysics*, ed. E. Vangioni-Flam, J. Audouze, M. Chasse, J. P. Chieze, and J. Tran Thanh Van (Gif sur Avette: Editions Frontières), p. 419.
Bao, Z. Y. and Käppeler, F. 1987, *At. Data and Nucl. Data Tables*, **36**, 411.
Bidelman, W. P. and Keenan, P. C. 1951, *Ap. J.*, **114**, 473.
Böhm-Vitense, E. 1980, *Ap. J. (Letters)*, **239**, L79.
Böhm-Vitense, E., Nemec, J. and Proffitt, C. 1984, *Ap. J.*, **278**, 726.
Bond, H. E. 1974, *Ap. J.*, **194**, 95.
_____. 1984, in *Future of Ultraviolet Astronomy Based on Six Years of IUE Research*,
Bond, H. E., Liller, W. and Mannery, E. J. 1978, *Ap. J.*, **223**, 252.
Boothroyd, A. I. 1987, Ph.D. Thesis, California Institute of Technology.
Catchpole, R. M. and Feast, M. W. 1985, in *Cool Stars with Excesses of Heavy Elements*, ed. M. Jaschek and P. C. Keenan (Dordrecht: Reidel), p. 113.
Clayton, D. D. and Ward, R. A. 1974, *Ap. J.*, **193**, 397.
Clegg, R. E. S., Lambert, D. L. and Bell, R. A. 1979, *Ap. J.*, **234**, 188.
Cosner, K., Iben, I., Jr., and Truran, J. W. 1980, *Ap. J. (Letters)*, **238**, L91.
Dominy, J. F. 1984, *Ap. J. Suppl.*, **55**, 27.
Dominy, J. F. and Lambert, D. L. 1983, *Ap. J.*, **270**, 180.
Halbwachs, J. L. 1985, in *Cool Stars with Excesses of Heavy Elements*, ed. M. Jaschek and P. C. Keenan (Dordrecht: Reidel), p. 337.
Harris, M. J., Lambert, D. L. and Smith, V. V. 1985, *Ap. J.*, **292**, 620.
_____. 1987, *Ap. J.*, in press.
Iben, I., Jr., and Tutukov, A. V. 1985, *Ap. J. Suppl.*, **58**, 661.'
Käppeler, F. 1986, in *Nucleosynthesis and Its Implications on Nuclear and Particle Physics*, ed. J. Audouze and N. Mathieu (Dordrecht: Reidel), p. 253.
Keenan, P. C. 1942, *Ap. J.*, **96**, 101.
Kovács, N. 1983, *Astr. Ap.*, **124**, 63.
Krishnaswamy, K. and Sneden, C. 1985, *Pub. A. S. P.*, **97**, 407.
Lambert, D. L. 1985, in *Cool Stars with Excesses of Heavy Elements*, ed. M. Jaschek and P. C. Keenan (Dordrecht: Reidel), p. 191
Lambert, D. L., Gustafsson, B., Eriksson, K. and Hinkle, K. H. 1986, *Ap. J. Suppl.*, **62**, 373.
Lambert, D. L. and Ries, L. M. 1981, *Ap. J.*, **248**, 228.
Luck, R. G. and Bond, H. E. 1982, *Ap. J.*, **259**, 792.
McClure, R. D. 1983, *Ap. J.*, **268**, 264.
_____. 1984, *Pub. A. S. P.*, **96**, 117.
_____. 1985, in *Cool Stars with Excesses of Heavy Elements*, ed. M. Jaschek and P. C. Keenan (Dordrecht: Reidel), p. 315.
McClure, R. D., Fletcher, J. M., and Nemec, J. M. 1980, *Ap. J. (Letters)*, **238**, L35.
McWilliam, A. and Lambert, D. L. 1987, *M.N.R.A.S.*, in press.

Malaney, R. A. 1986, *M.N.R.A.S.*, **223**, 683.
———. 1987a, *Ap. J.*, **321**, 832.
———. 1987b, in Proc. 2nd IAP Rencontre on *Nucl. Astrophys*, in press.
Malaney, R. A. and Boothroyd, A. I. 1987, *Ap. J.*, **320**, 866.
Malaney, R. A. and Lambert, D. L. 1987, in Proc. ACS Symposium on *The Origin and Distribution of the Elements*, in press.
Mathews, G. J., Ward, R. A., Takahashi, K., and Howard, W. M. 1986, in *Nucleosynthesis and Its Implications on Nuclear and Particle Physics*, ed. J. Audouze and N. Mathieu (Dordrecht: Reidel), p. 277.
Merrill, P. W. 1952, *Ap. J.*, **116**, 21.
Meyer, F. and Meyer-Hofmeister, E. 1979, *Astr. Ap.*, **78**, 167.
Paczyński, B. 1976, in *IAU Symposium 73, Structure and Evolution of Close Binary Systems*, ed. P. Eggleton, S. Mitton and J. Whelan (Dordrecht: Reidel), p. 75.
Peery, B. F. and Beebe, R. F. 1970, *Ap. J.*, **160**, 619.
Pinsonneault, M. H., Sneden, C., and Smith, V. V. 1984, *Pub. A. S. P.*, **96**, 239.
Scalo, J. M. 1978, *Ap. J.*, **221**, 627.
Schadee, A. and Davis, D. N. 1968, *Ap. J.*, **152**, 169.
Smith, V. V. 1984, *Astr. Ap.*, **132**, 326.
———. 1987, in Proc. ACS Symposium on *The Origin and Distribution of the Elements*, in press.
Smith, V. V. and Lambert, D. L. 1984, *Pub. A. S. P.*, **96**, 226.
———. 1985, *Ap. J.*, **294**, 326.
———. 1986a, *Ap. J.*, **303**, 226.
———. 1986b, *Ap. J.*, **311**, 843.
———. 1987, *A. J.*, in press.
Smith, V. V., Lambert, D. L., and McWilliam, A. 1987, *Ap. J.*, in press.
Sneden, C. 1983, *Pub. A. S. P.*, **95**, 745.
Sneden, C., Lambert, D. L., and Pilachowski, C. A. 1981, *Ap. J.*, **247**, 1052.
Tomkin, J. and Lambert, D. L. 1979, *Ap. J.*, **227**, 209.
———. 1983, *Ap. J.*, **273**, 722.
———. 1986, *Ap. J.*, **311**, 819.
Torres-Peimbert, S. and Wallerstein, G. 1966, *Ap. J.*, **146**, 724.
Ulrich, R. K. 1973, in *Explosive Nucleosynthesis*, eds. D. N. Schramm and W. D. Arnett (Austin: Univ. of Texas Press), p. 139.
Utsumi, K. 1985, in *Cool Stars with Excesses of Heavy Elements*, ed. M. Jaschek and P. C. Keenan (Dordrecht: Reidel), p. 243.
Wallerstein, G. and Conti, P. S. 1969, *Ann. Rev. Astr. Ap.*, **7**, 99.
Warner, B. 1965, *M.N.R.A.S.*, **129**, 263.
Wood, P. R. 1985, in *Cool Stars with Excesses of Heavy Elements*, ed. M. Jaschek and P. C. Keenan (Dordrecht: Reidel), p. 357.
Yamashita, Y. 1972, *Ann. Tokyo. Astr. Obs.*, **13**, 169.
———. 1975, *Pub. A. S. J.*, **27**, 325.
Zook, A. C. 1978, *Ap. J. (Letters)*, **221**, 413.
———. 1985, *Ap. J.*, **289**, 356.

ABUNDANCE CLUES TO EARLY GALACTIC CHEMICAL EVOLUTION

James W. Truran
Max-Planck-Institut für Astrophysik
8046 Garching b. München, FRG
Dept. of Astronomy, University of Illinois
Urbana, Illinois 61801, USA

ABSTRACT. High S/N spectroscopic studies of the abundance patterns characterizing extremely metal-deficient halo field stars and globular cluster stars have served to provide significant clues to and increasingly stringent boundary conditions upon the chemical evolution of the halo population of our galaxy. Guided by our current knowledge of nucleosynthesis as a function of stellar mass occurring in stars and supernovae, we identify some interesting constraints that these combined observational and theoretical considerations impose upon theories of the early history of our galaxy.

1. INTRODUCTION

Abundance determinations from analyses of stellar spectra have, historically, played a critical and defining role in the development of nucleosynthesis theory. The early recognition of the fact that there exist halo stars in our galaxy with metal contents less than one hundredth that of solar system matter provided support for the view that heavy element synthesis has occurred predominantly in galactic sources. The detection of technetium (an element having no stable isotopes) in the atmospheres of red giant stars (Merrill 1952) further confirmed that nucleosynthesis mechanisms are indeed operating in stars, triggering an epoch of theoretical activity in stellar and supernova nucleosynthesis that has persisted over more than a quarter century. Recent observational studies utilizing high S/N spectroscopy are continuing in this tradition. Some observational and theoretical considerations relevant to the interpretation of the anomalous abundance patterns characterizing certain classes of peculiar red giants, carbon stars, barium stars and R Corona Borealis stars have been reviewed in the paper by D. Lambert in these proceedings. In this paper, we will concentrate on the anomalous abundance patterns observed in metal deficient field halo stars and globular cluster stars and examine their implications for models of stellar and supernova nucleosynthesis and galactic chemical evolution.

2. FIELD HALO STARS AND EARLY GALACTIC EVOLUTION

One area in which high S/N spectroscopic studies have yielded significant returns is that involving the abundances observed in extreme halo population stars and their implications for the early chemical and dynamical evolution of the Galaxy. We will briefly review both the predictions of nucleosynthesis theory and the existing observational data regarding extremely metal deficient stars and show how these serve to constrain modeling of stellar activity during the halo collapse phase (see also Truran 1983; Truran and Thielemann 1986; Truran 1987).

2.1 Nucleosynthesis Expectations

Nucleosynthesis predictions are now available for both the stable phases of stellar evolution and the matter ejected in supernova events (Truran 1984; Woosley 1987). For the purposes of this discussion, we confine our attention to the following interesting elements or classes of nucleosynthesis products: carbon, nitrogen, oxygen, the elements from neon to calcium, the iron-group nuclei, the s-process heavy elements, and the r-process heavy elements. We also identify the interesting ranges of stellar mass with which specific nucleosynthesis products are found to be associated: (1) the mass range $M \geq 10\ M_\odot$ of the massive star progenitors of Type II supernovae; (2) the mass range $1 \leq M \leq 10\ M_\odot$ of intermediate mass stars for which significant nucleosynthesis occurs during the asymptotic giant branch phase of evolution; and (3) Type I supernovae, which are believed to result from the evolution of intermediate mass stars in binary systems. We note a crucial distinction in the production timescales for these nucleosynthesis sources as defined by their corresponding stellar lifetimes: intermediate mass stars evolve on timescales $\tau \geq 10^8 - 10^9$ years, while massive stars $M \geq 10\ M_\odot$ evolve on timescales $\tau \leq 10^8$ years compatible with a halo collapse timescale.

Massive stars and associated Type II supernovae are known generally to synthesize nuclei from carbon to nickel (Woosley 1987). Detailed model predictions indicate, however, that carbon and the iron group elements are underproduced in such events relative to oxygen and the neon-to-calcium elements. This characteristic signature of nucleosynthesis in massive stars, as we shall see, is presumably reflected in the abundance patterns observed in metal deficient stars: $[O/Fe] \approx + 0.5$ and $[Mg,Si,Ca/Fe] \approx + 0.5$. We also assume that these massive stars provide the site of formation of the r-process heavy nuclei, this being associated perhaps with the ejection of highly neutronized matter from the vicinity of the mass cut in supernovae which leave neutron star remnants (Truran 1984; Hillebrandt 1978; Truran, Cowan and Cameron 1985).

Intermediate mass stars provide important contributions to heavy element abundances in the galaxy particularly as a consequence of the occurrence of thermal pulses in their helium burning shells on the asymptotic giant branch. Estimates of nucleosynthesis yields from asymptotic giant branch stars (Iben and Truran 1978; Renzini and Voli 1981) specifically suggest that significant production of carbon, nitrogen, and the s-process elements, in solar proportions, can be achieved in this environment. The longer timescales of evolution of intermediate mass stars suggest that their nucleosynthesis yields will begin to influence the interstellar gas abundances only relatively late in the halo collapse phase.

Type I supernovae are believed to be the major contributors to the abundances of the iron group nuclei in our galaxy. Calculations of explosive nucleosynthesis associated with carbon deflagration models of Type I supernovae (Thielemann, Nomoto, and Yokoi 1986) arising from binary evolution predict that sufficient iron-peak nuclei are formed to explain both the powering of the light curves of Type I supernovae by the decay of ^{56}Ni and ^{56}Co and the observed mass fraction of iron in galactic matter. The production timescale is expected to be compatible with the lifetimes of the intermediate mass stars which characterize these binary systems.

2.2 Composition Trends in Field Halo Stars

Existing observations of the abundances in the most metal deficient field halo stars have recently been reviewed by Spite and Spite (1985). Interesting trends from the point of view of nucleosynthesis may be briefly summarized as follows. For CNO nuclei, both [C/Fe] and [N/Fe] are found (Laird 1985) to be approximately constant and compatible with solar for a sample of disk and halo stars which span a range in [Fe/H] from -2.45 to +0.5. In contrast, high oxygen to iron ratios [O/Fe] $\approx$ + 0.5 are found to characterize the halo stars. Similar trends are evident for the intermediate mass elements Mg, Si, and Ca, which are found to be enriched relative to Fe by approximately 0.5 dex (Luck and Bond 1985; Gratton and Sneden 1987).

Further interesting trends have been identified in the heavy element region. The data clearly establishes the existence of deple-tions in the abundances of the designated s-process elements Sr and Ba, relative to iron, in stars of low Fe/H (Spite and Spite 1978; Luck and Bond 1985; Gratton and Sneden 1987). In point of fact, both theory (Truran 1981) and observation (Sneden and Pilachowski 1985) now seem strongly to suggest that these heavy element abundance patterns charac-teristic of extreme metal deficient stars are dominated by r-process contributions. On the theoretical side, it is recognized that it is a natural consequence of r-process synthesis that the resulting ratios Ba/Sr, Sr/Er and Ba/Er should deviate from those of solar system matter, in a manner compatible with those observed for metal deficient stars. The

heavy element abundance pattern for the extremely metal deficient
star HD 110184, determined from the high S/N spectroscopic analysis
of Sneden and Pilachowski (1985), indeed strongly suggests that the
r-process contribution dominates in this object. This is again
compatible with our assumption that massive stars represent a site
of r-process nucleosynthesis.

We note finally that the data also give evidence for the
presence of a mild odd-even effect involving the products of
explosive nucleosynthesis: specifically, elements containing odd
numbers of protons (e.g. Na, Al, and Sc) seem perhaps to show
somewhat greater relative deficiencies than neighboring even-Z
nuclei in extremely metal deficient stars (Luck and Bond 1985;
François 1986a,b; Gratton and Sneden 1987). Such an odd-even
effect in Z has indeed been predicted by calculations of explosive
nucleosynthesis (Truran and Arnett 1971) for stars of low
metallicity. Unfortunately, the uncertainties in the data are such
that a definite trend is not unambiguously established.

2.3 Discussion

The trends in elemental abundance patterns in the extreme halo
population stars in our galaxy are thus seen to be quite consistent
with the predictions of nucleosynthesis theory for the ejecta of
normal stars of masses $\geq 10\ M_\odot$ and associated Type II supernovae.
These massive stars, of lifetimes $\leq 10^8$ years compatible with a
halo collapse timescale, are the major galactic sources of oxygen,
the elements from neon to calcium and the r-process heavy elements,
while carbon, nitrogen, the iron peak nuclei, and the s-process
heavy elements are produced in stars of lower masses and longer
lifetimes. The "anomalous" abundance features characterizing
extremely metal deficient stars (e.g. the high O/Fe, Mg/Fe, Si/Fe,
and Ca/Fe ratios and the strikingly r-process-like heavy element
abundance pattern) may thus be understood in a straightforward
manner. One cannot strictly rule out contributions to the
abundances of the most extreme metal deficient halo stars from such
more exotic sources as supermassive stars or a Population III, but
neither is there any compelling evidence to suggest that such
sources may have contributed significantly.

3. GLOBULAR CLUSTER STARS

Observational studies utilizing high S/N spectroscopy now indicate
that globular cluster stars exhibit abundance patterns similar to
those of the extreme metal deficient field halo stars (Pilachowski,
Sneden, and Wallerstein 1983; Gratton, Quarta, and Ortolani 1987).
Significant observed trends include the following: (1)
intermediate mass elements such as magnesium, silicon, calcium, and
titanium are typically enriched relatively to iron by up to
approximately 0.5 dex for clusters with values of [Fe/H] ranging
from -2.2 to -0.9; (2) high O/Fe ratios are encountered in a
substantial fraction of the observed clusters; and (3) the heavy

element (A > 60) abundance patterns in globular cluster stars reveal anomalies in the ratios of Zr, Ba, and La to Fe similar to trends observed in the metal deficient field halo stars and again seem suggestive, in this writer's view, of an r-process origin. The situation with respect to the light odd-Z elements Na, Al, Sc and V can be quite complicated (François, Spite, and Spite 1987), but the evidence seems to suggest that these elements are relatively more abundant in the globular cluster stars than in the field stars.

The similarities in the abundance patterns characterizing globular cluster stars and extreme halo population field stars are strongly suggestive of a similar, if not common, nucleosynthesis origin. Their abundance distributions in both instances are quite consistent with the contamination of the gas from which they were formed by the ejecta of normal stars of masses ≥ 10 $M_\odot$, and associated Type II supernovae, on a timescale $\leq 10^8 - 10^9$ years. The question remains as to how this was accomplished. Possible explanations for these observed patterns include: (1) the primordial compositions of the gas from which both the globular cluster stars and the extreme metal deficient field halo stars formed reflect the contamination from a common earlier and elusive (Bond 1981) population III; (2) the first stellar generation was selectively polluted by massive stars and Type II supernovae formed first at the centers of massive collapsing clouds (Cayrel 1986); and (3) the globular cluster and extreme halo population stars were entirely *independently* contaminated by the ejecta of massive stars and Type II supernovae.

In this regard, the interesting question arises as to whether a self-enrichment model is possible for and appropriate to globular clusters. The physical conditions which are believed to have characterized the protogalactic gas (Fall and Rees 1985, 1988) imply a Jean's mass of the order of 10^6 $M_\odot$ consistent with expectations for protoglobular clusters. Cayrel (1986) argues that it is quite natural to expect star formation early in the history of the galaxy to have occurred in such objects. Could some of these clouds have survived to evolve to the globular clusters we observe today? Here again, perhaps, high S/N spectroscopic studies can provide critical input.

Specifically, an interesting possible test of whether a self-enrichment model is appropriate to globular clusters might be provided by a careful study of the abundance patterns in the most metal-rich clusters of $Z > 0.1$ $Z_\odot$. Field stars of such metallicity exhibit patterns of abundances relative to iron which are consistent with those of solar system matter, since the nucleosynthesis contributions from the intermediate mass stars of relatively longer lifetimes have by this time enriched the interstellar medium in carbon, nitrogen, s-process elements, and iron-peak nuclei. Self-enrichment of a globular cluster must necessarily be realized on a much shorter timescale, hence the abundance patterns of stars in even the most metal-rich clusters should reflect only the contributions from the more massive stars.

If self-enrichment indeed occurred for the globular clusters, one might then expect even the metal-rich clusters to exhibit, to some degree, both the high O/Fe, Mg/Fe, Si/Fe, and Ca/Fe ratios and the r-process heavy element abundance pattern which are found to characterize the most metal-deficient field halo stars.

4. CONCLUDING REMARKS

My purpose in this paper has been to illustrate the manner in which increasingly sensitive spectroscopic studies can impact on one active and exciting area of astrophysical research. The abundance data thus obtained for both the extreme metal-deficient halo population stars and the stars in globular clusters is providing critical constraints on the nature of the stellar and supernova activity and associated nucleosynthesis contributions during the earliest phases of evolution of our galaxy. This allows at least a consistent model for the contamination of the gas in heavy nucleosynthesis products on a halo collapse (dynamical) timescale. We can anticipate that further detailed studies of this nature of the compositions of globular cluster stars, halo stars, and early disk population stars will provide important clues to the solutions of some of the many challenging problems which remain.

5. ACKNOWLEDGEMENTS

The author wishes to express his thanks to the Alexander von Humboldt Foundation for support by a U.S. Senior scientist Award and to Professor R. Kippenhahn for the hospitality of the Max-Planck-Institut für Astrophysik, Garching bei München. This research was supported in part by the United States National Science Foundation under grant AST 86-11500.

6. REFERENCES

Bond, H. E. 1981, *Astrophys. J.*, **248**, 606.
Cayrel, R. 1986, *Astr. Astrophys.*, **168**, 81.
Fall, S.M. and Rees, M.J. 1985, *Astrophys. J.*, **298**, 18.
Fall, S.M. and Rees, M.J. 1988, in *Globular Cluster Systems in Galaxies*, ed. J.E. Grindlay and A.G.D. Philip (Dordrecht: Reidel).
François, P. 1986a, *Astr. Astrophys.*, **160**, 264.
François, P. 1986b, *Astr. Astrophys.*, **165**, 183.
François, P., Spite, M., and Spite, F. 1987, *Astr. Astrophys.*, in press.
Gratton, R.G., Quarta, M.L., and Ortolani, S. 1987, *Astr. Astrophys.*,
    in press.
Gratton, R.G. and Sneden, C. 1987, *Astr. Astrophys.*, **178**, 179.
Hillebrandt, W. 1978, *Space Sci. Rev.*, **21**, 639.

Iben, I., Jr. and Truran, J.W. 1978, *Astrophys. J.*, **220**, 980.
Laird, J.B. 1985, *Astrophys. J.*, **289**, 556.
Luck, R.E. and Bond, H.E. 1985, *Astrophys. J.*, **292**, 559.
Merrill, P.W. 1952, *Science*, **115**, 484.
Pilachowski, C.A., Sneden, C., and Wallerstein, G. 1983, *Astrophys. J. Suppl.*, **52**, 241.
Renzini, A. and Voli, M. 1981, *Astr. Astrophys.*, **94**, 175.
Sneden, C. and Pilachowski, C.A. 1985, *Astrophys. J. Letters*, **288**, L55.
Spite, M. and Spite, F. 1978, *Astr. Astrophys.*, **67**, 23.
Spite, M. and Spite, F. 1985, *Ann. Rev. Astr. Ap.*, **23**, 225.
Thielemann, F.-K., Nomoto, K., and Yokoi, K. 1986, *Astr. Astrophys.*, **158**, 17.
Truran, J.W. 1981, *Astr. Astrophys.*, **97**, 391.
Truran, J.W. 1983, *Mem. S.A. Ita.*, **54**, 113.
Truran, J.W. 1984, *Ann. Rev. Nucl. Part. Sci.*, **34**, 53.
Truran, J.W. 1987, in *Relativistic Astrophysics*, ed. M.P. Ulmer (Singapore: World Scientific), p. 430.
Truran, J.W. and Arnett, W.D. 1971, *Astrophys. Space Sci.*, **11**, 430.
Truran, J.W., Cowan, J.J., and Cameron, A.G.W. 1985, in *Nuclear Astrophysics*, ed. W. Hillebrandt (Munich: Max-Planck Publication MPA 199), p. 81.
Truran, J.W. and Thielemann, F.-K. 1987, in *Stellar Populations*, ed. C.A. Norman, A. Renzini, and M. Tosi (Cambridge: Cambridge University Press), p. 149.
Woosley, S.E. 1987, *Saas Fee Lecture Notes*, preprint.

DISCUSSION

M. SPITE May I remark that aluminum becomes underdeficient relative to iron as soon as the resonance aluminum lines are used to determine this abundance.

MAGAIN If one accepts that the resonance lines give a wrong abundance, one has to invoke non-LTE effects (or so). If non-LTE play a role, I have shown yesterday that they may also affect excited lines, and that nothing might be left of the odd-even effect. Maybe one should not invoke non-LTE effects only in the cases when observation disagrees with theory...

BERYLLIUM ABUNDANCES OF F DWARFS

Kent G. Budge
Ann Merchant Boesgaard

Palomar Observatory and Department of Astronomy
California Institute of Technology

John Varsik

Institute for Astronomy
University of Hawaii

ABSTRACT. We have obtained spectra near 3130 Å of 23 F and G main-sequence stars at 0.05 Å px^{-1}. We have made preliminary estimates of the equivalent widths for the Be II doublet at 3130.4 and 3131.1 Å by means of a comparison procedure that matches each star against a standard star of similar spectral type with known Be abundance. It appears that stars that are deficient in Be are very uncommon. The halo dwarf, HD 76932, clearly has Be.

1. INTRODUCTION

The beryllium abundance of a main-sequence star is an indication of the depth and extent of convective mixing, since Be is destroyed at moderately low temperatures ($\sim 3.5 \times 10^6$ K). In particular, one might expect a subset of the known Li-deficient stars to be Be-deficient as well. The cosmic Be abundance is also of interest, since Be is formed exclusively by spallation and is therefore a measure of cosmic-ray activity during the lifetime of the galaxy.

2. OBSERVATIONS

We have obtained spectra for a number of F main-sequence stars from various populations, including stars in the Hyades Li gap (Boesgaard and Trippico 1986) and the halo dwarf HD 76932. Observations were made at CFHT on the nights of October 10 and December 10-11, 1986. Spectra were obtained at the coudé focus using the ultraviolet optical train and a TI CCD as detector. The dispersion was .05 Å/pixel and the signal-to-noise ratio was typically ~ 100 as opposed to ~ 20 for IUE. Although progress has been made in increasing the blue sensitivity of CCD detectors, exposures of ~ 2 hours were required to obtain this signal-to-noise ratio for the fainter Hyades stars.

3. REDUCTION OF DATA

The spectra have been flat-fielded and dark-subtracted and cosmic ray signatures removed by an interpolation process from adjoining pixels. The spectral region near the Be doublet is heavily line-blanketed, making it difficult to determine the true continuum level. This line-blanketing also results in numerous blends, which are further accentuated by the rapid rotation of many stars of interest in this temperature range.

To solve these problems, a "differential spectroscopy" procedure is being developed. This procedure includes the following steps: First, a cross-correlation algorithm is used to match the wavelength scale of the object spectrum to a standard spectrum. The object spectrum is then multiplied by a low-order polynomial (usually linear) selected by a least-squares algorithm to give the best fit between the object and standard spectra in the region immediately around the Be doublet. Finally, the standard spectrum is subtracted from the object spectrum. The resulting difference spectrum shows any differences between the object and standard spectra; furthermore, if the continuum level is known for the standard, the equivalent width of the Be doublet may be obtained. This procedure is adapted to rapidly-rotating stars by convolving the standard spectrum with a rotation profile of the correct width.

Figure 1 shows the results of this procedure as applied to the object star ι Psc (heavy line at top) with the standard star γ^1 Del (light line at top). The difference spectrum, showing the signature of Be, appears at the botton of the figure.

This procedure reduces the problems with close blends and, assuming the continuum level has been carefully determined for the standard stars, avoids the necessity of determining the continuum level for the object stars. However, it requires that standard stars of known Be abundance be available. We have selected Procyon (α CMa, F5IV-V) as our primary standard. Although this star has spectral peculiarities arising from its location slightly above the main sequence, its spectrum shows no detectable Be, a result consistent with a probable history of mass transfer between it and its white dwarf companion. In addition, Procyon has a very small value of $v \sin i$. Our secondary standards include γ^1 Del (F7V), ξ Peg (F7V), and χ^1 Ori (G0V), all of which have small values of $v \sin i$. All of these stars have normal Be abundance except for γ^1 Del, which appears to be somewhat deficient in Be.

4. RESULTS

Our preliminary results indicate that Be deficiency is a much rarer phenomenon than Li deficiency (Boesgaard 1976). This is not particularly surprising when one considers that Be is destroyed at temperatures of $\sim 3.5 \times 10^6$K while Li is destroyed at only $\sim 2.5 \times 10^6$K. We find no evidence for Be deficiency in the Hyades Li-gap stars, although the rapid rotation of these stars makes the measurement particulary uncertain. We also find that HD 76932, the halo dwarf, is definitely overabundant in Be, with a near-solar abundance of this element despite an overall low metallicity (Figure 2). This suggests that much of the galactic Be was formed early, but measurements are needed for more halo dwarfs.

This work was supported by NSF grants AST-8216192 and RII-8521715 and a Guggenheim Fellowship to AMB and by an NSF Graduate Fellowship to KGB.

REFERENCES

Boesgaard, A.M. 1976, *Astrophys. J.*, **210**, 466.
Boesgaard, A.M., and Trippico, M.J. 1986, *Astrophys. J.*, **302**, L49.

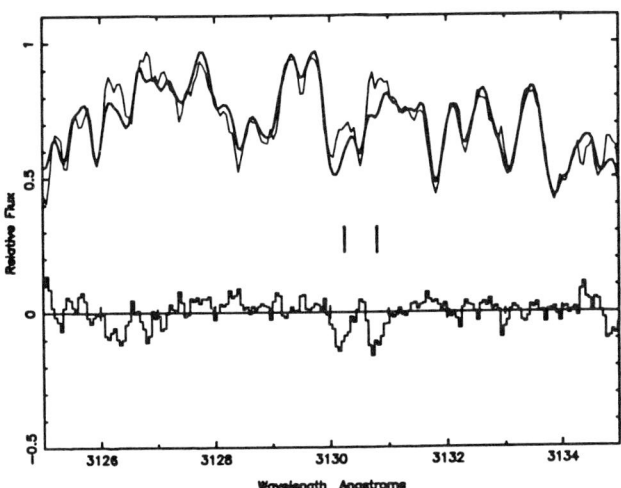

Figure 1. ι Psc vs. γ^1 Del. The Be signature is seen clearly in the lower spectrum.

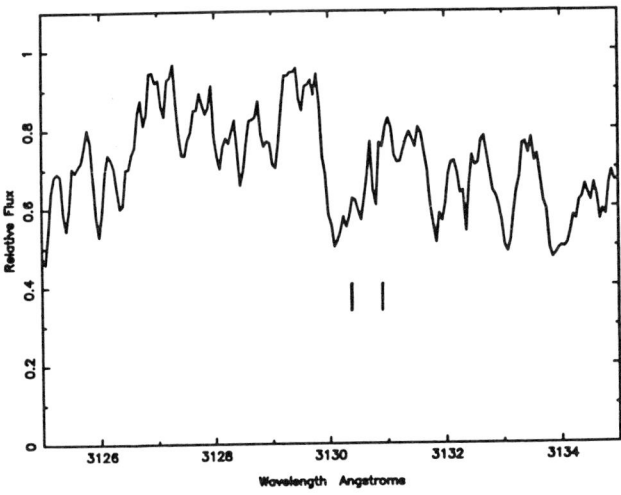

Figure 2. The halo dwarf HD 76932

DISCUSSION

VAUCLAIR Can you give approximate values of the beryllium abundances in stars on each side of the lithium gap in the Hyades, and the uncertainty on these values ?

BUDGE Our numbers are very rough at this stage ; I wouldn't claim better than a factor of two uncertainty. However, Be/H in VB14 is around 1.2×10^{-11}; VB37 is possibly somewhat depleted, perhaps Be/H$\approx 4 \times 10^{-12}$. Other Hyades stars all fall in this range.

SODERBLOM Have you Be observations of other halo dwarfs to see how ubiquitous their high Be is ?

BUDGE Unfortunately, no. We do have some Be observations of the old disk, which show roughly solar Be, but no other halo dwarfs.

THE $^{12}C/^{13}C$ RATIO IN UNEVOLVED COOL STARS

Y. Chmielewski[1] and D.L. Lambert[2]

[1] Observatoire de Genève, 1290 Sauverny, Switzerland
[2] Department of Astronomy, University of Texas
Austin, TX 78712, USA

ABSTRACT. We show that the carbon isotope ratio $^{12}C/^{13}C$ in the atmosphere of dwarf stars can be determined with reasonable accuracy from high resolution, high signal-to-noise ratio observations of the CH G-band in their spectra. Lines suitable for this purpose are selected from consideration of the solar case, for which $^{12}C/^{13}C = 89$ is derived. A preliminary analysis of these features in the spectra of μ Her, δ Eri and τ Cet yields $^{12}C/^{13}C$ values of 84, 80 and 150 respectively.

1. INTEREST OF $^{12}C/^{13}C$ RATIO DETERMINATION IN DWARF AND SUBGIANT STARS

The knowledge of the $^{12}C/^{13}C$ ratio in dwarf and subgiant stars can provide valuable constraints in the study of
- <u>Chemical evolution models of the Galaxy</u>. $^{12}C/^{13}C$ should decrease in the interstellar medium (ISM) as the Galaxy ages. In the solar system $^{12}C/^{13}C = 89$. Analyses of $^{12}C/^{13}C$ in the present ISM in the solar vicinity yield values between 40 and 70, which suggests that it is indeed richer in ^{13}C than the solar system. The value of the carbon isotope ratio in unevolved stars of different ages and kinematics should reflect its value in the I.S.M. at the epoch and location of their formation.
- <u>Internal structure of main sequence stars</u>. Standard models predict no deep mixing between the photosphere and the interior. But several non-standard models predict that deep mixing (e.g. induced by rapid rotation) might occur in some late-type main-sequence stars. According to other scenarios, low mass stars might experience severe mass loss in their early main-sequence lifetime, thus showing later ^{13}C enriched surface layers.
- <u>Subgiants</u>. The determination of the $^{12}C/^{13}C$ ratio in their atmospheres may help to map the onset of deep mixing as the stars evolve towards the giant branch. Otherwise, the above remarks on main-sequence stars apply to subgiants not evolved enough to have undergone mixing, with the advantage that they can be dated.

2. SELECTION OF SUITABLE ^{13}CH LINES FOR THE DETERMINATION OF THE ^{12}C/^{13}C RATIO

The value ^{12}C/^{13}C = 90 for the solar atmosphere has been derived by Hall (1973) and Harris et al. (1987) from observations of the CO infrared vibration-rotation bands. It is now firmly established.

The ratio has been determined for many normal giants by Lambert and collaborators from observations of the CN red system. For most of them a ratio around 20 - 25 is obtained which agrees beautifully with the values predicted by the standard theory of stellar evolution.

Sneden et al. (1986) determined the ^{12}C/^{13}C ratio in Pop. II giants : in these very metal-poor stars, CN becomes too weak and they had to use observations of lines of the CH G-band. The CH G-band had previously been used by Lambert and Dearborn (1972) and Krupp (1973) to derive the ^{12}C/^{13}C ratio in Arcturus.

In main-sequence stars CN also becomes too weak to be used. The CO infrared bands cannot be observed with sufficient resolution and S/N ratio. We therefore have to use observations of the CH G-bands.

These bands occur in the λ 4200-4400 Å region which is very crowded with absorption lines, implying that the continuum location and equivalent width measurements of blend-free lines are very difficult. We therefore have to use the spectrum synthesis technique. For this purpose a very detailed study of the CH bands in the solar spectrum has been carried out (Chmielewski, 1984).

Reconsideration of the solar case was all the more necessary since previous studies of the solar ^{12}C/^{13}C based on CH observations pointed to values of the ratio in the range 100-150 (see, e.g., Iversen, 1976).

All the potentially observable lines of ^{13}CH were considered in synthetic spectra of the Sun. Many of them are clearly blended with unidentified lines. The basic difficulty is that ^{12}C/^{13}C is likely to be high ($\simeq$ 100) for dwarf stars and ^{13}CH lines accordingly very weak, so that lines useful for giants with low ^{12}C/^{13}C turn out to be useless for unevolved stars.

After very careful examination, only three lines appear usable for the determination of the carbon isotope ratio : λ 4231.4 (0,0) $R_{2dc}(13)+R_{1dc}(13)$, λ 4366.2 (0,0)$P_{2cd}(12)$ + (1,1)$P_{1cd}(12)$, λ 4366.3 (0,0) λ $P_{1cd}(12)$. The line (0,0)$P_{1dc}(11)$ λ 4363.8 can also be used as a check.

In the solar spectrum all these lines are very weak and therefore very sensitive to the continuum location. The doublet at 4231.4 Å is the strongest and the one for which the continuum location can be best controlled : it will be given the highest weight. These lines fit very well the observed solar spectrum (Liège - Jungfraujoch Atlas) when they are synthesized with the ratio ^{12}C/^{13}C = 89.

3. SELECTION OF WEAK ^{12}CH LINES

Since the ^{12}C/^{13}C ratio is expected to be very high in unevolved stars, the ^{12}CH lines will be saturated and the calculation of their strength will be too model-dependent. In order to retain the advantage of differentiality of the isotope ratio determinations, we have to choose weak ^{12}CH lines to derive the abundance of ^{12}C which is to be compared with the ^{13}C abundance derived from the ^{13}CH lines.

Weak lines are sensitive to the choice of the continuum location. Ideally, the selected ^{12}CH weak lines should be observable on the same spectrograph scans as the selected ^{13}CH lines in order to keep the same continuum definition. Scans of a typical high resolution Reticon or CCD spectrograph have a width of about 15 to 30 Å.

Lines corresponding to these requirements are the following : CH (2,2) $R_{1cd}(19) + R_{2cd}(19)$ λ 4240.2; (2,2) $R_{1dc}(25)$ λ 4241.3; (2,2) $P_{1dc}(5)$ λ 4348.6 ; (1,1) $P_{2dc}(11)$ λ 4363.5 ; (1,1) $P_{1dc}(11)$ λ 4363.6 ; (1,1) $P_{2cd}(13)$ λ 4370.1.

Other useful weak lines are the (0,0) P(3) lines around 4329 Å.

4. PRELIMINARY RESULTS FOR THREE STARS

Preliminary results are given for μ Her, δ Eri and τ Cet, i.e. two subgiants and a metal-poor dwarf, all three belonging to the old disk population. The observations were obtained with the high resolution échelle spectrograph and Reticon detector attached to the 2.7 m telescope at the Mac Donald Observatory.

The model atmospheres used for the spectrum synthesis were scaled from the solar empirical model atmosphere of Holweger, with corrections for gravity and metallicity interpolated in the grid of theoretical model atmospheres of Gustafsson et al.. The f-values for the lines in the stellar synthetic spectra wereadjusted from fits to the corresponding regions in the solar spectrum. The model atmosphere parameters are given below, together with the resulting carbon isotope ratio :

Star	T_{eff}	log g	[M/H]	$^{12}C/^{13}C$
μ Her	5500	4.0	0	84
δ Eri	4940	3.75	-0.1	80
τ Cet	5350	4.5	-0.5	150

At a typical resolution of 100'000, it is found that an accuracy of ± 15 is obtained on the $^{12}C/^{13}C$ ratio if the signal-to-noise ratio of the observations is higher than 180 when $^{12}C/^{13}C$ is between 60 and 90, and higher than 380 when $^{12}C/^{13}C$ is between 90 and 150.

The result for the old dwarf star τ Cet is the first for an unevolved star other than the sun. Its markedly higher than solar value is an important clue to the existence of systematic trends of $^{12}C/^{13}C$ with age and/or galactocentric radius. More stars will soon be obseved in view of assessing such trends.

REFERENCES

Chmielewski, Y.: 1984, Astron. Astrophys. **133**, 83
Hall, D.N.B.: 1973, Ap. J. **182**, 977
Harris, M.J., Lambert, D.L., Goldman, A.: 1987, Monthly Notices Roy. Astron. Soc. **224**, 237
Iversen, O.: 1976, Inst. Theor. Astrophys. Blindern-Oslo, Rep. No. 45
Krupp, B.M.: 1973, Ph.D. Thesis, University of Maryland
Lambert, D.L., Dearborn, D.S.: 1972, Mem. Soc. Roy. Sci. Liège, 6e série, **3**, 147
Sneden, C., Pilachowski, C.A., VandenBerg, D.A.: 1986, Ap.J. **311**, 826

DISCUSSION

EDVARDSSON : I guess that you might be able to test your solar gf-values if you use also the solar flux spectrum, since molecular lines might be expected to be stronger near the solar limb. Since these stars are also not too different from the Sun, it might make the analysis somewhat more differential to use the flux spectrum also for the Sun.

CHMIELEWSKI : I agree that to make the analysis fully differential, I should have adjusted the f-values on spectra of the reflected sunlight secured with the same spectrograph as the stellar spectra. I chose however to use the Jungfraujoch Atlas intensity at the center of the solar disk because of its much higher resolution and S/N than that of the solar flux data available when I started this work. In particular the solar wavelengths of the components in close blends can be much better assessed at the center of the disk where the lines are less broadened by micro- and macro-turbulence.

BOESGAARD : What is the explanation for the high value of $^{12}C/^{13}C$ in τ Cet?

AUDOUZE : One should expect to determine large $^{12}C/^{13}C$ ratios in very old stars like τ Cet since ^{13}C should be of secondary origin. Do you also plan to observe $^{12}C/^{13}C$ ratios in young or very young stars? This would be of utmost importance for constraining models of chemical evolution of galaxies.

CHMIELEWSKI : τ Cet and δ Eri are both old disk stars with similar kinematics, i.e. a galactic orbit external relative to that of the Sun. Being more metal-poor, τ Cet should be somewhat older than δ Eri, but is it enough to account for the difference in $^{12}C/^{13}C$ between the two stars, unless one assumes that δ Eri, which is evolved, has aready undergone some mixing?
We have an observing run on the CAT + CES at ESO scheduled for August. In the sample which we intend to observe we have selected stars belonging to the halo, thick disk, old disk and young populations. It includes in particular a few stars belonging to the Hyades or Sirius groups.

THE $^{12}C/^{13}C$ RATIO IN BARIUM STARS

A. JORISSEN*
Institut d'Astronomie, d'Astrophysique et de Géophysique
Université Libre de Bruxelles C.P.165
50, av. F. Roosevelt
B-1050 Bruxelles
Belgium

ABSTRACT. It is suggested that the position of BaII stars with respect to normal red giants in the (log L, $^{12}C/^{13}C$) and ([C/Fe], $^{12}C/^{13}C$) diagrams supports the hypothesis that $^{13}C(\alpha,n)^{16}O$ was the neutron source responsible for the synthesis of the heavy elements now present in the BaII star envelopes.

It is well known that $^{12}C/^{13}C$ is a function of luminosity in first ascent giants, as a result of the first dredge-up (e.g. Lambert, 1976). The comparison between $^{12}C/^{13}C$ in BaII stars and in normal giants should thus take this luminosity effect into account. Fig.1 sums up all the data available for normal G-K giants and subgiants, extracted from Lambert (1976), Tomkin et al. (1976) and Lambert and Ries (1981). For the stars of these samples, the luminosity is derived mainly from the CaII K-line emission width (Wilson, 1976), although some other methods are also used. Concerning BaII stars, $^{12}C/^{13}C$ and luminosity are simultaneously available for only 6 *classical* BaII stars. As those data are scattered in the literature, they are summarized in Table I. Luminosities of BaII stars are derived from Eggen's (1972) (R-I, M_{bol}) relation. When available, the uncertainty on the $^{12}C/^{13}C$ ratio has been drawn in Fig.1, whereas the uncertainty on log $L/L_\odot$ is difficult to evaluate.

Let us consider the BaII star HD16458 (=HR774), the $^{12}C/^{13}C$ ratio of which appears to fall quite inside the range of $^{12}C/^{13}C$ ratios displayed by normal giants of the *same luminosity*[1]. In the ($^{12}C/^{13}C$, [C/Fe]) diagram (Fig. 2), this same star is found to have a strong C-overabundance (by a factor 2 to 5) with respect to normal red giants, although $^{12}C/^{13}C$ is normal!

---

[1] It should be mentioned, however, that values of the order of log $L/L_\odot$=2.25 to 2.65 have been proposed for that star by Böhm-Vitense et al. (1984) using spectroscopically derived gravities and effective temperatures, *assuming* a mass of 2.5 $M_\odot$.
*Boursier I.R.S.I.A. (Belgium).

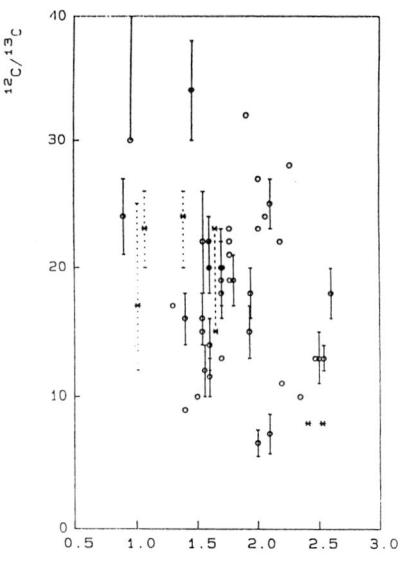

Fig.1: $^{12}C/^{13}C$ vs. luminosity

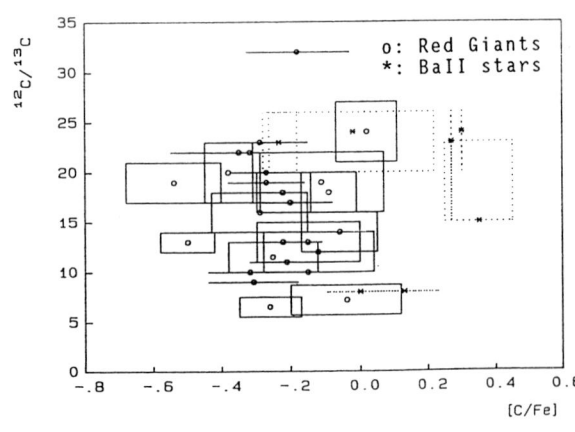

Fig.2: $^{12}C/^{13}C$ vs. $[C/Fe]_\odot$

TABLE I

Name	Sp.	M_{bol}	$logL/L_\odot$	$^{12}C/^{13}C$	$[C/Fe]_\odot$	$[Fe/H]_\odot$
HD 16458	K1IIIBa5$_k$	0.5$_c$	1.65$_c$	23$_{ab}$	.27$_a$	-.32$_a$
				15$_g$	.35±.10$_g$	-.30$_g$
HD 46407	K0IIIBa2$_k$	1.9$_c$	1.07$_c$	21$_a$	-.02±.04$_a$	.02$_a$
				23±3$_b$	+.27$_l$	-.42$_l$
HD 101013	K0IIIBa5$_k$	2.1$_c$	1.01$_c$	17±$^8_5$$_h$	-.50±.10$_{hi}$	.25$_i$
				13±1$_b$		
HD 116713	K1IIIBa3$_k$	1.1$_{cn}$	1.39$_c$	20$_a$	-.02±.24$_a$	+.15$_a$
			1.43$_n$	24±2$_b$	+.30$_l$	-.29$_l$
HD 121447	K7IIIBa5$_k$	-1.7$_c$	2.53$_c$	8$_g$	.00±.10$_g$	.05$_g$
HD 178717	K4IIIBa5$_k$	-1.4$_c$	2.41$_c$	8$_g$	.13±.10$_g$	-.18$_g$

References: a) Sneden et al., 1981; b) Tomkin and Lambert, 1979; c) Eggen, 1972; g) Smith, 1984; h) Harris et al., 1985; i) Williams, 1975; k) Lü et al., 1983; l) Kovacs, 1985; n) Dominy and Lambert, 1983.

In order to understand how this paradoxical situation can be accounted for, let us consider the contamination of a giant star convective envelope by material having undergone a s-type of processing by neutrons liberated by either $^{13}C(\alpha,n)^{16}O$ or $^{22}Ne(\alpha,n)^{25}Mg$. The value of $^{12}C/^{13}C$ after such a mixing, referred to as the final $(^{12}C/^{13}C)_F$ ratio, is given by

$$(^{12}C/^{13}C)_F = [(1-g)^{12}C_I + g^{12}C_P]/[(1-g)^{13}C_I + g^{13}C_P], \quad (1)$$

where g denotes the dilution factor of the s-processed material (abundances by number with subscripts P) into the envelope (abundances by number before contamination with subscripts I). Two special cases are of interest :

(i) if $(^{12}C/^{13}C)_P = (^{12}C/^{13}C)_I$, then $(^{12}C/^{13}C)_F = (^{12}C/^{13}C)_I$, (2)

whatever the total C abundance could be.

(ii) if the contaminating material is essentially devoid of ^{13}C (i.e. $^{13}C_P \approx 0$), then, assuming $1-g \approx 1$,

$$(^{12}C/^{13}C)_F \approx (^{12}C/^{13}C)_I \times (^{12}C_F/^{12}C_I). \quad (2')$$

Case (ii) is relevant if the contaminating material has been processed by $^{22}Ne(\alpha,n)^{25}Mg$, since the ^{13}C abundance is very low in typical He-burning environments and $^{12}C(n,\gamma)^{13}C(n,\gamma)^{14}C$ does not favor the production of ^{13}C. Assuming further that the envelope composition before contamination is representative of normal red giants, we conclude that the $^{12}C/^{13}C$ ratio now observed in BaII stars should be that of red giants scaled by the overabundance of carbon (^{12}C) if $^{22}Ne(\alpha,n)^{25}Mg$ is the active neutron source, as indicated by Eq.(2').

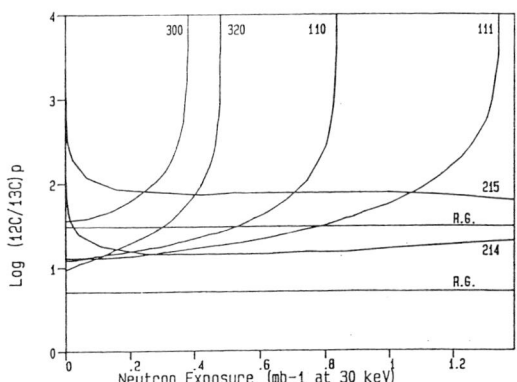

Fig.3: $(^{12}C/^{13}C)_P$ in the processed material as a function of the neutron exposure for the following sets of the parameters describing the operation of the $^{13}C(\alpha,n)^{16}O$ neutron source:
110: $T=1\times10^8K$, $^{12}C/p(0)=10.$, $\vartheta_{ing}=0$;
111: $T=1\times10^8K$, $^{12}C/p(0)=10.$, $\vartheta_{ing}=0$, $^{12}C/^4He(0)=1.$;
214: $T=2\times10^8K$, $^{12}C/p(0)=1.$, $\vartheta_{ing}=10$ $\vartheta_\alpha(^{13}C)$;
215: $T=2\times10^8K$, $^{12}C/p(0)=1.$, $\vartheta_{ing}=100$ $\vartheta_\alpha(^{13}C)$;
300: $T=3\times10^8K$, $^{12}C/p(0)=20.$, $\vartheta_{ing}=0$;
320: $T=3\times10^8K$, $^{12}C/p(0)=5.$, $\vartheta_{ing}=0.$,
with $\rho=500$ g/cm^3 and $^{12}C/^4He(0)=1/3$ (except for case 111).
ϑ_{ing} is the proton ingestion timescale and $\vartheta_\alpha(^{13}C)$ is the ^{13}C lifetime for α-captures. The two horizontal lines labelled R.G. border the domain $5 \leq ^{12}C/^{13}C \leq 30$ characterizing red giant values.

Notwithstanding their scarcity and uncertainties, the observational data appear to favor the situation described by Eq.(2) rather than that corresponding to Eq.(2'). This speaks against the $^{22}Ne(\alpha,n)^{25}Mg$ neutron source, which cannot account for the mixing case with $(^{12}C/^{13}C)_F = (^{12}C/^{13}C)_I$, and thus for Eq.(2). It now remains to evaluate the virtues of $^{13}C(\alpha,n)^{16}O$ in that respect. Fig.3 displays the $(^{12}C/^{13}C)_P$ ratio in the processed material during the nucleosynthesis event as a function of the neutron exposure φ at 30 keV, for various sets of the parameters describing the operation of the $^{13}C(\alpha,n)^{16}O$ neutron source. That neutron source operates when protons are ingested (with a timescale ϑ_{ing}) into a ^{4}He- and ^{12}C-rich layer [characterized by the initial $^{12}C/^4He(0)$ and $^{12}C/p(0)$ ratios; see Jorissen and Arnould, 1986, for more details about the parametrization], as a result of the chain $^{12}C(p,\gamma)^{13}N(\beta^+)^{13}C(\alpha,n)^{16}O$. It can be seen that many $(^{12}C/^{13}C)_P$ curves cross the $5 \leq (^{12}C/^{13}C)_I \leq 30$ domain, characterizing normal red giant envelopes. Thus, a freeze out of the matter processed at $\varphi < \varphi_{max}$ before complete ^{13}C burning is just what is needed in order to keep $(^{12}C/^{13}C)_P$ in a range typical for normal red giants. The curve corresponding to case 211 does not even require such a freeze out. Mixing of such a processed material in a giant envelope then keeps the $^{12}C/^{13}C$ ratio close to its initial value.

In conclusion, the method presented here could be useful for identifying the operation of the $^{13}C(\alpha,n)^{16}O$ neutron source in BaII stars, but the scatter in $^{12}C/^{13}C$, the uncertainties affecting the luminosity as well as the difficulty of distinguishing first ascent giants from core-He burning giants make it difficult to draw a definite conclusion.

REFERENCES:

Böhm-Vitense, E., Nemec, J., Proffitt, Ch.: 1984, *Astrophys.J.* 278, 726.
Dominy, J.F., Lambert, D.L.: 1983, *Astrophys.J.* 270, 180.
Eggen, O.J.: 1972, *Mon. Notices Roy. Astron. Soc.* 159, 403.
Harris, M.J., Lambert D.L., Smith, V.V.: 1985, *Astrophys.J.* 292, 620.
Jorissen, A., Arnould, M.: 1987, in *Advances in Nuclear Astrophysics*, eds. E. Vangioni-Flam, J. Audouze, M. Cassé, J.P. Chièze and J. Tran Thanh Van (Gif sur Yvette: Frontières), p. 419.
Kovacs, N.: 1985, *Astron.Astrophys.* 150, 232.
Lambert, D.L.: 1976, *Mém. Soc. Roy. Sc. Liège* IX, 6ᵉ Série, p.405.
Lambert, D.L., Ries, L.M.: 1981, *Astrophys.J.* 248, 228.
Lü, P.K., Dawson, D.W., Upgren, A.R., Weis, E.W.: 1983, *Astrophys.J.* 52, 169.
Smith, V.V.: 1984, *Astron.Astrophys.* 132, 326.
Sneden, C., Lambert, D.L., Pilachowski, C.A.: 1981, *Astrophys.J.* 247, 1052.
Tomkin, J., Luck, R.E., Lambert, D.L.: 1976, *Astrophys.J.* 210, 694.
Tomkin, J., Lambert D.L.: 1979, *Astrophys.J.* 227, 209.
Williams, P.M.: 1975, *Monthly Notices Roy. Astron. Soc.* 170, 343.
Wilson, O.C.: 1976, *Astrophys.J.* 205, 823.

COSMOLOGICAL SIGNATURES IN HIGH S:N SPECTRA

J.E. Beckman[1] and B.E.J. Pagel[2]
1. Instituto de Astrofisica de Canarias, 38200-La Laguna, Tenerife, Spain.
2. Royal Greenwich Observatory, Herstmonceux Castle, Hailsham, East Sussex BN27 1RP, United Kingdom.

ABSTRACT. We review briefly the observational status of abundance determinations for those elements produced in the first seconds of a "standard" Big Bang Universe, showing that although [^{4}He] and [D] appear not to be mutually compatible there is still room for reconciliation via astration of D, for values consistent with estimated [^{7}Li]. New limits to production cross-sections for ^{7}Li appear to limit N_ν (the number of light neutrino types) to no more than 3. Non-standard models with fluctuations are briefly considered; they are attractive because they use $\Omega = 1$ and may also yield early C, N, O in significant quantities. The true value of ^{7}Li [10^{-9} or 10^{-10} w.r.t. H] is the critical test here. We show where spectra of improved S:N are needed to progress in constraining cosmological models.

1. INTRODUCTION

One triumph of hot Big Bang model universes is their ability to account for the observed set of abundances of light elements, notably ^{2}H and ^{4}He, but also ^{7}Li and ^{3}He, over a range of nine orders of magnitude (^{4}He:^{7}Li). Global agreement is felt to be satisfactory enough to use the "standard" model (see Yang et al., 1984) to derive the universal photon:baryon ratio and to limit N_ν, the number of light neutrino types more stringently (see e.g. Rebolo et al., 1988) than the limits from the experimental "width" of the Z measured at CERN.

Referring the reader to the recent review by Boesgaard and Steigman (1985) here we concentrate on only two topics: possible discrepancies between the standard model and observations, and those cases where S:N ratio, rather than interpretation, limits our progress in grasping the physics.

2. ARE THE OBSERVED ABUNDANCES CONSISTENT WITH THE STANDARD MODEL?

(a) ^{4}He. For details of the derivation of the primordial ^{4}He abundance we refer the reader to the recent review by Pagel (1987) and the proceedings of the ESO workshop (Shaver et al., 1983). The method favoured by Pagel uses emission from HII regions of external galaxies (e.g. Vilchez et al., 1987) in the $\lambda 4417$ Å and $\lambda 5876$ Å He recombination lines. Limitations are the interpretation of observed fluxes to give relative abundances, and the required extrapolation to low metallicity. Recent work (Pagel et al., 1986) has stressed the reliability of the use of nitrogen (rather than only oxygen) as a measure of a nuclide produced in primary processes against which ^{4}He can be extrapolated to zero metallicity. The most recent observed value for Y_p is 23.2 (± 0.4)% by number. For ^{4}He the primordial abundance depends on η, and also on N_ν and the neutron half-life τ_n. Using a reasonable modern estimate for τ_n of 10.5 (± 0.1) min, we find, from ^{4}He alone (see Yang et al., 1984):

$$N_\nu = 4: \quad 0.8 \times 10^{-10} \leq \eta \leq 1.3 \times 10^{-10}; \quad 10^{-10} \leq \eta \leq 3 \times 10^{-10} : N_\nu = 3$$

$$2.5 \times 10^{-10} \leq \eta \leq 10 \times 10^{-10} : N_\nu = 2$$

We will see below when discussing Li that a value of $N_\nu = 4$ appears to be excluded, and $N_\nu > 4$ is certainly rule out.

(b) D+ ^{3}He. It is difficult to use primordial D directly as a check on η, because D is so readily destroyed by astration.

Interstellar determinations, using the Ly series of deuterium, or from molecules give $5 \times 10^{-6} < [D] < 2 \times 10^{-5}$, but even 2×10^{-5} must be taken as a lower limit, because D is so fragile that no production, only destruction is expected within stars. Using $[D] = 2 \times 10^{-5}$, these standard models (Yang et al., 1984) gives $\eta = 4 \times 10^{-10}$, which is incompatible with the ^{4}He data and 3 light neutrino types. Either $N_\nu = 2$ (the neutrino would have mass), or the standard model fails or the true $[D]$ is larger. A value for $[D]$ of 2×10^{-4} is needed for agreement with ^{4}He and ^{7}Li estimates. Direct determination of primordial D via the Lyα forest in QSO's offers a solution but is at present balked by inadequate S:N (Carswell et al., 1986).

An indirect method for D is to subtract the pre-solar ^{3}He abundance measured in carbonaceous chondrites: 1.5×10^{-5} (Eberhardt, 1978) from the ^{3}He abundance in the solar wind: 4×10^{-5} (Black, 1972), or in a prominence: 4×10^{-5} (Hall, 1975). The difference is supposedly attributable to D burning to ^{3}He in the sun, and yield $[D]_{protosolar} = 2.5 \times 10^{-5}$, agreeing with proto-solar D in planetary atmospheres (Encrenaz and Combes, 1982). The problem is that $[D]_{protosolar} / [D]_{primordial}$ could be as high as 40 (Pagel, 1986; Larson, 1086) depending on galactic chemical evolution.

To improve reliability, Yang et al. (1984) suggested $[D+^3He]$ instead of $[D]$ as a test abundance, assuming that all astrated D becomes ^{3}He, a fraction, g, of which survives astration. Assuming g=0.25, Delbourgo-Salvador et al. (1985) using a specific galactic model, obtain $[D]_p < 8.6 \times 10^{-5}$. This is still incompatible with the range permitted by ^{4}He (unless $N_\nu = 2$), but the gap is small, and with present errors in observations, and uncertainties in chemical evolution, does not at this point rule out the standard model. Observational estimate of Y have tended to decrease, thus widening the gap between (^{4}He) and (D+^{3}He), but the total light element picture can still accommodate the standard model, as we consider below.

(c) ^{7}Li. The sound use of ^{7}Li as a primordial probe dates from the elegant work by the Spites (1982) on subdwarfs. Their original value log $N(Li)_p$ 2.08 has been confirmed. although slightly raised to 2.20 by more recent observations (Beckman et al., 1986; Rebolo et al., 1987 a; Hobbs and Duncan, 1987). Three arguments support this value as primordial. (a) The uniformity of log N(Li) with Teff between Teff=6300 K and 5700 K; (b) the uniformity of log N(Li) with metallicity over the wide range -1.5 > [Fe/H] > -3.5, and (c) the uniformity of log N(Li) for halo dwarfs with sites of origin separated by up to several kpc.

Nevertheless two question marks can be placed on the use of log N(Li)= 2.2 as a primordial value: (a) was ^{7}Li produced in a significant quantity between the Big Bang and the formation of Pop II ? and (b) have Pop II subdwarfs suffered significant ^{7}Li depletion? In a recent article Rebolo et al. (1987 b) give negative answers to both questions. Against extra ^{7}Li prior to Pop II they point out: (i) observational limits of ^{6}Li (Maurice et al., 1984) and to ^{9}Be (Rebolo et al., 1987 c) in these stars, excluding virtually all non-primordial sources of ^{7}Li; (ii) the uniformity of ^{7}Li over a hundred-fold metallicity range below [Fe/H] =1.5 showing no tendency for much ^{7}Li to be produced in this early phase of chemical galactic evolution. Observations expecially of ^{9}Be and ^{6}Li, at higher S:N are needed to clinch these arguments. Against depletion, Rebolo et al. (1987 b) compare the log N(Li) v. Teff curves for "EMD" stars (with [Fe/H] ≠1.5) with those of open clusters, showing that non-diffusive depletion could not have occurred in the EMD group. Diffusive depletion is not ruled out, but the observed "plateau" in log N(Li) v. Teff between 6300 K and 5700 K in EMD stars does not conform to any theoretical accounts of diffusion effects (Michaud, 1984). Taking a value for log N(Li) of 2.2 we find, with observational (Rebolo et al., 1987) and theoretical (Kawano, 1987) uncertainties accounted for:

$$1.5 \times 10^{-10} \leq \eta \leq 6.5 \times 10^{-10}$$

Comparing this range with the ranges infered from ^{4}He we see that the maximum number of light neutrino types permitted in a standard model is 3.

3. NON-STANDARD MODELS

Cosmological models have recently been produced in order to fit the inflationary premise that Ω_0 =1 with a purely baryonic density. these either invoke ad hoc baryon density fluctuations (Applegate et al., 1987) or fluctuations arising from the quark-hadron phase transition (Alcock et al., 1987). Although these variants are able to reconcile the apparent ^{4}He-D discrepancy, they "overproduce" ^{7}Li by at least a factor 10. This is one reason for perhaps wishing to reconsider whether the arguments of those who prefer to identify the present-day Pop I ^{7}Li abundance of log N(Li)$_p$ ~3 with the primordial value may be justified. The evidence cited above, plus the trend of ^{7}Li to rise in measured stellar abundance linearly with metallicity (Rebolo et al., 1987 a) points towards log N(Li)$_p$ ~ 2, but the question is still open. It is unlikely that higher S:N spectra of ^{7}Li or ^{4}He will help to decide whether non-standard models are required. The key datum is now primordial [D] and hence the Ly-α forest observations mentioned above.

Table 1
Light element abundances with cosmological signatures: requirements of high S:N spectra

Nuclide	Measurement	S:N (and spectral resolution) required	Significance
D	D/H in Ly-α forest (QSO forest)	>100 ($\lambda/\Delta\lambda \geqslant 10.000$)	Direct non-astrated chemistry free primordial D.
^{3}He	Subdwarfs; emission at 10830 equivalent of ^{3}He.	>150 ($\lambda/\Delta\lambda > 50.000$)	Nearest data to primordial ^{3}He (very difficult measurement.
^{6}Li	Subdwarfs	>200 ($\lambda/\Delta\lambda > 30.000$)	Limits non-primordial sources of ^{7}Li.
	ISM	>300 ($\lambda/\Delta\lambda > 50.000$)	Examine formation rate of ^{6}Li in galaxy (^{7}Li data will emerge automatically).
^{9}Be	Subdwarfs	$\geqslant$200 ($\lambda/\Delta\lambda > 50.000$)	Limit non-primordial ^{7}Li. Explore early galactic evolution.
^{10}B	Subdwarfs	>200 ($\lambda/\Delta\lambda > 20.000$)	Limit non-primordial ^{7}Li.
O C N	Subdwarfs	>200 ($\lambda/\Delta\lambda > 20.000$)	Test of possible light metal production in Big Bang; early galactic evolution.

A further feature of the fluctuation models is that they can yield quantities of C, N, O which approach measurability limits. Since the finite metallicities of the earliest stars are a puzzle, this feature may prove attractive, but there is no prima facie evidence (Cayrel, 1986) that any phenomenon other than a "normal" massive early stellar population is required. High S:N data on early chemical evolution i.e. of weak C N O in Pop II would certainly be useful in this context.

4. CONCLUSIONS

Table 1 lists the requirements of spectral resolution and S:N for nuclides relevant to constraining cosmological models, where improved S:N is important. The isotopes ^{6}Li, ^{9}Be and ^{10}B constrain early galactic ^{7}Li production and high S:N is necessary because of their low abundances (similarly for very early C, N, O). However imprecision of the neutron half-life, ulterior knowledge of the number of neutrino types, understanding of D astration and Li depletion and model-dependent abundance determinations are at present the principal impediments to further advance.

References

Alcock, C., Fuller, M., Mathews, G.J.: 1987, (Preprint).
Applegate, J.H., Hogan, C.I., and Scherrer, R.J.: 1987, Phys. Rev. D. (submitted).
Beckman, J., Rebolo, R., and Molaro, P.: 1986, in "Advances in Nuclear Astrophysics" Eds. Vangioni-Flam, Audouze et al., p. 29.
Black, D.G.: 1972, Geochim. et Cosmochim. Acta **36**, 347.
Boesgaard, A., Steigman, C.: 1985, Ann. Rev. Astron. Astrophys. **23**, 319.
Carswell, R.F., Irwin, M.J., Webb, J.K., Baldwin, J.A., Atwood, B., Robertson, J.C., and Cayrel, R.: 1986, Astron. Astrophys. **168**, 81.
Cayrel, R.: 1986, Astron. Astrophys. **168**, 181.
Eberhardt, P.: 1978, Proc. 9th Lunar Plan. Sci. Conf. p. 1027.
Kawano, L., Schramm, D.N., Steigman, G.: 1987, (Preprint).
Encrenaz, T., and Combes, M.: 1982, Icarus **52**, 54.
Pagel, B.E.J.: 1986, Phil. Trans. Roy. Soc. **320**, 557.
Pagel, B.E.J.: 1987, Lectures to ESO-CERN School; Astroparticle physics, A. de Rojula, D. Nanopoulos, P.A. Shaver (Eds) Singapore. World Scientific Publishing Company Ltd.
Pagel, B.E.J., Terlevich, R., and Melnick, J.: 1986, P.A.S.P. **98**, 1005.
Rebolo, R., Beckman, J.E., Molaro, P.: 1988 (This volume).
Rebolo, R., Molaro, P., and Beckman, J.E.: 1987 a, Astron. Astrophys. (In Press).
Rebolo, R., Molaro, P., Abia, C., Beckman, J.E.: 1987 b, Astron. Astrophys. (In Press).
Shaver, P.A., Kunth, D., and Kjar, K (Eds): 1983, Primordial Helium, ESO Garching.
Spite, F., and Spite, M.: 1982, Astron. Astrophys. **115**, 357.
Vilchez, J.M., Pagel, B.E.J., Terlevich, R.J., and Melnick, J.: 1987 (In preparation).
Yang, J., Turner, M.S., Steigman, G., Schramm, D.N., and Olive, K.A.: 1984, Astrophysics. J. **281**, 493.
Michaud, G., Fontaine, G., and Beaudet, G.: 1984, Astrophys. J. **282**, 206.

CHEMICAL EVOLUTION AND PRIMORDIAL NUCLEOSYNTHESIS

J.Audouze[1,2] and E.Vangioni-Flam[1]
1 Institut d'Astrophysique du CNRS, Paris–France
2 Laboratoire René Bernas, Orsay–France

ABSTRACT. This contribution starts with a very brief updated report on our present knowledge concerning the primordial abundances of the lightest elements (D, ^{3}He, ^{4}He and ^{7}Li). From this information it is claimed here that specific models of chemical evolution able to account for a thorough destruction of D during the galactic history should be involved to reconcile the ratios : $\eta = \eta_B/\eta_\gamma$ (baryonic density relative to background photon density) predicted respectively from the primordial ^{4}He (Y) abundance and from the primordial D (and ^{3}He) abundance. Among different possibilities, it is shown that galactic evolution models implying variations of the role of star formation (SFR) are more successful for that goal than models implying bimodal star formation. From this analysis it is also argued that contrary to a common belief, a large D destruction rate over the galactic history implies very specific conditions concerning the chemical evolution of our galaxy.

1. INTRODUCTION

The goal of this contribution is first to advocate that the primordial abundances of the lightest elements (D, ^{3}He, ^{4}He and ^{7}Li) which are presently available lead to somewhat discordant predictions regarding the baryonic density of the Universe when the simplest standard assumptions are adopted for both the models of early nucleosynthesis and of the chemical evolution of our Galaxy. In order to reconcile such predictions in the frame of the simple "canonical" Big Bang model of nucleosynthesis, we felt compelled (see eg Delbourgo- Salvador etal,1985 and Vangioni-Flam and Audouze,1987(VFA)) to propose specific models of galactic evolution leading to a thorough D destruction during the history of our Galaxy. The second goal of this paper is to present a short account of a promising hypothesis : the model of galactic evolution when the rate of star formation (SFR) is supposed to vary with time and which is analysed in detail in VFA. After a very brief review of the primordial abundances of D, ^{3}He, ^{4}He and ^{7}Li (section 2), the problem concerning the discordant baryonic densities deduced from the ^{4}He abundance (Y) on one hand and from D and ^{3}He on the other hand, is presented in section 3. The specific models of the galactic evolution of D are presented in section 4 and the conclusions of this analysis in section 5.

2. THE PRIMORDIAL ABUNDANCES OF THE LIGHTEST ELEMENTS (D, ^{3}He, ^{4}He AND ^{7}Li)

Given the cosmological implications of the primordial abundances of the lightest elements (D, ^{3}He, ^{4}He and ^{7}Li), they are reviewed in many recent papers, see eg Boesgaard and Steigman,1985 and Audouze,1987. An update of the present determination of these abundances is given in some detail in Audouze and Vangioni-Flam, 1988 and in Audouze, Spite and Spite, 1988. Table 1 represents the summary of their review (see also Beckman and Pagel, this volume for further details.)

TABLE 1

Primordial Abundances of D, ^{3}He, ^{4}He and ^{7}Li

Element	Primordial	Old and low Z Objects	Solar System	Present Interstellar medium
(Age, Gyr)	> 12	10-12	4-6	0
D	$3 \cdot 10^{-5} - 10^{-4}$	$(2-8) \cdot 10^{-5}$	$(2\pm1) \cdot 10^{-5}$	$(1\pm0.3) \cdot 10^{-5}$
^{3}He	$3 \cdot 10^{-5} - 10^{-4}$		$(2\pm1) \cdot 10^{-5}$	$10^{-5} - 1.5 \cdot 10^{-4}$
D+^{3}He	$6 \cdot 10^{-5} - 2 \cdot 10^{-4}$		$(2-4) \cdot 10^{-5}$	$2 \cdot 10^{-5} - 1.6 \cdot 10^{-4}$
^{4}He(Y)	0.24 ± 0.01	$(0.24\pm0.01)+AZ$	0.24-0.28	0.24-0.30
^{7}Li	$(1\pm0.3) \cdot 10^{-10}(s^2)$ $(2-9) \cdot 10^{-10}(DH)$	$8 \cdot 10^{-11} - 2.5 \cdot 10^{-1}$	10^{-9}	10^{-9}

3. SOME DISCORDANCES IN THE "CANONICAL MODEL OF EARLY NUCLEOSYNTHESIS"

The remarkable agreement between the primordial abundances of the lightest elements and those computed in the frame of the standard (simplest) models of Big Bang, (see eg Boesgaard and Steigman,1985) is rightly considered as one of the major successes of this cosmological model. As shown especially in the review, the primordial nucleosynthesis leads to two most exciting constraints (i) on the overall baryonic density which is such that $\Omega_B \sim 0.1$ (where Ω_B is the baryonic cosmological parameter) ; this means that the baryonic density represents only $\sim 10\%$ of the critical density above which the Universe would be closed . In the sequel we will use either Ω_B or $\eta_B = \eta_B/\eta_\gamma$ where η_B and η_γ are respectively the baryon and the background photon densities. Ω_B is related to η_B by the relation $\Omega_B = 3.53 \; 10^{-3} \; \eta_{10} \; h^{-2} \; (T/2.7)^3$ where $\eta_{10} = 10^{10} \eta_B$; $h = (H_o/100)$ where H_o is the Hubble constant expressed in km s^{-1}Mpc^{-1} and T is the temperature of the background radiation ; (ii) on the maximum number of neutrino (lepton) families which is about 3 to 4 (remember that three lepton families have already been discovered). This implies that there is no much room for further discoveries of new lepton flavours in the frame of this standard model.

Given the importance of such conclusions, they should be scrutinized very precisely. As discussed at some length eg in Audouze 1987 we claim that the primordial abundance of (D+^{3}He) used in Boesgaard and Steigman (1985) and the primordial abundance of ^{4}He (Y_p) are only consistent for an <u>extremely narrow</u> range of η_{10} ($\eta_{10} = 3.2 \pm 0.2$). This conclusion has also been reached by Pagel (1986). This narrow range corresponds to $\Omega_b = 0.01$ to 0.04: the width of that range is only due to the fact that the "battle" between the $H_o = 50$ km s^{-1} Mpc^{-1} proponents and those who claim that $H_o = 100$ km s^{-1} Mpc^{-1} is not yet ended. One should note also that this very restricted agreement may disappear if Y_p is found to be as low as 0. 23 (Beckman and Pagel, this conference).

In the analysis of Yang <u>et al</u> 1984, on which the Boesgaard and Steigman (1985) review is based, the primordial (D+^{3}He/H)$\leq 10^{-4}$ value is determined by assuming that the chemical evolution of our Galaxy is properly described by the simplest models such as those of Audouze and Tinsley (1974). In such models the overall D destruction is limited in such a way that $D_{primordial}/D_{present} \leq 3-4$.

This is why we felt compelled to examine the possibility of considering models of chemical evolution leading to higher primordial values of D consistent with lower values of ^{4}He.

4. SPECIFIC MODELS OF GALACTIC EVOLUTION LEADING TO A THOROUGH DESTRUCTION OF D

Up to a quite recent past, we (Delbourgo-Salvador et al 1985) have proposed two types of models able to destroy thoroughly D during the galactic history that we will not review again here : they are (i) an inflow of processed (D free) material inside the solar vicinity and (ii) the ejection of D free material by stellar winds during the premain sequence. Following our analysis (VFA), we will consider here two specific models related to recent ideas put forward eg by Larson 1986, Wyse and Silk (WS) 1987, Audouze et al 1987, concerning both varying rate of star formation (SFR) and bimodal SFR. The first model hereafter referred to as model II has been designed to study the effect of a large SFR at the beginning of the galactic history. Model I is the standard one (Audouze 1987) with SFR $\psi(t)$ proportional to the gas density σ such that ν the constant of proportionality is $\nu=0.3$ Gyr^{-1} and a monomodal IMF $\Phi(m)$ $\alpha m^{(-x+1)}$ with x=2 and $0.4 \leq m/Mo \leq 100$. In model II, the SFR parameter $\Psi(t)$ has been modified on the following way : $\Psi_2(t) = \nu_2$ for t< τGyr and $\Psi_1(t) = \nu_1 \sigma$ for t> τ(Gyr). In this model we have conserved the classical IMF with x=2 for the whole stellar mass spectrum 0.4–100Mo. The parameters ν_2 and τ are such that $0.4 \leq \nu_1 \leq 2$ and $2 \geq \tau \geq 0.5$Gyr while $\nu_1 = 0.3$.

Model III has been designed to study the effect of bimodal SFR. It follows the WS prescriptions: the number of stars formed in the mass interval dm during a time dt is :

$$d^2 n(m,t) = [\Phi_1(m)/\Psi_1(t) + \Phi_2(t)\Psi_2(t)]dmdt \qquad (1)$$

The first term at the right hand side of (1) represents a constant mode of star formation from 0.4 to 100 Mo such as $\Psi_1(t)=\nu_1$, the second term represents a massive mass mode from the lower mass limit m_{L2}(1 to 2 Mo) to 100 Mo such that $\Psi_2(t)=\nu_2.e^{-t/\tau}$. The corresponding IMF has the same slope x=2 but different lower mass limits. The parameters ν_1, ν_2 and τ are $\nu_1=0.06$, $\nu_2=0.3$ and $\tau=2$ Gyr. Finally for reasons which are discussed below, we have also considered model IV which is similar to model III but where the first constant SFR term in the right hand side of equation (1) is replaced by $\Psi_1 t_1 = \nu_1 \sigma$ i.e. decreasing proportionally to the gas density.

Table 2 provides an account of the galactic evolution corresponding to these three different models, Figures 1, 2 and 3 display respectively the evolution of the normalized gas density (σ), the overall metallicity (z) and the D mass abundance (X_D).

To sum up the discussion of these models which can be found in VFA the following points should be stressed :

a) Model II (with varying SFR) is the model which is the best suited to account (at the same time for the evolution of σ, Z and a large X_D destruction. This is because gas processing is due to stars of any mass. Moreover as shown by Andreani et al 1988 that type of evolution model is also able to reproduce most satisfactorily the evolution of heavier elements (N, O,Fe,the s process and the r process elements). The only price to pay is that this model may not be satisfactory enough to reproduce properly the overall present luminosity function

Table 2

	Primordial Phase $t \simeq 100$ s		$T_0 = 12.5$ Gyr			
	Models IId, IVa,b,c	Observational constraints	Model IId	Model IVa	Model IVb	Model IVc
$\sigma = \frac{M_{gas}}{M_{total}}$	1	0.05 to 0.1	0.05	0.06	0.05	0.06
H	0.76	~ 0.70	0.69	0.6	0.63	0.62
D	10^{-4}	$5\ 10^{-6} \leq D \leq 1.5\ 10^{-5}$	$0.15\ 10^{-4}$	$0.1\ 10^{-4}$	$0.1\ 10^{-4}$	$0.1\ 10^{-4}$
$\frac{D_{prim}}{D_{pres}}$	1		~ 7	~ 10	~ 10	~ 10
^{3}He	$5\ 10^{-5}$	$(2-7)\ 10^{-5}$	$6.6\ 10^{-5}$	$3\ 10^{-5}$	$3.9\ 10^{-5}$	$3.5\ 10^{-5}$
^{4}He	0.24	~ 0.28	0.29	0.33	0.33	0.33
Z	0	~ 0.02	0.02	0.06	0.04	0.05

Resulting values of two models: Model IId, where $\psi_1(t) = \nu_1\sigma$, $\nu_1 = 0.3$, $t \geq 1$ Gyr, $\psi_2(t) = \nu_2$, $\nu_2 = 0.8$, $t \leq 1$ Gyr, Model IVa, where $\psi_1(t) = \nu_1\sigma$, $\nu_1 = 0.3$, $\psi_2(t) = \nu_2 e^{-t/2}$, $\nu_2 = 0.3$, the IMF parameters are $x=2$ and $m_{L2} = 2$ M$_\odot$. Model IVb has the same parameters than Model IVa but with an IMF slope $x=2.2$. Model IVc has the same parameters than Model IVa but with a low mass limit for the massive mode $m_{L2}=1.5$ M$_\odot$.

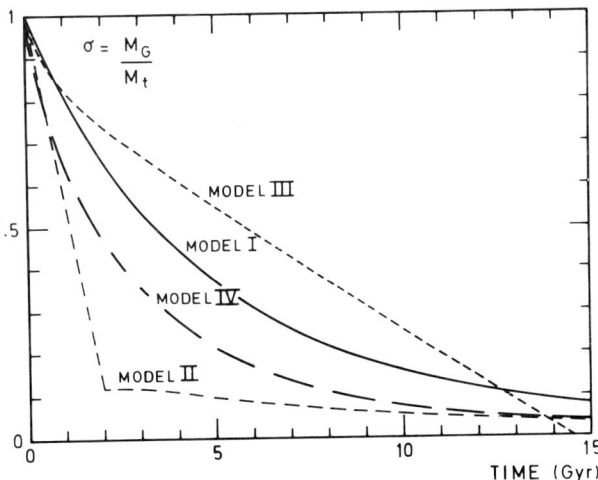

Figure 1. The evolution of normalized surface gas density ($\sigma = M_{gas}/M_{total}$) is presented for 4 models : — Model I, standard rate of star formation : $\psi(t)=0.3\ \sigma\ \phi(m) \propto m^{-3}$ $0.4 \leq m/M_\odot \leq 100$ — Model II, two regimes in the rate of star formation : $\psi_2(t)=0.5$ if $t \leq 2$ Gyr $\psi_1(t)=0.3\ \sigma$ if $t \geq 2$ Gyr $\phi(m) \propto m^{-3}$ $0.4 \leq m/M_\odot \leq 100$ — Model III, bimodal rate of star formation $\psi_2(t) = 0.3e^{-t/2}$ $\phi_2 \propto m^{-3}$ $2 \leq m/M_\odot \leq 100$ $\psi_1(t) = 0.06$ $\phi_1 \propto m^{-3}$ $0.4 \leq m/M_\odot \leq 100$ — Model IV, $\psi_2(t) = 0.3e^{-t/2}$ $\phi_2 \propto m^{-3.2}$ $2 < m/M_\odot < 100$ $\psi_1(t) = 0.3\sigma$ $\phi_1 \propto m^{-3.2}$ $0.4 < m/M_\odot < 100$

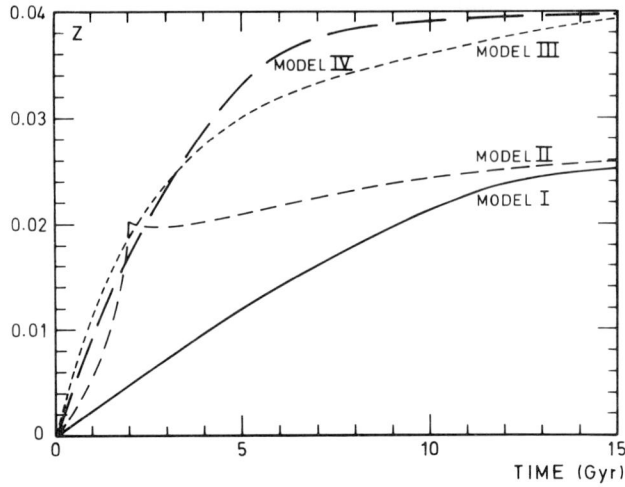

Figure 2. The evolution of metallicity Z for the same models as in figure 1.

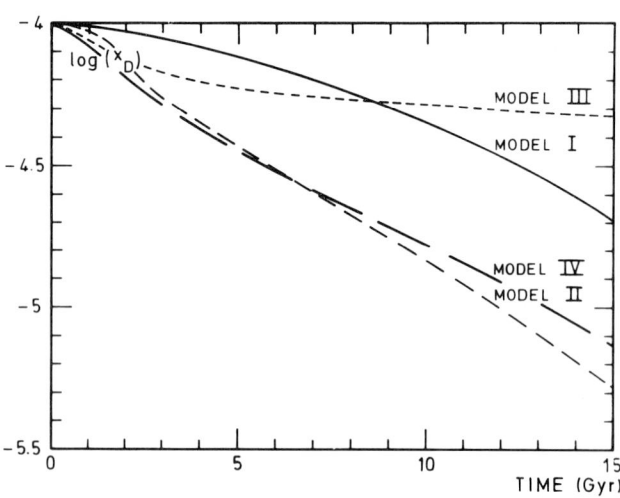

Figure 3. The evolution of deuterium (in mass) for the same models as in figures 1 and 2.

of stars. We (VFA) have proposed a slight modification of the parameters of this model ($\nu_2=0.8$ and $\tau=1$) which satisfy the Scalo (1986) SFR limits imposed by the stellar luminosity function. But this modification decreases the D destruction ($D_{primordial}/Dpresent=7$ instead of 10-13).

b) Model III is indeed well suited to reproduce satisfactorily the evolution of the luminosity function (WS) but leads to an overmetallicity (too high Z) and a much too low D destruction. As discussed thoroughly in VFA, an only formation of <u>massive</u> stars at the very beginning of the galactic evolution does not imply necessarily a large D destruction while low mass stars evolve slowly and eject D free material and few metals (since D is destroyed in <u>any</u> stellar environment), by contrast high mass stars evolve quickly but destroy D at the expense of rejecting metal rich stars. This is why in evolution models like model III SFR has to be strongly limited in order to avoid a too large overproduction of metals.

c) Since type II models are successful in accounting for the D destruction but not as good regarding the present luminosity stellar function while it is reverse for type III models, we have considered model IV which is an hybrid of models II and III : this model satisfies quite properly the D and stellar luminosity requirements but still leads to a metallicity (too high by a factor 2) and to an ^{4}He(Y) abundance of 0.33 which is also much too high.

5. SUMMARY AND CONCLUSIONS

The chemical evolution of the lightest elements produced by the Big Bang nucleosynthesis is indeed not the only exciting problem which one has to adress itself now : the early nucleosynthesis possibly affected by the quark- hadron phase transition is presently inspiring almost all the specialists involved in this field (see eg Applegate <u>et al</u>,1987, Alcock <u>et al</u>,1987, Audouze <u>et al</u>,1988 and Reeves,1988) . Nevertheless this contribution devoted to these evolutionary aspects gives us an opportunity to draw the attention of the interested reader on the complementarity which exists between the early nucleosynthesis (and therefore its implications on cosmology and particle physics). As discussed eg by Boesgaard and Steigman 1985, Audouze 1987, and Reeves <u>et al</u> 1988 standard Big Bang nucleosynthesis leading to an upper limit of $\Omega_B \leq 0.1$ seems still to be the most secure hypothesis. We claim here that in order to alleviate any difficulty which could arise from some discordance between the predictions on Ω_B coming from ^{4}He on one hand and (D+^{3}He) on the other hand, D has to be thoroughly destroyed during the galactic history. In order to achieve that goal some <u>specific</u> evolution models are required the most promising being those assuming time varying SFR (although some "fine tuning" are required in order to avoid any difficulty which could arise from the stellar luminosity function). Such time varying SFR (models may also be found useful in other contexts such as the evolution of heavier elements (Andreani <u>et al</u>

1988). Evolution models implying a bimodal IMF such as those of WS, are not successful in that respect since they are unable to lead to a thorough D destruction during the galactic history. Finally, one should note that only very specific evolution models are able to lead to such a D galactic destruction. Although this element is extremely sensitive to any extraction process its overall destruction is contingent to conditions on evolutionary parameters such as SFR. The apparent simplicity of the standard Big Bang nucleosynthesis and its quite remarkable consequences on the baryonic density of the Universe and the maximum number of families of relativistic particles (such as neutrinos) may require some somewhat contrived galactic evolution models.

ACKNOWLEDGEMENTS

This contribution has been efficiently typed by M.C.Pantalacci. The research reported here has been supported in part by PICS 18.

REFERENCES

Alcock,C.R., Fuller,G.M. and Mathews,G.J., 1987, Ap.J. (in press)

Andreani,P., Vangioni-Flam,E. and Audouze,J., 1988, Ap.J. (submitted)

Applegate,J.H., Hogan,C.T. and Scherrer,R.J., 1987, Phys.Rev. D 35,1151

Audouze,J., 1987, in Observational Cosmology, eds A.Hewitt et al, Reidel-Dordrecht, p.89

Audouze,J., Delbourgo-Salvador,P. and Salati,P.,1988, eds J.Audouze and F. Melchiorri, Varenna (in press)

Audouze,J., Delbourgo-Salvador,P. and Vangioni-Flam,E., 1987, in Advances in Nuclear Astrophysics, eds E.Vangioni-Flam et al, Editions Frontières p.47

Audouze,J., Spite,F. and Spite M., 1988, Phys. Reports (in preparation)

Audouze,J. and Tinsley,B.M., 1974, Ap.J. $\underline{192}$, 487

Audouze,J. and Vangioni-Flam,E., 1988, in Varenna proceedings, eds J.Audouze and F.Melchiorri (in press)

Boesgaard,A.M. and Steigman,G., 1985, Ann. Rev. Astron. Astroph., $\underline{23}$,319

Cameron,A.G.W., 1982, Essays in Nuclear Astrophysics, eds C.A.Barnes et al, Cambridge Univ. Press, p.23

Delbourgo-Salvador,P., Gry,C., Malinie,G. and Audouze,J., 1985, Astron. Astroph., $\underline{150}$, 53

Larson,R.B., 1986, MNRAS, $\underline{218}$, 409

Meyer,JP., 1985, Ap.J. suppl., $\underline{41}$, 513

Peimbert,M., 1986, in Star forming regions, eds M.Peimbert and J.Jugaku, Reidel-Dordrecht

Reeves,H., 1988,"Confrontation between observations and theories in Cosmology", in Varenna proceedings, eds J.Audouze and F.Melchiorri,(in press)

Reeves,H., Delbourgo–Salvador,P., Salati,P. and Audouze,J., 1988, Phys. Rev. $\underline{D}$, (submitted)
Rood,R.T., Bania, T.M., Wilson,T.L., 1984, Ap.J., $\underline{280}$, 629
Sanders,D.B., Solomon, P.M., Scoville, N.Z., 1984, Astron. Astroph., $\underline{276}$, 182
Scalo, J.M., 1986, Fundam. Cosm. Phys., $\underline{11}$, 1
Vangioni–Flam,E. and Audouze,J., 1987, Astron. Astroph., (in press)
Vidal-Madjar,A.,1987, in Space Astronomy and Solar System Exploration, ESA SP 268, p.73
Wyse,R. and Silk,J., 1987, Ap.J. Letters, $\underline{313}$, L11

DISCUSSION

Gustafsson : You honted at the possibility to use heavier element abundances for proving or disproving the bi–modal star–formation hypothesis. What particular suggestions would you make?

Audouze : Our current analysis (P.Andreani, E.Vangioni-Flam and myself) concerning the chemical evolution of heavier elements shows clearly that models with varying rate of star formation are more successful than those assuming a bimodal initial mass function.

Beckman : It appears that we are left with the choice of fine-tuning the galaxy or fine -tuning the Universe!

Audouze : I do not agree entirely with this statement. It seems more satisfactory to find out a general solution for the early universe to the expense of fine- tuning the galaxy. But for me it is still an open question.

Boesgaard :
Your models can predict the Li astration also. How you done that?
Audouze :
We should indeed determine as precisely as possible the chemical evolution of ^7Li by taking into account both destruction and formation processes. We plan to study this problem in a very near future. However we know already especially from the work I did with Beatrice Tinsley some years ago that ^7Li will be much less affected than D by destruction processes.

CONCLUDING SUGGESTIONS FOR FUTURE HIGH S/N SPECTROSCOPIC WORK ON STARS

P. E. Nissen
Institute of Astronomy
University of Aarhus
DK-8000 Aarhus C, Denmark

ABSTRACT. A summary of the main subjects discussed at this symposium is given, and some suggestions for future high signal-to-noise spectroscopic work, which could have a major impact on our understanding of the physics of stars and on astrophysics in general are presented.

1. INTRODUCTION

Let me begin this concluding talk with a few remarks on the interpretation of the term *very high S/N spectroscopy* that appears in the title of this symposium. I think it should be reserved to spectroscopy with a S/N of 300 or higher at resolutions R $\geq$ 20,000. In many works presented at this meeting a S/N of 100 only has been reached, which in some cases was not sufficient to obtain unambiguous results. Another problem is that even if the S/N is very high the astrophysical results derived from high-resolution spectroscopy may be severely affected by *systematic errors* as emphasized in the paper by D. F. Gray. Examples are: unknown amounts of scattered light, uncertainties in the instrumental profile used in the reduction, non-linearity of the detector, and errors in the flat fielding. In order to determine the size of these systematic errors there is a need for a more careful comparison of spectra obtained with different instruments. Maybe one should set up a list of standard stars and standard spectral regions in order to be able to make such comparisons in an effective way.

2. INSTRUMENTATION

About one day of the symposium was spent on instrumentation for high S/N spectroscopy. It is evident from papers published and from the contributions at this symposium that three high-resolution instruments with RETICON detectors (the McDonald Spectrometer, the CFTH Coudé Spectrograph, and the ESO CAT/CES) have been the basis for the rapid development of stellar spectroscopy during the last 10 years. However, these instruments do not represent the final state of art. Several problems have been encountered and improvements could be made in the following respects:

i) *Efficiency* of telescope, instrument and detector which according to the S/N obtained for a given stellar magnitude and exposure time is not higher than about 1%. One obvious way to improve the efficiency is to apply image slicers.

ii) *Spectral coverage* which is limited to a range of 50-100Å as determined by the length of the RETICON. With 2-dimensional CCD detectors it should be possible to cover larger spectral ranges. Furthermore, some spectrometers should be optimized for observations in the near UV, 3000 - 4000Å. This spectral region is very important for the study of early-type stars and very metal-poor stars, and some interesting elements like beryllium have spectral lines suitable for abundance determinations only in this spectral region (see paper by K. G. Budge et al.).

iii) *Flat fielding* which is not always as accurate as one would like due to different illumination of the optics by stars and flat-field lamps. In particular this is a great problem in connection with interference fringing in the near infrared for some CCD detectors. One possible solution is to feed the spectrometer through optical fibers.

iv) *Wavelength stability* that needs to be of the order of a few m/s in some research programs, e.g. the search for planets around stars or studies of stellar oscillations. Fiber optical scrambling may be one possible solution.

v) *Detectors* that should be made larger and more efficient especially in the near UV. An annoying problem is cosmic ray events, which produce disturbing spikes and limit the S/N severely for long exposures. Protection of instruments by heavy concrete should be investigated as a possible cure of the problem.

It is promising that many new high-resolution spectrometers have come into operation recently or are being built. Some of the improvements suggested above have already been incorporated. Thus S. S. Vogt reported on the Lick Observatory Echelle Spectrometer that seems to reach an overall efficiency as high as 4%. J. S. Yang gave new ideas for improving the efficiency, and H. Mandel as well as P. Felenbok and J. Guerin reported on fiber linked spectrometers.

A particular rapid instrumental development is taking place in the field of high-resolution infrared spectroscopy. S. T. Ridgway and K. Hinkle discussed new infrared detectors, which will make it possible to reach nearly as faint magnitudes in the infrared as in the visual spectral region. Fourier Transform Spectrometers remain an interesting alternative both in the visual and the infrared for the very highest resolutions or in cases where a very clean instrumental profile is needed.

In the future it is to be expected that the instrumental development will be very much influenced by the construction of very large telescopes. One of the major reasons for building VLT's is the possibility to extend high-resolution, high S/N spectroscopy to interesting classes of stars that cannot be reached by present days telescopes. However, it should be realized that it is very difficult to construct spectrometers for the VLT's with the same efficiency as for smaller telescopes. Detectors and gratings have to be scaled up in proportion to the diameter of the telescope in order to maintain spectral coverage and efficiency at a given resolution. Thus, the VLT spectrometers become very bulky and expensive. For example, an amount of 25 million DM is foreseen for the construction of Coudéroom, beam combination and a high-resolution spectrometer for the ESO 16 m VLT. Clearly, it is a challenge for astronomers interested in high-resolution spectroscopy to ensure that this amount of money is spent in the best possible way.

3. PHYSICS OF STELLAR ATMOSPHERES

About two days of the symposium were spent on the structure and physics of stellar atmospheres. This is a field, where the impact of the new generation of high-resolution spectrometers has been particular strong. A wealth of innovative research has been carried out during the last few years and was presented or reviewed at this symposium. Let me just mention a few highlights. The works of D. Dravins and D. F. Gray on line profiles and bisectors have given completely new information on velocity fields and granulation of stellar atmospheres. Spectroscopy of lines formed high up in the stellar atmospheres has led to a much better understanding of chromospheres, coronas and circumstellar envelopes not the least for T Tauri stars (see papers by I. Appenzeller and G. Basri). The impressive work of D. Baade on line profile variations has given new insight into the atmospheric structure and velocity fields of early-type stars. Doppler imaging is now applied for most spectral types and gives fairly detailed information on the distribution of stellar spots. Very accurate measurements of line profiles have made it possible to determine both the magnetic field strength in active regions and the fraction of stellar surface covered by the fields (see e.g. S. H. Saar's paper).

These new results are not only important for a better understanding of stellar atmospheres, but they are also of fundamental importance in a wider context. Thus, we need better models of stellar atmospheres to derive accurate chemical abundances, and we need information on stellar parameters like rotation and magnetic field strength in order to study the details of stellar evolution.

4. CHEMICAL COMPOSITION OF STARS

The last two days of the symposium were spent on stellar abundances. This is a field with a long tradition but relatively slow progress in the sixties and the seventies. Conflicting results on the details of the abundance patterns in stars were published and the distrust to the results obtained was widely spread. This was clearly expressed by two solar physicists in a paper entitled "*Can astrophysical abundances be taken seriously?*" (Worrall and Wilson, 1972). They questioned the reality of abundance differences at a level of a factor of four for the following three reasons: i) Published equivalent widths often disagreed with more than 50%. ii) An unphysical parameter (the microturbulence) was introduced in order to eliminate the dependence of the derived abundances on line strength. iii) The derived abundances were based on the LTE assumption.

How is the situation today after a decade of further work on stellar abundances based on spectra obtained with the new generation of spectrometers? Concerning the first problem there is no doubt that great improvements have been achieved. Equivalent widths are now measured with an accuracy of ±3 mÅ for the range 10 < W < 50 mÅ. Comparison of several sets of independent observations in the paper by D. Duncan confirms this. Thus, equivalent width measurements are accurate enough to allow abundance determinations to an accuracy of ±0.05 dex, if no other error sources were present.

Concerning the next problem the situation has also been much improved. The hydrodynamical models of the solar atmosphere by Nordlund (1979) show that not only the shape but also the *strength* of spectral lines are affected by the velocity field. To a high degree of accuracy this effect can be described by a single parameter, the microturbulence, as shown in the work of Simmons and Blackwell (1982). Furthermore, in the case of weak lines, W < 30 mÅ, the derived abundance is practically independent of the value of the microturbulence parameter adopted. Modern abundance determinations are often based on such faint lines.

There is however one snag that makes the derived abundances more uncertain than ±0.05 dex for some stars. The method of Simmons and Blackwell can only be applied if oscillator strengths are known to an accuracy of a few percent. Very few lines have been measured to that level of accuracy and therefore oscillator strengths derived from the solar spectrum are often used. This works well if the lines are weak in the solar spectrum as shown in the paper by R. J. Rutten. However, weak lines in e.g. metal-poor stars are rather strong in the solar spectrum. Thus, the derived gf-values are affected by uncertainties in the solar microturbulence and in the model atmosphere used for the sun. Because of this problem the derived abundances for metal-poor stars and for stars with T_{eff} and g much different from the solar values are often more uncertain than ±0.05 dex. Clearly, there is a great need for improved experimental oscillator strengths as also stressed in the paper by M. C. E. Huber. Physicists doing such measurements should be supported and encouraged by astrophysicists to continue and extend their work.

In connection with the third problem raised by Worrall and Wilson - the assumption of LTE - severe uncertainties still exist. As discussed by B. Gustafsson and as shown by P. Magain the derived abundances in metal-poor stars show a significant trend with exitation potential if LTE is assumed. The effect may be as large as 0.2 to 0.4 dex. This means that the interesting trends of various abundance ratios as a function of [Fe/H] derived by many participants in this meeting may be a product of the assumption of LTE. Obviously, there is a great need for non-LTE computations of spectral lines and fortunately this is possible with present days computers. Important work is being done by groups in Kiel, Munich, Uppsala and in the United States (see papers by D. Gigas, U. Heber et al., B. Gustafsson, and R. C. Peterson). Hopefully, the situation concerning non-LTE effects will become more clear within the next few years, but progress is considerably hampered by the lack of atomic data - in particular collisional cross sections.

In addition to the non-LTE problems inadequate model atmospheres also lead to considerable errors in the abundances derived. As reported by B. Gustafsson the solar iron abundance derived from the hydrodynamical model of Nordlund (1984) is significantly lower than the value derived from traditional homogeneous models. This is a serious problem that should be further investigated.

We conclude that errors in stellar abundances at present are dominated by errors in atomic data, in the models, and in the computation of line spectra. These effects could be responsible for some of the derived trends of abundance ratios as a function of [Fe/H], which are very interesting for studies of nucleosynthesis and galactic chemical evolution (see papers by J. W. Truran as well as E. Vangioni-Flam and J. Audouze). On the other hand very reliable results may be obtained if one is working differentially such that the above mentioned errors cancel. A good example is the study of Li abundance differences among F and G stars in open clusters (Boesgaard and Tripicco, 1986).

5. STELLAR PARAMETERS

It is well known that errors in the stellar atmospheric parameters, T_{eff} and g, introduce rather large errors in the abundances derived. Also in connection with studies of stellar structure and evolution there is a great interest in accurate values of these two parameters. As reviewed by B. Gustafsson the accuracy in the determination of T_{eff} and g has been much improved during the last few years. As shown in the paper by Perrin et al. the profile of H_α can be used to determine T_{eff} values to an accuracy of ± 35 K for solar-type stars. From Balmer lines in O and B stars one can determine log g with errors less than ± 0.15 dex (see paper by U. Heber et al.), and for late-type stars the surface gravity may be determined from the wings of strong, pressure-broadened lines. The method was demonstrated at this symposium in papers by J. Drake and B. Edvardsson. For very high quality line profiles the accuracy may be as good as ± 0.05 dex in log g (Smith et al. 1986).

As emphasized by W. Däppen at this meeting and as discussed in details by Gough (1987), accurate values of T_{eff}, g and luminosity (to be determined when parallaxes from HIPPARCOS become available) are needed to make a fundamental new investigation of stellar interiors by measuring the frequencies of stellar oscillations. The first (still controversial) results for Procyon, αCen and ε Eridani were reported on by E. Fossat. More convincing detection of oscillations of Pollux and Aldebaran was shown by P. H. Smith and R. S. McMillan. Clearly, these very interesting attempts to measure stellar oscillations should be continued with improved techniques (see paper by S. Frandsen).

6. SOME SUGGESTIONS FOR HIGH S/N SPECTROSCOPIC OBSERVING PROGRAMS.

In this section I would like to give just a few specific suggestions for high S/N spectroscopic observations that could have a major impact on our knowledge of stellar physics and chemical abundances of stars. These examples are listed according to the S/N and resolution that are needed to obtain good results.

6.1. S/N≃500, R≃100,000

Work on stellar granulation and convection is of fundamental importance for a better modelling of stars. Until now the relevant observations of bisectors have been carried out only for the very brightest stars. Extension to fainter stars representing all spectral types and to Pop. II stars is extremely important. Probably a VLT is needed to reach the Pop. II stars.

Detection and measurement of frequencies of stellar oscillations is a similar demanding field of utmost importance. Improved techniques will be necessary and a VLT will probably be helpful.

6.2. $S/N \simeq 300$, $R \simeq 100,000$

This is the kind of S/N and resolution needed to get accurate information on stellar rotation, turbulence, activity and magnetic fields or isotopic abundances. Limiting magnitudes are V = 6-7 with present days telescopes and spectrometers, e.g. the ESO CAT/CES. With larger existing telescopes and spectrometers with improved efficiency one should be able to reach V = 9-10 and with a VLT one may reach V = 12-13.

An interesting problem to be solved is the influence of parameters like rotation and magnetic fields on the global structure and chemical composition of stars. There is growing evidence that stellar colours and the strength of weak spectral lines are influenced by variations in other parameters than T_{eff}, g and chemical composition. Apparent abundance variations has been found among the Hyades stars (Cayrel et al. 1985). Unevolved Hyades F-type stars deviate very significantly from ZAMS field stars in the Strömgren c_1 - (b-y) diagram (the so-called "Hyades anomaly", Strömgren et al. 1982). Some open clusters, e.g. NGC752 (Twarog 1983) and NGC3680 (Nissen 1987) have a bimodal turnoff in the CM diagram. Only through a detailed study of the rotation, turbulence, activity and magnetic fields we can hope to find out which effects are responsible for these anomalies.

The study of isotope ratios is another field, where it would be extremely important to continue and extend very high S/N spectroscopy. As reviewed by D. Lambert isotope ratios in late-type giants give information about late stages of stellar evolution, whereas isotope ratios in dwarf stars provide information on the chemical evolution of our galaxy. For the latest spectral types, M, N, and C stars, very complicated spectra consisting of molecular bands are observed. Only recently it has been possible to model the atmospheric structure of these stars, synthesize their spectra and determine abundances and isotope ratios (e.g. Lambert et al. 1986). Extension of this kind of work to a substantial number of open and globular cluster giants and Pop. II field dwarfs would be much interesting.

6.3. $S/N \simeq 200$, $R \simeq 50,000$

This is the typical S/N and resolution needed to determine accurate abundances of most elements. Present limiting magnitudes are V = 11-12 with 3-4 m telescopes. In the future it should be possible to reach V = 14-15 with a VLT.

One of the best examples of the impact of high S/N spectroscopy on astrophysics is coming from recent work on lithium abundances in F and G stars. The striking similarity of the Li/H ratio in Pop. II dwarfs found by Spite and Spite (1982) is of great interest for cosmology, and the discovery of the dip in Li-abundance around T_{eff} = 6600 K in Hyades stars by Boesgaard and Tripicco (1986) together with the smooth decrease of Li/H for

$T_{eff} < 6000$ K (Cayrel et al. 1984) provide very interesting information on the internal structure of dwarf stars. Still, the mechanism of Li destruction is not well understood and it is unclear whether the Pop. II Li-abundance or the 10 times higher young Pop. I Li-abundance represent the primordial Big-Bang abundance. Much work is required to solve these problems, especially on Li-abundances in open clusters and field stars with different ages, rotation, metal abundance and activity. Also, the apparent constancy of the Li-abundance in Pop. II stars should be carefully tested, and stars further away in our galaxy should be observed.

Interesting work has been done in recent years on the trends of different abundance ratios in F and G stars near the sun. In view of the importance of such information for nucleosynthesis theories and models of chemical evolution it is evident that these investigations should be continued. Of particular interest is the scatter and possible discontinuities in the trends. Even more important would it be to extend the trends to a large group of very metal-poor stars, [Fe/H] < -3.0. For this purpose a VLT may be needed.

6.4. S/N ≃ 100, R ≃ 20,000

With this rather low resolution, accurate information can still be obtained on the parameters T_{eff} and g as well as abundances of some elements e.g. C, N, and O in early-type stars and O, Fe in late-type stars. Present days limiting magnitudes with 3-4 m telescopes are V = 13-15. With a VLT V = 18 may be reached.

At this symposium impressive spectra of giants in globular clusters and of supergiants in the Magellanic Clouds were shown (see papers by M. Spite et al., V. V. Smith, and M. Bessell). These works should be continued and extended to fainter luminosity classes in order to obtain more reliable abundances.

CNO abundances in main sequence B-type stars can give new information on the formation and evolution of stellar associations and on chemical gradients in our galaxy. The comparison with the chemical composition of HII regions provides an important consistency check. So far such work has been performed for B-stars ranging in galactocentric distance from 6 to 16 kpc (Gehren et al. 1985). With a VLT the investigations could be extended to the Magellanic Clouds or other nearby dwarf galaxies. New information on the primordial helium abundance may be obtained as well as information concerning the early galaxy formation epochs in the Universe.

Super-metal-rich stars were not discussed very much at this meeting. Yet there is no doubt that stars with [Fe/H] > 0.3 do exist. A good example is μLeo with [Fe/H] ≃ 0.5 (Branch et al. 1978). It is likely that such stars exist in large numbers and with [Fe/H] up to 1.0 in the galactic bulge (Whitford and Rich 1983). Determination of abundances in a substantial number of such stars seen through Baade's window could provide information on a stage of chemical evolution that is completely unknown today. However, a VLT is needed because the K giants in Baade's window are as faint as V = 17-18.

Finally, let me just mention the late stages of stellar evolution, blue subdwarfs and white dwarfs. As shown in the papers by U. Heber and J. L. Greenstein much new information has been gained by means of the new generation of spectrographs. Undoubtly, this type of work will continue to be a good example of the importance of very high S/N spectroscopy at modest resolutions.

7. CONCLUSION

The progress in high resolution, high S/N spectroscopy has been remarkable during the last decade but, as we have seen, much work remains to be done. Instrumentation at existing telescopes can still be improved significantly and the VLT's will give new possibilities in a few years. Many interesting observing programs are possible to carry out; just a few examples have been mentioned here. However, as I have tried to emphasize, it is also very important to improve physical theories and models used to interpret the observations. Otherwise we will not learn very much or we will get the wrong information even if the S/N of the spectra is very high.

REFERENCES

Boesgaard, A. M., Tripicco, M. J.: 1986, *Astrophys. J. Letters* **302**, L49

Branch, D., Bonnell, J., Tomkin, J.: 1978, *Astrophys. J.* **225**, 902

Cayrel, R., Cayrel de Strobel, G., Campbell, B., Däppen, W.: 1984, *Astrophys. J.* **283**, 205

Cayrel, R., Cayrel de Strobel, G., Campbell, B.: 1985, *Astron. Astrophys.* **146**, 249

Gehren, T., Nissen, P. E., Kudritzki, R. P., Butler, K.: 1985, *Proc. ESO workshop "Production and Distribution of C, N, O elements"*, Eds. E. J. Danziger, F. Matteucci and K. Kjär, p.171

Gough, D.: 1987, *NATURE* **326**, 257

Lambert, D. L., Gustafsson, B., Eriksson, K., Hinkle, K. H.: 1986, *Astrophys. J. Suppl.* **62**, 373

Nissen, P. E.: 1987, submitted to *Astron. Astrophys.*

Nordlund, Å.: 1979, *Proc. IAU Coll. 51 "Stellar Turbulence"*, Eds. D. F. Gray and J. L. Linsky in "Lecture Notes in Physics" vol. 114, p. 213

Nordlund, Å.: 1984, *Proc. National Solar Observatory Conference "Small-Scale Dynamic Processes in Quiet Stellar Atmospheres"*, Ed. S. L. Keil, Sunspot, New Mexico, p.181

Smith, G., Edvardsson, B., Frisk, U.: 1986, *Astron. Astrophys.* **165**, 126

Spite, F., Spite, M.: 1982, *Astron. Astrophys.* **115**, 357

Strömgren, B., Olsen, E. H., Gustafsson, B.: 1982, *Publ.Astr.Soc.Pac.* **94**, 5

Twarog, B. A.: 1983, *Astrophys. J.* **267**, 207

Whitford, A. E., Rich, R. M.: 1983, *Astrophys. J.* **274**, 723

Worrall, G., Wilson, A. M.: 1972, *NATURE* **236**, 15

FINAL REMARKS ON IAU SYMPOSIUM 132

FURENLID Without breaking the orderly progression of events during these last moments of IAU Symposium No. 132, I would like to insert a few words. We have just finished an excellent Symposium, and at this point there is no more any discussion or arguments when we all join in gratitude to the supporting staff at the meeting as well as to the scientific and local organizing committees. The Symposium has for me been like a school, where I have found myself —except for 11 too short minutes— on the appropriate side of the teacher's desk, the student side. The teachers were outstanding ! Our special thanks go to the Principal of this school, Dr. Giusa Cayrel, and as a small token of gratitude I would like to present her with a copy of the solar flux atlas. Dr. Cayrel, you are dear to us and we thank you very much !

-INDEX-

abundance (see chemical composition, isotopic abundances, light elements abundance (He,Li,Be,B),
        surface abundance distribution, stratification of the elements)

age	213 474
asymmetries (see lines : profile)	
atmospheric parameters	334 345 387 433 437 445 528
atomic data	361
Barium stars	541 563 593
chemical composition	333 345 387 389 395 409 411 421 425 429 433 437 441 449 453 459 477 485 493 501 515 519 521 525 537 541 545 551 577

      (see also: isotopic abundances, light elements abundance (He,Li,Be,B),
         (surface abundance distribution, stratification of the elements)

chromosphere	55 87 149 223 283 287
clusters [see open clusters or globular clusters]	
cool stars (see late-type stars)	
convection	239 249 381 463
corona	123 164
cosmological models	597 605
data analysis	345 355
detectors	
CCD detectors	16 45 49
Digicon detector	35
infra-red detectors	61
Photocounting array	23
Reticon detector	39 79
2D Frutti detector	559
diffusion	176 202 235 389 463
disk Population (see Population I)	
Doppler imaging	193 199 223 253 287
early type stars	143
echelle-spectrograph	1 9 16 28 61
emission lines	99 283

envelopes (stellar)	135 196
evolution (stellar)	95 209 211 455 563
evolution (galactic)	522 603
fiber spectrograph	9 18 31
flux (stellar,solar)	153 435
Fourier transform spectroscopy	71 185 249 425
flare	231
globular clusters	493 525 537 541 580
granulation	188 239
halo stars (see Population II stars)	
Herbig stars (see pre-main-sequence stars)	
high velocity stars (see Population II stars)	
Hyades see open clusters	
IMF (initial mass function)	456
infra-red spectroscopy	71 109 139 435
internal stellar structure	273 279 589
isotopic abundances	462 568 589 593
late-type stars	249 295 301 325 411 459 477 589
lines : profile	127 131 135 139 143 150 153 163 169 193 241 407
lines : broadening	186
light elements abundance (He,Li,Be,B)	273 279 373 381 401 407 409 463 469 473 507 511 585 597 603
macroturbulence	186
microturbulence	335 415 429 433 437 495 528 546
mass loss	127 217 555
Magellanic clouds	545 551 555 557 559
magnetic field	87 108 175 295 301 309 313 317 321 325 329
magnetic stars	180 199 317 321
metal-poor (see Population II stars)	
Mira variables	146
model atmospheres	117 127 333 411 531
non-LTE effects	117 139 369 387 389 395 411 445 485 493
nuclear processes	441 501 517 521 539 567 577 593

open clusters [Pleiades, UMa, Coma, Hyades]	274 401 445 453 463 469 473
oscillator strengths	361 367
oscillations (see : seismology)	
polarimetry	317
Population I stars	378 411 441
Population II stars	276 374 421 477 485 493 501 507 511 515 521 577
pre-main-sequence stars	87 95 99 105 109
pulsating stars	143
pulsations	193 217
radial velocity	79 83
rotation	95 105 469 473
rotational modulation	105 131
shock-wave (in stars)	143 169 205
seismology (stellar)	27 83 209 211 291
solar-type stars (see G stars)	
solar analogs	429
spots	253
supergiants	135 559
stars (different types):	
A stars [Ae, Am, Ap]	199 235 309 313 317 321 329 401
B stars [Be, Bp]	131 217 329 389 559
F stars	273 531 585
G stars	301 381 429 433 531
K stars	291 301 433 519 531
O stars [Of...]	117 123 127 389 555 557
β Cephei stars	169
ζ Aur stars	55
R CrB stars	205
RS Cvn stars	231
UU Her stars	407
S Dor stars	557

    (see also : Barium stars, early type stars, late type stars, magnetic stars, Mira variables, pre-main-sequence stars, pulsating stars, white dwarfs).

stratification of the elements   202 235
surface abundance distribution   309 321 329
turbulence (see velocity fields, macroturbulence, microturbulence)
T Tauri stars (see pre-main-sequence stars)
ultra-violet   50 163 223 231 555
variability   100 109 131 169 193 217 287 551
velocity fields   151 186 283
winds (stellar)   88 123
white dwarfs   175 573
Zeeman effect   295 301 313 329